Engineering Fundamentals of the Internal Combustion Engine

SECOND EDITION

Engineering Fundamentals of the Internal Combustion Engine

SECOND EDITION

Willard W. Pulkrabek
University of Wisconsin—Platteville

Prentice-Hall of India Private Limited
New Delhi - 110 001
2006

This Indian Reprint—Rs. 295.00
(Original U.S. Edition—Rs. 6428.00)

ENGINEERING FUNDAMENTALS OF THE INTERNAL COMBUSTION ENGINE, 2nd Ed.
by Willard W. Pulkrabek

ISBN-81-203-3031-5

Published by Asoke K. Ghosh, Prentice-Hall of India Private Limited, M-97, Connaught Circus, New Delhi-110001 and Printed by Mohan Makhijani at Rekha Printers Private Limited, New Delhi-110020.

Contents

Preface

The goal of the second edition of this book is the same as that of the first edition, with updated material in several areas that reflects the ever-advancing technology of internal combustion engines. This book was written to be used as an applied thermoscience textbook in a one-semester, college-level, undergraduate engineering course on internal combustion engines. It provides the material needed for a basic understanding of the operation of internal combustion engines. Students are assumed to have knowledge of fundamental thermodynamics, heat transfer, and fluid mechanics as a prerequisite to get maximum benefit from the text. This book can also be used for self-study or as a reference book in the field of engines.

Contents include the fundamentals of most types of internal combustion engines, with a major emphasis on reciprocating engines. Both spark ignition and compression ignition engines are covered, as are those operating on four-stroke cycles and on two-stroke cycles, and ranging in size from small model airplane engines to the largest stationary engines. Rocket engines and jet engines are not included. Because of the large number of engines that are used in automobiles and other vehicles, a major emphasis is placed on these.

The book is divided into 11 chapters. Chapters 1 and 2 give an introduction, terminology, definitions, and basic operating characteristics. These are followed in Chapter 3 with detailed analysis of basic engine cycles. Chapter 4 reviews fundamental thermochemistry as it applies to engine operation and engine fuels. Chapters 5 through 9 follow the air–fuel charge as it passes sequentially through an engine, including intake, motion within a cylinder, combustion, exhaust, and emissions. Engine heat transfer, friction, and lubrication are covered in Chapters 10 and 11. Each chapter includes solved example problems and historical notes, followed by a set of unsolved review problems. Also included at the end of each chapter are open-ended problems that require limited design application, in keeping with the modern engineering education trend of emphasizing design through the entire curriculum. These design problems can be used as minor weekly exercises or as major group projects.

Fueled by intensive commercial competition and increasingly strict government regulations on emissions and safety, the field of engine technology is constantly changing. It is difficult to stay knowledgeable about all advancements in engine design, materials, controls, and fuel development that we experience at an ever-increasing rate. During the years as the outline for this text evolved, continuous changes were required as new developments occurred. Those advancements that are covered in this book include the Miller cycle, lean burn engines, hybrid vehicles, 42-volt electrical systems, variable valve timing, fuel cell technology, gasoline direct injection, variable compression ratios, cylinder cutout, thermal storage, etc. Advancements and technological changes will continue to occur, and further updating of this text will be required periodically.

Information in the book represents an accumulation of general material collected by the author over a period of years while teaching courses and working in research and development in the field of internal combustion engines in the Mechanical Engineering Department of the University of Wisconsin—Platteville. During this time, information has been collected from many sources, including conferences, newspapers, personal communication, books, technical periodicals, research, product literature, television, etc. This information became the basis for the outline and notes used in the teaching of a class about internal combustion engines. These class notes, in turn, evolved into the general outline for this textbook. A list of references from the technical literature from which specific information for this book was taken is included at the back of the book.* Several references were of special importance in the development of these notes and are suggested for additional reading and more in–depth study. For keeping up with information about the latest research and development in automobile and internal combustion engine technology, publications by SAE International (Society of Automotive Engineers) are highly recommended; reference [11] is very good. For general information about most engine subjects, references [40, 58, 93, 100, 116] are recommended. On certain subjects, some of these references go into too much depth to be manageable in a one-semester course. Some of the information is slightly out of date, but overall, these are very informative references. For historical information about engines and automobiles in general, references [29, 45, 97, 102] are suggested. General data, formulas, and principles of engineering thermodynamics and heat transfer are used at various places throughout the text. Most undergraduate textbooks on these subjects would supply the needed information. References [63, 90] were used by the author.

In preparing this second edition, the author has followed the suggestions for improvements made by reviewers, when possible. Following the suggestion made by several reviewers, many more data dealing with real engines are included, using the actual figures from technical literature when possible.

In keeping with the trend of the world, the international system of units (SI) are used throughout the book, often supplemented with English units. Most research and development of engines is done using SI units, and this is reflected in the technical literature. The consumer market, however, still commonly uses English units, such as horsepower, miles per gallon, and cubic inch displacement, particularly in the automobile market. A conversion table of SI and English units for common parameters used in engine work is included in the appendix at the back of the book.

The reader should remember that many of the mathematical formulas used in this book and other technical literature are only models of what is actually happening. Often these are simplified models of very complex occurrences (e.g., chemical reactions, engine

*Numbers in brackets [ ] refer to references found in the back of the book.

cycles, flow in cylinders, etc.). These models often allow us to relate important variables and solve many technical problems, but should not be used beyond their intended range.

We have entered an exciting revolutionary period of internal combustion engine technology, largely brought about by the use of computer controls. For over 100 years, most combustion engines operated on the same basic four-stroke Otto or Diesel cycles. These engines had a fixed displacement, a fixed compression ratio, and fixed valve actuation controlled with a camshaft. The major improvement experienced during this time was higher thermal efficiency, brought about by better fuels and higher compression ratios. Now at the start of the twenty-first century, we suddenly have engines with variable displacement, variable compression ratios, variable valve control, and other technological advancements; and we have computer power to control these variables. The challenge to today's automobile and engine engineers is to develop new engine concepts that utilize and optimize these new technologies. We must understand the basic concepts, but we must not limit ourselves to evolutionary improvements of these concepts.

The author would like to express his gratitude to the many people who have influenced him and helped in the writing of this book. First, I thank Dorothy, who was always there, along with John, Tim, Becky, and Chad. I thank the people who reviewed the original book manuscript of the first edition, and those who reviewed the book in preparation of this second edition. The suggestions for additions and improvements that these people made have resulted in a better text. Although I have never met them, I am indebted to authors J. B. Heywood, C. R. Ferguson, E. F. Obert, and R. Stone. The books these men have written about internal combustion engines have certainly influenced the content of this textbook. I thank my father, who many years ago introduced me to the field of automobiles and generated a lifelong interest. I thank Earl of Capital City Auto Electric for carrying on the tradition.

WILLARD W. PULKRABEK

University of Wisconsin–Platteville

Acknowledgments

The author wishes to thank and acknowledge the following organizations for permission to reproduce photographs, drawings, and tables from their publications in this text: Auburn-Cord-Duesenberg Museum, ASME International, Briggs and Stratton, Carnot Press, Cummins Inc., Curtiss-Wright, Fairbanks Morse Engine Division of Coltec Industries, Ford Motor Company, General Motors, Harley Davidson, the Institute of Mechanical Engineers, MIT Press, Pearson Prentice Hall (Pearson Education, Inc.), SAE International, The Combustion Institute, Tuescher Photography.

Notation

Notation, symbols, and abbreviations used in this text. Common units are given in brackets, both SI and English.

A	Cross-section area of flow [cm^2] [$in.^2$]
A_c	Flow area of fuel capillary tube [mm^2] [$in.^2$]
A_{cc}	Surface area of combustion chamber [cm^2] [$in.^2$]
A_{ch}	Cylinder head surface area [cm^2] [$in.^2$]
A_{ex}	Area of exhaust valve [cm^2] [$in.^2$]
A_i	Area of intake valve [cm^2] [$in.^2$]
A_p	Piston face area of flat-faced piston, or cross-section area of cylinder [cm^2] [$in.^2$]
A_t	Throat area of carburetor [cm^2] [$in.^2$]
AF	Air–fuel ratio [kg_a/kg_f] [lbm_a/lbm_f]
AKI	Anti-knock index
AON	Aviation octane number
B	Cylinder bore [cm] [in.]
C	Constant
C_D	Discharge coefficient
C_{Dc}	Discharge coefficient of capillary tube
C_{Dt}	Discharge coefficient of carburetor throat
CI	Cetane index
CN	Cetane number
Eff	Aftercooler effectiveness
EGR	Exhaust gas recycle [%]
F	Force [N] [lbf]
F_f	Friction force [N] [lbf]
F_r	Force of connecting rod [N] [lbf]
F_x	Forces in the X direction [N] [lbf]
F_y	Forces in the Y direction [N] [lbf]
F_{1-2}	View factor
FA	Fuel–air ratio [kg_f/kg_a] [lbm_f/lbm_a]
FS	Fuel sensitivity

I Moment of inertia [kg-m^2] [lbm-ft^2]
ID Ignition delay [sec]
K_e Chemical equilibrium constant
M Molecular weight (molar mass) [kg/kgmole] [lbm/lbmmole]
MON Motor octane number
N Engine speed [RPM]
N Number of moles
N_c Number of cylinders
N_v Moles of vapor
Nu Nusselt number
ON Octane number
P Pressure [kPa] [atm] [psi]
P_a Air pressure [kPa] [atm] [psi]
P_{ex} Exhaust pressure [kPa] [atm] [psi]
P_{EVO} Pressure when the exhaust valve opens [kPa] [psi]
P_f Fuel pressure [kPa] [atm] [psi]
P_i Intake pressure [kPa] [atm] [psi]
P_{inj} Injection pressure [kPa] [atm] [psi]
P_o Standard pressure [kPa] [atm] [psi]
P_t Pressure in carburetor throat [kPa] [atm] [psi]
P_v Vapor pressure [kPa] [atm] [psi]
Q Heat transfer [kJ] [BTU]
$\dot{Q}$ Heat transfer rate [kW] [hp] [BTU/sec]
Q_{HHV} Higher heating value [kJ/kg] [BTU/lbm]
Q_{HV} Heating value of fuel [kJ/kg] [BTU/lbm]
Q_{LHV} Lower heating value [kJ/kg] [BTU/lbm]
R Ratio of connecting rod length to crank offset
R Gas constant [kJ/kg-K] [ft-lbf/lbm-°R] [BTU/lbm-°R]
Re Reynolds number
RON Research octane number
S Stroke length [cm] [in.]
S_g Specific gravity
SOF Soluble organic fraction
SR Swirl ratio
T Temperature [°C] [K] [°F] [°R]
T_a Temperature of air [°C] [K] [°F] [°R]
T_c Temperature of coolant [°C] [K] [°F] [°R]
T_{EVO} Temperature when the exhaust valve opens [°C] [K] [°F] [°R]
T_{ex} Temperature of exhaust [°C] [K] [°F] [°R]
T_g Temperature of gas [°C] [K] [°F] [°R]
T_i Intake temperature [°C] [K] [°F] [°R]
T_m Temperature of mixture [°C] [K] [°F] [°R]

T_{mp} Midpoint boiling temperature in °F [°F]
T_o Standard temperature [°C] [°F]
T_w Wall temperature [°C] [K] [°F] [°R]
TR Tumble ratio
U_p Piston speed [m/sec] [ft/sec]
$\overline{U}_p$ Average piston speed [m/sec] [ft/sec]
V Cylinder volume [L] [cm^3] [in.^{3}]
V_{BDC} Cylinder volume at bottom dead center [L] [cm^3] [in.^{3}]
V_c Clearance volume [L] [cm^3] [in.^{3}]
V_d Displacement volume [L] [cm^3] [in.^{3}]
V_{TDC} Cylinder volume at top dead center [L] [cm^3] [in.^{3}]
W Work [kJ] [ft-lbf] [BTU]
W_b Brake work [kJ] [ft-lbf] [BTU]
W_f Friction work [kJ] [ft-lbf] [BTU]
W_i Indicated work [kJ] [ft-lbf] [BTU]
$\dot{W}$ Power [kW] [hp]
$\dot{W}_b$ Brake power [kW] [hp]
$\dot{W}_c$ Power to drive compressor [kW] [hp]
$\dot{W}_f$ Friction power [kW] [hp]
$\dot{W}_i$ Indicated power [kW] [hp]
$\dot{W}_m$ Power to motor engine [kW] [hp]
$\dot{W}_{sc}$ Power to drive supercharger [kW] [hp]
$\dot{W}_t$ Turbine power [kW] [hp]
a Crank offset (crank radius) [cm] [in.]
c Speed of sound [m/sec] [ft/sec]
c_{ex} Speed of sound at exhaust conditions [m/sec] [ft/sec]
c_i Speed of sound at inlet conditions [m/sec] [ft/sec]
c_o Speed of sound at ambient conditions [m/sec] [ft/sec]
c_p Specific heat at constant pressure [kJ/kg-K] [BTU/lbm-°R]
c_v Specific heat at constant volume [kJ/kg-K] [BTU/lbm-°R]
d_v Valve diameter [cm] [in.]
dB Decibel
g Acceleration due to gravity [m/sec^2] [ft/sec^2]
h Height differential in fuel capillary tube [cm] [in.]
h Convection heat transfer coefficient [kW/m^2-K] [BTU/hr-ft^2-°R]
h Specific enthalpy [kJ/kg] [BTU/lbm]
h_a Specific enthalpy of air [kJ/kg] [BTU/lbm]
h_c Convection heat transfer coefficient on the coolant side [kW/m^2-K] [BTU/hr-ft^2-°R]
h_{ex} Specific enthalpy of exhaust [kJ/kg] [BTU/lbm]
h_g Convection heat transfer coefficient on the gas side [kW/m^2-K] [BTU/hr-ft^2-°R]
h_m Specific enthalpy of mixture [kJ/kg] [BTU/lbm]
h_f° Enthalpy of formation [kJ/kgmole] [BTU/lbmmole]

Δh	Change in enthalpy from standard conditions [kJ/kgmole]
k	Ratio of specific heats
k	Thermal conductivity [kW/m-K] [BTU/hr-ft-°R]
k_g	Thermal conductivity of gas [kW/m-K] [BTU/hr-ft-°R]
l	Valve lift [cm] [in.]
m	Mass [kg] [lbm]
m_a	Mass of air [kg] [lbm]
m_{ex}	Mass of exhaust [kg] [lbm]
m_f	Mass of fuel [kg] [lbm]
m_m	Mass of gas mixture [kg] [lbm]
m_{mi}	Mass of mixture ingested [kg] [lbm]
m_{mt}	Mass of mixture trapped [kg] [lbm]
m_{tc}	Mass of total charge trapped [kg] [lbm]
$\dot{m}$	Mass flow rate [kg/sec] [lbm/sec]
$\dot{m}_a$	Mass flow rate of air [kg/sec] [lbm/sec]
$\dot{m}_{EGR}$	Mass flow rate of exhaust gas recycle [kg/sec] [lbm/sec]
$\dot{m}_f$	Mass flow rate of fuel [kg/sec] [lbm/sec]
$\dot{m}_i$	Mass flow into the cylinder [kg/sec] [lbm/sec]
mep	Mean effective pressure [kPa] [atm] [psi]
n	Number of revolutions per cycle
q	Heat transfer per unit mass [kJ/kg] [BTU/lbm]
q	Heat transfer per unit area [kJ/m^2] [BTU/ft^2]
$\dot{q}$	Heat transfer rate per unit mass [kW/kg] [BTU/hr-lbm]
$\dot{q}$	Heat transfer rate per unit area [kW/m^2] [BTU/hr-ft^2]
r	Connecting rod length [cm] [in.]
r_c	Compression ratio
r_e	Expansion ratio
rh	Relative humidity [%]
s	Distance between wrist pin and crankshaft axis [cm] [in.]
t	Time [sec]
u	Specific internal energy [kJ/kg] [BTU/lbm]
u_t	Swirl tangential speed [m/sec] [ft/sec]
v	Specific volume [m^3/kg] [ft^3/lbm]
v_{BDC}	Specific volume at bottom dead center [m^3/kg] [ft^3/lbm]
v_{ex}	Specific volume of exhaust [m^3/kg] [ft^3/lbm]
v_{TDC}	Specific volume at top dead center [m^3/kg] [ft^3/lbm]
w	Specific work [kJ/kg] [ft-lbf/lbm] [BTU/lbm]
w_b	Brake-specific work [kJ/kg] [ft-lbf/lbm] [BTU/lbm]
w_f	Friction-specific work [kJ/kg] [ft-lbf/lbm] [BTU/lbm]
w_i	Indicated-specific work [kJ/kg] [ft-lbf/lbm] [BTU/lbm]
x	Distance [cm] [m] [in.] [ft]
x_{ex}	Fraction of exhaust

x_r	Exhaust residual
x_v	Mole fraction of water vapor
α	Pressure ratio
α	Ratio of valve areas
β	Cutoff ratio
Γ	Angular momentum [kg-m^2/sec] [lbm-ft^2/sec]
ε_g	Emissivity of gas
ε_w	Emissivity of wall
η_c	Combustion efficiency [%]
η_f	Fuel conversion efficiency [%]
η_m	Mechanical efficiency [%]
η_s	Isentropic efficiency [%]
η_t	Thermal efficiency [%]
η_v	Volumetric efficiency of the engine [%]
θ	Crank angle measured from TDC [°]
λ	Lambda value
λ_{ce}	Charging efficiency
λ_{dr}	Delivery ratio
λ_{rc}	Relative charge
λ_{se}	Scavenging efficiency
λ_{te}	Trapping efficiency
μ	Dynamic viscosity [kg/m-sec] [lbm/ft-sec]
μ_g	Dynamic viscosity of gas [kg/m-sec] [lbm/ft-sec]
ν	Stoichiometric coefficients
ρ	Density [kg/m^3] [lbm/ft^3]
ρ_a	Density of air [kg/m^3] [lbm/ft^3]
ρ_o	Density of air at standard conditions [kg/m^3] [lbm/ft^3]
ρ_f	Density of fuel [kg/m^3] [lbm/ft^3]
σ	Stefan–Boltzmann constant [W/m^2-K^4] [BTU/hr-ft^2-°R^4]
τ	Torque [N-m] [lbf-ft]
τ_s	Shear force per unit area [N/m^2] [lbf/ft^2]
ϕ	Equivalence ratio
ϕ	Angle between connecting rod and centerline of the cylinder
ω	Angular velocity of swirl [rev/sec]
ω_t	Angular velocity of tumble [rev/sec]
ω_v	Specific humidity [kg$_v$/kg$_a$] [grains$_v$/lbm$_a$]

CHAPTER 1

Introduction

"... the modern high-class automobile is in the main satisfactory and can be relied upon if the driver is thoroughly cognizant of its workings. It is needless to say that self-propelling vehicles, like other machines, will never do as much for one who does not understand them as for one who does."

The Endurance Test

The Automobile Magazine (October 1901)

This chapter introduces and defines the internal combustion engine. It lists ways of classifying engines and terminology used in engine technology. Descriptions are given of many common engine components and of basic four-stroke and two-stroke cycles for both spark ignition and compression ignition engines.

1.1 INTRODUCTION

The **internal combustion (IC) engine** is a heat engine that converts chemical energy in a fuel into mechanical energy, usually made available on a rotating output shaft. Chemical energy of the fuel is first converted to thermal energy by means of combustion or oxidation with air inside the engine. This thermal energy raises the temperature and pressure of the gases within the engine, and the high-pressure gas then expands against the mechanical mechanisms of the engine. This expansion is converted by the mechanical linkages of the engine to a rotating crankshaft, which is the output of the engine. The crankshaft, in turn, is connected to a transmission or power train to transmit the rotating mechanical energy to the desired final use. For engines, this will often be the propulsion of a vehicle (i.e., automobile, truck, locomotive, marine vessel, or airplane). Other applications include stationary engines to drive generators or pumps, and portable engines for things like chain saws and lawn mowers.

Most internal combustion engines are **reciprocating engines** having pistons that reciprocate back and forth in cylinders internally within the engine. This book concentrates on the thermodynamic study of this type of engine. Other types of IC engines also exist in much fewer numbers, one important one being the rotary engine [104]. These engines will be given brief coverage. Engine types not covered by the book include steam engines and gas turbine engines, which are better classified as **external combustion engines** (i.e., combustion takes place outside the mechanical engine system). Also not included in the book, although they could be classified as internal combustion engines, are rocket engines, jet engines, and firearms.

Reciprocating engines can have one cylinder or many cylinders, up to 20 or more. The cylinders can be arranged in many different geometric configurations. Sizes range from small model airplane engines with power output on the order of 100 watts to large multicylinder stationary engines that produce thousands of kilowatts per cylinder.

There are so many different engine manufacturers, past, present, and future, that produce and have produced engines which differ in size, geometry, style, and operating characteristics, that no absolute limit can be stated for any range of engine characteristics (i.e., size, number of cylinders, strokes in a cycle, etc.). This book will work within normal characteristic ranges of engine geometries and operating parameters, but there can always be exceptions to these.

Early development of modern internal combustion engines occurred in the latter half of the 1800s and coincided with the development of the automobile. History records earlier examples of crude internal combustion engines and self-propelled road vehicles dating back as far as the 1600s [29]. Most of these early vehicles were steam-driven prototypes which never became practical operating vehicles. Technology, roads, materials, and fuels were not yet adequately developed. Very early examples of heat engines, including both internal combustion and external combustion, used gun powder and other solid, liquid, and gaseous fuels. Major development of the modern steam engine and, consequently, the railroad locomotive occurred in the latter half of the 1700s and early 1800s. By the 1820s and 1830s, railroads were present in several countries around the world.

HISTORIC—ATMOSPHERIC ENGINES

Most of the very earliest internal combustion engines of the 17th and 18th centuries can be classified as **atmospheric engines**. These were large engines with a single piston and cylinder, the cylinder being open on the end. Combustion was initiated in the open cylinder using any of the various fuels that were available. Gunpowder was often used as the fuel. Immediately after combustion, the cylinder would be full of hot exhaust gas at atmospheric pressure. At this time, the cylinder end was closed and the trapped gas was allowed to cool. As the gas cooled, it created a vacuum within the cylinder. This caused a pressure differential across the piston, atmospheric pressure on one side and a vacuum on the other. As the piston moved because of this pressure differential, it would do work by being connected to an external system, such as raising a weight [29].

Some early steam engines were also atmospheric engines. Instead of combustion, the open cylinder was filled with hot steam. The end was then closed and the steam was allowed to cool and condense. This created the necessary vacuum.

FIGURE 1-1

The Charter Engine made in 1893 at the Beloit works of Fairbanks, Morse & Company was one of the first successful gasoline engines offered for sale in the United States. Printed with permission, Fairbanks Morse Engine Division, Coltec Industries.

In addition to a great amount of experimentation and development in Europe and the United States during the middle and latter half of the 1800s, two other technological occurrences during this time stimulated the emergence of the internal combustion engine. In 1859, the discovery of crude oil in Pennsylvania finally made available the development of reliable fuels that could be used in these newly developed engines. Up to this time, the lack of good, consistent fuels was a major drawback in engine development. Fuels like whale oil, coal gas, mineral oils, coal, and gun powder, which were available before this time, were less than ideal for engine use and development. It still took many years before products of the petroleum industry evolved from the first crude oil to **gasoline**, the automobile fuel of the 20th century. However, improved hydrocarbon products began to appear as early as the 1860s and gasoline, lubricating oils, and the internal combustion engine evolved together.

The second technological invention that stimulated the development of the internal combustion engine was the pneumatic rubber tire, which was first marketed by John B. Dunlop in 1888 [141]. This invention made the automobile much more practical and desirable and thus generated a large market for propulsion systems, including the internal combustion engine.

During the early years of the automobile, the internal combustion engine competed with electricity and steam engines as the basic means of propulsion. Early in the 20th century, electricity and steam faded from the automobile picture—electricity because of the limited range it provided and steam because of the long start-up time

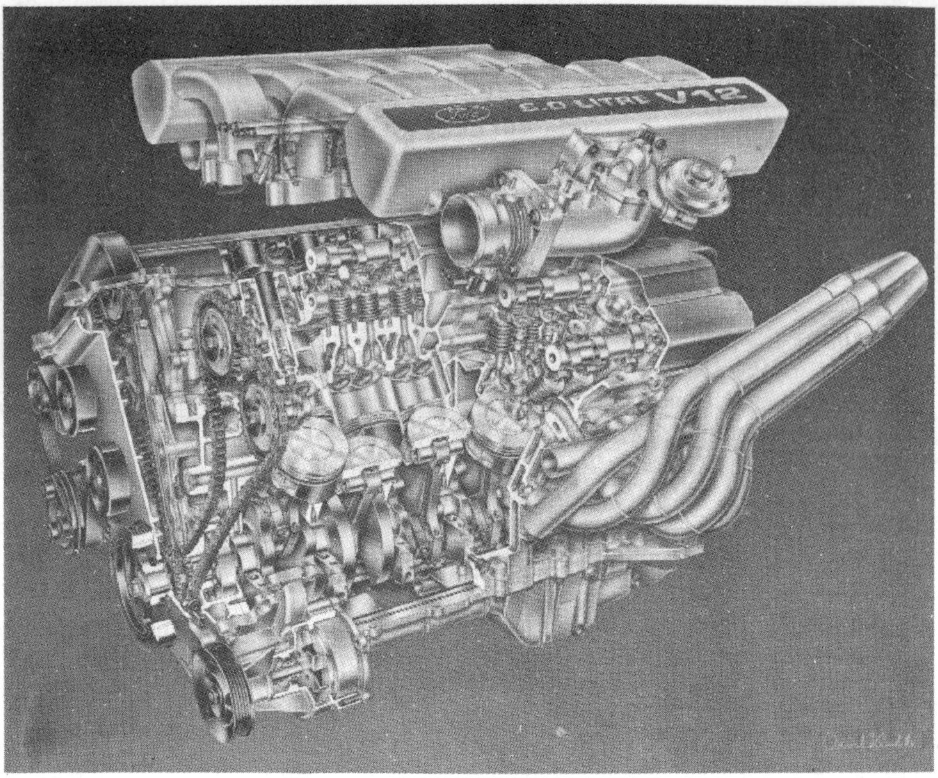

FIGURE 1-2

Cutaway view of Ford four-stroke cycle, spark-ignition, 367 in.^{3} (6.0 liter) V12 engine, which was used in luxury/sports cars of mid 1990s. The 60° V12 engine was made of aluminum alloy, with electronic fuel injection. Maximum brake power was 313 hp at 5350 RPM (233 kW), with maximum torque of 353 lbf-ft at 3750 RPM (478 N-m). Courtesy of Ford Motor Company.

needed. Thus, the 20th century was the period of the internal combustion engine and the automobile powered by the internal combustion engine. Now, at the start of the 21st century, the internal combustion engine is again being challenged by electricity and other forms of propulsion systems for automobiles and other applications. What goes around comes around.

1.2 EARLY HISTORY

During the second half of the 19th century, many different styles of internal combustion engines were built and tested. Reference [29] is suggested as a good history of this period. These engines operated with variable success and dependability using many different mechanical systems and engine cycles.

The first fairly practical engine was invented by J.J.E. Lenoir (1822–1900) and appeared on the scene about 1860 (Fig. 3-20). During the next decade, several hundred of these engines were built with power up to about 4.5 kW (6 hp) and mechanical efficiency up to 5%. The Lenoir engine cycle is described in Section 3-13. In 1867 the Otto–Langen engine, with efficiency improved to about 11%, was first introduced, and several thousand of these were produced during the next decade. This was a type of atmospheric engine with the power stroke propelled by atmospheric pressure acting against a vacuum. Nicolaus A. Otto (1832–1891) and Eugen Langen (1833–1895) were two of many engine inventors of this period.

During this time, engines operating on the same basic four-stroke cycle as the modern automobile engine began to evolve as the best design. Although many people were working on four-stroke cycle design, Otto was given credit when his prototype engine was built in 1876.

In the 1880s the internal combustion engine first appeared in automobiles [45]. Also in this decade the two-stroke cycle engine became practical and was manufactured in large numbers.

By 1892, Rudolf Diesel (1858–1913) had perfected his compression ignition engine into basically the same diesel engine known today. This was after years of development work, which included the use of solid fuel in his early experimental engines. Early compression ignition engines were noisy, large, slow, single-cylinder engines. They were, however, generally more efficient than spark ignition engines. It wasn't until the 1920s that multicylinder compression ignition engines were made small enough to be used with automobiles and trucks.

1.3 ENGINE CLASSIFICATIONS

Internal combustion engines can be classified in a number of different ways:

1. Types of Ignition

- **(a) Spark Ignition (SI).** An SI engine starts the combustion process in each cycle by use of a spark plug. The spark plug gives a high-voltage electrical discharge between two electrodes which ignites the air–fuel mixture in the combustion chamber surrounding the plug. In early engine development, before the invention of the electric spark plug, many forms of torch holes were used to initiate combustion from an external flame.
- **(b) Compression Ignition (CI).** The combustion process in a CI engine starts when the air–fuel mixture self-ignites due to high temperature in the combustion chamber caused by high compression.

2. Engine Cycle

- **(a) Four-Stroke Cycle.** A four-stroke cycle has four piston movements over two engine revolutions for each cycle.
- **(b) Two-Stroke Cycle.** A two-stroke cycle has two piston movements over one revolution for each cycle.

  Three-stroke cycles and six-stroke cycles were also tried in early engine development [29].

FIGURE 1-3

Engine of Ford Model A automobile built from 1928 to 1932. Four-cylinder L-head engine had displacement of 195 $in.^3$ (3.20 L), with 3.875 in. bore (9.84 cm), 4.125 in. stroke (10.48 cm), and 4.22:1 compression ratio. The water-cooled engine had three main bearings, with brake power rated as 40 hp at 2200 RPM (29.8 kW), and torque of 128 lbf-ft at 1000 RPM (173 N-m). There was no fuel pump with fuel gravity fed to a downdraft carburetor [165, 169]. Figure courtesy of Ford Motor Company.

3. Valve Location (see Fig. 1-4)
   - (a) **Valves in head (overhead valve)**, also called **I Head engine**.
   - (b) **Valves in block (flat head)**, also called **L Head engine**. Some historic engines with valves in block had the intake valve on one side of the cylinder and the exhaust valve on the other side. These were called **T Head engines**.
   - (c) One valve in head (usually intake) and one in block, also called **F Head engine**; this is much less common.
4. Basic Design
   - (a) **Reciprocating.** Engine has one or more cylinders in which pistons reciprocate back and forth. The combustion chamber is located in the closed end of each cylinder. Power is delivered to a rotating output crankshaft by mechanical linkage with the pistons.

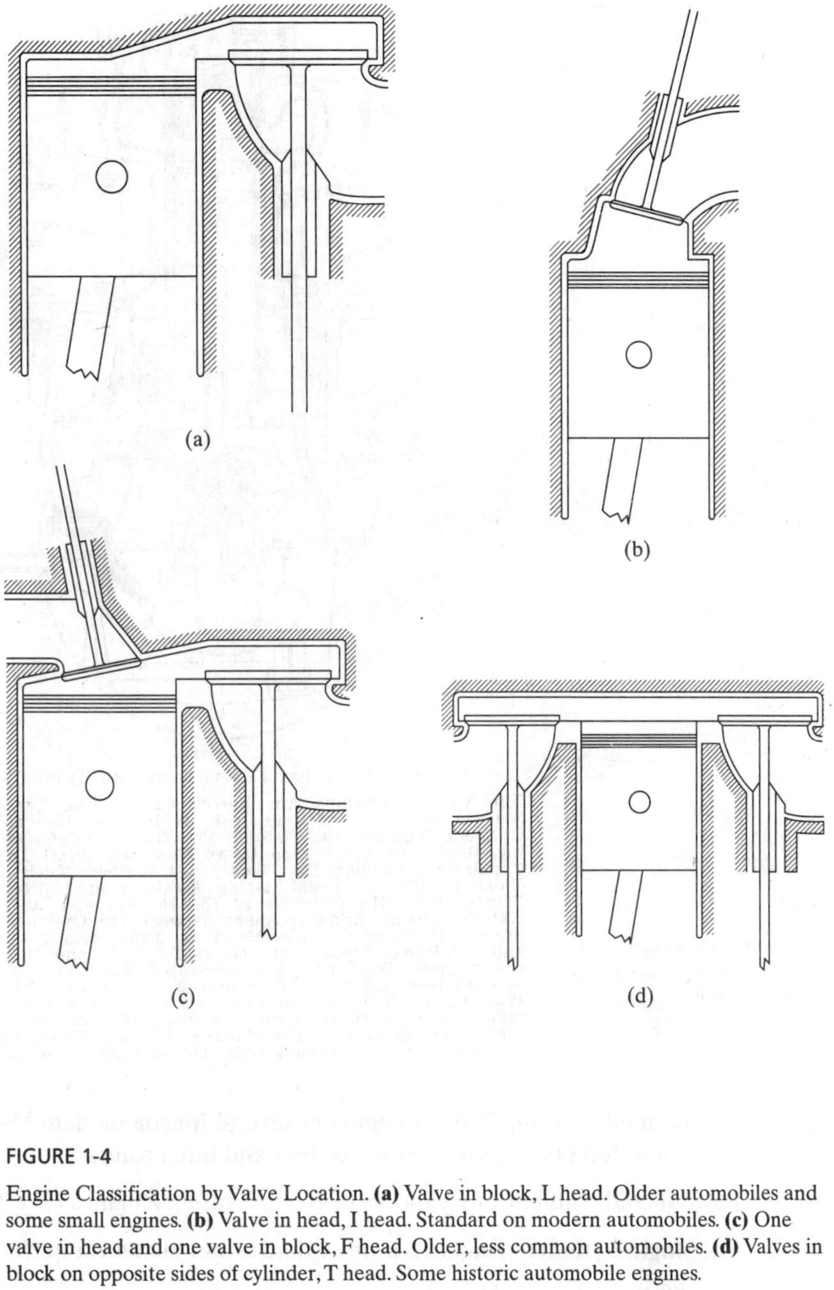

FIGURE 1-4

Engine Classification by Valve Location. **(a)** Valve in block, L head. Older automobiles and some small engines. **(b)** Valve in head, I head. Standard on modern automobiles. **(c)** One valve in head and one valve in block, F head. Older, less common automobiles. **(d)** Valves in block on opposite sides of cylinder, T head. Some historic automobile engines.

(b) Rotary. Engine is made of a block (stator) built around a large nonconcentric rotor and crankshaft (Figs. 2-22, 2-23). The combustion chambers are built into the nonrotating block. A number of experimental engines have been tested using this concept, but the only design that has ever become common in an

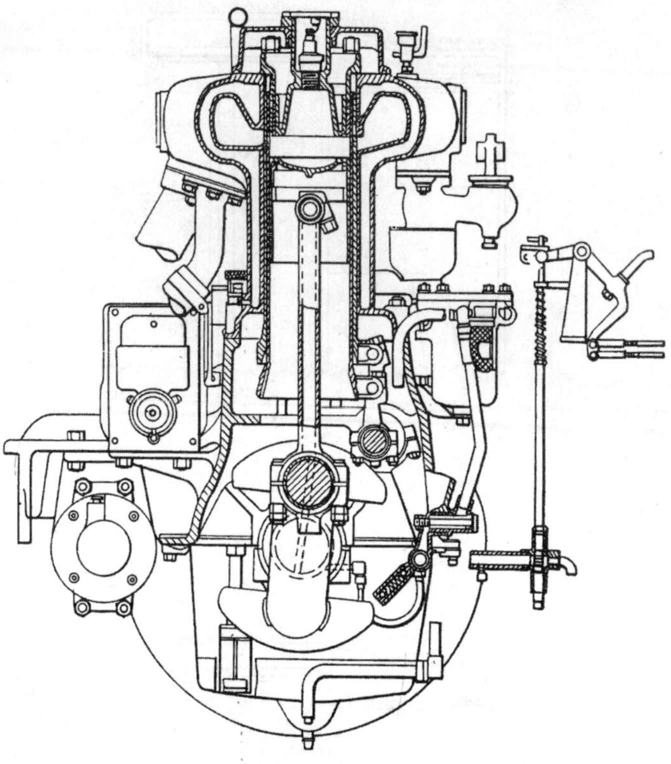

PRINCIPAL FEATURES OF THE KNIGHT ENGINE

The Valve Functions Are Performed by Two Concentric, Ported Sleeves, Generally of Cast Iron, Which Are Inserted between the Cylinder-Wall and the Piston. The Sleeves Are Given a Reciprocating Motion by Connection to an Eccentric Shaft Driven from the Crankshaft through the Usual 2 to 1 Gear, Their Stroke, in the Older Designs at Least, Being Either 1 or 1⅛ In. The Sleeves Project from the Cylinder at the Bottom and, at the Top, They Extend into an Annular Space between the Cylinder-Wall and the Special Form of Cylinder-Head So That, during the Compression and the Power Strokes, the Gases Do Not Come into Contact with the Cylinder-Wall But Are Separated Therefrom by Two Layers of Cast Iron and Two Films of Lubricating Oil. The Cylinder, As Well As Each Sleeve, Is Provided with an Exhaust-Port on One Side and with an Inlet-Port on the Opposite Side. The Passage for Either the Inlet or the Exhaust Is Open When All Three of the Ports on the Particular Side Are in Register with Each Other

FIGURE 1-5

Sectional view of Willy-Knight sleeve valve engine of 1926. Reprinted with permission from © 1995 Automotive Engineering magazine, SAE International.

automobile is the Wankel engine in several Mazda models. Mazda builds rotary automobile engines with one, two, and three rotors.

5. Position and Number of Cylinders of Reciprocating Engines (Fig. 1-7)

   (a) Single Cylinder. Engine has one cylinder and piston connected to the crankshaft.

   (b) In-Line. Cylinders are positioned in a straight line, one behind the other along the length of the crankshaft. They can consist of 2 to 11 cylinders or possibly more. In-line four-cylinder engines are very common for automobile and other applications. In-line six and eight cylinders are historically common automobile engines. In-line engines are sometimes called **straight** (e.g., straight six or straight eight).

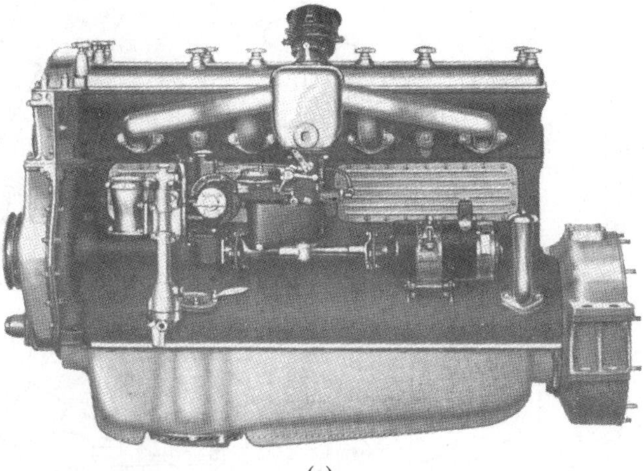

(a)

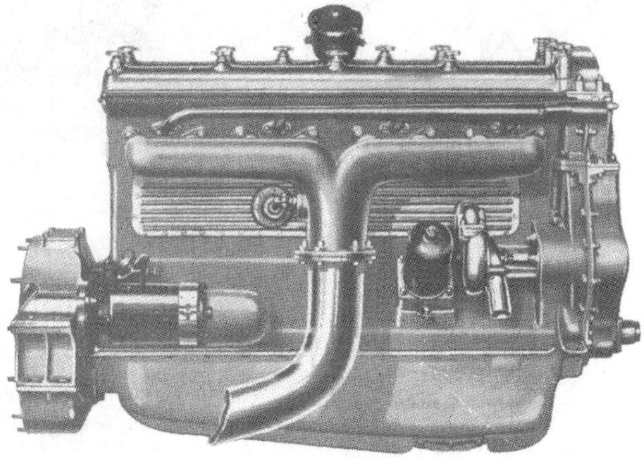

(b)

FIGURE 1-6

Duesenberg Model J in-line, eight-cylinder, spark-ignition automobile engine of the 1930s; in the author's opinion, the greatest automobile engine relative to its time and competition that there ever was or ever will be (powering Duesenbergs, the greatest automobiles ever). The engine was designed by the Duesenberg Company, but was manufactured by Lycoming Aircraft Engine Company. The engine had hemispherical combustion chambers with total displacement of 420 in.^{3} (6.88 L), $3\frac{3}{4}$ in. bore (9.53 cm), $4\frac{3}{4}$ in. stroke (12.07 cm), 5.2:1 compression ratio, and produced 265 bhp (198 kW) at 4200 RPM. When supercharged as the Model SJ it produced 320 bhp (239 kW) at 4000 RPM. Twin chain-driven overhead camshafts operated four valves per cylinder. Intake valves had $1\frac{1}{2}$ in. (3.81 cm) diameter and 0.35 in. (0.89 cm) lift, exhaust valves had $1\frac{7}{8}$ in. (4.76 cm) diameter and 0.36 in. (0.91 cm) lift. Aluminum pistons each had three compression rings and one oil ring. The alloy steel crankshaft had five main bearings, was balanced both dynamically and statically, and had a mercury-filled vibration damper. The lubrication system consisted of a gear pump in the oil pan, three filters, crankcase ventilation, and a finned aluminum oil pan for cooling. The seven gallon cooling system had a water pump, thermostatically controlled radiator shutters, and a belt-driven four-bladed fan. Ignition was six volt with Delco Remy coil and distributor. The Model J was equipped with a Schebler updraft carburetor, while the Model SJ had a Stromberg downdraft carburetor and a 12 in. centrifugal supercharger. In addition to luxurious automobiles, Duesenbergs were also very successful as racing cars in Grand Prix and track racing. The Duesenberg Automobile company went out of business in 1937, a victim of the Economic Depression. The slang expression "Its a Dusey," meaning it's very good, is a reference to the Duesenberg automobile. Printed with permission from Auburn–Cord–Duesenberg Museum, Auburn, Indiana, [147].

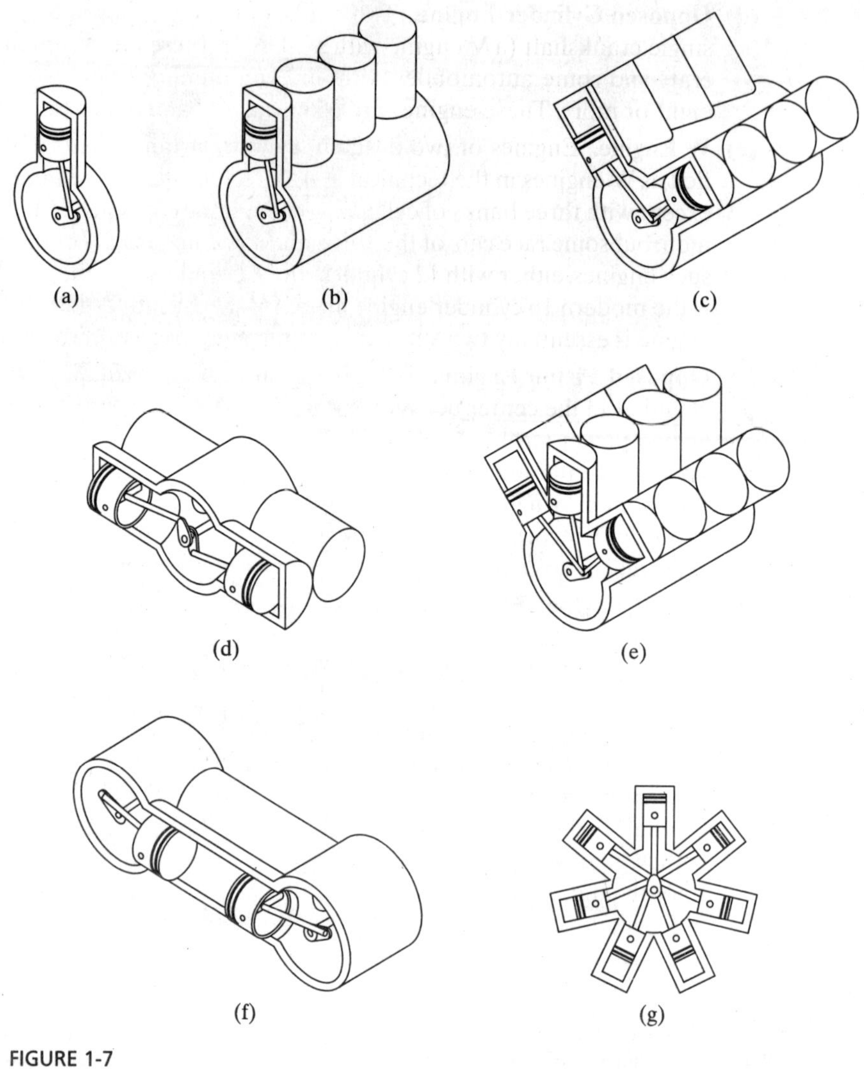

FIGURE 1-7

Engine Classification by Cylinder Arrangement. **(a)** Single cylinder. **(b)** In-line, or straight. **(c)** V engine. **(d)** Opposed cylinder. **(e)** W engine. **(f)** Opposed piston. **(g)** Radial.

(c) V Engine. Two banks of cylinders at an angle with each other along a single crankshaft, allowing for a shorter engine block. The angle between the banks of cylinders can be anywhere from 15° to 120°, with 60°–90° being common. V engines usually have even numbers of cylinders from 2 to 20 or more. V6s and V8s are common automobile engines, with V12s and V16s (historic) found in some luxury and high-performance vehicles. Large ship and stationary engines have anywhere from 8 to 20 cylinders. Volkswagen has a V5 on the market with two cylinders slightly out of line (15°) with the other three so that the cylinders can be moved closer together to shorten the engine block. Honda makes a true V5 motorcycle engine.

(d) **Opposed Cylinder Engine.** Two banks of cylinders opposite each other on a single crankshaft (a V engine with a 180° V). These are common on small aircraft and some automobiles with an even number of cylinders from two to eight or more. These engines are often called **flat** engines (e.g., flat four).

(e) **W Engine.** Engines of two different cylinder arrangements have been classified as W engines in the technical literature. One type is the same as a V engine except with three banks of cylinders on the same crankshaft. They are not common, but some race cars of the 1930s and some luxury cars of the late 1990s had such engines, either with 12 cylinders or 18 cylinders. Another type of W engine is the modern 16 cylinder engine made for the Bugatti automobile (W16). This engine is essentially two V8 engines connected together on a single crankshaft.

(f) **Opposed Piston Engine.** Two pistons in each cylinder with the combustion chamber in the center between the pistons. A single combustion process causes two power strokes at the same time, with each piston being pushed away from the center and delivering power to a separate crankshaft at each end of the cylinder. Engine output is either on two rotating crankshafts or on one crankshaft incorporating a complex mechanical linkage. These engines are generally of large displacement, used for powerplants, ships, or submarines.

(g) **Radial Engine.** Engine with pistons positioned in a circular plane around a central crankshaft. The connecting rods of the pistons are connected to a master rod which, in turn, is connected to the crankshaft (Fig. 1-12). A bank of cylinders on a radial engine almost always has an odd number of cylinders ranging from 3 to 13 or more. Operating on a four-stroke cycle, every other cylinder fires and has a power stroke as the crankshaft rotates, giving a smooth operation. Many medium and large size propeller-driven aircraft use radial engines. For large aircraft, two or more banks of cylinders are mounted together, one behind the other on a single crankshaft, making one powerful smooth engine. Very large ship engines exist with up to 54 cylinders, six banks of 9 cylinders each. In the early part of the 20th century there were a few experimental radial aircraft engines that had an even number of cylinders (4 to 12). These engines operated on a two-stroke cycle and never became standard.

HISTORIC—RADIAL ENGINES

There are a few historical examples of radial engines being mounted with the crankshaft fastened to the vehicle while the heavy bank of radial cylinders rotated around the stationary crank. This was done to enhance the cooling of the cylinders on these air cooled engines. The Sopwith Camel, a very successful World War I fighter aircraft, had the engine so mounted with the propeller fastened to the rotating bank of cylinders. The gyroscopic forces generated by the large rotating engine mass allowed these planes to do some maneuvers that were not possible with other airplanes, and restricted them from some other maneuvers. Snoopy has been flying a Sopwith Camel in his battles with the Red Baron for many years.

The little-known early Adams–Farwell automobiles had three- and five-cylinder radial engines rotating in a horizontal plane with the stationary crankshaft mounted vertically. The gyroscopic effects must have given these automobiles unique steering characteristics [45].

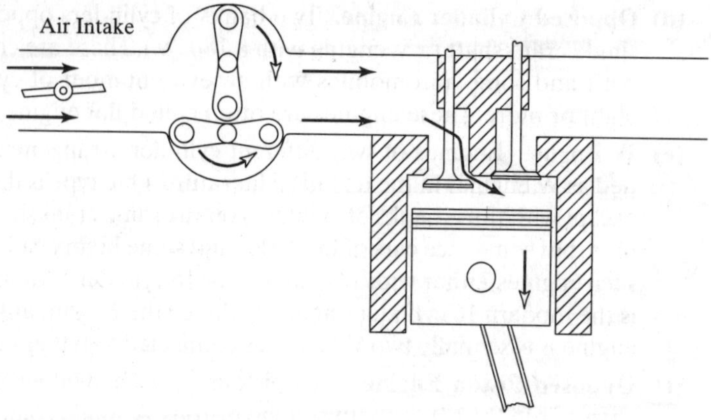

FIGURE 1-8

Supercharger used to increase inlet air pressure to engine. Compressor is driven off engine crankshaft, which gives fast response to speed changes but adds parasitic load to engine.

6. Air Intake Process
   (a) **Naturally Aspirated.** No intake air pressure boost system.
   (b) **Supercharged.** Intake air pressure increased with the compressor driven off of the engine crankshaft (Fig. 1-8).
   (c) **Turbocharged.** Intake air pressure increased with the turbine–compressor driven by the engine exhaust gases (Fig. 1-9).
   (d) **Crankcase Compressed.** Two-stroke cycle engine which uses the crankcase as the intake air compressor. Limited development work has also been done on design and construction of four-stroke cycle engines with crankcase compression.
7. Method of Fuel Input for Spark Ignition Engines
   (a) **Carbureted.**
   (b) **Multipoint Port Fuel Injection.** One or more injectors at each cylinder intake.
   (c) **Throttle Body Fuel Injection.** Injectors upstream in intake manifold.
   (d) **Gasoline Direct Injection.** Injectors mounted in combustion chambers with injection directly into cylinders.
8. Method of Fuel Input for Compression Ignition Engines
   (a) **Direct Injection.** Fuel injected into main combustion chamber.
   (b) **Indirect Injection.** Fuel injected into secondary combustion chamber.
   (c) **Homogeneous Charge Compression Ignition.** Some fuel added during intake stroke.

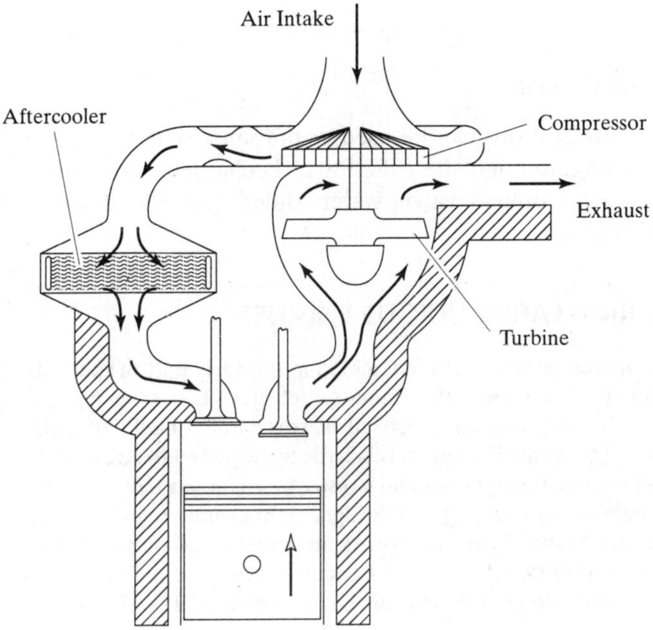

FIGURE 1-9

Turbocharger used to increase inlet air pressure to engine. Turbine that drives compressor is powered by exhaust flow from engine. This adds no load to the engine but results in turbo lag, a slower response to engine speed changes.

9. Fuel Used

(a) **Gasoline.**

(b) **Diesel Oil or Fuel Oil.**

(c) **Gas, Natural Gas, Methane.**

(d) **LPG.**

(e) **Alcohol—Ethyl, Methyl.**

(f) **Dual Fuel.** There are a number of engines that use a combination of two or more fuels. Some, usually large, CI engines use a combination of natural gas and diesel fuel. These are attractive in developing third-world countries because of the high cost of diesel fuel. Combined gasoline–alcohol fuels are becoming more common as an alternative to straight gasoline automobile engine fuel.

(g) **Gasohol.** Common fuel consisting of 90% gasoline and 10% alcohol.

10. Application

(a) **Automobile, Truck, Bus.**

(b) **Locomotive.**

(c) **Stationary.**

(d) **Marine.**

(e) **Aircraft.**

(f) **Small Portable, Chain Saw, Model Airplane.**

11. Type of Cooling

(a) Air Cooled.

(b) Liquid Cooled, Water Cooled.

Several or all of these classifications can be used at the same time to identify a given engine. Thus, a modern engine might be called a turbocharged, reciprocating, spark ignition, four-stroke cycle, overhead valve, water-cooled, gasoline, multipoint fuel-injected, V8 automobile engine.

HISTORIC—LARGE UNIQUE ENGINES

Several large very powerful unique engines were assembled in the early part of the 20th century, mostly for experimental tanks and other military vehicles [227]. One type of engine, made by the Italian company Bugatti, consisted of two parallel banks of eight cylinders each, constitúting a U16 engine. The two crankshafts were geared to one central shaft output. At least one engine of this type was later used in a Duesenberg race car. Bugatti also made an H32 engine by connecting four straight eight engines together, with each engine making up one arm of the H (up and down), and one central output crankshaft. A German engine consisted of two inverted V12 engines connected together on a single crankshaft (an M24 engine?). None of these engines saw commercial production.

FIGURE 1-10

General Motors 6.6 liter, V8, compression ignition, LB7 "Duramax 6600 Turbo Diesel Engine" used in 2002 model trucks. The engine has four overhead valves per cylinder, compression ratio of 17.5:1, bore of 10.3 cm (4.06 in.), stroke of 9.90 cm (3.90 in.), and a total displacement of 6599 cm^3 (403 $in.^3$). High pressure injection with common rail is used. Copyright General Motors Corporation, used with permission.

1.4 TERMINOLOGY AND ABBREVIATIONS

The following terms and abbreviations are commonly used in engine technology literature and will be used throughout this book. These should be learned to assure maximum understanding of the following chapters.

Internal Combustion (IC)

Spark Ignition (SI) An engine in which the combustion process in each cycle is started by use of a spark plug.

Compression Ignition (CI) An engine in which the combustion process starts when the air–fuel mixture self-ignites due to high temperature in the combustion chamber caused by high compression. CI engines are often called **diesel** engines, especially in the nontechnical community.

Top-Dead-Center (TDC) Position of the piston when it stops at the furthest point away from the crankshaft. *Top* because this position is at the top of most engines (not always), and *dead* because the piston stops at this point. Because in some engines top-dead-center is not at the top of the engine (e.g., horizontally opposed engines, radial engines, etc.), some sources call this position **Head-End-Dead-Center** (HEDC). Some sources call this position **Top-Center** (TC). When an occurrence in a cycle happens

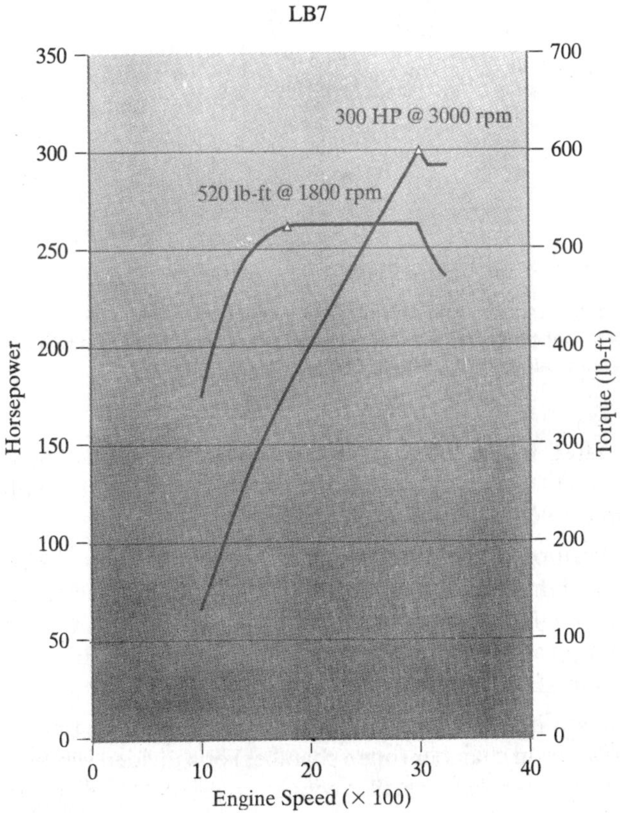

FIGURE 1-11

Power and torque curves of General Motors "Duramax 6600 Turbo Diesel" truck engine shown in Fig. 1-10. The engine has a brake maximum power rating of 300 hp (224 kW) at 3000 RPM, and a maximum torque of 520 lbf-ft (705 N-m) at 1800 RPM. Maximum speed is governed at 3250 RPM. Copyright General Motors Corporation, used with permission.

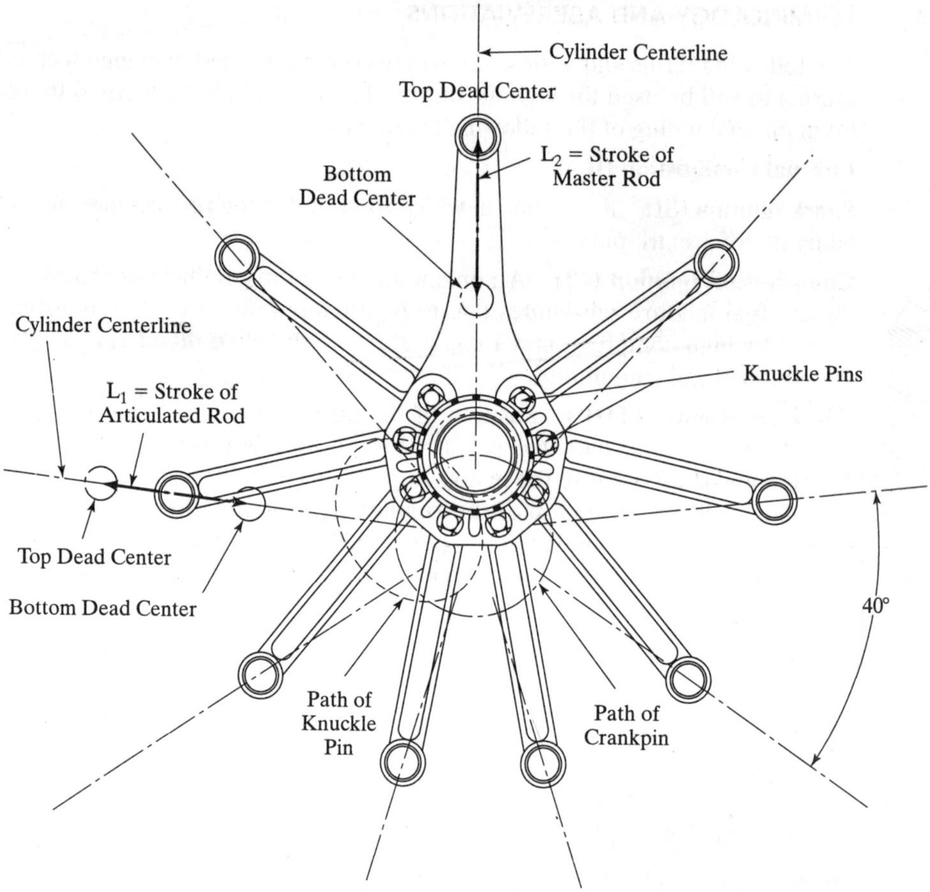

FIGURE 1-12

Connecting rod assembly for a nine-cylinder radial engine. The piston of number one cylinder is connected to the crankshaft by a master connecting rod. All other cylinders have connecting rods which fasten to the master rod. Printed with permission, Curtiss Wright Corporation.

before TDC, it is often abbreviated bTDC or bTC. When the occurrence happens after TDC, it will be abbreviated aTDC or aTC. When the piston is at TDC, the volume in the cylinder is a minimum called the *clearance volume*.

Bottom-Dead-Center (BDC) Position of the piston when it stops at the point closest to the crankshaft. Some sources call this **Crank-End-Dead-Center** (CEDC) because it is not always at the bottom of the engine. Some sources call this point **Bottom-Center** (BC). During an engine cycle things can happen before bottom-dead-center, bBDC or bBC, and after bottom-dead-center, aBDC or aBC.

Direct Injection (DI) Fuel injection into the main combustion chamber of an engine. Engines have either one main combustion chamber (open chamber) or a divided combustion chamber made up of a main chamber and a smaller connected secondary chamber.

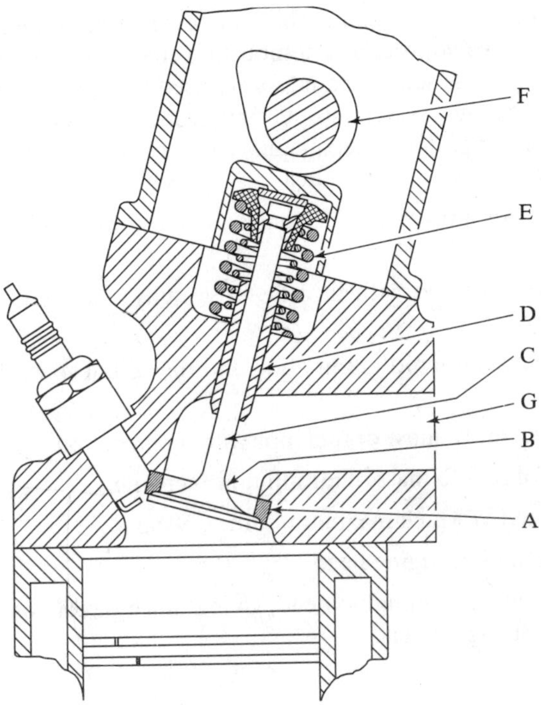

FIGURE 1-13

Poppet valve is spring loaded closed, and pushed open by cam action at proper time in cycle. Most automobile engines and other reciprocating engines use poppet valves. Much less common are sleeve valves and rotary valves. Components include: (A) valve seat, (B) head, (C) stem, (D) guide, (E) spring, (F) camshaft, (G) manifold.

Indirect Injection (IDI) Fuel injection into the secondary chamber of an engine with a divided combustion chamber.

Bore Diameter of the cylinder or diameter of the piston face, which is the same minus a very small clearance.

Stroke Movement distance of the piston from one extreme position to the other: TDC to BDC or BDC to TDC.

Clearance Volume Minimum volume in the combustion chamber with piston at TDC.

Displacement or Displacement Volume Volume displaced by the piston as it travels through one stroke. Displacement can be given for one cylinder or for the entire engine (one cylinder times number of cylinders). Some literature calls this ***swept volume***.

Gasoline Direct Injection (GDI) Spark ignition engine with fuel injectors mounted in combustion chambers. Gasoline fuel is injected directly into cylinders during compression stroke.

Homogeneous Charge Compression Ignition (HCCI) Compression-ignition engine operating with a homogeneous air–fuel charge instead of the diffusion combustion mixture normally used in CI engines.

Smart Engine Engine with computer controls that regulate operating characteristics such as air–fuel ratio, ignition timing, valve timing, exhaust control, intake tuning, etc. Computer inputs come from electronic, mechanical, thermal, and chemical sensors located throughout the engine. Computers in some automobiles are even programmed to adjust engine operation for things like valve wear and combustion chamber deposit buildup as the engine ages. In automobiles, the same computers are used to make *smart cars* by controlling the steering, brakes, exhaust system, suspension, seats, anti-theft systems, sound-entertainment systems, shifting, doors, repair analysis, navigation, noise suppression, environment, comfort, etc.

Engine Management System (EMS) Computer and electronics used to control smart engines.

Wide-Open Throttle (WOT) Engine operated with throttle valve fully open when maximum power and/or speed is desired.

Ignition Delay (ID) Time interval between ignition initiation and the actual start of combustion.

Air–Fuel Ratio (AF) Ratio of mass of air to mass of fuel input into engine.

Fuel–Air Ratio (FA) Ratio of mass of fuel to mass of air input into engine.

Brake Maximum Torque (BMT) Speed at which maximum torque occurs.

Overhead Valve (OHV) Valves mounted in engine head.

Overhead Cam (OHC) Camshaft mounted in engine head, giving more direct control of valves which are also mounted in engine head.

Fuel Injected (FI)

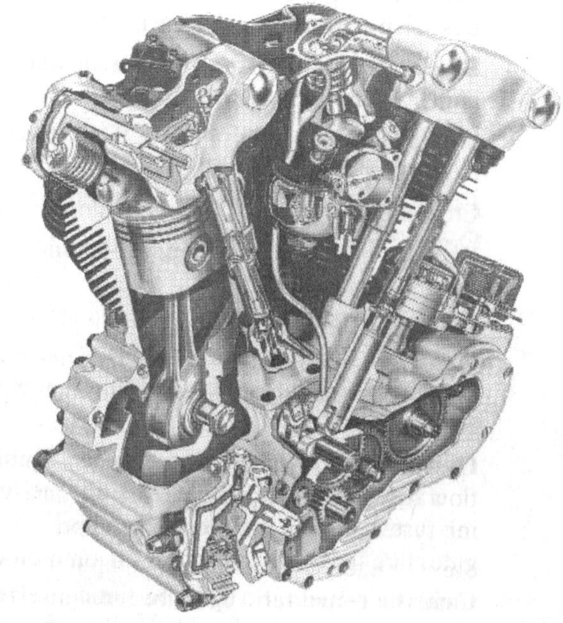

FIGURE 1-14

Harley–Davidson two-cylinder, air-cooled, overhead valve "Knucklehead" motorcycle engine first introduced in 1936. The 45° V engine had displacement of 60 cubic inches with 3.3125 inch bore and 3.500 inch stroke. Operating on a four-stroke cycle with a compression ratio of 7:1 the engine was rated at 40 bhp at 4800 RPM. Ignition was by Harley–Davidson generator–battery system. Photograph courtesy of the Harley–Davidson Juneau Avenue Archives. All rights reserved. Copyright Harley–Davidson.

FIGURE 1-15

Harley–Davidson motorcycle of 1936 powered by "Knucklehead" engine shown in Fig. 1-14. The motorcycle had a rated top speed of 90–95 MPH with a fuel economy of 35–50 MPG. Photograph courtesy of the Harley–Davidson Juneau Avenue Archives. All rights reserved. Copyright Harley–Davidson.

1.5 ENGINE COMPONENTS

The following is a list of major components found in most reciprocating internal combustion engines (see Fig. 1-16).

Block Body of engine containing the cylinders, made of cast iron or aluminum. In many older engines, the valves and valve ports were contained in the block. The block of water-cooled engines includes a water jacket cast around the cylinders. On air-cooled engines, the exterior surface of the block has cooling fins.

Camshaft Rotating shaft used to push open valves at the proper time in the engine cycle, either directly or through mechanical or hydraulic linkage (push rods, rocker arms, tappets). Most modern automobile engines have one or more camshafts mounted in the engine head (overhead cam). Most older engines had camshafts in the crankcase. Camshafts are generally made of forged steel or cast iron and are driven off the crankshaft by means of a belt or chain (timing chain). To reduce weight, some cams are made from a hollow shaft with the cam lobes press-fit on. In four-stroke cycle engines, the camshaft rotates at half engine speed.

Carburetor Venturi flow device that meters the proper amount of fuel into the air flow by means of a pressure differential. For many decades it was the basic fuel metering system on all automobile (and other) engines. It is still used on low-cost small engines like lawn mowers, but is uncommon on new automobiles.

Catalytic converter Chamber mounted in exhaust flow containing catalytic material that promotes reduction of emissions by chemical reaction.

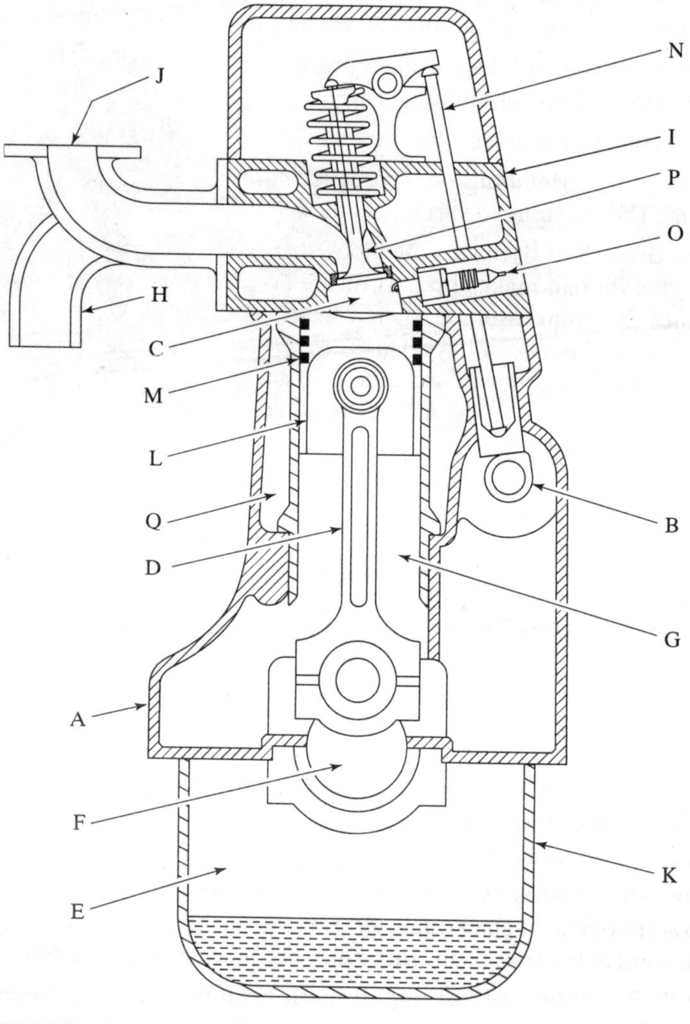

FIGURE 1-16

Cross-section of four-stroke cycle SI engine showing engine components: (A) block, (B) camshaft, (C) combustion chamber, (D) connecting rod, (E) crankcase, (F) crankshaft, (G) cylinder, (H) exhaust manifold, (I) head, (J) intake manifold, (K) oil pan, (L) piston, (M) piston rings, (N) push rod, (O) spark plug, (P) valve, and (Q) water jacket.

Choke Butterfly valve at carburetor intake, used to create rich fuel–air mixture in intake system for cold weather starting.

Combustion chamber The end of the cylinder between the head and the piston face where combustion occurs. The size of the combustion chamber continuously changes from a minimum volume when the piston is at TDC to a maximum when the piston is at BDC. The term "cylinder" is sometimes synonymous with "combustion chamber" (e.g., "the engine was firing on all cylinders"). Some engines have *open* combustion chambers which consist of one chamber for each cylinder. Other engines have *divided* chambers which consist of dual chambers on each cylinder connected by an orifice passage.

Connecting rod Linkage connecting piston with rotating crankshaft, usually made of steel, alloy forging, or aluminum.

Connecting rod bearing Bearing where connecting rod fastens to crankshaft.

Cooling fins Metal fins on the outside surfaces of cylinders and head of an air-cooled engine. These extended surfaces cool the cylinders by conduction and convection.

Crankcase Part of the engine block surrounding the rotating crankshaft. In many engines the oil pan makes up part of the crankcase housing. In some high-performance engines the crankcase is designed with "windows" between the piston bays to allow freer air flow between bays. This is to reduce air pressure buildup on the backside of the pistons during power and intake strokes.

Crankshaft Rotating shaft through which engine work output is supplied to external systems. The crankshaft is connected to the engine block with the *main bearings*. It is rotated by the reciprocating pistons through connecting rods connected to the crankshaft, offset from the axis of rotation. This offset is sometimes called *crank throw* or *crank radius*. Most crankshafts are made of forged steel, while some are made of cast iron.

Cylinders The circular cylinders in the engine block in which the pistons reciprocate back and forth. The walls of the cylinder have highly polished hard surfaces. Cylinders may be machined directly in the engine block, or a hard metal (drawn steel) sleeve may be pressed into the softer metal block. Sleeves may be dry sleeves, which do not contact the liquid in the water jacket, or wet sleeves, which form part of the water jacket. In a few engines, the cylinder walls are given a knurled surface to help hold a lubricant film on the walls. In some very rare cases, the cross section of the cylinder is not round.

Exhaust manifold Piping system that carries exhaust gases away from the engine cylinders, usually made of cast iron.

Exhaust system Flow system for removing exhaust gases from the cylinders, treating them, and exhausting them to the surroundings. It consists of an exhaust manifold that carries the exhaust gases away from the engine, a thermal or catalytic converter to reduce emissions, a muffler to reduce engine noise, and a tailpipe to carry the exhaust gases away from the passenger compartment.

Fan Most engines have an engine-driven fan to increase air flow through the radiator and through the engine compartment, which increases waste heat removal from the engine. Fans can be driven mechanically or electrically, and can run continuously or be used only when needed.

Flywheel Rotating mass with a large moment of inertia connected to the crankshaft of the engine. The purpose of the flywheel is to store energy and furnish a large angular momentum that keeps the engine rotating between power strokes and smooths out engine operation. After the electrical systems of automobiles evolve from 12 volts to 42 volts in the next decades the engine starter and generator will be built as part of a multipurpose flywheel. On some aircraft engines the propeller serves as the flywheel, as does the rotating blade on many lawn mowers.

Fuel injector A pressurized nozzle that sprays fuel into the incoming air on SI engines or into the cylinder on CI engines. On SI engines, fuel injectors are located at the intake valve ports on multipoint port injection systems, upstream at the intake manifold inlet on throttle body injection systems, and in the combustion chambers on direct injection systems.

Fuel pump Electrically or mechanically driven pump to supply fuel from the fuel tank (reservoir) to the engine. Many modern automobiles have an electric fuel pump mounted submerged in the fuel tank. Most small engines and some early automobiles had no fuel pump, relying on gravity feed.

HISTORIC—FUEL PUMPS

Lacking a fuel pump, it was necessary to back Model T Fords (1909–1927) up steep hills because of the location of the fuel tank relative to the engine.

Glow plug Small electrical resistance heater mounted inside the combustion chamber of many CI engines, used to preheat the chamber so that combustion will occur when first starting a cold engine. The glow plug is turned off after the engine is started.

Head The piece that closes the end of the cylinders, usually containing part of the clearance volume of the combustion chamber. The head is usually cast iron or aluminum, and bolts to the engine block. In some less common engines, the head is one piece with the block. The head contains the spark plugs in SI engines and the fuel injectors in CI engines and some SI engines. Most modern engines have the valves in the head, and many have the camshaft(s) positioned there also (overhead valves and overhead cam).

Head gasket Gasket that serves as a sealant between the engine block and the head where they bolt together. The gaskets are usually made in sandwich construction of metal and composite materials. Some engines use liquid head gaskets.

Intake manifold Piping system that delivers incoming air to the cylinders, usually made of cast metal, plastic, or composite material. In most SI engines, fuel is added to the air in the intake manifold system either by fuel injectors or with a carburetor. Some intake manifolds are heated to enhance fuel evaporation. The individual pipe to a single cylinder is called a *runner*.

Jake brake System that allows the engine to assist in slowing down a vehicle, usually large trucks. Actuation of the exhaust valves is modified so that some kinetic energy of the vehicle is absorbed by compressing air in the engine cylinders.

Main bearing The bearings connected to the engine block in which the crankshaft rotates. The maximum number of main bearings would be equal to the number of pistons plus one, or one between each set of pistons plus the two ends. On some less powerful engines, the number of main bearings is less than this maximum.

Oil pan Oil reservoir usually bolted to the bottom of the engine block, making up part of the crankcase. Acts as the oil sump for most engines.

Oil pump Pump used to distribute oil from the oil sump to required lubrication points. The oil pump can be electrically driven, but is most commonly mechanically driven by the engine. Some small engines do not have an oil pump and are lubricated by splash distribution.

Oil sump Reservoir for the oil system of the engine, commonly part of the crankcase. Some automobile engines with overhead camshafts have a secondary oil sump in the engine head to supply the cam and valve mechanism. Some engines (aircraft, race cars, etc.) have a separate closed reservoir called a *dry sump*.

Piston The cylindrically shaped mass that reciprocates back and forth in the cylinder, transmitting the pressure forces in the combustion chamber to the rotating crankshaft. The top of the piston is called the *crown* and the sides are called the *skirt*. The face on the crown makes up one wall of the combustion chamber and may be a flat or highly contoured surface. Some pistons contain an indented bowl in the crown, which makes up a large percent of the clearance volume. Pistons are made of cast iron, steel, or aluminum. Iron and steel pistons can have sharper corners because of their higher strength. They also have lower thermal expansion, which allows for tighter tolerances and less crevice volume. Aluminum pistons are lighter and have less mass inertia. Sometimes synthetic or composite materials are used for the body of the piston, with only the crown made of metal. Some pistons have a ceramic coating on the face.

Piston rings Metal rings that fit into circumferential grooves around the piston and form a sliding surface against the cylinder walls. Near the top of the piston are usually two or more compression rings made with highly polished surfaces. The purpose of these rings is to form a seal between the piston and cylinder walls and to restrict the high-pressure gases in the combustion chamber from leaking past the piston into the crankcase (blowby). Below the compression rings on the piston is at least one oil ring, which assists in lubricating the cylinder walls and scrapes away excess oil to reduce oil consumption.

Push rods Mechanical linkage between the camshaft and valves on overhead valve engines with the camshaft in the crankcase. Many push rods have oil passages through their length as part of a pressurized lubrication system.

Radiator Liquid-to-air heat exchanger of honeycomb construction used to remove heat from the engine coolant after the engine has been cooled. The radiator is usually mounted in front of the engine in the flow of air as the automobile moves forward. An engine-driven or electric fan is often used to increase air flow through the radiator.

Spark plug Electrical device used to initiate combustion in an SI engine by creating a high-voltage discharge spark across an electrode gap. Spark plugs are usually made of metal surrounded with ceramic insulation. Some modern spark plugs have built-in pressure sensors that supply one of the inputs into engine control.

Speed control–cruise control Automatic electric–mechanical control system that keeps the automobile operating at a constant speed by controlling engine speed.

Starter Several methods are used to start IC engines. Most are started by use of an electric motor (starter) geared to the engine flywheel. Energy is supplied from an electric battery. When automobiles change to 42-volt electrical systems, the starter and generator will become part of the flywheel.

On some very large engines, such as those found in large tractors and construction equipment, electric starters have inadequate power, and small IC engines are used as starters for the large IC engines. First the small engine is started with the normal electric motor, and then the small engine engages gearing on the flywheel of the large engine, turning it until the large engine starts.

Early aircraft engines were often started by hand spinning the propeller, which also served as the engine flywheel. Many small engines on lawn mowers and similar equipment are hand started by pulling a rope wrapped around a pulley connected to the crankshaft.

Compressed air is used to start some large engines. Cylinder release valves are opened, which keeps the pressure from increasing in the compression strokes. Compressed air is then introduced into the cylinders, which rotates the engine in a *freewheeling* mode. When rotating inertia is established, the release valves are closed and the engine is fired.

HISTORIC—STARTERS

Early automobile engines were started with hand cranks that connected with the crankshaft of the engine. This was a difficult and dangerous process, sometimes resulting in broken fingers and arms when the engine would fire and snap back the hand crank. The first electric starters appeared on the 1912 Cadillac automobiles, invented by C. Kettering, who was motivated when his friend was killed in the process of hand starting an automobile [45].

Supercharger Mechanical compressor powered off of the crankshaft, used to compress incoming air of the engine.

Throttle Butterfly valve mounted at the upstream end of the intake system, used to control the amount of air flow into an SI engine. Some small engines and stationary constant-speed engines have no throttle.

Turbocharger Turbine-compressor used to compress incoming air into the engine. The turbine is powered by the exhaust flow of the engine and thus takes very little useful work from the engine.

Valves Used to allow flow into and out of the cylinder at the proper time in the cycle. Most engines use *poppet valves*, which are spring loaded closed and pushed open by camshaft action (Fig. 1-13). Valves are mostly made of forged steel. Surfaces against which valves close are called **valve seats** and are made of hardened steel or ceramic. *Rotary valves* and *sleeve valves* are sometimes used, but are much less common. Many two-stroke cycle engines have *ports* (slots) in the side of the cylinder walls instead of mechanical valves.

Water jacket System of liquid flow passages surrounding the cylinders, usually constructed as part of the engine block and head. Engine coolant flows through the water jacket and keeps the cylinder walls from overheating. The coolant is usually a water–ethylene glycol mixture.

Water pump Pump used to circulate engine coolant through the engine and radiator. It is usually mechanically run off of the engine.

Wrist pin Pin fastening the connecting rod to the piston (also called *piston pin*).

1.6 BASIC ENGINE CYCLES

Most internal combustion engines, both spark ignition and compression ignition, operate on either a four-stroke cycle or a two-stroke cycle. These basic cycles are fairly standard for all engines, with only slight variations found in individual designs.

Four-Stroke SI Engine Cycle (Fig. 1-17)

1. *First Stroke: Intake Stroke or Induction* The piston travels from TDC to BDC with the intake valve open and exhaust valve closed. This creates an increasing volume in the combustion chamber, which in turn creates a vacuum. The resulting pressure differential through the intake system from atmospheric pressure on the outside to the vacuum on the inside causes air to be pushed into the cylinder. As the air passes through the intake system, fuel is added to it in the desired amount by means of fuel injectors or a carburetor.

2. *Second Stroke: Compression Stroke* When the piston reaches BDC, the intake valve closes and the piston travels back to TDC with all the valves closed. This compresses the air–fuel mixture, raising both the pressure and the temperature in the cylinder. The finite time required to close the intake valve means that actual compression doesn't start until sometime aBDC. Near the end of the compression stroke, the spark plug is fired and combustion is initiated.

3. *Combustion* Combustion of the air–fuel mixture occurs in a very short but finite length of time with the piston near TDC (i.e., nearly constant-volume combustion). It starts near the end of the compression stroke slightly bTDC and lasts into the power stroke slightly aTDC. Combustion changes the composition of the gas mixture to that of exhaust products and increases the temperature in the cylinder to a very high peak value. This, in turn, raises the pressure in the cylinder to a very high peak value.

4. *Third Stroke: Expansion Stroke or Power Stroke* With all valves closed, the high pressure created by the combustion process pushes the piston away from TDC. This is the stroke which produces the work output of the engine cycle. As the piston travels from TDC to BDC, cylinder volume is increased, causing pressure and temperature to drop.

5. *Exhaust Blowdown* Late in the power stroke, the exhaust valve is opened and exhaust blowdown occurs. Pressure and temperature in the cylinder are still high relative to the surroundings at this point, and a pressure differential is created through the exhaust system which is open to atmospheric pressure. This pressure differential causes much of the hot exhaust gas to be pushed out of the cylinder and through the exhaust system when the piston is near BDC. This exhaust gas carries away a high amount of enthalpy, which lowers the cycle thermal efficiency. Opening the exhaust valve before BDC reduces the work obtained during the power stroke but is required because of the finite time needed for exhaust blowdown.

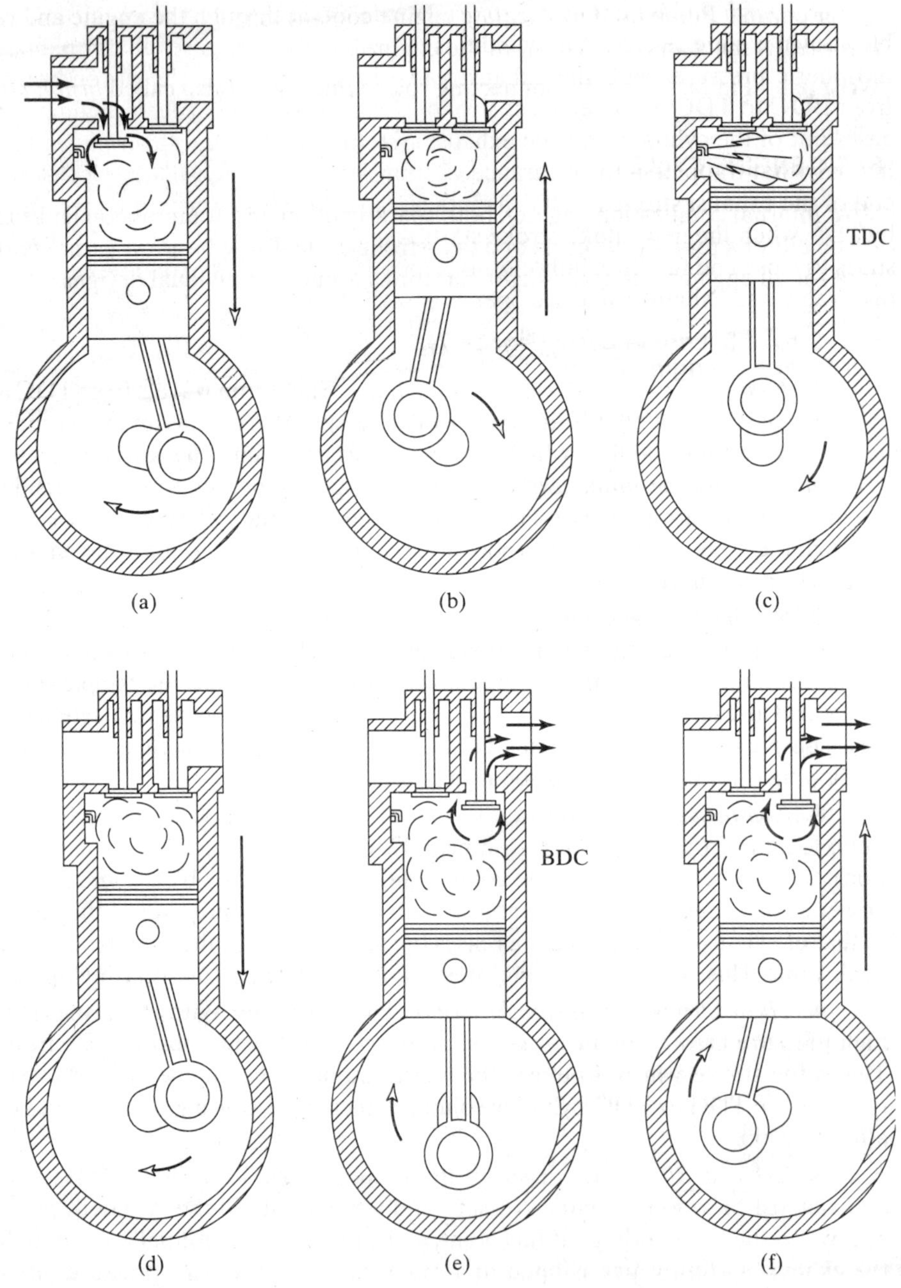

FIGURE 1-17

Four-stroke SI engine operating cycle. **(a)** Intake stroke. Ingress of air–fuel as piston moves from TDC to BDC. **(b)** Compression stroke. Piston moves from BDC to TDC. Spark ignition occurs near end of compression stroke. **(c)** Combustion at almost constant volume near TDC. **(d)** Power or expansion stroke. High cylinder pressure pushes piston from TDC towards BDC. **(e)** Exhaust blowdown when exhaust valve opens near end of expansion stroke. **(f)** Exhaust stroke. Remaining exhaust pushed from cylinder as piston moves from BDC to TDC.

6. *Fourth Stroke: Exhaust Stroke* By the time the piston reaches BDC, exhaust blowdown is complete, but the cylinder is still full of exhaust gases at approximately atmospheric pressure. With the exhaust valve remaining open, the piston now travels from BDC to TDC in the exhaust stroke. This pushes most of the remaining exhaust gases out of the cylinder into the exhaust system at about atmospheric pressure, leaving only that trapped in the clearance volume when the piston reaches TDC. Near the end of the exhaust stroke bTDC, the intake valve starts to open, so that it is fully open by TDC when the new intake stroke starts the next cycle. Near TDC the exhaust valve starts to close and finally is fully closed sometime aTDC. This period when both the intake valve and exhaust valve are open is called **valve overlap**.

Four-Stroke CI Engine Cycle

1. *First Stroke: Intake Stroke* The same as the intake stroke in an SI engine with one major difference: no fuel is added to the incoming air.
2. *Second Stroke: Compression Stroke* The same as in an SI engine except that only air is compressed and compression is to higher pressures and temperature. Late in the compression stroke fuel is injected directly into the combustion chamber, where it mixes with the very hot air. This causes the fuel to evaporate and self-ignite, causing combustion to start.
3. *Combustion* Combustion is fully developed by TDC and continues at about constant pressure until fuel injection is complete and the piston has started towards BDC.
4. *Third Stroke: Power Stroke* The power stroke continues as combustion ends and the piston travels towards BDC.
5. *Exhaust Blowdown* Same as with an SI engine.
6. *Fourth Stroke: Exhaust Stroke* Same as with an SI engine.

HISTORIC—DIESEL AIRCRAFT ENGINES

A number of attempts have been made to develop a CI engine for aircraft [120, 227]. In 1927 Detroit Diesel and Stinson Aircraft Companies, working together, built a nine-cylinder radial CI diesel engine for a midsize airplane. The engine showed promise, with good fuel economy in flight, but failed in becoming a commercial product due mainly to limited development priority. The Economic Depression of 1929 ended work on the project. The engine is now on display at the Henry Ford Museum in Dearborn, Michigan.

Two-Stroke SI Engine Cycle (Fig. 1-18)

1. *Combustion* With the piston at TDC combustion occurs very quickly, raising the temperature and pressure to peak values, almost at constant volume.
2. *First Stroke: Expansion Stroke or Power Stroke* Very high pressure created by the combustion process forces the piston down in the power stroke. The expanding volume of the combustion chamber causes pressure and temperature to decrease as the piston travels towards BDC.

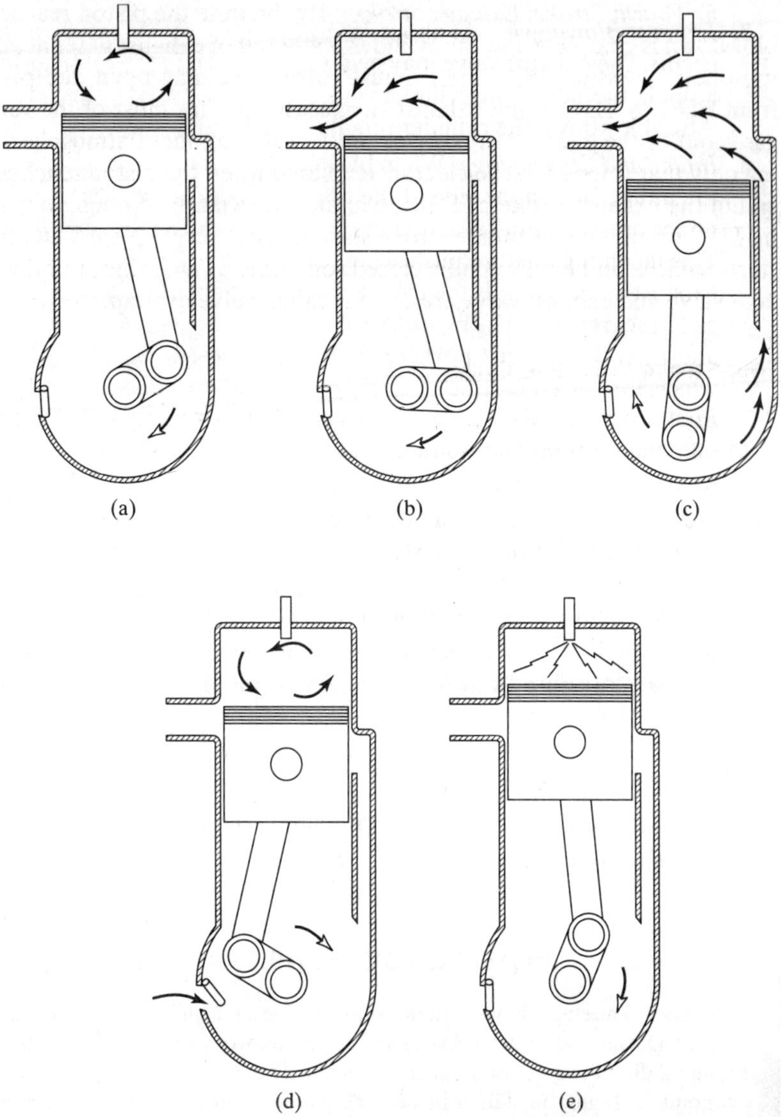

FIGURE 1-18

Two-stroke SI engine operating cycle with crankcase compression. **(a)** Power or expansion stroke. High cylinder pressure pushes piston from TDC towards BDC with all ports closed. Air in crankcase is compressed by downward motion of piston. **(b)** Exhaust blowdown when exhaust port opens near end of power stroke. **(c)** Cylinder scavenging when intake port opens and airfuel is forced into cylinder under pressure. Intake mixture pushes some of the remaining exhaust out the open exhaust port. Scavenging lasts until piston passes BDC and closes intake and exhaust ports. **(d)** Compression stroke. Piston moves from BDC to TDC with all ports closed. Intake air fills crankcase. Spark ignition occurs near end of compression stroke. **(e)** Combustion at almost constant volume near TDC.

3. *Exhaust Blowdown* At about 75° bBDC, the exhaust valve opens and blowdown occurs. The exhaust valve may be a poppet valve in the cylinder head, or it may be a slot in the side of the cylinder which is uncovered as the piston approaches BDC. After blowdown the cylinder remains filled with exhaust gas at lower pressure.
4. *Intake and Scavenging* When blowdown is nearly complete, at about 50° bBDC, the intake slot on the side of the cylinder is uncovered and intake air–fuel enters under pressure. Fuel is added to the air with either a carburetor or fuel injection. This incoming mixture pushes much of the remaining exhaust gases out the open exhaust valve and fills the cylinder with a combustible air–fuel mixture, a process called **scavenging.** The piston passes BDC and very quickly covers the intake port and then the exhaust port (or the exhaust valve closes). The higher pressure at which the air enters the cylinder is established in one of two ways. Large two-stroke cycle engines generally have a supercharger, while small engines will intake the air through the crankcase. On these engines the crankcase is designed to serve as a compressor in addition to serving its normal function.
5. *Second Stroke: Compression Stroke* With all valves (or ports) closed, the piston travels towards TDC and compresses the air–fuel mixture to a higher pressure and temperature. Near the end of the compression stroke, the spark plug is fired; by the time the piston gets to TDC, combustion occurs and the next engine cycle begins.

Two-Stroke CI Engine Cycle

The two-stroke cycle for a CI engine is similar to that of the SI engine, except for two changes. No fuel is added to the incoming air, so that compression is done on air only. Instead of a spark plug, a fuel injector is located in the cylinder. Near the end of the compression stroke, fuel is injected into the hot compressed air and combustion is initiated by self-ignition.

FIGURE 1-19

Ford Motor Company 1999 spark-ignition, four-stroke cycle V8 engine. Engine has bore of 9.02 cm (3.55 in.), stroke of 9.00 cm (3.54 in.), displacement of 4.60 L (281 in.^{3}), and compression ratio of 9.85:1. Maximum brake power is 205 kW at 5750 RPM (275 hp), with maximum torque of 373 N-m at 4750 RPM (275 lbf-ft). Courtesy of Ford Motor Company.

FIGURE 1-20

Ford 3.0 liter Vulcan V6, spark ignition, four-stroke cycle engine. This was the standard engine of the 1996 Ford Taurus and Mercury Sable automobiles. It is rated at 108 kW at 5250 RPM and develops 230 N-m of torque at 3250 RPM. Courtesy Ford Motor Company.

1.7 HYBRID VEHICLES

At the beginning of the 21st century a great amount of research and development is being put into a new type of vehicle utilizing an internal combustion engine. This is the **hybrid automobile** (or truck), which uses both an electric motor and an internal combustion engine to provide power for propelling the vehicle. The goals of this type of vehicle are to provide better fuel economy with fewer emissions.

Several different methods and motor–engine combinations have been and are being tried. The most common scenario is to use the electric motor(s) as the main driving force, with the IC engine running in steady-state operation to keep the batteries charged. The batteries then supply the electric energy to the one or more motors. Usually the combustion engine is connected such that it can also be used to help propel the vehicle directly when more power is needed than can be supplied by the motors.

It is accepted that without a connected IC engine, electric motors and batteries by themselves do not provide enough range and power under some conditions (e.g., higher speeds or going up a hill). This may change as fuel cell technology improves. On the other hand, the combination of an IC engine and an electric motor can produce enough power and range, while using less fuel and producing less emissions. A combustion engine by itself must be large enough to produce maximum power when needed, which is many times greater than the average power being used under most conditions. When utilized in conjunction with a generator–battery–motor system, a much smaller engine can be used and operated at near steady-state conditions. This allows the engine to be made smaller and optimized for one operating condition, resulting in a very efficient unit. When the batteries are fully charged and the electric motor has enough power to supply driving needs, the engine can be turned off, resulting in additional savings.

A number of different ways of combining an IC engine with an electric motor(s) are being tried. Some vehicles have electric drive on the front wheels and the combustion engine connected to the rear wheels and generator. Some vehicles incorporate the motor and engine connected in series, using one or both as needed. The extreme example of this is extending the engine crankshaft such that it also becomes the rotating output shaft of the motor. Other vehicles use a parallel connection of the motor and engine, with various ways of using one or both under different driving conditions (Fig. 1-21). Most systems allow the vehicle to be driven totally by the motor, totally by the engine, or a combination of both. With input from many sensors, the onboard computer determines the most efficient way of operation.

The internal combustion engines in most of the early hybrid automobiles are gasoline fueled SI engines with two to four cylinders, and having a displacement of about one liter. This probably will evolve to compression-ignition engines because of the higher thermal efficiency these offer. This will happen as CI (diesel) engines become cleaner burning and a more complete fuel distribution system is established. Future *advanced hybrids* may include fuel cells, gas turbines, Stirling cycle engines, etc.

In the 1990s the Partnership for a New Generation of Vehicles (PNGV), a cooperative research effort between industry and government, set standards for a *supercar*, the goals of which are shown in Table 1-1. A hybrid vehicle may be the only practical hope of ever meeting these standards.

Positive characteristics of a hybrid automobile:

1. Better fuel mileage. With a smaller engine operating at lean-burn steady-state conditions, the engine can be optimized for minimum fuel usage. As CI engines are phased in, the fuel savings will become even greater because of the higher thermal efficiency. Present hybrid automobiles available to the public obtain 50 to 80 miles per gallon (mpg) with gasoline (Fig 1-22).
2. Fewer emissions. Engines that run at one speed steady-state can be optimized to run much cleaner. Present available hybrid vehicles meet all present and most future pollution standards.
3. Combustion engines can be shut off when not needed or when the vehicle is momentarily stopped. When the vehicle is then again started, the startup can be very smooth with the help of the electric motor. Test automobiles powered totally by IC engines have been developed to run on this on–off mode of operation. However, without a large electric motor drive to help in startup, these vehicles suffer in starting time lapse and probable long-term starter deterioration.
4. The electric drive motors can be built as dual motor–generator units. This allows for some of the kinetic energy of the moving vehicle to be recovered when the vehicle is slowed down or stopped. The regained energy is used by the generator to recharge the battery.

Negative characteristics of a hybrid automobile:

1. High cost at present. The combination of many more components and lower sales volume make these vehicles much more costly to manufacture. Toyota expects to sell about 300,000 hybrid vehicles in year 2005 [218].
2. Vehicles must carry dual weight of two power units, engine and motor. Often only one of these units will be in use, with the other unit adding to the dead weight of the vehicle.

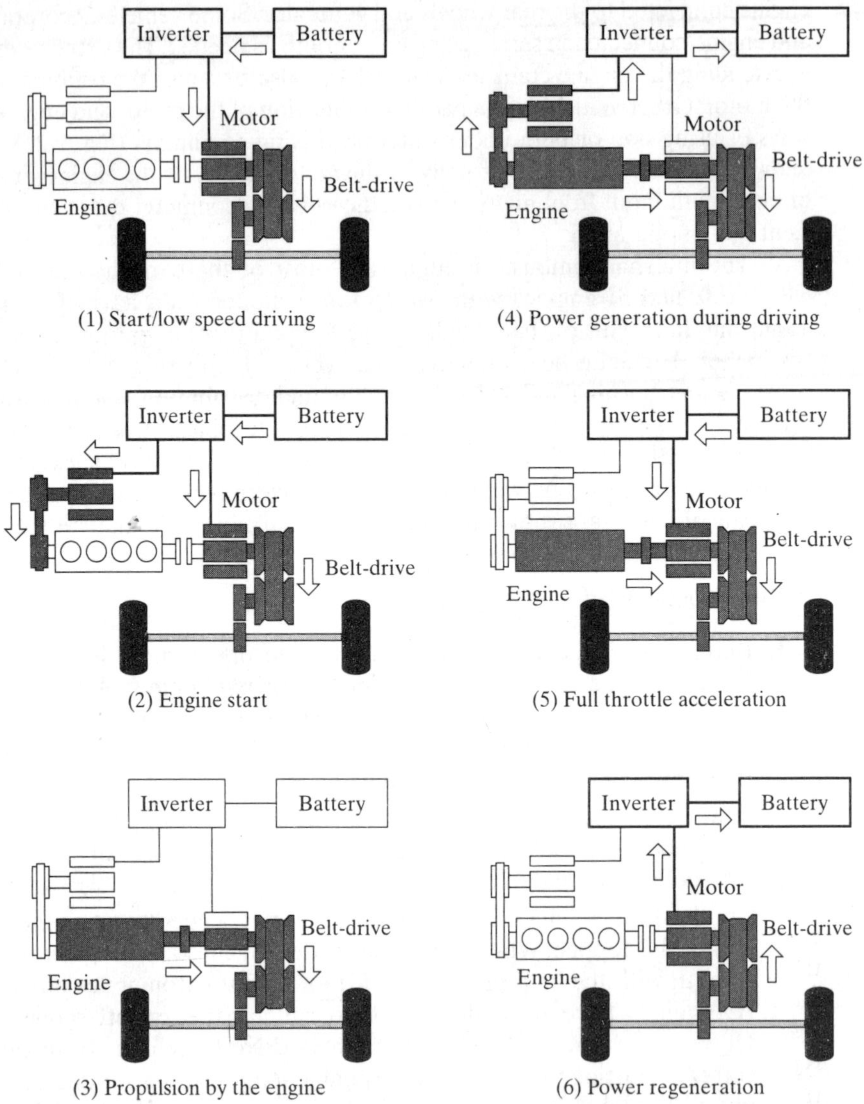

FIGURE 1-21

Various operational modes of a hybrid automobile. Reprinted with permission from SAE Paper No. 2000-01-0992 © 2000, SAE International, [204].

3. Any battery system will have a negative environmental impact when the batteries are used up and discarded. Better battery technology is needed.
4. Air conditioning and other auxiliary power requirements are more difficult to satisfy with an electrical system. The IC engine could not be turned off when auxiliary power is required. When the engine is turned off the engine coolant system would have to be circulated by electrical energy.

TABLE 1-1 PNGV Standards for Supercar

1.	Up to 80 miles per equivalent gallon of gasoline
2.	Acceleration 0–60 mph in 12 seconds
3.	Family sedan accommodating 5–6 passengers
4.	Exhaust emissions 0.125/1.7/0.2 gm/mile HC/CO/NOx at 100,000 miles
5.	Luggage capacity of 475 liters
6.	Range of at least 380 miles
7.	Ride, handling, noise, vibration, comfort equivalent to 1994 family sedan
8.	Total cost of ownership equivalent to 1994 family sedan (adjusted for economics)
9.	Minimum useful life of 100,000 miles
10.	Meet present and future safety standards
11.	Be at least 80% recyclable
12.	Concept vehicle demonstration by 2001
13.	Production prototype ready by 2004

HISTORIC—HYBRID-POWERED AUTOMOBILES

In 1916–1917 a Woods Dual Power automobile could be purchased for $2650. This vehicle had both an IC engine and an electric motor with regenerative braking [139].

1.8 FUEL CELL VEHICLES

Some hybrid vehicles in the future may replace the IC engine–generator system with a **fuel cell** stack to supply electricity to the batteries. Or, more likely, fuel cells will supply electricity directly to the drive motors of the vehicle, without the need for batteries or a combustion engine [162, 187, 188, 206]. The use of fuel cells to supply power for propelling vehicles is attractive in that it is a more efficient method of converting chemical energy of the fuel to useful power output. It also produces far fewer harmful emissions than a combustion engine, with water vapor being the major exhaust. Fuel cells generate electricity by reversing the electrolysis process of water. This can be done in one of several ways, a common type being a *proton exchange membrane cell (PEM)*. With this type of fuel cell, the unit is fueled with hydrogen H_2, and oxygen O_2 (air). As these fuels pass through the special membrane of the cell they chemically combine forming water vapor and low voltage electricity. The cells are *stacked* in series to create high enough voltage to supply the 60 to 90 kW of power needed to propel an automobile. At the beginning of the 21st century several major problem obstacles remain, and must be solved before fuel cell technology challenges the dominance of the internal combustion engine for vehicle propulsion. These problems include weight, heat rejection, freezing, cost, safety, startup time, refueling, and fuel storage.

The greatest technical problem is onboard storage of the hydrogen fuel, and the corresponding refueling of the vehicle. Hydrogen can be stored as a cryogenic liquid, a compressed gas, or as a solid in a metal hydride compound. None of these methods is ideal and major development work is needed before any of these become practical (see Section 4-6). To circumvent the fuel storage problem, methods of generating the

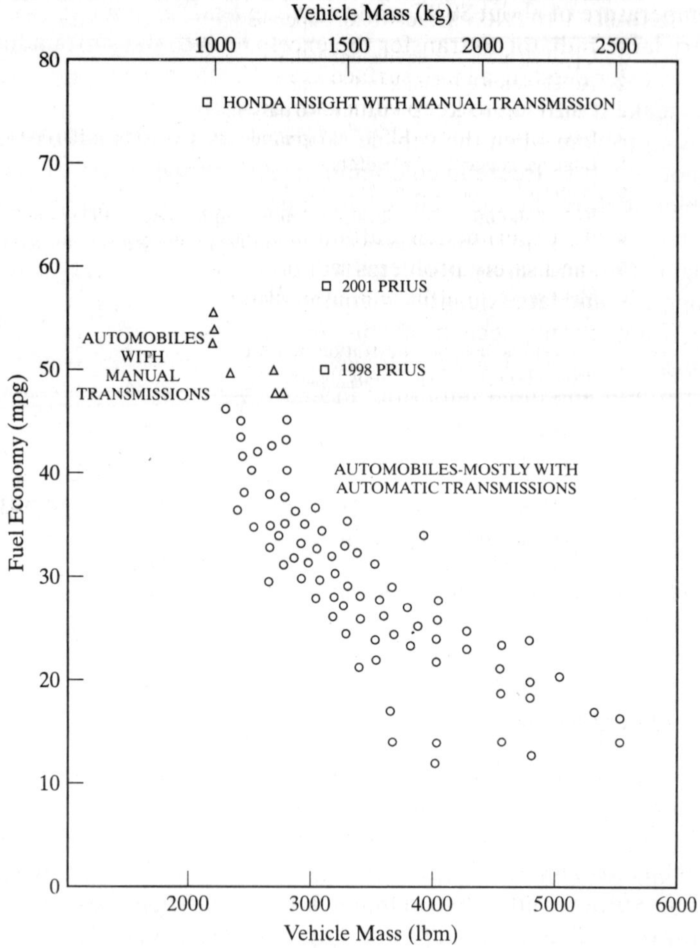

FIGURE 1-22
Fuel economy as a function of vehicle mass for automobiles powered with internal combustion engines. Also shown are data for three hybrid automobiles, the Honda Insight and two models of the Toyota Prius. Adapted from [228].

hydrogen onboard as it is needed seem to be the practical solution. This is done using methanol, gasoline, or some other hydrocarbon fuel to supply the needed hydrogen. These fuels, which are easily stored and refueled as a liquid, are passed through a catalytic *reformer* that converts them to H_2, CO, and CO_2. The H_2 is sent to the fuel cell as needed, while the CO and CO_2 are exhausted to the atmosphere after the CO is converted to CO_2. Methanol is the best of these fuels and supplies the most hydrogen, but there are other problems in using methanol (see Section 4-6).

The various components (membrane, reformer, etc.) of a fuel cell operate at different temperatures, creating unique cooling demands. Many PEM systems operate at a temperature of about 80°C, while creating a fair amount of excess heat. This requires a very large radiator to transfer the excess heat to the surroundings, low temperature heat transfer needing a large surface area. The large air supply openings for these radiators make it difficult to keep vehicle drag coefficients low. In northern climates, freezing is a problem when the vehicle is turned off. Water is generated in the fuel cell and cannot be left to freeze in cold temperatures. When cold, there is a startup time delay problem before the fuel cell reaches operating temperature, usually on the order of seconds. Some experimental automobiles are equipped with a battery for startup. Weight, cost, and safety problems will be improved upon by evolution, additional development, and large quantity manufacturing.

There are no mechanical components in a reformer–fuel cell system so there are no mechanical breakdowns. However, there is chemical cycling and thermal cycling wear. Sulfur and other impurities in fuel and air contaminate cell membranes, reformers, and other components. Evidence and predictions are that, with cleaner fuel and adequate filtration, fuel cells will last the lifetime of an automobile.

1.9 ENGINE EMISSIONS AND AIR POLLUTION

The exhaust of automobiles is one of the major contributors to the world's air pollution problem. Recent research and development has made major reductions in engine emissions, but a growing population and a greater number of automobiles means that the problem will exist for many years to come.

During the first half of the 1900s, automobile emissions were not recognized as a problem, mainly due to the lower number of vehicles. As the number of automobiles grew along with more power plants, home furnaces, and the population in general, air pollution became an ever-increasing problem. During the 1940s, the problem was first seen in the Los Angeles area due to the high density of people and automobiles, as well as unique weather conditions. By the 1970s, air pollution was recognized as a major problem in most cities of the United States as well as in many large urban areas around the world.

Laws were passed in the United States and in other industrialized countries which limit the amount of various exhaust emissions that are allowed. This put a major restriction on automobile engine development during the 1980s and 1990s. Although harmful emissions produced by engines have been reduced by over 90% since the 1940s, they are still a major environmental problem.

Four major emissions produced by internal combustion engines are hydrocarbons (HC), carbon monoxide (CO), oxides of nitrogen (NOx), and solid particulates (part). Hydrocarbons are fuel and lubrication particles that did not get burned and smaller non-equilibrium particles of partially burned fuel. Carbon monoxide occurs when insufficient oxygen is present to fully convert all carbon to CO_2 or when incomplete air–fuel mixing occurs due to the very short engine cycle time. Oxides of nitrogen are created in an engine when high combustion temperatures cause some normally stable N_2 to dissociate into monatomic nitrogen N, which then combines with reacting oxygen. Particulates are solid carbon particles formed in compression ignition engines and are seen as black smoke in the exhaust of these engines. Other emissions found in the exhaust of engines include aldehydes, sulfur, lead, phosphorus, etc.

Two methods are used to reduce harmful engine emissions. One is to improve the technology of engines and fuels so that better combustion occurs and fewer emissions are generated. The second method is aftertreatment of the exhaust gases. This is done by using thermal converters or catalytic converters that promote chemical reactions in the exhaust flow. These chemical reactions convert the harmful emissions to acceptable CO_2, H_2O, and N_2.

In Chapter 2, methods of classifying emissions will be introduced. Chapter 9 studies emissions and aftertreatment methods in detail.

PROBLEMS

1.1 List five differences between SI engines and CI engines.

1.2 A four-stroke cycle engine may or may not have a pressure boost (supercharger, turbocharger) in the intake system. Why must a two-stroke cycle engine always have an intake pressure boost?

1.3 List two advantages of a two-stroke cycle engine over a four-stroke cycle engine. List two advantages of a four-stroke cycle engine over a two-stroke cycle engine.

1.4 **(a)** Why do most very small engines operate on a two-stroke cycle?
(b) Why do most very large engines operate on a two-stroke cycle?
(c) Why do most automobile engines operate on a four-stroke cycle?
(d) Why would it be desirable to operate automobile engines on a two-stroke cycle?

1.5 A single-cylinder vertical atmospheric engine with a 1.2 m bore and a piston of 2700 kg mass is used to lift a weight. Pressure in the cylinder after combustion and cooling is 22 kPa, while ambient pressure is 98 kPa. Assume piston motion is frictionless.
Calculate:

(a) Mass that can be lifted if the vacuum is at the top of the cylinder and the piston moves up. [kg]
(b) Mass that can be lifted if the vacuum is at the bottom of the cylinder and the piston moves down. [kg]

1.6 An early atmospheric engine has a single horizontal cylinder with a 3.2-ft bore, a 9.0-ft stroke, and no clearance volume. After a charge of gunpowder is set off in the open cylinder, the conditions in the cylinder are ambient pressure and a temperature of 540°F. The piston is now locked in position and the end of the cylinder is closed. After cooling to ambient temperature, the piston is unlocked and allowed to move. The power stroke is at constant temperature and lasts until pressure equilibrium is obtained. Assume the gas in the cylinder is air and piston motion is frictionless. Ambient conditions are 70°F and 14.7 psia.
Calculate:

(a) Possible lifting force at start of power stroke. [lbf]
(b) Length of effective power stroke. [ft]
(c) Cylinder volume at end of power stroke. [ft^3]

1.7 Two automobile engines have the same total displacement volume and the same total power produced within the cylinders.

List the possible advantages of:

(a) A V6 over a straight six.
(b) A V8 over a V6.
(c) A V6 over a V8.
(d) An opposed cylinder four over a straight four.
(e) An in-line six over an in-line four.

1.8 A nine cylinder, four-stroke cycle, radial SI engine operates at 900 RPM. Calculate:

(a) How often ignition occurs, in degrees of engine rotation.
(b) How many power strokes per revolution.
(c) How many power strokes per second.

1.9 A family wishes to buy a new midsize automobile. The two options they consider are (a) a standard automobile with an SI engine that gets 31 miles per gallon fuel economy (mpg) and costs \$18,000; and (b) a hybrid automobile that gets 82 mpg and costs \$32,000. On average, the family drives 16,000 miles each year, and gasoline costs \$1.65 per gallon. Calculate:

(a) Amount of gasoline each vehicle would use each year. [gal]
(b) Gasoline cost savings of hybrid over standard. [\$/year]
(c) Time it would take to make up the difference in vehicle cost, with fuel savings. Disregard any interest on car loans and time-worth of money. [months]

DESIGN PROBLEMS

1.1D Design a single-cylinder atmospheric engine capable of lifting a mass of 1000 kg to a height of three meters. Assume reasonable values of cylinder temperature and pressure after combustion. Decide which direction the cylinder will move, and give the bore, piston travel distance, mass of piston, piston material, and clearance volume. Give a sketch of the mechanical linkage to lift the mass.

1.2D Design an alternate fuel engine to be used in a large truck by designating all engine classifications used in Section 1-3.

1.3D Design a four-stroke cycle for an SI engine using crankcase compression. Draw schematics of the six basic processes: intake, compression, combustion, expansion, blowdown, and exhaust. Describe fully the intake of air, fuel, and oil.

CHAPTER 2

Operating Characteristics

"Electric, hydrocarbon, and steam power are already successfully established in use. But their specialization is only beginning. Each has highways and byways to be explored, theories to test and sift, and principles to be demonstrated. This important work falls not only to the professional designer and builder, but to all whose interest prompts either suggestion or co-operative effort."

The Place of the Automobile

by *Robert Bruce* (1900)

This chapter examines the operating characteristics of reciprocating internal combustion engines. These include the mechanical output parameters of work, torque, and power; the input requirements of air, fuel, and combustion; efficiencies; and emission measurements of engine exhaust.

2.1 ENGINE PARAMETERS

For an engine with bore B (see Fig. 2-1), crank offset a, stroke length S, turning at an engine speed of N,

$$S = 2a \tag{2-1}$$

The average piston speed is

$$\overline{U}_p = 2SN \tag{2-2}$$

N is generally given in RPM (revolutions per minute), $\overline{U}_p$ in m/sec (ft/sec), and B, a, and S in m or cm (ft or in.).

The maximum average piston speed for all engines will normally be in the range of 5 to 20 m/sec (15 to 65 ft/sec), with large diesel engines on the low end and high-performance automobile engines on the high end. There are two reasons why engines

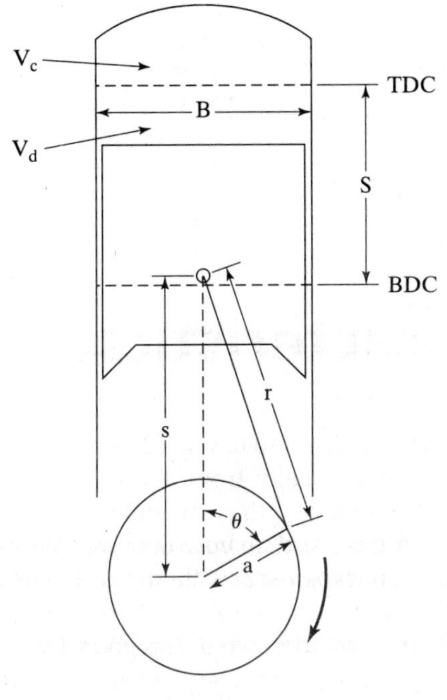

FIGURE 2-1

Piston and cylinder geometry of reciprocating engine. B = bore; S = stroke; r = connecting rod length; a = crank offset; s = piston position; θ = crank angle; V_c = clearance volume; V_d = displacement volume.

operate in this range. First, this is about the safe limit that can be tolerated by material strength of the engine components. For each revolution of the engine, each piston is accelerated twice from stop to maximum speed and back to stop. At a typical engine speed of 3000 RPM, each revolution lasts 0.02 sec (0.005 sec at 12,000 RPM). If engines operated at higher speeds, there would be a danger of material failure in the pistons and connecting rods as the pistons are accelerated and decelerated during each stroke. From Eq. (2-2), it can be seen that this range of acceptable piston speeds also defines a range of acceptable engine speeds, depending on engine size. There is a strong inverse correlation between engine size and operating speed. Very large engines with bore sizes on the order of 0.5 m (1.6 ft) typically operate in the 200 to 400 RPM range, while the very smallest engines (model airplane) with bores on the order of 1 cm (0.4 in.) operate at speeds of 12,000 RPM and higher. Race cars give us an example of engines being run at speeds above this safe range (e.g., Indianapolis 500 Race). These engines are generally operated at piston speeds up to 35 m/sec and engine speeds upwards of 14,000 RPM. Even though these engines receive far better care than the average automobile engine, a percentage of them experience failure within a few hundred miles. Table 2-1 gives representative values of engine speeds and operating variables for engines of various sizes. Automobile engines usually operate in a speed range of 500 to 5000 RPM, with a cruising speed of about 2500 RPM.

The second reason why maximum average piston speed is limited is because of the gas flow into and out of the cylinders. Piston speed determines the instantaneous flow rate of air–fuel into the cylinder during intake and exhaust flow out of the cylinder during the exhaust stroke. Higher piston speeds would require larger valves to allow for higher flow rates. In most engines, valves are at a maximum size with no room for enlargement.

TABLE 2-1 Typical Engine Operating Parameters

	Model Airplane Two-Stroke Cycle	Automobile Four-Stroke Cycle	Large Stationary Two-Stroke Cycle
Bore (cm)	2.00	9.42	50.0
Stroke (cm)	2.04	9.89	161
Displacement/cyl (L)	0.0066	0.69	316
Speed (RPM)	13,000	5200	125
Power/cyl (kW)	0.72	35	311
Average Piston Speed (m/sec)	8.84	17.1	6.71
Power/Displacement (kW/L)	109	50.7	0.98
bmep (kPa)	503	1170	472

Bore sizes of engines range from 0.5 m down to 0.5 cm (20 in. to 0.2 in.). The ratio of bore to stroke, B/S, for small engines is usually from 0.8 to 1.2. An engine with $B = S$ is often called a *square* engine. If stroke length is longer than bore diameter the engine is *under square*, and if stroke length is less than bore diameter the engine is *over square*. Very large engines are always under square, with stroke lengths up to four times bore diameter.

The distance between the crank axis and wrist pin axis is given by

$$s = a \cos \theta + \sqrt{r^2 - a^2 \sin^2 \theta} \tag{2-3}$$

where

a = crankshaft offset
r = connecting rod length
θ = crank angle, which is measured from the cylinder centerline and is zero when the piston is at TDC

When s is differentiated with respect to time, the instantaneous piston speed U_p is obtained:

$$U_p = ds/dt \tag{2-4}$$

The ratio of instantaneous piston speed divided by the average piston speed can then be written as

$$U_p/\bar{U}_p = (\pi/2) \sin \theta [1 + (\cos \theta / \sqrt{R^2 - \sin^2 \theta})] \tag{2-5}$$

where

$$R = r/a \tag{2-6}$$

R is the ratio of connecting rod length to crank offset and usually has values of 3 to 4 for small engines, increasing to 5 to 10 for the largest engines. Figure 2-2 shows the effect of R on piston speed.

Displacement, or displacement volume, V_d, is the volume displaced by the piston as it travels from BDC to TDC:

$$V_d = V_{BDC} - V_{TDC} \tag{2-7}$$

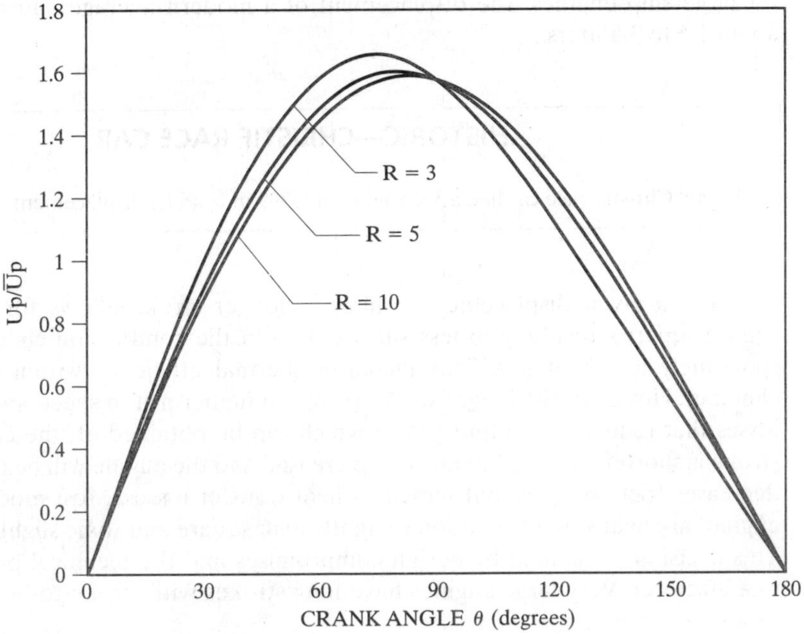

FIGURE 2-2

Instantaneous piston speed relative to average piston speed as a function of crank angle for various R values, where $R = r/a$, r = connecting rod length, a = crankshaft offset.

Some books call this swept volume. Displacement can be given for one cylinder or for the entire engine. For one cylinder,

$$V_d = (\pi/4)B^2S \tag{2-8}$$

For an engine with N_c cylinders,

$$V_d = N_c(\pi/4)B^2S \tag{2-9}$$

where

B = cylinder bore

S = stroke

N_c = number of engine cylinders

Engine displacements can be given in m^3, cm^3, $in.^3$, and, most commonly, in liters (L).

$$1\ L = 10^{-3}\ m^3 = 10^3\ cm^3 = 61.2\ in.^3$$

Typical values for engine displacement range from 0.1 cm^3 (0.0061 $in.^3$) for small model airplanes, to about 8 L (490 $in.^3$) for large automobiles, to much larger numbers for large ship engines. The displacement of a modern average automobile engine is about 1.5 to 3.5 liters.

HISTORIC—CHRISTIE RACE CAR

A 1908 Christie race car had a V4 engine of 2799 $in.^3$ (46 L) displacement [45].

For a given displacement volume, a longer stroke allows for a smaller bore (under square), resulting in less surface area in the combustion chamber and correspondingly less heat loss. This increases thermal efficiency within the combustion chamber. However, the longer stroke results in higher piston speed and higher friction losses that reduce the output power which can be obtained off the crankshaft. If the stroke is shortened, the bore must be increased and the engine will be over square. This decreases friction losses but increases heat transfer losses. Most modern automobile engines are near square, with some slightly over square and some slightly under square. This decision is dictated by design compromises and the technical philosophy of the manufacturer. Very large engines have long strokes with stroke-to-bore ratios as high as 4:1.

Minimum cylinder volume occurs when the piston is at TDC and is called the clearance volume V_c. We have

$$V_c = V_{TDC} \tag{2-10}$$

$$V_{BDC} = V_c + V_d \tag{2-11}$$

HISTORIC—SMALL HIGH-SPEED ENGINES

Engines for model airplanes and boats have been built with total displacement volumes as low as 0.075 cm^3 (0.0046 $in.^3$). Some of these engines, which are commercially available, can operate at speeds up to 38,000 RPM and have power output on the order of 0.15 to 1.5 kW (0.2 to 2.0 hp). It is interesting that the average piston speeds of these engines at high speed still fall within the general range of 5 to 20 m/sec.

The compression ratio of an engine is defined as

$$r_c = V_{BDC}/V_{TDC} = (V_c + V_d)/V_c = v_{BDC}/v_{TDC} \tag{2-12}$$

Modern spark ignition (SI) engines have compression ratios of 8 to 11, while compression ignition (CI) engines have compression ratios in the range 12 to 24. Engines with superchargers or turbochargers usually have lower compression ratios than naturally aspirated engines. Because of limitations in engine materials, technology, and fuel quality, very early engines had low compression ratios, on the order of 2 to 3.

FIGURE 2-3

Fairbanks Morse 10 cylinder PC4.2 diesel ship engine, capable of producing over 8000 brake horsepower (5970 kW). Printed with permission, Fairbanks Morse Engine Division, Coltec Industries.

FIGURE 2-4

Cox air-cooled, single cylinder, two-stroke cycle model airplane engine. Engine has displacement of 0.01 cubic inches (0.164 cm^3). Photo by Tuescher Photography, Platteville, Wisconsin.

Figure 2-5 shows how values of r_c increased over time to the 8–11 range used on modern spark ignition automobile engines. This limit of 8 to 11 is imposed mainly by gasoline fuel properties (see Section 4-4) and force limitations allowable in smaller high-speed engines.

Various attempts have been made to develop engines with a variable compression ratio. One unsuccessful early attempt used a split piston that expanded due to changing hydraulic pressure caused by engine speed and load. Three recent engine systems with a capability of changing compression ratio as the engine is running have been introduced. One system rotates the top of the engine block slightly to change the size of the clearance volume and compression ratio. Another system does this by using a pivot lever arm connected between the connecting rod and crankshaft. A third system uses secondary pistons in divided combustion chambers and operates on an Alvar cycle. These systems are examined in Chapter 7.

HISTORIC—CRANKSHAFTS

One interesting four-cylinder engine made by Austin of England had only two main bearings, one on each end of the crankshaft. To counteract the effect of crankshaft bending during operation, the two center cylinders were given a slightly higher static compression ratio than the two end cylinders [11].

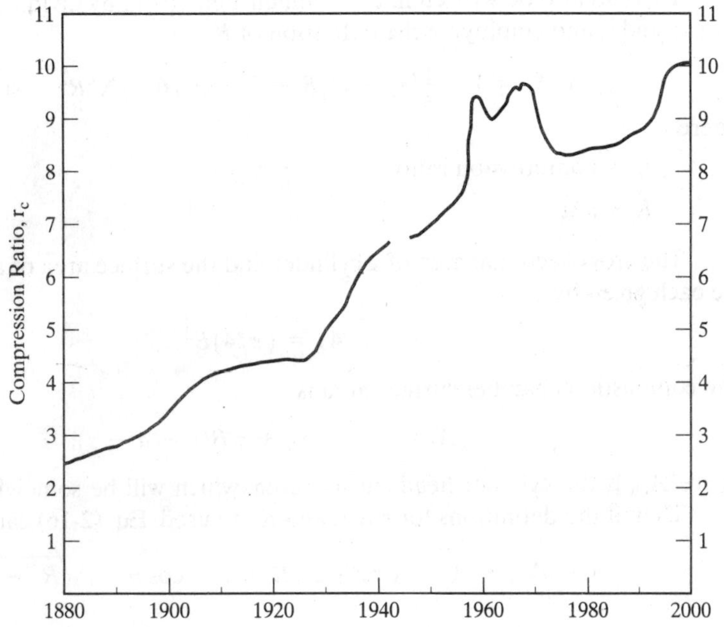

FIGURE 2-5

Average compression ratio of American spark ignition automobile engines as a function of year. During the first 40 years, compression ratios slowly increased from 2.5 to 4.5, limited mainly by low octane numbers of the available fuels. In 1923 TEL was introduced as a fuel additive and this was followed by a rapid increase in compression ratios. No automobiles were manufactured during 1942–1945 when production was converted to war vehicles during World War II. A rapid rise in compression ratios occurred during the 1950s when muscle cars became popular. During the 1970s TEL was phased out as a fuel additive, pollution laws were enacted, and gasoline became expensive due to an oil embargo imposed by some oil producing countries. These factors resulted in lower compression ratios during this time. In the 1980s and 1990s better fuels and combustion chamber technology allowed for higher compression ratios. Adapted from [5].

The cylinder volume at any crank angle is

$$V = V_c + (\pi B^2/4)(r + a - s) \tag{2-13}$$

where

V_c = clearance volume
B = bore
r = connecting rod length
a = crank offset
s = piston position shown in Fig. 2-1

This can also be written in a nondimensional form by dividing by V_c, substituting for r, a, and s, and employing the definition of R.

$$V/V_c = 1 + \tfrac{1}{2}(r_c - 1)[R + 1 - \cos\theta - \sqrt{R^2 - \sin^2\theta}] \tag{2-14}$$

where

r_c = compression ratio

$R = r/a$

The cross-sectional area of a cylinder and the surface area of a flat-topped piston are each given by

$$A_p = (\pi/4)B^2 \tag{2-15}$$

The combustion chamber surface area is

$$A = A_{ch} + A_p + \pi B(r + a - s) \tag{2-16}$$

where A_{ch} is the cylinder head surface area, which will be somewhat larger than A_p.

Then if the definitions for r, a, s, and R are used, Eq. (2-16) can be rewritten as

$$A = A_{ch} + A_p + (\pi BS/2)[R + 1 - \cos\theta - \sqrt{R^2 - \sin^2\theta}] \tag{2-17}$$

Example Problem 2-1

John's automobile has a three-liter SI V6 engine that operates on a four-stroke cycle at 3600 RPM. The compression ratio is 9.5, the length of the connecting rods is 16.6 cm, and the engine is square $(B = S)$. At this speed, combustion ends at 20° aTDC.
Calculate:

1. cylinder bore and stroke length
2. average piston speed
3. clearance volume of one cylinder
4. piston speed at the end of combustion
5. distance the piston has traveled from TDC at the end of combustion
6. volume in the combustion chamber at the end of combustion

(1) For one cylinder, using Eq. (2-8) with $S = B$ yields

$$V_d = V_{total}/6 = 3\text{L}/6 = 0.5\text{ L} = 0.0005\text{ m}^3 = (\pi/4)B^2S = (\pi/4)B^3$$

$$\underline{B = 0.0860\text{ m} = 8.60\text{ cm} = S}$$

(2) Use Eq. (2-2) to find the average piston speed:

$$\bar{U}_p = 2SN = (2\text{ strokes/rev})(0.0860\text{ m/stroke})(3600/60\text{ rev/sec})$$

$$\underline{= 10.32\text{ m/sec}}$$

(3) Use Eq. (2-12) to find the clearance volume of one cylinder:

$$r_c = 9.5 = (V_d + V_c)/V_c = (0.0005 + V_c)/V_c$$

$$\underline{V_c = 0.000059\text{ m}^3 = 59\text{ cm}^3}$$

(4) Crank offset, $a = S/2 = 0.0430 \text{ m} = 4.30 \text{ cm}$

$$R = r/a = 16.6 \text{ cm}/4.30 \text{ cm} = 3.86$$

Use Eq. (2-5) to find the instantaneous piston speed:

$$\begin{aligned} U_p/\overline{U}_p &= (\pi/2) \sin\theta[1 + (\cos\theta/\sqrt{R^2 - \sin^2\theta})] \\ &= (\pi/2) \sin(20°)\{1 + [\cos(20°)/\sqrt{(3.86)^2 - \sin^2(20°)}]\} \\ &= 0.668 \\ U_p &= 0.668\,\overline{U}_p = (0.668)(10.32 \text{ m/sec}) = \underline{6.89 \text{ m/sec}} \end{aligned}$$

(5) Use Eq. (2-3) to find the piston position:

$$\begin{aligned} s &= a\cos\theta + \sqrt{r^2 - a^2\sin^2\theta} \\ &= (0.0430 \text{ m})\cos(20°) + \sqrt{(0.166 \text{ m})^2 - (0.0430 \text{ m})^2\sin^2(20°)} \\ &= 0.206 \text{ m} \end{aligned}$$

The distance from TDC is

$$\begin{aligned} x &= r + a - s = (0.166 \text{ m}) + (0.043 \text{ m}) - (0.206 \text{ m}) \\ &= \underline{0.003 \text{ m} = 0.3 \text{ cm}} \end{aligned}$$

(6) Use Eq. (2-14) to find the instantaneous volume:

$$\begin{aligned} V/V_c &= 1 + \tfrac{1}{2}(r_c - 1)[R + 1 - \cos\theta - \sqrt{R^2 - \sin^2\theta}] \\ &= 1 + \tfrac{1}{2}(9.5 - 1)[3.86 + 1 - \cos(20°) - \sqrt{(3.86)^2 - \sin^2(20°)}] \\ &= 1.32 \\ V &= 1.32\,V_c = (1.32)(59 \text{ cm}^3) = \underline{77.9 \text{ cm}^3 = 0.0000779 \text{ m}^3} \end{aligned}$$

These results indicate that, during combustion, the volume in the combustion chamber has increased by only a very small amount and shows that combustion in an SI engine occurs at almost constant volume at TDC.

2.2 WORK

Work is the output of any heat engine, and in a reciprocating IC engine this work is generated by the gases in the combustion chamber of the cylinder. Work is the result of a force acting through a distance. That is, the force due to gas pressure on the moving piston generates the work in an IC engine cycle.

$$W = \int F\,dx = \int P A_p\,dx \qquad \text{(2-18)}$$

where

P = pressure in combustion chamber
A_p = area against which the pressure acts (i.e., the piston face)
x = distance the piston moves

and

$$A_p\,dx = dV \tag{2-19}$$

dV is the differential volume displaced by the piston as it travels a distance dx, so the work done can be written

$$W = \int P\,dV \tag{2-20}$$

Figure 2-6, which plots the engine cycle on P–V coordinates, is often called an indicator diagram. Early indicator diagrams were generated by mechanical plotters linked directly to the engine. Modern P–V indicator diagrams are generated on an oscilloscope using a pressure transducer mounted in the combustion chamber and an electronic position sensor mounted on the piston or crankshaft.

Because engines are often multicylinder, it is convenient to analyze engine cycles per unit mass of gas m within the cylinder. To do so, volume V is replaced with specific volume v and work is replaced with specific work:

$$w = W/m \qquad v = V/m \tag{2-21}$$

$$w = \int P\,dv \tag{2-22}$$

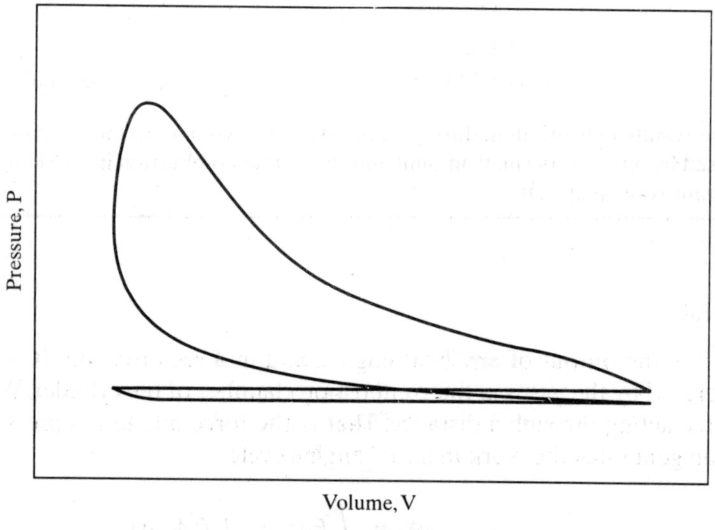

FIGURE 2-6

Indicator diagram for a typical four stroke cycle SI engine at WOT. An indicator diagram plots cylinder pressure as a function of combustion chamber volume over a 720° cycle. The diagram is generated on an oscilloscope using a pressure transducer mounted in the combustion chamber and a position sensor mounted on the piston or crankshaft.

FIGURE 2-7

2002 General Motors L47 Oldsmobile engine. The spark ignition 90° V8 engine has a displacement of 4.0 L (245 in.^{3}) and produces 250 hp (186 kW) of brake power. Copyright General Motors Corporation, used with permission.

The specific work w is equal to the area under the process lines on the P–v coordinates of Fig. 2-9.

If P represents the pressure inside the cylinder combustion chamber, then Eq. (2-22) and the areas shown in Fig. 2-9 give the work inside the combustion chamber. This is called **indicated work**. Work delivered by the crankshaft is less than indicated work, due to mechanical friction and parasitic loads of the engine. Parasitic loads include the oil pump, supercharger, air conditioner compressor, alternator, etc. Actual work available at the crankshaft is called **brake work,**

$$w_b = w_i - w_f \tag{2-23}$$

where

w_i = indicated specific work generated inside combustion chamber
w_f = specific work lost due to friction and parasitic loads

Units of specific work will be kJ/kg or BTU/lbm.

The upper loop of the engine cycle in Fig. 2-9 consists of the compression and power strokes where output work is generated and is called the **gross indicated work** (areas A and C in Fig. 2-9). The lower loop, which includes the intake and exhaust strokes, is called **pump work** and absorbs work from the engine (areas B and C). **Net indicated work** is

$$w_{net} = w_{gross} + w_{pump} \tag{2-24}$$

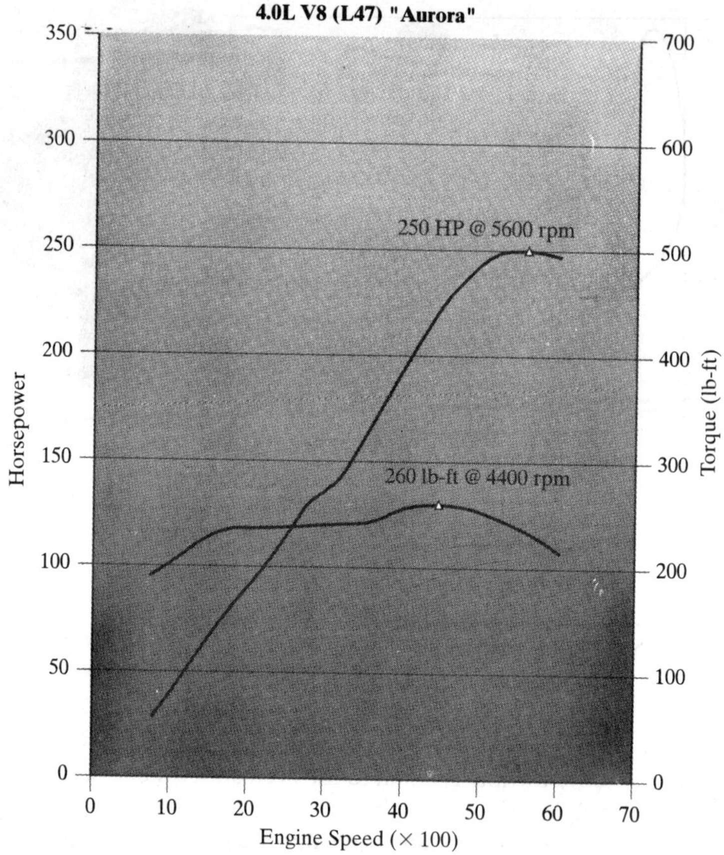

FIGURE 2-8

Power and torque curves of General Motors L47 Aurora engine shown in Fig. 2-7. In stock form the engine produces a maximum brake power of 250 hp at 5600 RPM (186 kW), and maximum torque of 260 lbf-ft at 4400 RPM (352 N-m). In modified racing form this engine produced 675 hp (503 kW), and generated a phenomenal record in the Indy Racing League (IRL) during the years 1997–2001. Race cars with this engine had the pole position in 51 out of 51 races and it won 49 of these races including 5 out of 5 of the Indianapolis 500 races. It had the fastest lap in 31 consecutive IRL races [159]. Copyright General Motors Corporation, used with permission.

Pump work w_{pump} is negative for engines without superchargers, so

$$w_{net} = (\text{Area A}) - (\text{Area B}) \tag{2-25}$$

Engines with superchargers or turbochargers can have intake pressure greater than exhaust pressure, giving a positive pump work (Fig. 2-10). When this occurs,

$$w_{net} = (\text{Area A}) + (\text{Area B}) \tag{2-26}$$

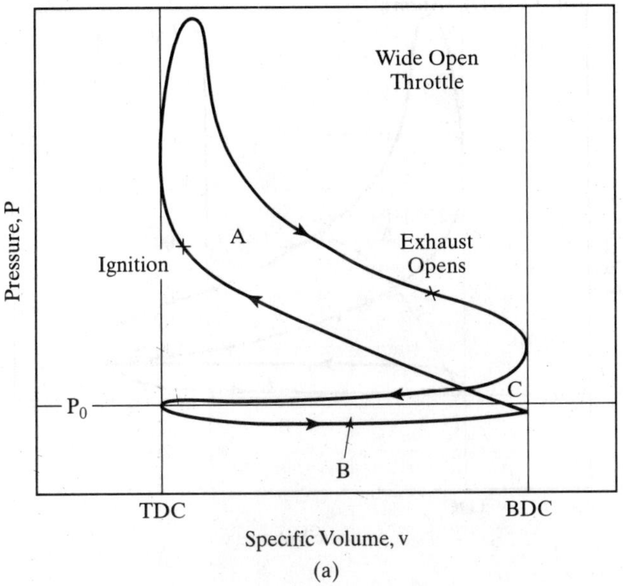

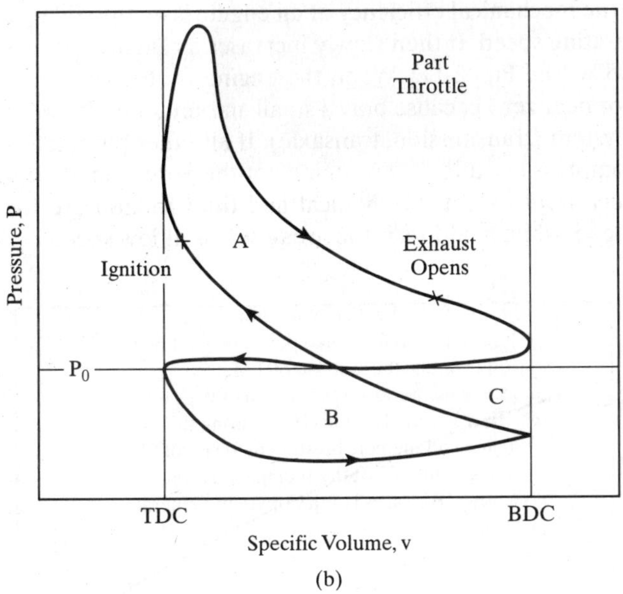

FIGURE 2-9

Four-stroke cycle of typical SI engine plotted on P-v coordinates at (a) wide open throttle, and (b) part throttle. The upper loop consists of the compression stroke and power stroke and the area represents gross indicated work. The lower loop represents negative work of the intake stroke and exhaust stroke. This is called indicated pump work.

Superchargers increase net indicated work but add to the friction work of the engine since they are driven by the crankshaft.

The ratio of brake work at the crankshaft to indicated work in the combustion chamber defines the **mechanical efficiency** of an engine:

$$\eta_m = w_b/w_i = W_b/W_i \tag{2-27}$$

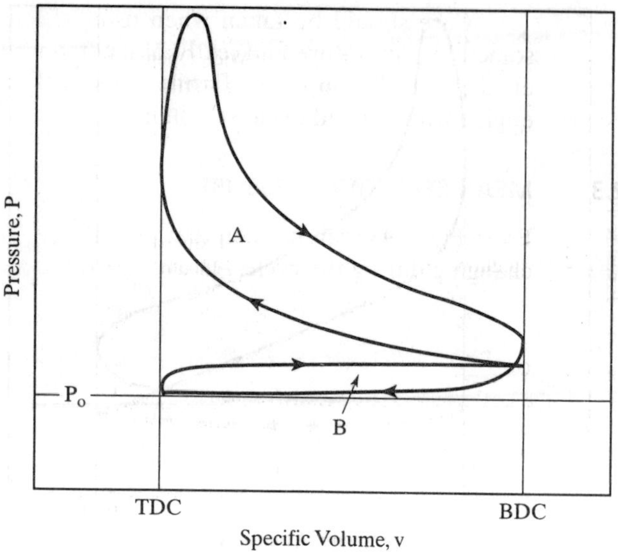

FIGURE 2-10
Four-stroke cycle of an SI engine equipped with a supercharger or turbocharger, plotted on P-v coordinates. For this cycle, intake pressure is greater than exhaust pressure and the pump work loop represents positive work.

Excluding parasitic loads, the mechanical efficiency of an engine is on the order of 55% to 60% at high engine operating speed. It then slowly increases as engine speed is decreased to the order of 85–95% (see Fig. 2-11). When the engine is at idle the mechanical efficiency falls to zero or near zero because only a small amount of brake work is being absorbed in the drive system (transmission, transaxle). If all other parameters are kept constant, neither the compression ratio of the engine nor the bore diameter affect mechanical efficiency to any great extent. Mechanical and fluid friction are the greatest energy losses at high speed, while heat loss is the greatest loss at low speed.

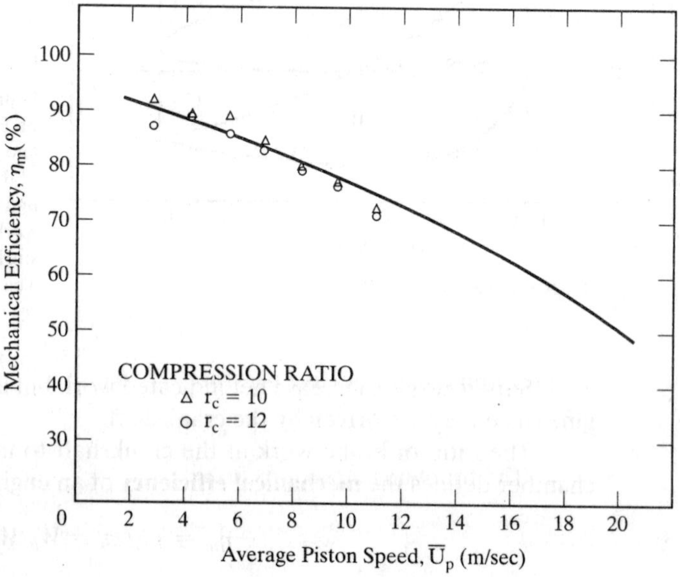

FIGURE 2-11
Mechanical efficiency of reciprocating internal combustion engines as a function of average piston speed. Data points and curve adapted from references [93, 197].

Care should be taken when using the terms "gross work" and "net work". In some older literature and textbooks, net work (or net power) meant the output of an engine with all components, while gross work (or gross power) meant the output of the engine with fan and exhaust system removed.

2.3 MEAN EFFECTIVE PRESSURE

From Fig. 2-9 it can be seen that pressure in the cylinder of an engine is continuously changing during the cycle. An average or **mean effective pressure** (mep) is defined by

$$w = (\text{mep})\Delta v \tag{2-28}$$

or

$$\text{mep} = w/\Delta v = W/V_d \tag{2-29}$$

$$\Delta v = v_{\text{BDC}} - v_{\text{TDC}} \tag{2-30}$$

where

W = work of one cycle
w = specific work of one cycle
V_d = displacement volume

Mean effective pressure is a good parameter for comparing engines with regard to design or output because it is independent of both engine size and speed. If torque is used for engine comparison, a larger engine will always look better. If power is used as the comparison, speed becomes very important.

Various mean effective pressures can be defined by using different work terms in Eq. (2-29). If brake work is used, **brake mean effective pressure** is obtained:

$$\text{bmep} = w_b/\Delta v \tag{2-31}$$

Indicated work gives **indicated mean effective pressure**.

$$\text{imep} = w_i/\Delta v \tag{2-32}$$

The imep can further be divided into gross indicated mean effective pressure and net indicated mean effective pressure:

$$(\text{imep})_{\text{gross}} = (w_i)_{\text{gross}}/\Delta v \tag{2-33}$$

$$(\text{imep})_{\text{net}} = (w_i)_{\text{net}}/\Delta v \tag{2-34}$$

Pump mean effective pressure (which can have negative values) is given by

$$\text{pmep} = w_{\text{pump}}/\Delta v \tag{2-35}$$

and **friction mean effective pressure** is given by

$$\text{fmep} = w_f/\Delta v \tag{2-36}$$

The following equations relate some of the previous definitions:

$$\text{nmep} = \text{gmep} + \text{pmep} \qquad \text{(a)}$$

$$\text{bmep} = \text{nmep} - \text{fmep} \qquad \text{(b)}$$

$$\text{bmep} = \eta_m \,\text{imep} \qquad \text{(c)}$$
$$\text{bmep} = \text{imep} - \text{fmep} \qquad \text{(d)} \qquad \text{(2-37)}$$

where

nmep = net mean effective pressure
η_m = mechanical efficiency of engine

Typical maximum values of bmep for naturally aspirated SI engines are in the range of 850 to 1050 kPa (120 to 150 psi). For CI engines, typical maximum values are 700 to 900 kPa (100 to 130 psi) for naturally aspirated engines and 1000 to 1200 kPa (145 to 175 psi) for turbocharged engines [58].

2.4 TORQUE AND POWER

Torque is a good indicator of an engine's ability to do work. It is defined as force acting at a moment distance and has units of N-m or lbf-ft. Torque τ is related to work by

$$2\pi\tau = W_b = (\text{bmep})\, V_d/n \qquad \text{(2-38)}$$

where

W_b = brake work of one revolution
V_d = displacement volume
n = number of revolutions per cycle

For a two-stroke cycle engine with one cycle for each revolution,

$$2\pi\tau = W_b = (\text{bmep})V_d \qquad \text{(2-39)}$$
$$\tau = (\text{bmep})V_d/2\pi \qquad \text{two-stroke cycle} \qquad \text{(2-40)}$$

For a four-stroke cycle engine that takes two revolutions per cycle,

$$\tau = (\text{bmep})V_d/4\pi \qquad \text{four-stroke cycle} \qquad \text{(2-41)}$$

In these equations, bmep and brake work W_b are used because torque is measured off the output crankshaft.

Most modern automobile engines have maximum torque per displacement in the range of 80 to 110 N-m/L (1 to 1.3 lbf-ft/in.^{3}), with some as high as 140 N-m/L. This gives maximum torques of 200 to 400 N-m (150 to 300 lbf-ft), usually at engine speeds around 4000 to 6000 RPM. The point of maximum torque is called maximum brake torque speed (MBT). A major goal in the design of a modern automobile engine is to *flatten* the torque-versus-speed curve as shown in Fig. 2-12, and to have high torque at both high and low speed. CI engines generally have higher torque than SI engines. Large engines often have very high torque values with MBT at relatively low speed.

Power is defined as the rate of work of the engine. If n = number of revolutions per cycle and N = engine speed, then

$$\dot{W} = WN/n \qquad \text{(2-42)}$$
$$\dot{W} = 2\pi N\tau \qquad \text{(2-43)}$$
$$\dot{W} = (1/2n)(\text{mep})A_p\overline{U}_p \qquad \text{(2-44)}$$

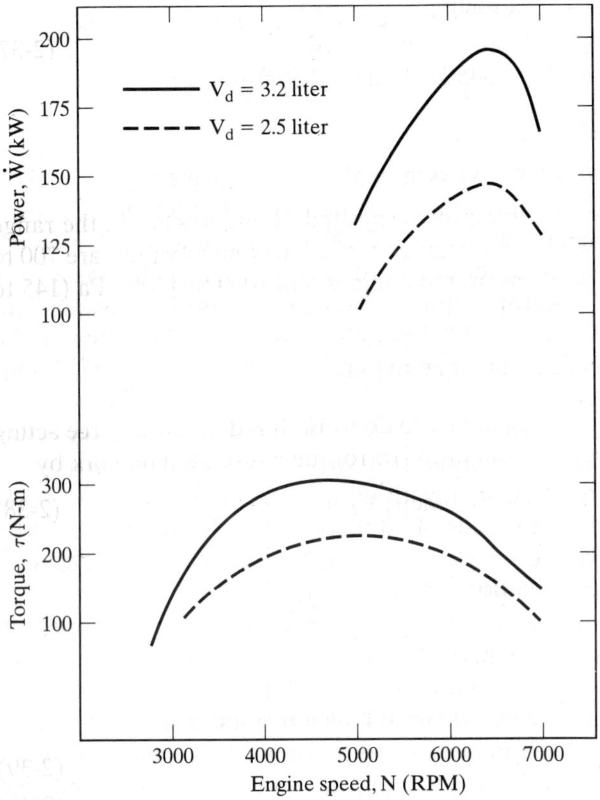

FIGURE 2-12

Brake power and torque of a typical automobile reciprocating engine as a function of engine speed and displacement. Speed at which peak torque occurs is called maximum brake torque (MBT) (or maximum best torque). Indicated power increases with speed while brake power increases to a maximum and then decreases. This is because friction increases with engine speed to a higher power and becomes dominant at higher speeds. Figs. 2-8, 2-14, 2-16, 2-20, and 2-24 show power and torque curves for specific engines.

$$\dot{W} = (\text{mep})A_p\overline{U}_p/4 \quad \text{four-stroke cycle} \tag{2-45}$$

$$\dot{W} = (\text{mep})A_p\overline{U}_p/2 \quad \text{two-stroke cycle} \tag{2-46}$$

where

W = work per cycle

A_p = piston face area of all pistons

$\overline{U}_p$ = average piston speed

Depending upon which definition of work or mep is used in Eqs. (2-42)–(2-46), power can be defined as brake power, net indicated power, gross indicated power, pumping power, and even friction power. Also,

$$\dot{W}_b = \eta_m \dot{W}_i \tag{2-47}$$

$$(\dot{W}_i)_{\text{net}} = (\dot{W}_i)_{\text{gross}} - (\dot{W}_i)_{\text{pump}} \tag{2-48}$$

$$\dot{W}_b = \dot{W}_i - \dot{W}_f \tag{2-49}$$

where η_m is the mechanical efficiency of the engine.

Power is normally measured in kW, but horsepower (hp) is still common:

$$1 \text{ hp} = 0.7457 \text{ kW} = 2545 \text{ BTU/hr} = 550 \text{ ft-lbf/sec}$$
$$1\text{kW} = 1.341 \text{ hp} \tag{2-50}$$

Engine power can range from a few watts in small model airplane engines to thousands of kW per cylinder in large multiple-cylinder stationary and ship engines. There is a large commercial market for engines in the 1.5- to 5-kW (2–7 hp) range for lawn mowers, chain saws, snowblowers, etc. Power for outboard motors (engines) for small boats typically ranges from 2 to 40 kW (3–50 hp), with much larger ones available. Modern automobile engines range mostly from 40 to 220 kW (50–300 hp). It is interesting to note that a modern midsize aerodynamic automobile requires only about 5 to 6 kW (7–8 hp) to cruise at 55 mph on level roadway.

Many large ship and stationary engines, with 4 to 20 cylinders, have brake power output in the range of 500 to 3000 kW/cylinder (670 to 4000 hp/cylinder) at speeds of 500 to 1000 RPM [228]. The largest engines, with up to 20 cylinders, can have brake power up to 60,000 kW, operating at speeds of 70 to 140 RPM. In the year 2002 the largest, most powerful, two-stroke cycle engine was claimed to be the MAN B & W 12-cylinder K98MC-C, which produced 68,640 kW (92,046 hp) [212].

Both torque and power are functions of engine speed. At low speed, torque increases as engine speed increases. As engine speed increases further, torque reaches a maximum and then decreases as shown in Figs. 2-8 and 2-12. Torque decreases because the engine is unable to ingest a full charge of air at higher speeds. Indicated power increases with speed, while brake power increases to a maximum and then decreases at higher speeds. This is because friction losses increase with speed and become the dominant factor at very high speeds. For many automobile engines, maximum brake power occurs at about 6000 to 7000 RPM, about one and a half times the speed of maximum torque.

Greater power can be generated by increasing displacement, mep, and/or speed. Increased displacement increases engine mass and takes up space, both of which are contrary to automobile design trends. For this reason, most modern engines are smaller but run at higher speeds, and are often turbocharged or supercharged to increase mep.

HISTORIC—DRAG COEFFICIENTS OF AUTOMOBILES

Only about 5 or 6 kW (7 or 8 hp) of power is needed to overcome air resistance on modern midsize automobiles traveling at 55 mph on level roadway. This is largely due to modern aerodynamic design which has reduced the drag coefficient of many automobiles to the 0.25 to 0.30 range. Wind tunnels large enough to test drag on full-size automobiles were first used in the 1930s. For some reason one automobile was tested with the vehicle facing backwards, and it was found that there was less air drag in reverse. This prompted additional testing of more vehicles, and it was found that many car models of that period had a lower drag coefficient in reverse than that facing forward.

Other ways which are sometimes used to classify engines are as follows:

specific power	$SP = \dot{W}_b/A_p$	(2-51)
output per displacement	$OPD = \dot{W}_b/V_d$	(2-52)
specific volume	$SV = V_d/\dot{W}_b$	(2-53)
specific weight	$SW = (\text{engine weight})/\dot{W}_b$	(2-54)

where

$\dot{W}_b$ = brake power
A_p = piston face area of all pistons
V_d = displacement volume

These parameters are important for engines used in transportation vehicles such as boats, automobiles, and especially airplanes, where keeping weight to a minimum is necessary. For large stationary engines, weight is not as important.

Modern automobile engines usually have brake power output per displacement in the range of 40 to 80 kW/L. The Honda eight-valve-per-cylinder V4 motorcycle engine generates about 130 kW/L, an extreme example of a high-performance racing engine [22]. One main reason for continued development to return to two-stroke cycle automobile engines is that they have up to 40% greater power output per unit weight.

HISTORIC—EIGHT-VALVES-PER-CYLINDER MOTORCYCLE ENGINE

In the early 1990s, Honda produced a racing motorcycle with a V4 engine of which each cylinder had four intake valves and four exhaust valves. The engine was developed by modifying a V8 engine so that the motorcycle could be raced under rules restricting engines to four cylinders. A four-valve-per-cylinder V8 engine block was modified by removing the metal between each set of two cylinders. Special pistons were built to fit into the resulting nonround, oblong cylinders. This resulted in each cylinder having eight valves and a piston with two connecting rods using a common piston pin.

The final product was a very fast, very expensive motorcycle with an aluminum block, 90° V4 engine having a displacement of 748 cm^3. It produced 96 kW at 14,000 RPM and maximum torque of 71 N-m at 11,600 RPM [22, 143].

2.5 DYNAMOMETERS

Dynamometers are used to measure torque and power over the engine operating ranges of speed and load. They do this by using various methods to absorb the energy output of the engine, all of which eventually ends up as heat.

Some dynamometers absorb energy in a mechanical friction brake (prony brake). These are the simplest dynamometers but are not as flexible and accurate as others at higher energy levels.

Fluid or hydraulic dynamometers absorb engine energy in water or oil pumped through orifices or dissipated with viscous losses in a rotor–stator combination. Large

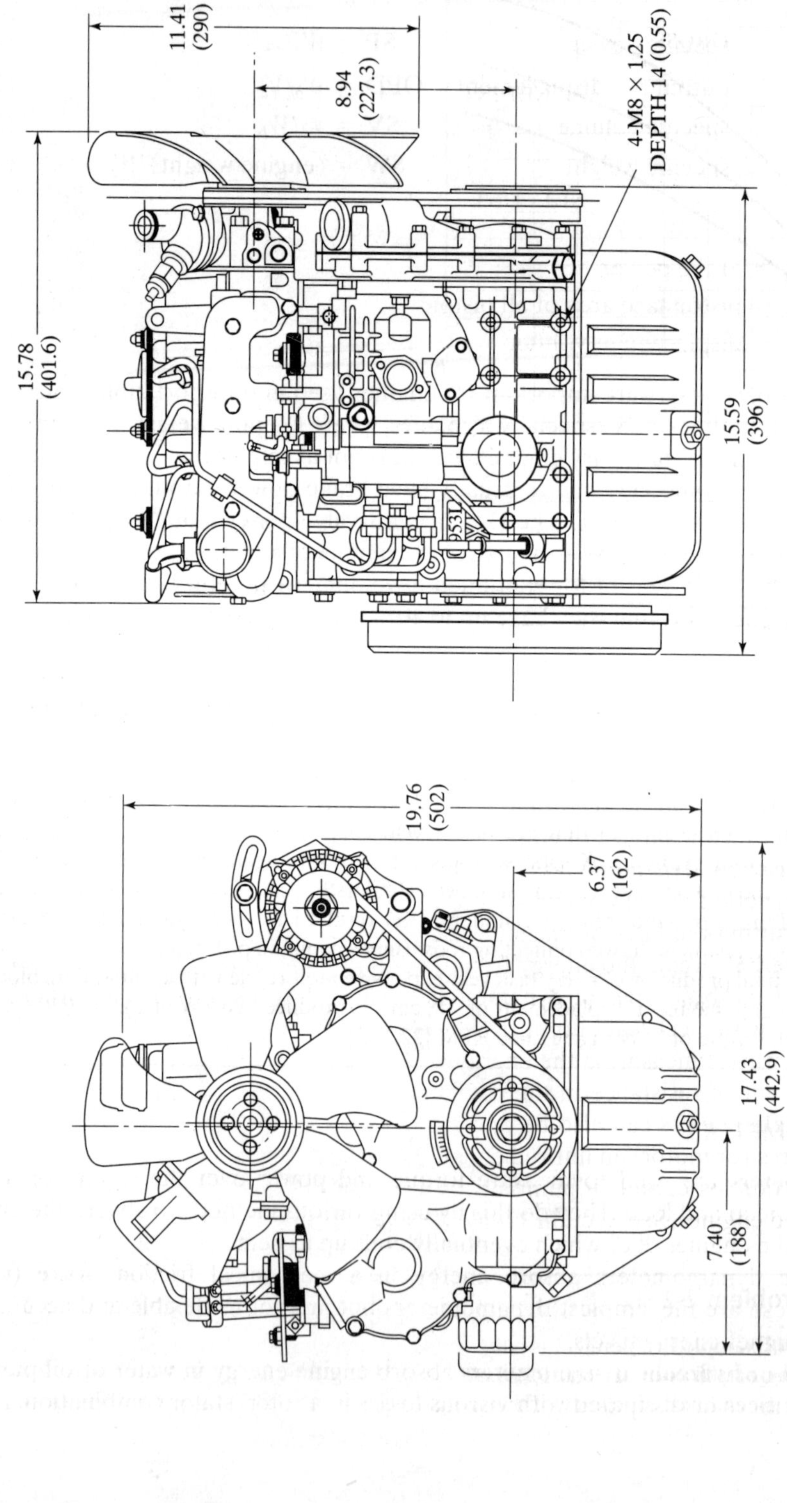

FIGURE 2-13
Vanguard 700D three-cylinder, in-line, four-stroke cycle, CI engine made by Briggs & Stratton Corporation. The engine has a bore of 6.8 cm (2.68 in.), stroke of 6.4 cm (2.52 in.), and a total displacement of 697 cm^3 (42.5 $in.^3$). Dimensions are in inches and (mm). Printed with permission of Briggs & Stratton Corporation.

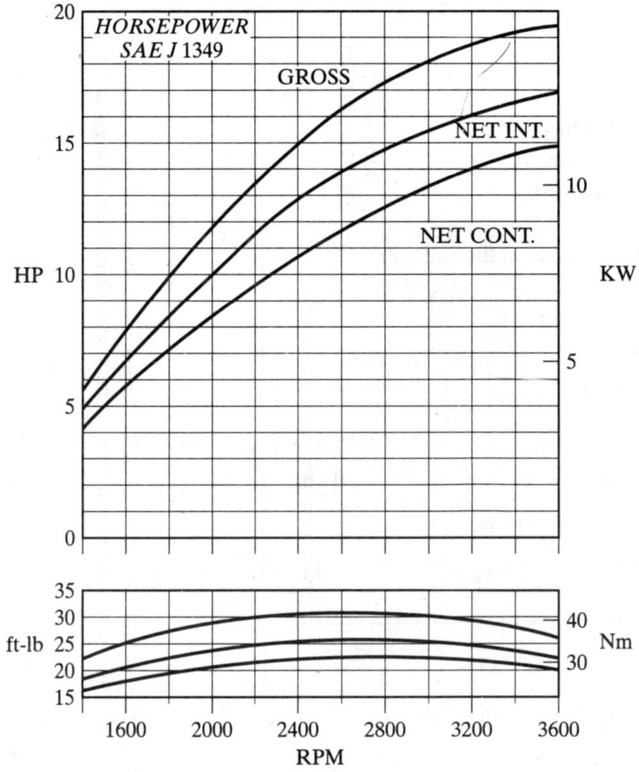

FIGURE 2-14

Power and torque curves of Briggs & Stratton Vanguard 700D CI engine shown in Fig. 2-13. Printed with permission of Briggs & Stratton Corporation.

amounts of energy can be absorbed in this manner, making this an attractive type of dynamometer for the largest of engines.

Eddy current dynamometers use a disk, driven by the engine being tested, rotating in a magnetic field of controlled strength. The rotating disk acts as an electrical conductor cutting the lines of magnetic flux and producing eddy currents in the disk. With no external circuit, the energy from the induced currents is absorbed in the disk.

One of the best types of dynamometer is the **electric dynamometer**, which absorbs energy with electrical output from a connected generator. In addition to having an accurate way of measuring the energy absorbed, the load is easily varied by changing the amount of resistance in the circuit connected to the generator output. Many electric dynamometers can also be operated in reverse, with the generator used as a motor to drive (or motor) an unfired engine. This allows the engine to be tested for mechanical friction losses and air pumping losses, quantities that are hard to measure on a running fired engine. (See Section 11-2.)

Example Problem 2-2

The engine in Example Problem 2-1 is connected to a dynamometer which gives a brake output torque reading of 205 N-m at 3600 RPM. At this speed air enters the cylinders at 85 kPa and 60°C, and the mechanical efficiency of the engine is 85%.

Calculate:

1. brake power
2. indicated power
3. brake mean effective pressure
4. indicated mean effective pressure
5. friction mean effective pressure
6. power lost to friction
7. brake work per unit mass of gas in the cylinder
8. brake specific power
9. brake output per displacement
10. engine specific volume

(1) Use Eq. (2-43) to find brake power:

$$\dot{W}_b = 2\pi N\tau = (2\pi \text{ radians/rev})(3600/60 \text{ rev/sec})(205 \text{ N-m})$$
$$= 77{,}300 \text{ N-m/sec} = \underline{77.3 \text{ kW} = 104 \text{ hp}}$$

(2) Use Eq. (2-47) to find indicated power:

$$\dot{W}_i = \dot{W}_b/\eta_m = (77.3 \text{ kW})/(0.85) = \underline{90.9 \text{ kW} = 122 \text{ hp}}$$

(3) Use Eq. (2-41) to find the brake mean effective pressure:

$$\text{bmep} = 4\pi\tau/V_d = (4\pi \text{ radians/cycle})(205 \text{ N-m})/(0.003 \text{ m}^3/\text{cycle})$$
$$= \underline{859{,}000 \text{ N/m}^2 = 859 \text{ kPa} = 125 \text{ psia}}$$

(4) Equation (2-37c) gives indicated mean effective pressure:

$$\text{imep} = \text{bmep}/\eta_m = (859 \text{ kPa})/(0.85) = \underline{1010 \text{ kPa} = 146 \text{ psia}}$$

(5) Equation (2-37d) is used to calculate friction mean effective pressure:

$$\text{fmep} = \text{imep} - \text{bmep} = 1010 - 859 = \underline{151 \text{ kPa} = 22 \text{ psia}}$$

(6) Equations (2-15) and (2-44) are used to find friction power lost:

$$A_p = (\pi/4)B^2 = (\pi/4)(0.086 \text{ m})^2 = 0.00581 \text{ m}^2 \text{ for one cylinder}$$
$$\dot{W}_f = (1/2\text{n})(\text{fmep})A_p\bar{U}_p$$
$$= (1/4)(151 \text{ kPa})(0.00581 \text{ m}^2/\text{cyl})(10.32 \text{ m/sec})(6 \text{ cyl})$$
$$= \underline{13.6 \text{ kW} = 18 \text{ hp}}$$

Alternatively, the friction power lost can be obtained from Eq. (2-49):

$$\dot{W}_f = \dot{W}_i - \dot{W}_b = 90.9 - 77.3 = 13.6 \text{ kW}$$

(7) First, brake work is found for one cylinder for one cycle, using Eq. (2-29):

$$W_b = (\text{bmep})V_d = (859 \text{ kPa})(0.0005 \text{ m}^3) = 0.43 \text{ kJ}$$

It can be assumed that the gas entering the cylinders at BDC is air:

$$m_a = PV_{BDC}/RT = P(V_d + V_c)/RT$$
$$= (85 \text{ kPa})(0.0005 + 0.000059)\text{m}^3/(0.287 \text{ kJ/kg-K})(333 \text{ K})$$
$$= 0.00050 \text{ kg}$$

Brake specific work per unit mass is

$$w_b = W_b/m_a = (0.43\text{ kJ})/(0.00050\text{ kg}) \underline{= 860\text{ kJ/kg} = 370\text{ BTU/lbm}}$$

(8) Equation (2-51) gives brake specific power:

$$\text{BSP} = \dot{W}_b/A_p = (77.3\text{ kW})/[(\pi/4)(0.086\text{ m})^2(6\text{ cylinders})]$$
$$\underline{= 2220\text{ kW/m}^2 = 0.2220\text{ kW/cm}^2 = 1.92\text{ hp/in.}^2}$$

(9) Equation (2-52) gives brake output per displacement:

$$\text{BOPD} = \dot{W}_b/V_d = (77.3\text{ kW})/(3\text{ L})$$
$$\underline{= 25.8\text{ kW/L} = 35\text{ hp/L} = 0.567\text{ hp/in.}^3}$$

(10) Equation (2-53) gives engine specific volume:

$$\text{BSV} = V_d/\dot{W}_b = 1/\text{BOPD} = 1/25.8$$
$$\underline{= 0.0388\text{ L/kW} = 0.0286\text{ L/hp} = 1.76\text{ in.}^3\text{/hp}}$$

Example Problem 2-3

When a three-cylinder, four-stroke cycle, SI engine, operating at 4000 RPM is connected to an eddy current dynamometer, 70.4 kW of power is dissipated by the dynamometer. The engine has a total displacement volume of 2.4 liters and a mechanical efficiency of 82% at 4000 RPM. Because of heat and mechanical losses, the dynamometer has an efficiency of 93%. η_{dyno} = (power recorded by dynamometer)/(actual power from engine).
Calculate:

1. power lost to friction in engine
2. brake mean effective pressure
3. engine torque at 4000 RPM
4. engine specific volume

(1) Brake power:

$$\dot{W}_b = (70.4\text{ kW})/(0.93) = 75.7\text{ kW} = 101.5\text{ hp}$$

Indicated power is obtained using Eq. (2-47):

$$\dot{W}_i = \dot{W}_b/\eta_n = (75.7\text{ kW})/(0.82) = 92.3\text{ kW} = 123.8\text{ hp}$$

Eq. (2-49) gives power lost to engine friction:

$$\dot{W}_f = \dot{W}_i - \dot{W}_b = (92.3\text{ kW}) - (75.7\text{ kW}) \underline{= 16.6\text{ kW} = 22.3\text{ hp}}$$

(2) The brake mean effective pressure is obtained by combining Eqs. (2-29) & (2-42):

$$\text{bmep} = W_b/V_d = [\dot{W}_b/(N/n)]/V_d$$
$$= \{(75.7\text{ kW})/[(4000/60\text{ rev/sec})/(2\text{ rev/cycle})]\}/(0.0024\text{ m}^3\text{/cycle})$$
$$\underline{= 946\text{ kPa} = 137\text{ psia}}$$

or using Eq. (2-88):

$$\text{bmep} = [(1000)(75.7)(2)]/[(2.4)(4000/60)] \underline{= 946\text{ kPa}}$$

(3) The engine torque using Eq. (2-43):

$$\tau = \dot{W}_b/2\pi N = (75.7\text{ kJ/sec})/[(2\pi\text{ radians/rev})(4000/60\text{ rev/sec})]$$
$$= \underline{181\text{ N-m} = 134\text{ lbf-ft}}$$

or using Eq. (2-76):

$$\tau = [(159.2)(75.7)]/(4000/60) \underline{= 181\text{ N-m}}$$

(4) Eq. (2-53) gives engine specific volume:

$$\text{SV} = V_d/\dot{W}_b = (2.4\text{ L})/(75.7\text{ kW}) \underline{= 0.0317\text{ L/kW}}$$

2.6 AIR–FUEL RATIO AND FUEL–AIR RATIO

Energy input to an engine Q_{in} comes from the combustion of a hydrocarbon fuel. Air is used to supply the oxygen needed for this chemical reaction. For combustion reaction to occur, the proper relative amounts of air (oxygen) and fuel must be present.

Air–fuel ratio (AF) and fuel–air ratio (FA) are parameters used to describe the mixture ratio. We have

$$\text{AF} = m_a/m_f = \dot{m}_a/\dot{m}_f \qquad (2\text{-}55)$$
$$\text{FA} = m_f/m_a = \dot{m}_f/\dot{m}_a = 1/\text{AF} \qquad (2\text{-}56)$$

where

m_a = mass of air
$\dot{m}_a$ = mass flow rate of air

FIGURE 2-15

Cummins QSK60-2700 sixteen-cylinder, four-stroke cycle, CI, V16 engine of 60.2 liter displacement (3,672 in.^{3}). Engine has bore of 15.9 cm (6.26 in.), stroke of 19.0 cm (7.48 in.), oil system capacity of 281 L (74.2 gal), coolant capacity of 170 L (45 gal), and wet mass of 9,305 kg (20,514 lbm). Using No. 2 diesel fuel, it is advertized as having fuel economy of 206 gm/kW-hr at 1900 RPM (0.339 lbm/bhp-hr). The oil management system "converts used engine oil into fuel, eliminating oil changes for up to 4,000 hours of operation." Printed with permission, Cummins, Inc.

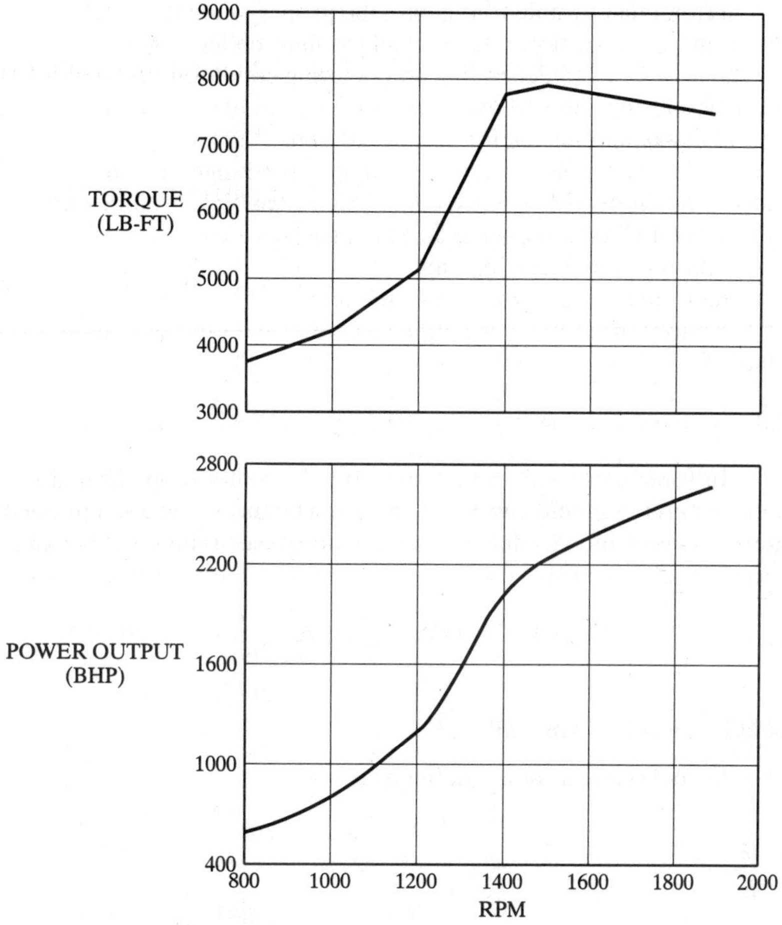

FIGURE 2-16

Power and torque curves for Cummins QSK60-2700 engine shown in Fig. 2-15. Engine has maximum power of 2013 kW at 1900 RPM (2700 hp), and maximum torque of 10,623 N-m at 1500 RPM (7840 lbf-ft). Printed with permission, Cummins, Inc.

m_f = mass of fuel
$\dot{m}_f$ = mass flow rate of fuel

The ideal or stoichiometric AF for many gasoline-type hydrocarbon fuels is very close to 15:1, with combustion possible for values in the range of 6 to 25. AF less than 6 is too *rich* to sustain combustion and AF greater than 25 is too *lean*. A vehicle will often be operated with a rich mixture when accelerating or starting cold, rich mixtures having better ignition. When cruising at light load, vehicles are often operated lean to save fuel. When this is done, it is often necessary to have a small fuel rich zone around the sparkplug to assure good ignition. The fuel input system of an engine, fuel injectors or carburetor, must be able to regulate the proper amount of fuel for any operating condition and given air flow rate. Normal gasoline-fueled engines usually have AF input in the range of 12 to 18 depending on operating conditions at the time (e.g., accelerating, cruising, starting, etc.). SI *lean-burn* engines can have AF as high as 25 to 40, but need special intake and mixing for proper ignition.

CI engines typically have AF input in the range of 18 to 70, which appears to be outside the limits within which combustion is possible. Combustion occurs because the cylinder of a CI engine, unlike an SI engine, has a very nonhomogeneous air–fuel mixture, with reaction occurring only in those regions in which a combustible mixture exists, other regions being too rich or too lean.

Equivalence ratio ϕ is defined as the actual ratio of fuel–air to ideal or stoichiometric fuel–air:

$$\phi = (\mathrm{FA})_{\mathrm{act}}/(\mathrm{FA})_{\mathrm{stoich}} = (\mathrm{AF})_{\mathrm{stoich}}/(\mathrm{AF})_{\mathrm{act}} \tag{2-57}$$

In some cases, AF and FA are given as molar ratios. This is much less common and AF and FA should always be considered mass ratios unless otherwise noted. Some literature use lambda value instead of equivalence ratio, lambda value being the reciprocal of the equivalence ratio:

$$\lambda = 1/\phi = (\mathrm{FA})_{\mathrm{stoich}}/(\mathrm{FA})_{\mathrm{act}} = (\mathrm{AF})_{\mathrm{act}}/(\mathrm{AF})_{\mathrm{stoich}} \tag{2-58}$$

2.7 SPECIFIC FUEL CONSUMPTION

Specific fuel consumption is defined as

$$\mathrm{sfc} = \dot{m}_f/\dot{W} \tag{2-59}$$

where

$\dot{m}_f$ = rate of fuel flow into engine

$\dot{W}$ = engine power

Brake power gives the **brake specific fuel consumption**:

$$\mathrm{bsfc} = \dot{m}_f/\dot{W}_b \tag{2-60}$$

Indicated power gives **indicated specific fuel consumption**:

$$\mathrm{isfc} = \dot{m}_f/\dot{W}_i \tag{2-61}$$

Other examples of specific fuel consumption parameters can be defined as follows:

fsfc = friction specific fuel consumption
igsfc = indicated gross specific fuel consumption
insfc = indicated net specific fuel consumption
psfc = pumping specific fuel consumption

It also follows that

$$\eta_m = \dot{W}_b/\dot{W}_i = (\dot{m}_f/\dot{W}_i)/(\dot{m}_f/\dot{W}_b) = (\text{isfc})/(\text{bsfc}) \tag{2-62}$$

where

$$\eta_m = \text{mechanical efficiency of the engine}$$

Brake specific fuel consumption decreases as engine speed increases, reaches a minimum, and then increases at high speeds (Fig. 2-17). Fuel consumption increases at high speed because of greater friction losses. At low engine speed, the longer time per cycle allows more heat loss and fuel consumption goes up. Figure 2-18 shows how bsfc also depends on compression ratio and fuel equivalence ratio. It decreases with higher compression ratio due to higher thermal efficiency. It is lowest when combustion occurs in a mixture with a fuel equivalence ratio near one, ($\phi = 1$). The further from stoichiometric combustion, either rich or lean, the higher will be the fuel consumption.

Brake specific fuel consumption generally decreases with engine size, being best (lowest) for very large engines (see Fig. 2-19).

Specific fuel consumption is generally given in units of gm/kW-hr or lbm/hp-hr. For transportation vehicles it is common to use **fuel economy** in terms of distance traveled per unit of fuel, such as miles per gallon (mpg). In SI units it is common to use the inverse of this, with (L/100 km) being a common unit. To decrease air pollution and depletion of fossil fuels, laws have been enacted requiring better vehicle fuel economy. Since the early 1970s, when most automobiles got less than 15 mpg (15.7 L/100 km)

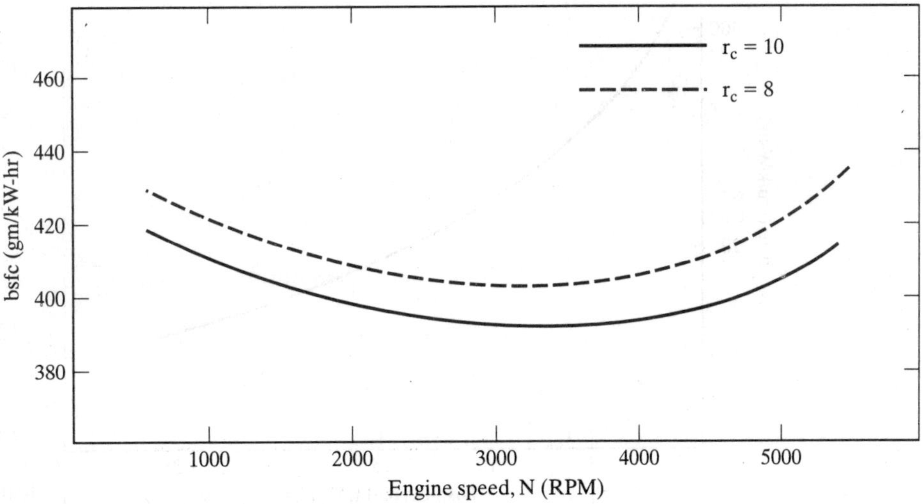

FIGURE 2-17

Brake specific fuel consumption as a function of engine speed. Fuel consumption decreases as engine speed increases due to the shorter time for heat loss during each cycle. At higher engine speeds fuel consumption again increases because of high friction losses. As compression ratio is increased fuel consumption decreases due to greater thermal efficiency.

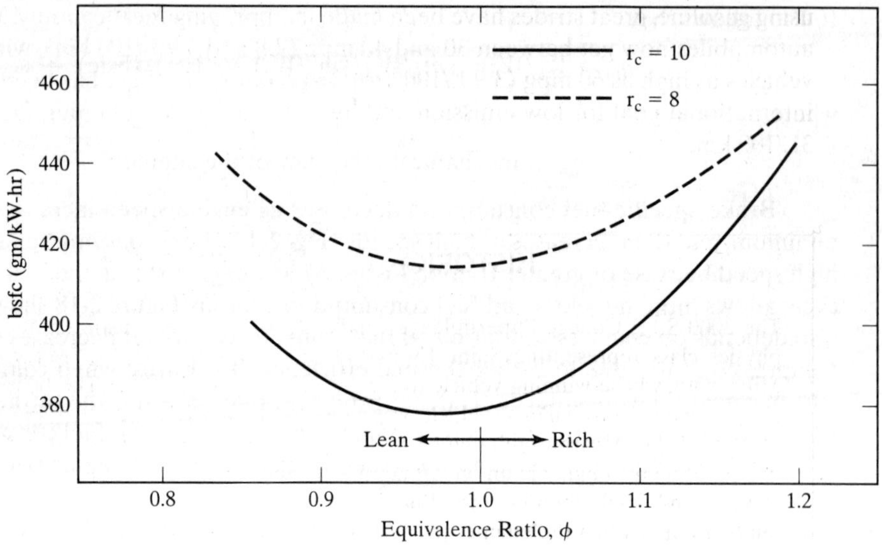

FIGURE 2-18

Brake specific fuel consumption as a function of fuel equivalence ratio. Consumption is minimum at a slightly lean condition, increasing with both richer and leaner mixtures.

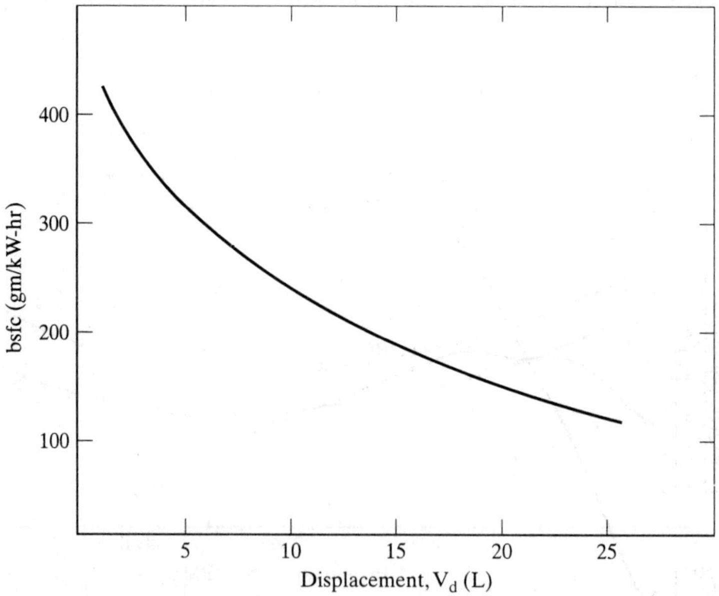

FIGURE 2-19

Brake specific fuel consumption as a function of engine displacement. Generally, average fuel consumption is less with larger engines. One reason for this is less heat loss due to the higher volume-to-surface-area ratio of the combustion chamber in a large engine. Also, larger engines operate at lower speeds which reduces friction losses. Adapted from [123].

using gasoline, great strides have been made in improving fuel economy. Many modern automobiles now get between 30 and 40 mpg (7.8 and 5.9 L/100 km), with some small vehicles as high as 60 mpg (3.9 L/100 km). In recent years, there has been an unwritten international goal for low-emission and hybrid vehicles of obtaining fuel economy of 3L/100 km.

HISTORIC—1322-MPG VEHICLE

The 2000 SAE College Supermileage Challenge was won by a team of students from a physics class representing Saint Thomas Academy High School of Mandota Heights, Minnesota, whose winning vehicle averaged 1131 mpg of gasoline. The light-weight, single-passenger, aerodynamic vehicle was powered by a 3.5 hp, single-cylinder, Briggs & Stratton, L-head, carbureted engine of 90 cm^3 displacement. High mileage was obtained by operating the engine in an on–off mode. The engine would accelerate the vehicle up to 25 mph, and would then be turned off. The vehicle was then allowed to coast until speed fell to 10 mph when the engine was again started. This method of operation satisfied the requirement of a 10 mph average over the closed course. The same vehicle was then used in the Minnesota Technology Education Association (MTEA) Supermileage Challenge, where it obtained a mileage of 1322 mpg using a fuel of 90% gasoline and 10% ethanol.

In 1999, the average vehicle fuel consumption worldwide was 27.5 mpg (8.6 L/100 km).

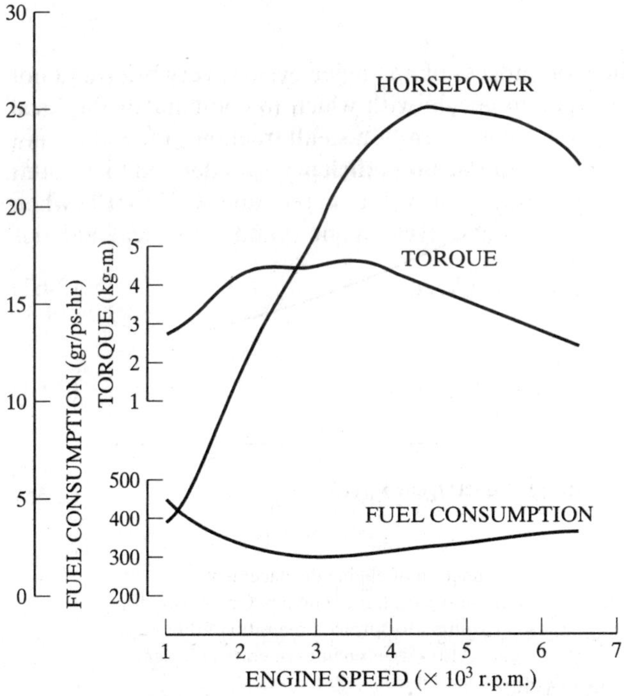

FIGURE 2-20

Power, brake specific fuel consumption, and torque as functions of engine speed for Suzuki three-cylinder, two-stroke cycle, minicar engine of 0.45 L displacement. Maximum brake power is 25 hp at 4500 RPM, with maximum torque of 46 N-m at 3500 RPM. Reprinted with permission from SAE Paper No. 770766 © 1977, SAE International, [231].

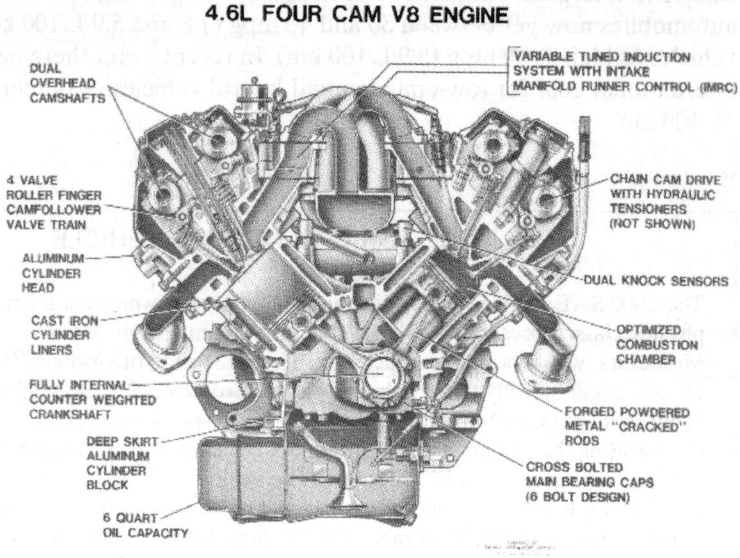

FIGURE 2-21

Cutaway view of Ford spark-ignition, four-stroke cycle, V8 engine showing main components. The engine has a displacement of 4.6 liters, four valves per cylinder, and two camshafts for each bank of cylinders. Courtesy of Ford Motor Company.

2.8 ENGINE EFFICIENCIES

The time available for the combustion process of an engine cycle is very brief, and not all fuel molecules may find an oxygen molecule with which to combine, or the local temperature may not favor a reaction. Consequently, a small fraction of fuel does not react and exits with the exhaust flow. A **combustion efficiency** η_c is defined to account for the fraction of fuel that burns. Typically, η_c has values in the range 0.95 to 0.98 when an engine is operating properly. For one engine cycle in one cylinder, the heat added is

$$Q_{\text{in}} = m_f Q_{\text{HV}} \eta_c \tag{2-63}$$

For steady state,

$$\dot{Q}_{\text{in}} = \dot{m}_f Q_{\text{HV}} \eta_c \tag{2-64}$$

and **thermal efficiency** is

$$\eta_t = W/Q_{\text{in}} = \dot{W}/\dot{Q}_{\text{in}} = \dot{W}/\dot{m}_f Q_{\text{HV}} \eta_c = \eta_f/\eta_c \tag{2-65}$$

where

W = work of one cycle
$\dot{W}$ = power
m_f = mass of fuel for one cycle

$\dot{m}_f$ = mass flow rate of fuel
Q_{HV} = heating value of fuel
η_f = fuel conversion efficiency (see Eq. 2-67)

Thermal efficiency can be given as indicated or brake, depending on whether indicated power or brake power is used in Eq. (2-65). It follows that engine mechanical efficiency is given by

$$\eta_m = (\eta_t)_b/(\eta_t)_i \tag{2-66}$$

Engines can have indicated thermal efficiencies in the range of 40% to 50%, with brake thermal efficiency usually about 30%. Some large, slow CI engines can have brake thermal efficiencies greater than 50%.

Fuel conversion efficiency is defined as

$$\eta_f = W/m_f Q_{HV} = \dot{W}/\dot{m}_f Q_{HV} \tag{2-67}$$

$$\eta_f = 1/(\text{sfc})Q_{HV} \tag{2-68}$$

For a single cycle of one cylinder the thermal efficiency can be written

$$\eta_t = W/m_f Q_{HV} \eta_c \tag{2-69}$$

This is the thermal efficiency introduced in basic thermodynamic textbooks, sometimes called *enthalpy efficiency*.

2.9 VOLUMETRIC EFFICIENCY

One of the most important processes that governs how much power and performance can be obtained from an engine is getting the maximum amount of air into the cylinder during each cycle. More air means more fuel can be burned and more energy can be converted to output power. Getting the relatively small volume of liquid fuel into the cylinder is much easier than getting the large volume of gaseous air needed to react with the fuel. Ideally, a mass of air equal to the density of atmospheric air times the displacement volume of the cylinder should be ingested for each cycle. However, because of the short cycle time available and the flow restrictions presented by the air cleaner, carburetor (if any), intake manifold, and intake valve(s), less than this ideal amount of air enters the cylinder. **Volumetric efficiency** is defined as

$$\eta_v = m_a/\rho_a V_d \tag{2-70}$$

$$\eta_v = n\dot{m}_a/\rho_a V_d N \tag{2-71}$$

where

m_a = mass of air into the engine (or cylinder) for one cycle
$\dot{m}_a$ = steady-state flow of air into the engine
ρ_a = air density evaluated at atmospheric conditions outside the engine
V_d = displacement volume
N = engine speed
n = number of revolutions per cycle

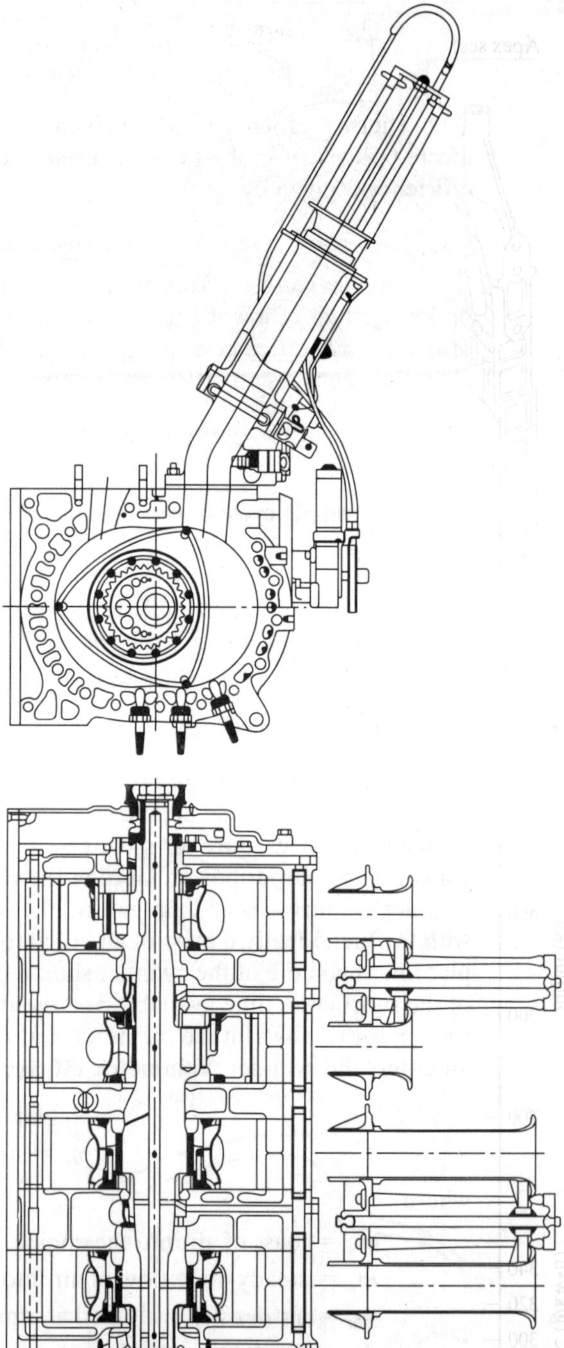

FIGURE 2-22

Cutaway view of Mazda four-rotor, R26B rotary engine, used in the race car which won the 1991 24-hour endurance race at Le Mans, France. The 2.62 liter, liquid cooled engine had a compression ratio of 10:1, three-spark-plug ignition, and variable length telescopic intake manifold. Operating parameters controlled by the engine management system (EMS) included injection timing, injection volume, ignition timing, and intake length of manifold runner. Reprinted with permission from SAE Paper No. 920309 © 1992, SAE International, [222].

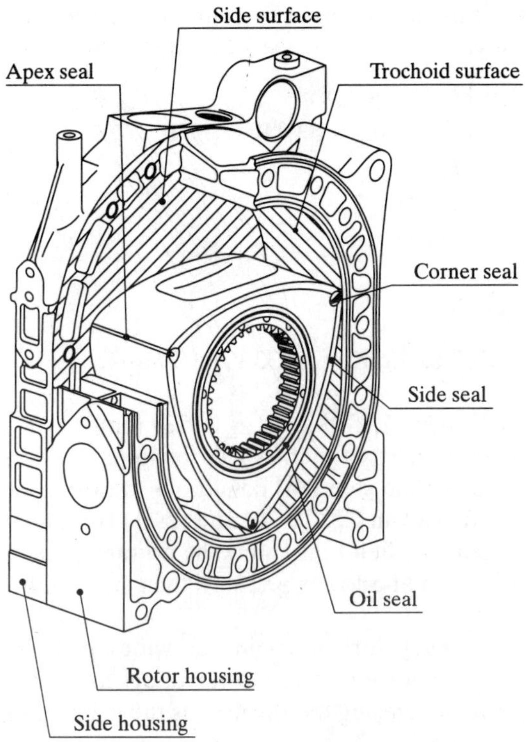

FIGURE 2-23

Cutaway view of rotor and combustion chamber of Mazda R26B rotary engine shown in Fig. 2-22. Reprinted with permission from SAE Paper No. 920309 © 1992, SAE International, [222].

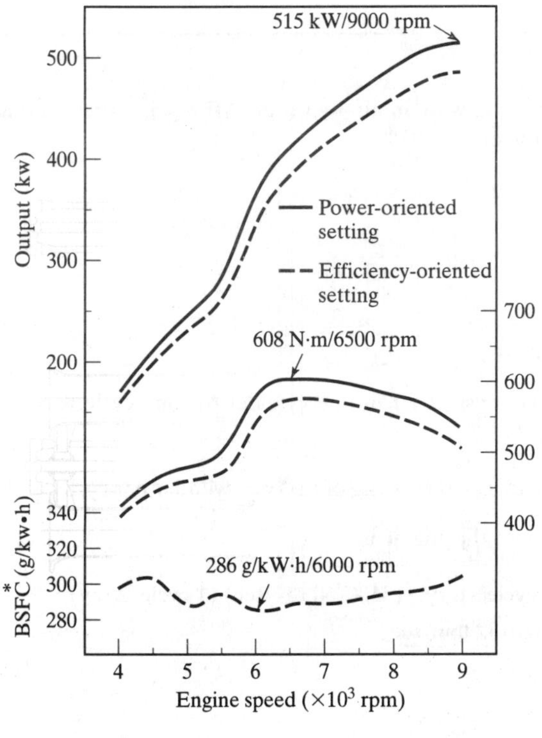

FIGURE 2-24

Power, torque, and brake specific fuel consumption curves of Mazda R26B rotary engine shown in Fig. 2-22. The engine produced maximum brake power of 515 kW at 9000 RPM (691 hp), maximum torque of 608 N-m at 6500 RPM (449 lbf-ft), and had a minimum bsfc of 286 gm/kW-hr at 6000 RPM. Reprinted with permission from SAE Paper No. 920309 © 1992, SAE International, [222].

Unless better values are known, standard values of surrounding air pressure and temperature can be used to find density.

$$P_o \text{ (standard)} = 101 \text{ kPa} = 14.7 \text{ psia}$$
$$T_o \text{ (standard)} = 298 \text{ K} = 25°\text{C} = 537°\text{R} = 77°\text{F}$$
$$\rho_a = P_o/RT_o \tag{2-72}$$

where

P_o = pressure of surrounding air
T_o = temperature of surrounding air
R = gas constant for air = 0.287 kJ/kg-K = 53.33 ft-lbf/lbm-°R

At standard conditions, the density of air $\rho_a = 1.181 \text{ kg/m}^3 = 0.0739 \text{ lbm/ft}^3$.

When volumetric efficiency is measured experimentally, corrections can be made for temperature and humidity when other than standard conditions are experienced.

Sometimes (less common) the air density in Eqs. (2-70) and (2-71) is evaluated at conditions in the intake manifold immediately before it enters the cylinder. The conditions at this point will usually be hotter and at a lower pressure than surrounding atmospheric conditions.

Typical values of volumetric efficiency for an engine at wide-open throttle (WOT) are in the range 75% to 90%, going down to much lower values as the throttle is closed. Restricting air flow into an engine (closing the throttle) is the primary means of power control for a spark ignition engine.

Example Problem 2-4

The engine in Example Problem 2-2 is running with an air–fuel ratio AF = 15, a fuel heating value of 44,000 kJ/kg, and a combustion efficiency of 97%.
Calculate:

1. rate of fuel flow into engine
2. brake thermal efficiency
3. indicated thermal efficiency
4. volumetric efficiency
5. brake specific fuel consumption

(1) From Example Problem 2-2, the mass of air in one cylinder for one cycle is m_a = 0.00050 kg. Then

$$m_f = m_a/AF = 0.00050/15 = 0.000033 \text{ kg of fuel per cylinder per cycle}$$

Therefore, the rate of fuel flow into the engine is

$$\dot{m}_f = (0.000033 \text{ kg/cyl-cycle})(6 \text{ cyl})\,(3600/60 \text{ rev/sec})\,(1 \text{ cycle/2 rev})$$
$$= \underline{0.0060 \text{ kg/sec} = 0.0132 \text{ lbm/sec}}$$

(2) Use Eq. (2-65) to find brake thermal efficiency:

$$(\eta_t)_b = \dot{W}_b/\dot{m}_f Q_{HV}\eta_c = (77.3\ \text{kW})/(0.0060\ \text{kg/sec})(44{,}000\ \text{kJ/kg})(0.97)$$
$$= \underline{0.302 = 30.2\%}$$

Or, using Eq. (2-68) for one cycle of one cylinder:

$$(\eta_t)_b = W_b/m_f Q_{HV}\eta_c = (0.43\ \text{kJ})/(0.000033\ \text{kg})(44{,}000\ \text{kJ/kg})(0.97)$$
$$= 0.302$$

(3) Indicated thermal efficiency using Eq. (2-65):

$$(\eta_t)_i = (\eta_t)_b/\eta_m = 0.302/0.85 = \underline{0.355 = 35.5\%}$$

(4) Use Eq. (2-70) with standard air density to calculate volumetric efficiency:

$$\eta_v = m_a/\rho_a V_d = (0.00050\ \text{kg})/(1.181\ \text{kg/m}^3)(0.0005\ \text{m}^3)$$
$$= \underline{0.847 = 84.7\%}$$

(5) Use Eq. (2-60) for brake specific fuel consumption:

$$\text{bsfc} = \dot{m}_f/\dot{W}_b = (0.0060\ \text{kg/sec})/(77.3\ \text{kW})$$
$$= 7.76 \times 10^{-5}\ \text{kg/kW-sec} = \underline{279\ \text{gm/kW-hr} = 0.459\ \text{lbm/hp-hr}}$$

2.10 EMISSIONS

The four main engine exhaust emissions that must be controlled are oxides of nitrogen (NOx), carbon monoxide (CO), hydrocarbons (HC), and solid particulates (part). Two common methods of measuring the amounts of these pollutants are **specific emissions** (SE) and the **emissions index** (EI). Specific emissions typically have units of gm/kW-hr, while the emissions index has units of emissions flow per fuel flow.

Specific Emissions:

$$\begin{aligned}(SE)_{NOx} &= \dot{m}_{NOx}/\dot{W}_b \\ (SE)_{CO} &= \dot{m}_{CO}/\dot{W}_b \\ (SE)_{HC} &= \dot{m}_{HC}/\dot{W}_b \\ (SE)_{part} &= \dot{m}_{part}/\dot{W}_b\end{aligned} \tag{2-73}$$

where

$\dot{m}$ = flow rate of emissions in gm/hr

$\dot{W}_b$ = brake power

Emissions Index:

$$\begin{aligned}(EI)_{NOx} &= \dot{m}_{NOx}[\text{gm/sec}]/\dot{m}_f[\text{kg/sec}] \\ (EI)_{CO} &= \dot{m}_{CO}[\text{gm/sec}]/\dot{m}_f[\text{kg/sec}] \\ (EI)_{HC} &= \dot{m}_{HC}[\text{gm/sec}]/\dot{m}_f[\text{kg/sec}] \\ (EI)_{part} &= \dot{m}_{part}[\text{gm/sec}]/\dot{m}_f[\text{kg/sec}]\end{aligned} \tag{2-74}$$

Example Problem 2-5

A 12-cylinder, two-stroke cycle CI engine produces 2440 kW of brake power at 550 RPM using stoichiometric light diesel fuel. The engine has bore of 24 cm, stroke of 32 cm, volumetric efficiency of 97%, mechanical efficiency of 88%, and combustion efficiency of 98%. Calculate:

1. mass flow rate of fuel into engine
2. brake specific fuel consumption
3. indicated specific fuel consumption
4. specific emissions of hydrocarbons due to unburned fuel
5. emissions index of hydrocarbons due to unburned fuel

(1) Equation (2-8) gives total displacement of the engine:

$$V_d = N_c(\pi/4)B^2S = (12\ \text{cylinders})(\pi/4)(0.24\ \text{m})^2(0.32\ \text{m}) = 0.1737\ \text{m}^3 = 173.7\ \text{L}$$

Equation (2-71) gives the air flow rate into the engine:

$$\begin{aligned}\dot{m}_a &= \eta_v \rho_a V_d N/n \\ &= (0.97)(1.181\ \text{kg/m}^3)(0.1737\ \text{m}^3\text{/cycle})(550/60\ \text{rev/sec})/(2\ \text{rev/cycle}) \\ &= 0.9120\ \text{kg/sec.}\end{aligned}$$

Equation (2-55) gives the mass flow rate of fuel into the engine:

$$\dot{m}_f = \dot{m}_a/\text{AF} = (0.9120\ \text{kg/sec})/(14.5) = \underline{0.0629\ \text{kg/sec} = 0.1387\ \text{lbm/sec}}$$

(2) Equation (2-60) gives brake specific fuel consumption:

$$\begin{aligned}\text{bsfc} &= m_f/W_b \\ &= [(0.0629\ \text{kg/sec})(3600\ \text{sec/hr})(1000\ \text{gm/kg})]/(2440\ \text{kW}) = \underline{92.8\ \text{gm/kW-hr}}\end{aligned}$$

(3) Equation (2-62) gives indicated specific fuel consumption:

$$\text{isfc} = \eta_m(\text{bsfc}) = (0.88)(92.8\ \text{gm/kW-hr}) = \underline{81.7\ \text{gm/kW-hr}}$$

(4) Mass flow rate of unburned fuel is

$$\begin{aligned}\dot{m}_{\text{unburned}} &= (1 - \eta_c)\dot{m}_f \\ &= (1 - 0.98)(0.0629\ \text{kg/sec}) = 0.001258\ \text{kg/sec} = 1.258\ \text{gm/sec}\end{aligned}$$

Equation (2-73) gives specific emissions of hydrocarbons from unburned fuel:

$$\begin{aligned}(\text{SE})_{\text{HC}} = \dot{m}_{\text{unburned}}/W_b &= [(1.258\ \text{gm/sec})(3600\ \text{sec/hr})]/(2440\ \text{kW}) \\ &= \underline{1.86\ \text{gm/kW-hr}}\end{aligned}$$

(5) Equation (2-74) gives the emissions index of hydrocarbons from unburned fuel:

$$(\text{EI})_{\text{HC}} = \dot{m}_{\text{unburned}}/\dot{m}_f = (1.258\ \text{gm/sec})/(0.0629\ \text{kg/sec}) = \underline{20.0\ \text{gm}_{\text{HC}}/\text{km}_{\text{f}}}$$

2.11 NOISE ABATEMENT

In recent years a lot of research and development has been directed towards reducing engine and exhaust noise. Although excessive noise is considered a pollution, total elimination of all noises is not always the goal of vehicle manufacturers. Some people consider some "sporty rumble" noise from an engine as desirable. The sound abatement systems on some models of European ultrasmall "city cars" are designed so that the vehicle sounds like an expensive luxury car. Several modern vehicles with nostalgic body designs reminiscent of the 1950s also have exhaust systems that are "tweaked" to sound like their 1950s counterparts (e.g., the rumbling of a Hollywood muffler). Motorcycle enthusiasts will sometimes reject new models because "they don't sound like a motorcycle." You could not sell a new Harley–Davidson cycle if it did not sound like a Harley–Davidson. A great effort was expended to make the new liquid-cooled Porsche engines sound like old air-cooled Porsche engines.

On the other hand, on many vehicles noise reduction has been so successful that some automobiles are now equipped with a safety switch on the starter. At idle speed, the engine is so quiet that the safety switch is required to keep drivers from trying to start the engine when it is already running.

2.12 42-VOLT ELECTRICAL SYSTEMS

In the early years of the 21st century, a revolutionary change will occur in the electrical systems of automobiles, a switch from 12 volts to 42 volts [221, 234]. This change is in response to the ever-increasing electrical power needs of the modern car, increased lighting, larger computer for controls, greater starter power for higher compression engines, air-conditioning, electrical accessaries, etc. To address this rising power need, automobile and component manufacturers spent billions of dollars in research and development during the 1990s and into the 2000s. The end result is a several-year changeover from 12 volts standard to 42 volts, starting in 2002. In the first years, prototype vehicles and a limited number of automobile models available to the public will have the higher voltage. Each year, more standard automobile models will have the new system, and in 10 years or so, it will be the industry standard.

Many factors were considered before international consensus agreed on 42 volts as the new standard, safety being a major consideration. The electrical industry considered anything over 60 volts potentially dangerous, with added requirements for wire insulation, connectors, etc. Years of electrical industry development had created great knowledge and a large line of products and standards in the 42 volt range; wires, relays, connectors, etc. In the old lower voltage systems, batteries were standard at 12 volts while alternators/generators operated at 14 volts. The new system will have 36-volt batteries and 42-volt generators. Wire sizes will be decreased, all mechanical relays will be replaced by solid-state switching, and many components will be of smaller size. The overall mass of the electrical system on a standard automobile is expected to be reduced by about 25%.

The first automobiles using the higher voltage will have two electrical systems, one of 42 volts and one of 12 volts. This is because some components, mainly the lights, operate better at the lower potential. The main electrical system will be 42 volts, with a transformer conversion down to 12 volts for that system. Most vehicles will have two batteries, one 36 volt and one 12 volt. It is uncertain whether the 12/14-volt systems will be phased out in the future.

The additional power available with a high voltage system opens up many possible uses of electrical components for engines and vehicles. Two main changes in engine operation which will quickly occur are the elimination of the camshaft, and the combining of the starter and generator with the flywheel. Other possibilities include beltless engines, better fuel injectors, electric water pumps, electric fuel pumps, electric oil pumps, quick defrost glass, windshield heaters, electric steering, heated catalysts, vehicle suspension control, electric particular filters for diesels, heated seats, greater entertainment systems, navigation systems, cellular equipment, high-tech security, electric brakes, climate control, power doors, etc.

One of the greater benefits of the higher voltage and power will be engine valve control using electromechanical actuators instead of a camshaft. Not only will this improve the mechanical efficiency of an engine, but it will provide variable valve control in both timing and lift. With more powerful computers of the EMS, complete control of valve timing and lift will allow the engine to operate much more efficiently over all speed and load conditions. Opening and closing of the valves can be much faster, and a soft closing is possible, which will allow the use of ceramic valves.

In most vehicles using the higher voltage system, the starter and generator will be built into a single unit along with the engine flywheel. This multipurpose flywheel will be mounted between the engine and transmission, much the same as standard flywheels. This will eliminate the need for a separate starter motor, and for a belt-driven generator. A flywheel-mounted starter will allow for very quick starting of a warmed engine—as quick as 0.3 seconds [234]. This will allow automobile engines to be turned off when stopped (e.g., at stop lights), saving fuel and reducing emissions. When the accelerator pedal is depressed the engine will then smoothly restart very quickly with the aid of the electric starter motor acting as a minihybrid. This type of unit will also save energy by recovering some of the vehicle kinetic energy when the vehicle slows or stops, energy that would normally be lost as heat in the break system. The flywheel-generator can recover some of this energy electrically and return it to the vehicle's battery. Creating a single unit that includes a starter and a generator was a major technological achievement, starters generally operating at low speed and generators only efficient at high speed.

An electric pump and controls in the engine cooling system will allow flow rates to be adjusted as needed, saving energy and eliminating the need for a thermostat. Engines will heat up quicker and passenger compartment heating can continue after the engine stops. Electric fuel pumps and oil pumps, along with sensors and controls, will allow more efficient use of these units. Better lubrication control (e.g., at cold startup) will reduce engine wear.

Electric braking and steering will eventually replace the systems now used. Electric brakes will be made safer by eliminating hydraulic systems. Eventually, electric steer-by-wire will probably eliminate the need for a steering column, giving greater space and flexibility for engine compartment design. The steering wheel could someday be replaced with a joystick. Eliminating all engine belts and using electricity for driving components only when needed (e.g., air conditioner pump, fan, etc.) will reduce noise and increase mechanical efficiency.

Both hybrid vehicles and all-electric fuel cell vehicles will operate more efficiently using 42 volts. Potential problems with the higher voltage include electrochemical corrosion, arcing, and jump-starting of automobiles.

2.13 VARIABLE DISPLACEMENT - CYLINDER CUTOUT

A large displacement SI engine becomes very inefficient when low output power is required. The throttle valve is partially closed, creating low inlet pressure and the resulting large pumping loss. Low inlet pressure reduces the pressure for the entire following cycle, resulting in poor combustion and a low imep. This, along with low engine speed, results in a very inefficient cycle. To compensate for this, several automobile manufacturers have developed engines that disconnect (cutout) half of the cylinders at low load, and run with only the remaining cylinders firing. This is usually done with large displacement V8s, which then run as four-cylinder engines at low load producing the same brake power output. Instead of a large engine running inefficiently at low speed, the unit runs efficiently as a smaller engine at higher speed.

When cylinder cutout is used, the valves are disconnected and fuel input and ignition to those cylinders are stopped. Typically, on a V8 engine, the two outer cylinders on one bank and the two inner cylinders on the other bank are disconnected. The EMS decides when cutout occurs, and then adjusts throttle, ignition timing, etc. for the new operating conditions. Early attempts to use cylinder cutout in the 1980s and 1990s generally gave less-than-satisfactory results due to inability of the system controls. Modern EMS systems now have the power and sophistication needed and some top-of-the-line automobiles (Mercedes) use cylinder cutout [190].

When the large engine is operated as a smaller four-cylinder engine, the throttle is opened, resulting in less pumping loss and higher engine speeds. The higher speed of a small engine operates closer to steady-state, and the higher cycle pressure allows for a leaner air–fuel ratio and a greater amount of exhaust gas recycling. All of this makes for greater efficiency, and fuel savings of 5% to 15% have been realized.

Mitsubishi has even developed a four-cylinder engine that cuts out cylinders 1 and 4, and runs as a two-cylinder when full power is not needed [206].

Yet another way of operating a large engine when only a light power output is needed is to convert from a four-stroke cycle to a six-stroke cycle. At present, no engine is known to operate in this manner, but it has been suggested for possible future development. With a 42-volt electrical system and complete variable valve control it would be possible to consider such an engine. After the exhaust stroke of a four-stroke cycle, two additional strokes could be added, with no fuel input and possibly all valves open. The engine could then run at a higher, more efficient speed, but still produce less brake output with a power stroke in each cylinder only on every third revolution.

Example Problem 2-6

A hybrid automobile with mass of 3200 lbm, traveling at 60 mph, slows to a stop. The automobile is equipped with a combined starter–generator–flywheel, and when slowing down, 58% of the kinetic energy of the vehicle is recovered as electrical energy in the battery. When the battery is being recharged with the vehicle's IC engine, there is a 28% efficiency of converting chemical energy in the fuel to electrical energy stored in the battery. The engine burns stoichiometric gasoline. Calculate:

1. electrical energy recovered in the battery by one slowdown of the automobile
2. mass of gasoline saved by recovering kinetic energy of one slowdown

(1) Kinetic energy of vehicle at 60 mph is

$$\text{KE} = mV^2/2g_c$$
$$= \frac{(3200\ \text{lbm})[(60\ \text{miles/hr})(5280\ \text{ft/mile})/(3600\ \text{sec/hr})]^2}{[(2)(32.2\ \text{lbm-ft/lbf-sec}^2)]}$$
$$= 384{,}800\ \text{ft-lbf} = 495\ \text{BTU}$$

Of this amount, 58% is recovered as stored electrical energy:

$$E = (0.58)(495\ \text{BTU}) = 287\ \text{BTU} = 303\ \text{kJ}$$

(2) The mass of gasoline needed to supply the same amount of electrical energy to the battery is

$$E = m_{\text{gasoline}} Q_{\text{LHV}} \eta_{\text{conversion}} = 287\ \text{BTU}$$
$$= m_{\text{gasoline}}(43{,}000\ \text{kJ/kg})[0.4299\ (\text{BTU/lbm})/(\text{kJ/kg})]$$
$$\times\ (0.28\ \text{conversion efficiency})$$
$$m_{\text{gasoline}} = 0.0554\ \text{lbm} = 0.025\ \text{kg}$$

2.14 CONCLUSIONS—WORKING EQUATIONS

In this chapter, equations relating the working parameters of engine operation have been developed, giving tools by which these parameters can be used for engine design and characterization. By combining earlier equations from the chapter, the following additional working equations are obtained. These are given as general equations and as specific equations to be used either with SI units or with English units. In the specific equations, units that must be used to satisfy the equality are given in brackets.

Torque:

$$\tau = \eta_f \eta_v V_d Q_{\text{HV}} \rho_a (\text{FA})/2\pi n \qquad (2\text{-}75)$$

For SI units,

$$\tau[\text{N-m}] = 159.2\ \dot{W}[\text{kW}]/\text{N}[\text{rev/sec}] \qquad (2\text{-}76)$$

For English units,

$$\tau[\text{lbf-ft}] = 5252\ \dot{W}[\text{hp}]/\text{N}[\text{RPM}] \qquad (2\text{-}77)$$

Power:

$$\dot{W}_b = \dot{m}_f/(\text{bsfc}) = (\text{FA})\dot{m}_a/(\text{bsfc}) \tag{2-78}$$

$$\dot{W}_b = \eta_f \eta_v N V_d Q_{\text{HV}} \rho_a (\text{FA})/n \tag{2-79}$$

For SI units,

$$\dot{W}_b[\text{kW}] = N[\text{rev/sec}]\tau[\text{N-m}]/159.2 \tag{2-80}$$

$$\dot{W}_b[\text{kW}] = \text{bmep}[\text{kPa}]V_d[\text{L}]N[\text{rev/sec}]/1000\, n[\text{rev/cycle}] \tag{2-81}$$

For English units,

$$\dot{W}_b[\text{hp}] = N[\text{RPM}]\tau[\text{lbf-ft}]/5252 \tag{2-82}$$

$$\dot{W}_b[\text{hp}] = \text{bmep}[\text{psia}]V_d[\text{in.}^3]N[\text{RPM}]/396{,}000\, n[\text{rev/cycle}] \tag{2-83}$$

Mechanical Efficiency:

$$\eta_m = \dot{W}_b/\dot{W}_i = \text{bmep}/\text{imep} = 1 - \dot{W}_f/\dot{W}_i \tag{2-84}$$

Mean Effective Pressure:

$$\text{bmep} = 2\pi n\tau/V_d \tag{2-85}$$

$$\text{mep} = n\dot{W}/V_d N \tag{2-86}$$

For SI units,

$$\text{bmep}[\text{kPa}] = 6.28\, n[\text{rev/cycle}]\tau[\text{N-m}]/V_d[\text{L}] \tag{2-87}$$

$$\text{mep}[\text{kPa}] = 1000\, \dot{W}[\text{kW}]n[\text{rev/cycle}]/V_d[\text{L}]N[\text{rev/sec}] \tag{2-88}$$

For English units,

$$\text{bmep}[\text{psia}] = 75.4\, n[\text{rev/cycle}]\tau[\text{lb}_f\text{-ft}]/V_d[\text{in.}^3] \tag{2-89}$$

$$\text{mep}[\text{psia}] = 396{,}000\, \dot{W}[\text{hp}]n[\text{rev/cycle}]/V_d[\text{in.}^3]N[\text{RPM}] \tag{2-90}$$

Specific Power:

$$\dot{W}/A_p = \eta_f \eta_v N S Q_{\text{HV}} \rho_a (\text{FA})/n \tag{2-91}$$

$$\dot{W}/A_p = \eta_f \eta_v \bar{U}_p Q_{\text{HV}} \rho_a (\text{FA})/2n \tag{2-92}$$

PROBLEMS

2.1 As Becky was driving "Old Betsy," the family station wagon, the engine finally quit, being worn out after 171,000 miles. It can be assumed that the average speed over its lifetime was 40 mph at an engine speed of 1700 RPM. The engine is a five-liter V8 operating on a four-stroke cycle.

Calculate:

(a) How many revolutions has the engine experienced?

(b) How many spark plug firings have occurred in the entire engine?

(c) How many intake strokes have occurred in one cylinder?

2.2 A four-cylinder, two-stroke cycle diesel engine with 10.9-cm bore and 12.6-cm stroke produces 88 kW of brake power at 2000 RPM. Compression ratio $r_c = 18:1$.
Calculate:

(a) Engine displacement. [cm^3, L]
(b) Brake mean effective pressure. [kPa]
(c) Torque. [N-m]
(d) Clearance volume of one cylinder. [cm^3]

2.3 A four-cylinder, 2.4-liter engine operates on a four-stroke cycle at 3200 RPM. The compression ratio is 9.4:1, the connecting rod length $r = 18$ cm, and the bore and stroke are related as $S = 1.06B$.
Calculate:

(a) Clearance volume of one cylinder in cm^3, L, and $in.^3$.
(b) Bore and stroke in cm and in.
(c) Average piston speed in m/sec and ft/sec.

2.4 What are the advantages of an over square engine? What are the advantages of an under square engine?

2.5 In Problem 2-3, what is the average piston speed and what is the piston speed when the crank angle $\theta = 90°$ aTDC? [m/sec]

2.6 A five-cylinder, 3.5-liter SI engine operates on a four-stroke cycle at 2500 RPM. At this condition, the mechanical efficiency of the engine is 62% and 1000 J of indicated work are produced each cycle in each cylinder.
Calculate:

(a) Indicated mean effective pressure. [kPa]
(b) Brake mean effective pressure. [kPa]
(c) Friction mean effective pressure. [kPa]
(d) Brake power in kW and hp.
(e) Torque. [N-m]

2.7 The engine operating at the conditions in Problem 2-6 is square, with $S = B$.
Calculate:

(a) Specific power. [kW/cm^2]
(b) Output per displacement. [kW/cm^3]
(c) Specific volume. [cm^3/kW]
(d) Power lost to friction in kW and hp.

2.8 The engine operating at the conditions in Example Problem 2-4 has a combustion efficiency of 97%.
Calculate:

(a) Rate of unburned hydrocarbon fuel that is expelled into the exhaust system. [kg/hr]
(b) Specific emission of HC. [(gm/kW-hr]
(c) Emission index of HC.

2.9 A construction vehicle has a diesel engine with eight cylinders of 5.375-inch bore and 8.0-inch stroke, operating on a four-stroke cycle. It delivers 152-shaft horsepower at 1000 RPM, with a mechanical efficiency of 0.60.

Calculate:

(a) Total engine displacement. [$in.^3$]
(b) Brake mean effective pressure. [psia]
(c) Torque at 1000 RPM. [lbf-ft]
(d) Indicated horsepower.
(e) Friction horsepower.

2.10 A 1500-cm^3, four-stroke cycle, four-cylinder CI engine, operating at 3000 RPM, produces 48 kW of brake power. Volumetric efficiency is 0.92 and air–fuel ratio AF = 21:1.

Calculate:

(a) Rate of air flow into engine. [kg/sec]
(b) Brake specific fuel consumption. [gm/kW-hr]
(c) Mass rate of exhaust flow. [kg/hr]
(d) Brake output per displacement. [kW/L]

2.11 A pickup truck has a five-liter four-stroke cycle, V6, SI engine operating at 2400 RPM. The compression ratio $r_c = 10.2:1$, the volumetric efficiency $\eta_v = 0.91$, and the bore and stroke are related as stroke $S = 0.92B$.

Calculate:

(a) Stroke length. [cm]
(b) Average piston speed. [m/sec]
(c) clearance volume of one cylinder. [cm^3]
(d) Air flow rate into engine. [kg/sec]

2.12 It takes a man 12.5 hours to complete a 500-mile trip in his automobile, during which 18 gallons of gasoline are consumed. During this trip the average emissions index for carbon monoxide is $(EI)_{CO} = 28$ (gm/sec)/(kg/sec). Density of liquid gasoline is 0.692 kg/L.

Calculate:

(a) Fuel economy in English units. [mpg]
(b) Fuel consumption rate using standard SI units of L/100 km.
(c) Amount of CO emitted to environment during trip. [kg]

2.13 A 5.6-liter V10 compression-ignition truck engine operates on a four-stroke cycle at 3600 RPM producing 162 kW of brake power. The bore and stroke of the engine are related as S = 1.12 B.

Calculate:

(a) Average piston speed. [m/sec]
(b) Torque. [N-m]
(c) Brake mean effective pressure. [kPa]

2.14 A 4.8-liter, spark-ignition, four-stroke cycle, V8 industrial engine operates 24 hours per day for five days at 2000 RPM using gasoline with AF = 14.6. The engine has a volumetric efficiency of 92%, with bore and stroke related as B = 1.06 S.

Calculate:

(a) Stroke length. [cm]
(b) Average piston speed. [m/sec]
(c) Number of times each spark plug has fired.
(d) Mass flow rate of air into engine. [kg/sec]
(e) Mass flow rate of fuel into engine. [kg/sec]

2.15 A small single-cylinder, two-stroke cycle SI engine operates at 8000 RPM with a volumetric efficiency of $\eta_v = 0.85$. The engine is square (bore = stroke) and has a displacement of 6.28 cm^3. The fuel–air ratio FA = 0.067.
Calculate:

(a) Average piston speed. [m/sec]
(b) Flow rate of air into engine. [kg/sec]
(c) Flow rate of fuel into engine. [kg/sec]
(d) Fuel input for one cycle. [kg/cycle]

2.16 A single-cylinder, four-stroke cycle CI engine with 12.9-cm bore and 18.0-cm stroke, operating at 800 RPM, uses 0.113 kg of fuel in four minutes while developing a torque of 76 N-m.
Calculate:

(a) Brake specific fuel consumption. [gm/kW-hr]
(b) Brake mean effective pressure. [kPa]
(c) Brake power. [kW]
(d) Specific power. [kW/cm^2]
(e) Output per displacement. [kW/L]
(f) Specific volume. [L/kW]

2.17 A 302-$in.^3$ displacement, V8, four-stroke cycle SI engine mounted on a hydraulic dynamometer has an output of 72 hp at 4050 RPM. Water absorbs the energy output of the engine as it flows through the dynamometer at a rate of 30 gallons per minute. The dynamometer has an efficiency of 93% and the water enters at a temperature of 46°F.
Calculate:

(a) Exit temperature of the water. [°F]
(b) Torque output of the engine at this condition. [lbf-ft]
(c) What is the bmep at this condition? [psia]

2.18 A 3.1-liter, four-cylinder, two-stroke cycle SI engine is mounted on an electrical generator dynamometer. When the engine is running at 1200 RPM, output from the 220-volt DC generator is 54.2 amps. The generator has an efficiency of 87%.
Calculate:

(a) Power output of the engine in kW and hp.
(b) Engine torque. [N-m]
(c) Engine bmep. [kPa]

2.19 An SI, six-liter, V8 race car engine operates at WOT on a four-stroke cycle at 6000 RPM using stoichiometric nitromethane. Fuel enters the engine at a rate of 0.198 kg/sec and combustion efficiency is 99%.

Calculate:

(a) Volumetric efficiency of engine. [%]
(b) Flow rate of air into engine. [kg/sec]
(c) Heat added per cycle per cylinder. [kJ]
(d) Chemical energy from unburned fuel in the exhaust. [kW]

2.20 A large V8 SI four-stroke cycle engine with a displacement of 4.6 liters is equipped with cylinder cutout, which converts the engine to a 2.3 liter V4 when less power is needed. At a speed of 1750 RPM the engine, as a V8, has a volumetric efficiency of 51%, a mechanical efficiency of 75%, an air–fuel ratio of 14.5, and produces 32.4 kW of brake power using gasoline. With cylinder cutout and operating at higher speed as a V4, the engine has a volumetric efficiency of 86%, a mechanical efficiency of 87%, and uses an air–fuel ratio of 18.2. Indicated thermal efficiency can be considered the same at all speeds, and combustion efficiency is 100%.

Calculate:

(a) Mass flow rate of air into V8 engine at 1750 RPM. [kg/sec]
(b) Mass flow rate of fuel into V8 engine at 1750 RPM. [kg/sec]
(c) The bsfc as V8 at 1750 RPM. [gm/kW-hr]
(d) Engine speed needed as a V4 to produce same brake power output. [RPM]
(e) The bsfc as V4 at higher speed. [gm/kW-hr]

2.21 A 1900-kg hybrid automobile which operates on ethanol fuel is equipped with a multipurpose motor-generator-flywheel. When the vehicle slows or stops, 51% of the kinetic energy is recovered as electrical energy in the battery. When the IC engine is used to recharge the battery there is a 24% efficiency of converting chemical energy in the fuel to electrical energy stored in the battery. The vehicle slows from 70 MPH to 20 MPH.

Calculate:

(a) Electrical energy recovered in battery. [kJ]
(b) Mass of fuel needed to store same amount of energy in battery. [kg]

DESIGN PROBLEMS

2.1D Design a six-liter race car engine that operates on a four-stroke cycle. Decide what the design speed will be, and then give the number of cylinders, bore, stroke, piston rod length, average piston speed, imep, brake torque, fuel used, AF, and brake power all at design speed. All parameter values should be within typical, reasonable ranges and should be consistent with the other values. State what assumptions you make (e.g., mechanical efficiency, volumetric efficiency, etc.)

2.2D Design a six-horsepower engine for a snowblower. Decide on the operating speed, number of strokes in cycle, carburetor or fuel injectors, and total displacement. Give the number of cylinders, bore, stroke, connecting rod length, average piston speed, brake torque, and brake power. What special considerations must be made, knowing that this engine must start in very cold weather? All parameter values should be within typical, reasonable ranges and should be consistent with the other values. State all assumptions you make.

2.3D Design a small four-stroke cycle Diesel engine to produce 50 kW of brake power at design speed when installed in a small pickup truck. Average piston speed should not exceed 8 m/sec at design conditions. Give the design speed, displacement, number of cylinders, bore, stroke, bmep, and torque. All parameter values should be within typical, reasonable ranges and should be consistent with the other values. State all assumptions you make.

CHAPTER 3

Engine Cycles

"Nature, in providing us with combustibles on all sides, has given us the power to produce, at all times and in all places, heat and the propelling power which is the result of it. To develop this power, to appropriate it to our uses, is the object of heat engines. The study of these engines is of the greatest interest, their importance is enormous, their use is continually increasing, and they seem destined to produce a great revolution in the civilized world."

On Heat Engines

by *Sadi Carnot* (1824)

This chapter presents the basic cycles used in reciprocating internal combustion engines, both four-stroke and two-stroke. The most common four-stroke SI and CI cycles are analyzed in detail using air-standard analysis. Lesser used cycles, including some historic cycles, are analyzed in less detail.

3.1 AIR-STANDARD CYCLES

The cycle experienced in the cylinder of an internal combustion engine is very complex. First, air (CI engine) or air mixed with fuel (SI engine) is ingested and mixed with the slight amount of exhaust residual remaining from the previous cycle. This mixture is then compressed and combusted, changing the composition to exhaust products consisting largely of CO_2, H_2O, and N_2 with many other lesser components. Then, after an expansion process, the exhaust valve is opened and this gas mixture is expelled to the surroundings. Thus, it is an open cycle with changing composition, a difficult system to analyze. To make the analysis of the engine cycle much more manageable, the real cycle is approximated with an ideal **air-standard cycle**, which differs from the actual cycle in the following ways:

1. The gas mixture in the cylinder is treated as air for the entire cycle, and property values of air are used in the analysis. This is a good approximation during the first half of the cycle, when most of the gas in the cylinder is air with only up to about 7% fuel vapor. Even in the second half of the cycle, when the gas composition is mostly CO_2, H_2O, and N_2, using air properties does not create large errors in the analysis. Air will be treated as an ideal gas with constant specific heats.
2. The real open cycle is changed into a closed cycle by assuming that the gases being exhausted are fed back into the intake system. This works with ideal air-standard cycles, as both intake gases and exhaust gases are air. Closing the cycle simplifies the analysis.
3. The combustion process is replaced with a heat addition term Q_{in} of equal energy value. Air alone cannot combust.
4. The open exhaust process, which carries a large amount of enthalpy out of the system, is replaced with a closed system heat rejection process Q_{out} of equal energy value.
5. Actual engine processes are approximated with ideal processes.

   (a) The almost-constant-pressure intake and exhaust strokes are assumed to be constant pressure. At WOT, the intake stroke is assumed to be at a pressure P_o of one atmosphere. At partially closed throttle or when supercharged, inlet pressure will be some constant value other than one atmosphere. The exhaust stroke pressure is assumed constant at one atmosphere.

   (b) Compression strokes and expansion strokes are approximated by isentropic processes. To be truly isentropic would require these strokes to be reversible and adiabatic. There is some friction between the piston and the cylinder walls, but because the surfaces are highly polished and lubricated, this friction is kept to a minimum and the processes are close to frictionless and reversible. If this were not true, automobile engines would wear out long before the 150–200 thousand mile lifetimes they now have if properly maintained. There is also fluid friction because of the gas motion within the cylinders during these strokes. This too is minimal. Heat transfer for any one stroke will be negligibly small due to the very short time involved for that single process. Thus, an almost reversible and almost adiabatic process can quite accurately be approximated with an isentropic process.

   (c) The combustion process is idealized by a constant-volume process (SI cycle), a constant-pressure process (CI cycle), or a combination of both (CI Dual cycle).

   (d) Exhaust blowdown is approximated by a constant-volume process.

   (e) All processes are considered reversible.

In air-standard cycles, air is considered an ideal gas such that the following ideal gas relationships can be used:

$$Pv = RT \tag{a}$$
$$PV = mRT \tag{b}$$
$$P = \rho RT \tag{c}$$
$$dh = c_p\, dT \tag{d}$$

$$du = c_v\, dT \qquad \text{(e)}$$
$$Pv^k = \text{constant} \quad \text{isentropic process} \qquad \text{(f)}$$
$$Tv^{k-1} = \text{constant} \quad \text{isentropic process} \qquad \text{(g)}$$
$$TP^{(1-k)/k} = \text{constant} \quad \text{isentropic process} \qquad \text{(h)}$$
$$w_{1-2} = (P_2v_2 - P_1v_1)/(1 - k) \quad \text{isentropic work in closed system}$$
$$= R(T_2 - T_1)/(1 - k) \qquad \text{(i)}$$
$$c = \sqrt{kRT} \quad \text{speed of sound} \qquad \text{(j)} \quad \text{(3-1)}$$

where

P = gas pressure in cylinder
V = volume in cylinder
v = specific volume of gas
R = gas constant of air
T = temperature
m = mass of gas in cylinder
ρ = density
h = specific enthalpy
u = specific internal energy
c_p, c_v = specific heats
$k = c_p/c_v$
w = specific work
c = speed of sound

In addition to these, the following variables are used in this chapter for cycle analysis:

AF = air–fuel ratio
$\dot{m}$ = mass flow rate
q = heat transfer per unit mass for one cycle
$\dot{q}$ = heat transfer rate per unit mass
Q = heat transfer for one cycle
$\dot{Q}$ = heat transfer rate
Q_{HV} = heating value of fuel
r_c = compression ratio
W = work for one cycle
$\dot{W}$ = power
η_c = combustion efficiency

Subscripts used include the following:

a = air
f = fuel
ex = exhaust
m = mixture of all gases

For thermodynamic analysis, the specific heats of air can be treated as functions of temperature, which they are, or they can be treated as constants, which simplifies calculations at a slight loss of accuracy. In this textbook, constant specific heat analysis will be used. Because of the high temperatures and large temperature range experienced during an engine cycle, the specific heats and ratios of specific heats k do vary by a fair amount (see Table A-1 in the Appendix). At the low-temperature end of a cycle during intake and start of compression, a value of $k = 1.4$ is correct. However, at the end of combustion the temperature has risen such that $k = 1.3$ would be more accurate. A constant average value between these extremes is found to give better results than a standard condition (25°C) value, as is often used in elementary thermodynamics textbooks.

An algebraic average gives $k = (k_1 + k_2)/2 = (1.40 + 1.30)/2 = 1.35$, as does a geometric average $k = \sqrt{k_1 k_2} = \sqrt{(1.40)(1.30)} = 1.35$.

When analyzing what occurs within engines during the operating cycle and exhaust flow, this book uses the following air property values:

$$
\begin{aligned}
c_p &= 1.108 \text{ kJ/kg-K} = 0.265 \text{ BTU/lbm-°R} \\
c_v &= 0.821 \text{ kJ/kg-K} = 0.196 \text{ BTU/lbm-°R} \\
k &= c_p/c_v = 1.108/0.821 = 1.35 \\
R &= c_p - c_v = 0.287 \text{ kJ/kg-K} \\
&= 0.069 \text{ BTU/lbm-°R} = 53.33 \text{ ft-lbf/lbm-°R}
\end{aligned}
$$

Air flow before it enters an engine is usually closer to standard temperature, and for these conditions a value of $k = 1.4$ is correct. This would include processes such as inlet flow in superchargers, turbochargers, and carburetors, and air flow through the engine radiator. For these conditions, the following air property values are used:

$$
\begin{aligned}
c_p &= 1.005 \text{ kJ/kg-K} = 0.240 \text{ BTU/lbm-°R} \\
c_v &= 0.718 \text{ kJ/kg-K} = 0.172 \text{ BTU/lbm-°R} \\
k &= c_p/c_v = 1.005/0.718 = 1.40 \\
R &= c_p - c_v = 0.287 \text{ kJ/kg-K}
\end{aligned}
$$

HISTORIC—SIX-STROKE CYCLES

During the second half of the 19th century, when development of the modern reciprocating internal combustion engine was in its early stages, many types of engines operating on many different cycles were tried. These included various two-, four-, and even six-stroke cycles. Six-stroke cycles were similar to four-stroke cycles with two added strokes for additional exhaust removal (i.e., three revolutions per cycle instead of two). With poor fuel quality, low compression ratios, and large clearance volumes, early engines had problems with excessive exhaust residual. After the exhaust stroke, an additional intake stroke was added which ingested only air. The air mixed with the exhaust residual and was then expelled with a second exhaust stroke. Compare this with the concept of EGR, which adds exhaust gas to the incoming air of all modern automobile engines [29].

3.2 OTTO CYCLE

The cycle of a four-stroke, SI, naturally aspirated engine at WOT is shown in Fig. 2-6. This is the cycle of many automobile engines and other four-stroke SI engines. For analysis, this cycle is approximated by the air-standard cycle shown in Fig. 3-1. This ideal cycle is called an **Otto cycle**, named after one of the early developers of this type of engine. The Otto cycle is the air-standard model of most four-stroke SI engines of the last 140 years, including many of today's automobile engines.

The intake stroke of the Otto cycle starts with the piston at TDC and is a constant-pressure process at an inlet pressure of one atmosphere (process 6-1 in Fig. 3-1). This is a good approximation to the inlet process of a real engine at WOT, which will actually be at a pressure slightly less than atmospheric due to pressure losses in the inlet air flow. The temperature of the air during the inlet stroke is increased as the air passes through the hot intake manifold. The temperature at point 1 will generally be on the order of 25° to 35°C hotter than the surrounding air temperature.

The second stroke of the cycle is the compression stroke, which in the Otto cycle is an isentropic compression from BDC to TDC (process 1-2). This is a good approximation to compression in a real engine, except for the very beginning and the very end of the stroke. In a real engine, the beginning of the stroke is affected by the intake valve not being fully closed until slightly after BDC. The end of compression is affected by the

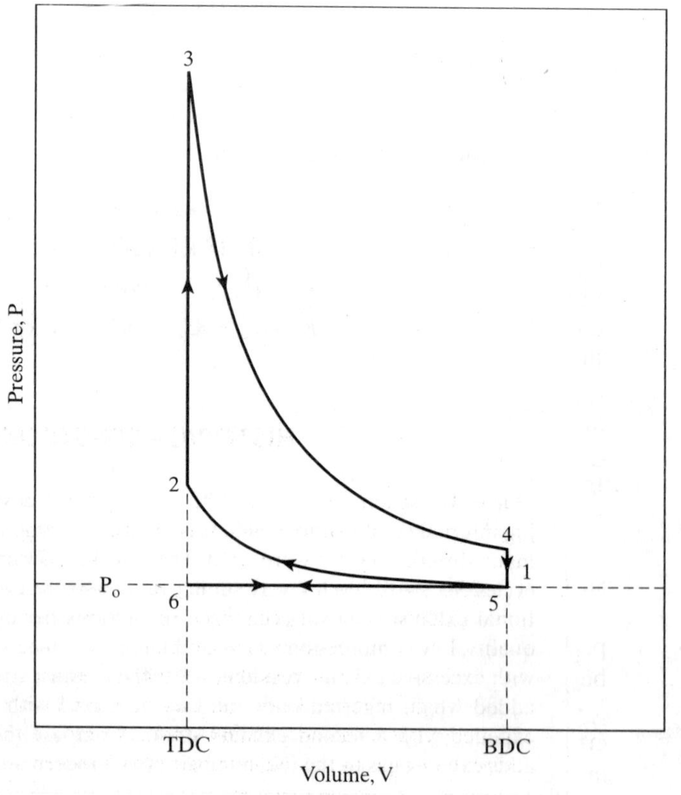

FIGURE 3-1

Ideal air-standard Otto cycle, 6-1-2-3-4-5-6, which approximates the four-stroke cycle of an SI engine on P–V coordinates.

firing of the spark plug before TDC. Not only is there an increase in pressure during the compression stroke, but the temperature within the cylinder is increased substantially due to compressive heating.

The compression stroke is followed by a constant-volume heat input process 2-3 at TDC. This replaces the combustion process of the real engine cycle, which occurs at close to constant-volume conditions. In a real engine combustion is started slightly bTDC, reaches its maximum speed near TDC, and is terminated a little aTDC. During combustion or heat input, a large amount of energy is added to the air within the cylinder. This energy raises the temperature of the air to very high values, giving peak cycle temperature at point 3. This increase in temperature during a closed constant-volume process results in a large pressure rise also. Thus, peak cycle pressure is also reached at point 3.

The very high pressure and enthalpy values within the system at TDC generate the power stroke (or expansion stroke) which follows combustion (process 3–4). High pressure on the piston face forces the piston back towards BDC and produces the work and power output of the engine. The power stroke of the real engine cycle is approximated with an isentropic process in the Otto cycle. This is a good approximation, subject to the same arguments as the compression stroke on being frictionless and adiabatic. In a real engine, the beginning of the power stroke is affected by the last part of the combustion process. The end of the power stroke is affected by the exhaust valve being opened before BDC. During the power stroke, values of both the temperature and pressure within the cylinder decrease as volume increases from TDC to BDC.

Near the end of the power stroke of a real engine cycle, the exhaust valve is opened and the cylinder experiences exhaust blowdown. A large amount of exhaust gas is expelled from the cylinder, reducing the pressure to that of the exhaust manifold. The exhaust valve is opened bBDC to allow for the finite time of blowdown to occur. It is desirable for blowdown to be complete by BDC so that there is no high pressure in the cylinder to resist the piston in the following exhaust stroke. Blowdown in a real engine is therefore almost, but not quite, constant volume. A large quantity of enthalpy is carried away with the exhaust gases, limiting the thermal efficiency of the engine. The Otto cycle replaces the exhaust blowdown open-system process of the real cycle with a constant-volume pressure reduction, closed-system process 4–5. Enthalpy loss during this process is replaced with heat rejection in the engine analysis. Pressure within the cylinder at the end of exhaust blowdown has been reduced to about one atmosphere, and the temperature has been substantially reduced by expansion cooling.

The last stroke of the four-stroke cycle now occurs as the piston travels from BDC to TDC. Process 5–6 is the exhaust stroke that occurs at a constant pressure of one atmosphere due to the open exhaust valve. This is a good approximation of the real exhaust stroke, which occurs at a pressure slightly higher than the surrounding pressure due to the small pressure drop across the exhaust valve and in the exhaust system.

At the end of the exhaust stroke, the engine has experienced two revolutions, the piston is again at TDC, the exhaust valve closes, the intake valve opens, and a new cycle begins.

When analyzing an Otto cycle, it is more convenient to work with specific properties by dividing by the mass of air within the cylinder. Figure 3-2 shows the Otto cycle in P–v and T–s coordinates. It is not uncommon to find the Otto cycle shown with

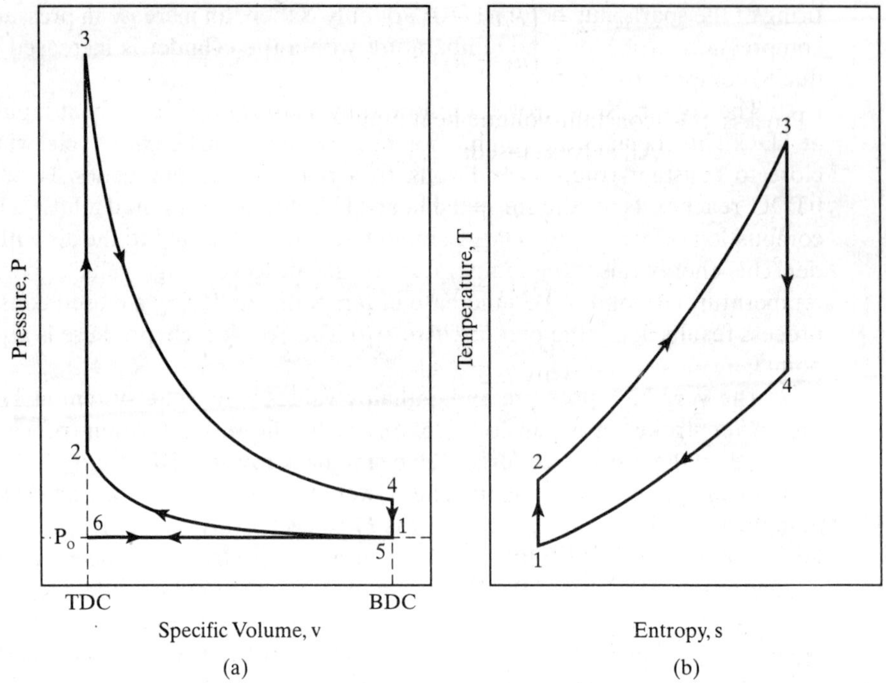

FIGURE 3-2

Otto cycle, 6-1-2-3-4-5-6, on (a) pressure-specific volume coordinates, and (b) temperature-entropy coordinates.

processes 6–1 and 5–6 left off the figure. The reasoning to justify this is that these two processes cancel each other thermodynamically and are not needed in analyzing the cycle.

Thermodynamic Analysis of Air-Standard Otto Cycle at WOT

Process 6-1—constant-pressure intake of air at P_o.

Intake valve open and exhaust valve closed:

$$P_1 = P_6 = P_o \tag{3-2}$$

$$w_{6-1} = P_o(v_1 - v_6) \tag{3-3}$$

Process 1-2—isentropic compression stroke.

All valves closed:

$$T_2 = T_1(v_1/v_2)^{k-1} = T_1(V_1/V_2)^{k-1} = T_1(r_c)^{k-1} \tag{3-4}$$

$$P_2 = P_1(v_1/v_2)^k = P_1(V_1/V_2)^k = P_1(r_c)^k \tag{3-5}$$

$$q_{1-2} = 0 \tag{3-6}$$

$$w_{1-2} = (P_2v_2 - P_1v_1)/(1 - k) = R(T_2 - T_1)/(1 - k) \tag{3-7}$$
$$= (u_1 - u_2) = c_v(T_1 - T_2)$$

Process 2-3—constant-volume heat input (combustion).
All valves closed:

$$v_3 = v_2 = v_{TDC} \tag{3-8}$$
$$w_{2-3} = 0 \tag{3-9}$$
$$Q_{2-3} = Q_{in} = m_f Q_{HV}\eta_c = m_m c_v(T_3 - T_2)$$
$$= (m_a + m_f)c_v(T_3 - T_2) \tag{3-10}$$
$$Q_{HV}\eta_c = (AF + 1)c_v(T_3 - T_2) \tag{3-11}$$
$$q_{2-3} = q_{in} = c_v(T_3 - T_2) = (u_3 - u_2) \tag{3-12}$$
$$T_3 = T_{max} \tag{3-13}$$
$$P_3 = P_{max} \tag{3-14}$$

Process 3-4—isentropic power or expansion stroke.
All valves closed:

$$q_{3-4} = 0 \tag{3-15}$$
$$T_4 = T_3(v_3/v_4)^{k-1} = T_3(V_3/V_4)^{k-1} = T_3(1/r_c)^{k-1} \tag{3-16}$$
$$P_4 = P_3(v_3/v_4)^k = P_3(V_3/V_4)^k = P_3(1/r_c)^k \tag{3-17}$$
$$w_{3-4} = (P_4v_4 - P_3v_3)/(1 - k) = R(T_4 - T_3)/(1 - k)$$
$$= (u_3 - u_4) = c_v(T_3 - T_4) \tag{3-18}$$

Process 4-5—constant-volume heat rejection (exhaust blowdown).
Exhaust valve open and intake valve closed:

$$v_5 = v_4 = v_1 = v_{BDC} \tag{3-19}$$
$$w_{4-5} = 0 \tag{3-20}$$
$$Q_{4-5} = Q_{out} = m_m c_v(T_5 - T_4) = m_m c_v(T_1 - T_4) \tag{3-21}$$
$$q_{4-5} = q_{out} = c_v(T_5 - T_4) = (u_5 - u_4) = c_v(T_1 - T_4) \tag{3-22}$$

Process 5-6—constant-pressure exhaust stroke at P_o.
Exhaust valve open and intake valve closed:

$$P_5 = P_6 = P_o \tag{3-23}$$
$$w_{5-6} = P_o(v_6 - v_5) = P_o(v_6 - v_1) \tag{3-24}$$

Thermal efficiency of Otto cycle:

$$(\eta_t)_{OTTO} = |w_{net}|/|q_{in}| = 1 - (|q_{out}|/|q_{in}|)$$
$$= 1 - [c_v(T_4 - T_1)/c_v(T_3 - T_2)]$$
$$= 1 - [(T_4 - T_1)/(T_3 - T_2)] \tag{3-25}$$

Only cycle temperatures need to be known to determine thermal efficiency. This can be simplified further by applying ideal gas relationships for the isentropic compression and expansion strokes and recognizing that $v_1 = v_4$ and $v_2 = v_3$:

$$(T_2/T_1) = (v_1/v_2)^{k-1} = (v_4/v_3)^{k-1} = (T_3/T_4) \tag{3-26}$$

Rearranging the temperature terms gives

$$T_4/T_1 = T_3/T_2 \tag{3-27}$$

Equation (3-25) can be rearranged to

$$(\eta_t)_{OTTO} = 1 - (T_1/T_2)\{[(T_4/T_1) - 1]/[(T_3/T_2) - 1]\} \tag{3-28}$$

Using Eq. (3-27) gives

$$(\eta_t)_{OTTO} = 1 - (T_1/T_2) \tag{3-29}$$

Combining this with Eq. (3-4)

$$(\eta_t)_{OTTO} = 1 - [1/(v_1/v_2)^{k-1}] \tag{3-30}$$

With $v_1/v_2 = r_c$, the compression ratio is

$$(\eta_t)_{OTTO} = 1 - (1/r_c)^{k-1} \tag{3-31}$$

Only the compression ratio is needed to determine the thermal efficiency of the Otto cycle at WOT. As the compression ratio goes up, the thermal efficiency goes up as seen in Fig. 3-3. This efficiency is the **indicated thermal efficiency**, as the heat transfer values are those to and from the air within the combustion chamber.

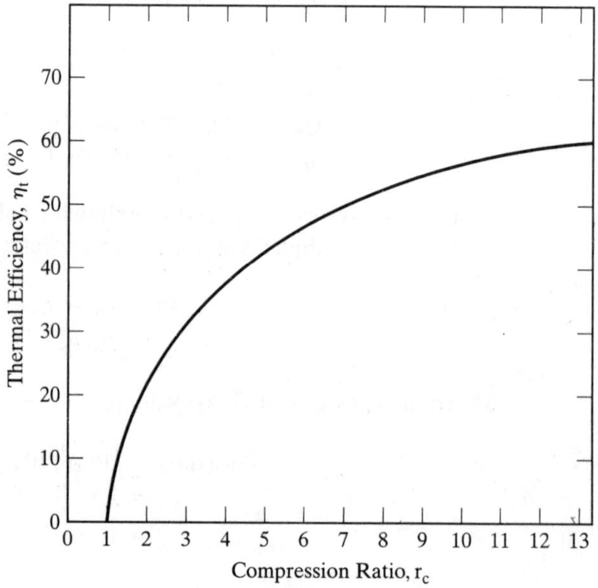

FIGURE 3-3

Indicated thermal efficiency as a function of compression ratio for SI engines operating at WOT on air-standard Otto cycle ($k = 1.35$).

Example Problem 3-1

A four-cylinder, 2.5-liter, SI automobile engine operates at WOT on a four-stroke air-standard Otto cycle at 3000 RPM. The engine has a compression ratio of 8.6:1, a mechanical efficiency of 86%, and a stroke-to-bore ratio $S/B = 1.025$. Fuel is isooctane with AF = 15, a heating value of 44,300 kJ/kg, and combustion efficiency $\eta_c = 100\%$. At the start of the compression stroke, conditions in the cylinder combustion chamber are 100 kPa and 60°C. It can be assumed that there is a 4% exhaust residual left over from the previous cycle.

Do a complete thermodynamic analysis of this engine.

For one cylinder, the displacement volume is

$$V_d = 2.5\ \text{liter}/4 = \underline{0.625\ \text{L} = 0.000625\ \text{m}^3}$$

Use Eq. (2-12) to find clearance volume:

$$r_c = V_1/V_2 = (V_c + V_d)/V_c = 8.6 = (V_c + 0.000625)/V_c$$
$$\underline{V_c = 0.0000822\ \text{m}^3 = 0.0822\ \text{L} = 82.2\ \text{cm}^3}$$

Use Eq. (2-8) to find bore and stroke:

$$V_d = (\pi/4)B^2S = (\pi/4)B^2(1.025B) = 0.000625\ \text{m}^3$$
$$\underline{B = 0.0919\ \text{m} = 9.19\ \text{cm}}$$
$$S = 1.025B = \underline{0.0942\ \text{m} = 9.42\ \text{cm}}$$

State 1:

$$\underline{T_1 = 60°\text{C} = 333\ \text{K}} \quad \text{given in problem statement}$$
$$\underline{P_1 = 100\ \text{kPa}} \quad \text{given}$$
$$V_1 = Vd + V_c = 0.000625 + 0.0000822 = \underline{0.000707\ \text{m}^3}$$

Mass of gas mixture in the cylinder can be calculated at State 1. The mass within the cylinder will then remain the same for the entire cycle.

$$m_m = P_1V_1/RT_1 = (100\ \text{kPa})(0.000707\ \text{m}^3)/(0.287\ \text{kJ/kg-K})(333\ \text{K})$$
$$= 0.000740\ \text{kg}$$

State 2: The compression stroke 1-2 is isentropic. Use Eqs. (3-4) and (3-5) to find the pressure and temperature:

$$P_2 = P_1(r_c)^k = (100\ \text{kPa})(8.6)^{1.35} = \underline{1826\ \text{kPa}}$$
$$T_2 = T_1(r_c)^{k-1} = (333\ \text{K})(8.6)^{0.35} = \underline{707\ \text{K} = 434°\text{C}}$$
$$V_2 = mRT_2/P_2 = (0.000740\ \text{kg})(0.287\ \text{kJ/kg-K})(707\ \text{K})/(1826\ \text{kPa})$$
$$= \underline{0.0000822\ \text{m}^3 = V_c}$$

This is the clearance volume of one cylinder, which agrees with the preceding. Another way of getting this value is to use Eq. (2-12):

$$V_2 = V_1/r_c = 0.000707\ \text{m}^3/8.6 = 0.0000822\ \text{m}^3$$

The mass of gas mixture m_m in the cylinder is made up of air m_a, fuel m_f, and exhaust residual m_{ex}:

mass of air	$m_a = (15/16)(0.96)(0.000740)$	$= 0.000666$ kg
mass of fuel	$m_f = (1/16)(0.96)(0.000740)$	$= 0.000044$ kg
mass of exhaust	$m_{ex} = (0.04)(0.000740)$	$= 0.000030$ kg
Total		$m_m = 0.000740$ kg

State 3: Use Eq. (3-10) to calculate the heat added during one cycle:

$$\begin{aligned} Q_{in} &= m_f Q_{HV} \eta_c = m_m c_v (T_3 - T_2) \\ &= (0.000044 \text{ kg})(44{,}300 \text{ kJ/kg})(1.00) \\ &= (0.000740 \text{ kg})(0.821 \text{ kJ/kg-K})(T_3 - 707 \text{ K}) \end{aligned}$$

Solving this for T_3

$$\underline{T_3 = 3915 \text{ K} = 3642°\text{C} = T_{max}}$$
$$V_3 = V_2 = \underline{0.0000822 \text{ m}^3}$$

For constant volume

$$P_3 = P_2(T_3/T_2) = (1826 \text{ kPa})(3915/707) = \underline{10{,}111 \text{ kPa} = P_{max}}$$

State 4: Power stroke 3-4 is isentropic. Use Eq. (3-16) and (3-17) to find temperature and pressure:

$$\begin{aligned} T_4 &= T_3(1/r_c)^{k-1} = (3915 \text{ K})(1/8.6)^{0.35} = \underline{1844 \text{ K} = 1571°\text{C}} \\ P_4 &= P_3(1/r_c)^k = (10{,}111 \text{ kPa})(1/8.6)^{1.35} = \underline{554 \text{ kPa}} \\ V_4 &= mRT_4/P_4 = (0.000740 \text{ kg})(0.287 \text{ kJ/kg-K})(1844 \text{ K})/(554 \text{ kPa}) \\ &= \underline{0.000707 \text{ m}^3 = V_1} \end{aligned}$$

This agrees with the value of V_1 found earlier.

Work produced in the isentropic power stroke for one cylinder during one cycle is

$$\begin{aligned} W_{3-4} &= mR(T_4 - T_3)/(1 - k) \\ &= (0.000740 \text{ kg})(0.287 \text{ kJ/kg-K})(1844 - 3915)\text{K}/(1 - 1.35) \\ &= 1.257 \text{ kJ} \end{aligned}$$

Work absorbed during the isentropic compression stroke for one cylinder during one cycle is

$$\begin{aligned} W_{1-2} &= mR(T_2 - T_1)/(1 - k) \\ &= (0.000740 \text{ kg})(0.287 \text{ kJ/kg-K})(707 - 333)\text{K}/(1 - 1.35) \\ &= -0.227 \text{ kJ} \end{aligned}$$

Work of the intake stroke is canceled by work of the exhaust stroke.

Net indicated work for one cylinder during one cycle is

$$W_{net} = W_{1-2} + W_{3-4} = (-0.227) + (+1.257) = \underline{+1.030 \text{ kJ}}$$

Use Eq. (3-10) to find heat added for one cylinder during one cycle:

$$Q_{in} = m_f Q_{HV} \eta_c = (0.000044 \text{ kg})(44{,}300 \text{ kJ/kg})(1.00) = \underline{1.949 \text{ kJ}}$$

Indicated thermal efficiency is

$$\eta_t = W_{net}/Q_{in} = 1.030/1.949 = \underline{0.529 = 52.9\%}$$

or, using Eqs. (3-29) and (3-31):

$$\begin{aligned} \eta_t &= 1 - (T_1/T_2) = 1 - (1/r_c)^{k-1} \\ &= 1 - (333/707) = 1 - (1/8.6)^{0.35} = 0.529 \end{aligned}$$

Equation (2-29) is used to find indicated mean effective pressure:

$$\text{imep} = W_{net}/(V_1 - V_2) = (1.030 \text{ kJ})/(0.000707 - 0.0000822)\text{m}^3 = \underline{1649 \text{ kPa}}$$

indicated power at 3000 RPM is obtained using Eq. (2-42):

$$\begin{aligned}\dot{W}_i &= WN/n \\ &= [(1.030 \text{ kJ/cyl-cycle})(3000/60 \text{ rev/sec})/(2 \text{ rev/cycle})](4 \text{ cyl}) \\ &= \underline{103 \text{ kW} = 138 \text{ hp}}\end{aligned}$$

Equation (2-2) is used to find mean piston speed:

$$\begin{aligned}\overline{U}_p &= 2SN = (2 \text{ strokes/rev})(0.0942 \text{ m/stroke})(3000/60 \text{ rev/sec}) \\ &= \underline{9.42 \text{ m/sec}}\end{aligned}$$

Equation (2-27) gives net brake work for one cylinder during one cycle:

$$W_b = \eta_m W_i = (0.86)(1.030 \text{ kJ}) = \underline{0.886 \text{ kJ}}$$

Brake power at 3000 RPM is

$$\begin{aligned}\dot{W}_b &= (3000/60 \text{ rev/sec})(0.5 \text{ cycle/rev})(0.886 \text{ kJ/cyl-cycle})(4 \text{ cyl}) \\ &= \underline{88.6 \text{ kW} = 119 \text{ hp}}\end{aligned}$$

or

$$\dot{W}_b = \eta_m \dot{W}_i = (0.86)(103 \text{ kW}) = 88.6 \text{ kW}$$

Torque is calculated using Eq. (2-43):

$$\begin{aligned}\tau &= \dot{W}_b/2\pi N = (88.6 \text{ kJ/sec})/(2\pi \text{radians/rev})(3000/60 \text{ rev/sec}) \\ &= \underline{0.282 \text{ kN-m} = 282 \text{ N-m}}\end{aligned}$$

Friction power lost is calculated using Eq. (2-49):

$$\dot{W}_f = \dot{W}_i - \dot{W}_b = 103 - 88.6 = \underline{14.4 \text{ kW} = 19.3 \text{ hp}}$$

Equation (2-37c) is used to find brake mean effective pressure:

$$\text{bmep} = \eta_m(\text{imep}) = (0.86)(1649 \text{ kPa}) = \underline{1418 \text{ kPa}}$$

This allows another way of finding torque using Eq. (2-41), which gives consistent results:

$$\tau = (\text{bmep})V_d/4\pi = (1418 \text{ kPa})(0.0025 \text{ m}^3)/4\pi = 0.282 \text{ kN-m}$$

Brake specific power is calculated using Eq. (2-51):

$$\text{BSP} = \dot{W}_b/A_p = (88.6 \text{ kW})/\{[(\pi/4)(9.19 \text{ cm})^2](4 \text{ cyl})\} = \underline{0.334 \text{ kW/cm}^2}$$

Output per displacement is found using Eq. (2-52):

$$\text{OPD} = \dot{W}_b/V_d = (88.6 \text{ kW})/(2.5 \text{ L}) = \underline{35.4 \text{ kW/L}}$$

Equation (2-60) is used to find brake specific fuel consumption:

$$\begin{aligned}\text{bsfc} &= \dot{m}_f/\dot{W}_b \\ &= (0.000044 \text{ kg/cyl-cycle})\ (50 \text{ rev/sec})\ (0.5 \text{ cycle/rev})\ (4 \text{ cyl})/(88.6 \text{ kW}) \\ &= \underline{0.000050 \text{ kg/sec/kW} = 180 \text{ gm/kW-hr}}\end{aligned}$$

Equation (2-70) is used to find volumetric efficiency using one cylinder and standard air density:

$$\begin{aligned}\eta_v &= m_a/\rho_a V_d = (0.000666 \text{ kg})/(1.181 \text{ kg/m}^3)(0.000625 \text{ m}^3) \\ &= \underline{0.902 = 90.2\%}\end{aligned}$$

3.3 REAL AIR–FUEL ENGINE CYCLES

The actual cycle experienced by an internal combustion engine is not, in the true sense, a thermodynamic cycle. An ideal air-standard thermodynamic cycle occurs on a closed system of constant composition. This is not what actually happens in an IC engine, and for this reason air-standard analysis gives, at best, only approximations to actual conditions and outputs. Major differences include the following:

1. Real engines operate on an open cycle with changing composition. Not only does the inlet gas composition differ from that of the gas which exits, but often the mass flow rate is not the same. Those engines that add fuel into the cylinders after air induction is complete (CI engines and some SI engines) change the amount of mass in the gas composition part way through the cycle. The gaseous mass exiting the engine in the exhaust is greater than the gaseous mass that entered in the induction process. This difference can be on the order of several percent. Other engines carry liquid fuel droplets with the inlet air that are idealized as part of the gaseous mass in air-standard analysis. During combustion, total mass remains about the same but molar quantity changes. Finally, there is a loss of mass during the cycle due to crevice flow and blowby past the pistons. Most of the crevice flow is a temporary loss of mass from the cylinder, but because it is greatest at the start of the power stroke, some output work is lost during expansion. Blowby can decrease the amount of mass in the cylinders by as much as 1% during compression and combustion. This is discussed in greater detail in Chapter 6.
2. Air-standard analysis treats the fluid flow through the entire engine as air and approximates air as an ideal gas. In a real engine inlet flow may be all air, or it may be air mixed with up to 7% fuel, either gaseous or as liquid droplets, or both. During combustion the composition is then changed to a gas mixture of mostly CO_2, H_2O, and N_2, with lesser amounts of CO and hydrocarbon vapor. In CI engines there will also be solid carbon particles in the combustion products gas mixture. Approximating exhaust products as air simplifies analysis but introduces some error.

   Even if all fluid in an engine cycle were air, some error would be introduced by assuming it to be an ideal gas with constant specific heats in air-standard analysis. At the low pressures of inlet and exhaust, air can accurately be treated as an ideal gas, but at the higher pressures during combustion, air will deviate from ideal gas behavior. A more serious error is introduced by assuming constant specific heats for the analysis. Specific heats of a gas have a fairly strong dependency on temperature and can vary as much as 30% in the temperature range of an engine (for air, $c_p = 1.004$ kJ/kg-K at 300 K and $c_p = 1.292$ kJ/kg-K at 3000 K [73]; see Review Problem 3-5).
3. There are heat losses during the cycle of a real engine that are neglected in air-standard analysis. Heat loss during combustion lowers the actual peak temperature and pressure from that predicted. The actual power stroke, therefore, starts at a lower pressure, and work output during expansion is decreased. Heat transfer continues during expansion, and this lowers the temperature and pressure below the ideal isentropic process towards the end of the power stroke. The result

of heat transfer is a lower indicated thermal efficiency than that predicted by air-standard analysis. Heat transfer is also present during compression which deviates the process from isentropic. However, this deviation is less than the deviation during the expansion stroke due to the lower temperatures at this time.

4. Combustion requires a short but finite time to occur, and heat addition is not instantaneous at TDC, as approximated in an Otto cycle. A fast but finite flame speed is desirable in an engine. This results in a finite rate of pressure rise in the cylinders, a steady force increase on the piston face, and a smooth engine cycle. A supersonic detonation would give almost instantaneous heat addition to a cycle, but would result in a rough cycle and quick engine destruction. Because of the finite time required, combustion is started before TDC and ends after TDC, not at constant volume as in air-standard analysis. By starting combustion bTDC, cylinder pressure increases late in the compression stroke, requiring greater negative work in that stroke. Because combustion is not completed until aTDC, some power is lost at the start of the expansion stroke (see Fig. 2-6). Another loss in the combustion process of an actual engine occurs because combustion efficiency is less than 100%. This happens because of less-than-perfect mixing, local variations in temperature and air–fuel due to turbulence, flame quenching, etc. SI engines will generally have a combustion efficiency of about 95%, while CI engines are generally about 98% efficient.

5. The blowdown process requires a finite real time and a finite cycle time, and does not occur at constant volume as in air-standard analysis. For this reason, the exhaust valve must open 40° to 60° bBDC, and some output work at the latter end of expansion is lost.

6. In an actual engine, the intake valve is not closed until after bottom-dead-center at the end of the intake stroke. Because of the flow restriction of the valve, air is still entering the cylinder at BDC, and volumetric efficiency would be lower if the valve closed here. Because of this, however, actual compression does not start at BDC but only after the inlet valve closes. With ignition then occurring before top-dead-center, temperature and pressure rise before combustion is less than predicted by air-standard analysis.

7. Engine valves require a finite time to actuate. Ideally, valves would open and close instantaneously, but this is not possible when using a camshaft. Cam profiles must allow for smooth interaction with the cam follower, and this results in fast but finite valve actuation. To assure that the intake valve is fully open at the start of the induction stroke, it must start to open before TDC. Likewise, the exhaust valve must remain fully open until the end of the exhaust stroke, with final closure occurring after TDC. The resulting valve overlap period causes a deviation from the ideal cycle. When electronic valve actuation replaces the use of camshafts, the time to open or close any valve will be greatly reduced.

8. Some error is introduced when the lower heating value of the fuel Q_{LHV} is used as the energy input to the cycle during combustion in air-standard analysis. Heating value of any fuel is calculated on conditions of 25°C in and 25°C out. This is not what happens in an engine cycle. Actual energy input during combustion in a real engine will be less than that predicted by Q_{LHV} (see Example Problem 4-4).

Because of these differences between real air-fuel cycles and ideal cycles, results from air-standard analysis will have errors and will deviate from actual conditions. Interestingly, however, the errors are not great, and property values of temperature and pressure are very representative of actual engine values, depending on the geometry and operating conditions of the real engine. By changing operating variables such as inlet temperature or pressure, compression ratio, peak temperature, etc., in Otto cycle analysis, good approximations can be obtained for output changes that will occur in a real engine as these variables are changed. Good approximation of power output, thermal efficiency, and mep can be expected.

Indicated thermal efficiency of a real four-stroke SI engine is always somewhat less than that predicted by air-standard Otto cycle analysis. This is due to the heat losses, friction, ignition timing, valve timing, finite time of combustion and blow-down, and deviation from ideal gas behavior of the real engine. Reference [120] shows that, over a large range of operating variables, the indicated thermal efficiency of an actual SI four-stroke cycle engine can be approximated by

$$(\eta_t)_{\text{actual}} \approx 0.85(\eta_t)_{\text{OTTO}} \tag{3-32}$$

This will be correct to within a few percent for large ranges of air–fuel equivalence ratio, ignition timing, engine speed, compression ratio, inlet pressure, exhaust pressure, and valve timing.

3.4 SI ENGINE CYCLE AT PART THROTTLE

When a four-stroke cycle SI engine is run at less than WOT conditions, air–fuel input is reduced by partially closing the throttle (butterfly valve) in the intake system. This creates a flow restriction and consequent pressure drop in the incoming air. Fuel input is then also reduced to match the reduction of air. Lower pressure in the intake manifold during the intake stroke and the resulting lower pressure in the cylinder at the start of the compression stroke are shown in Fig. 3-4. Although the air experiences an expansion cooling because of the pressure drop across the throttle valve, the temperature of the air entering the cylinders is about the same as at WOT because it first flows through the hot intake manifold.

Figure 3-4 shows that the net indicated work for the Otto cycle engine will be less at part throttle than at WOT. The upper loop of the cycle, made up of the compression and power strokes, represents positive work output, while the lower loop, consisting of the exhaust and intake strokes, is negative work absorbed off the rotating crankshaft. The more closed the throttle position, the lower will be the pressure during the intake stroke and the greater the negative pump work. Two main factors contribute to the reduced net work at part-throttle operation. The lower pressure at the start of compression results in lower pressures throughout the rest of the cycle except for the exhaust stroke. This lowers mep and net work. In addition, when less air is ingested into the cylinders during intake because of this lower pressure, fuel input by injectors or carburetor is also proportionally reduced. This results in less thermal energy from combustion in the cylinders and less resulting work out. It should be noted that although Q_{in} is reduced, the temperature rise in process 2-3 in Fig. 3-4 is about the same. This is because the mass of fuel and the mass of air being heated are both reduced by an equal proportion.

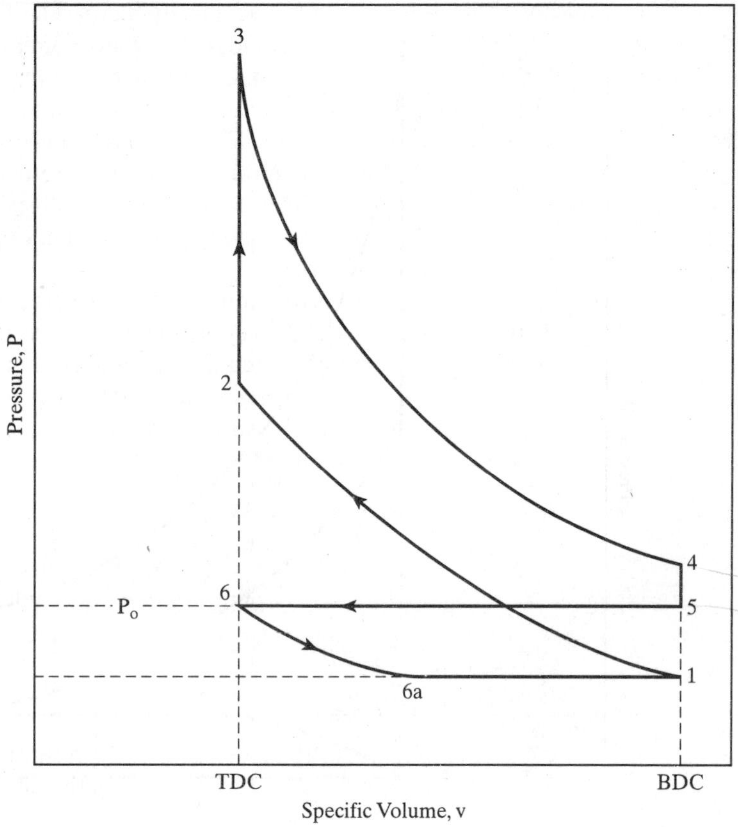

FIGURE 3-4

Four-stroke air-standard Otto cycle, 6-6a-1-2-3-4-5-6, for SI engine operating at part throttle.

If an engine is equipped with a supercharger or turbocharger the corresponding air-standard cycle is shown in Fig. 3-5, with intake pressure higher than atmospheric pressure. This results in more air and fuel in the combustion chamber during the cycle, and the resulting net indicated work is increased. Higher intake pressure increases all pressures thoughout the cycle, and increased air and fuel give greater Q_{in} in process 2-3. When air is compressed to a higher pressure by a supercharger or turbocharger, the temperature is also increased due to compressive heating. This would increase air temperature at the start of the compression stroke, which in turn raises all temperatures in the remaining cycle. This can cause self-ignition and knocking problems in the latter part of compression or during combustion. For this reason, engine compressors can be equipped with an aftercooler to again lower the compressed incoming air temperature. Aftercoolers are heat exchangers, which often use outside air as the cooling fluid. In principle, aftercoolers are desirable, but cost and space limitations often make them impractical on automobile engines. Instead, engines equipped with a supercharger or turbocharger will usually have a lower compression ratio to reduce

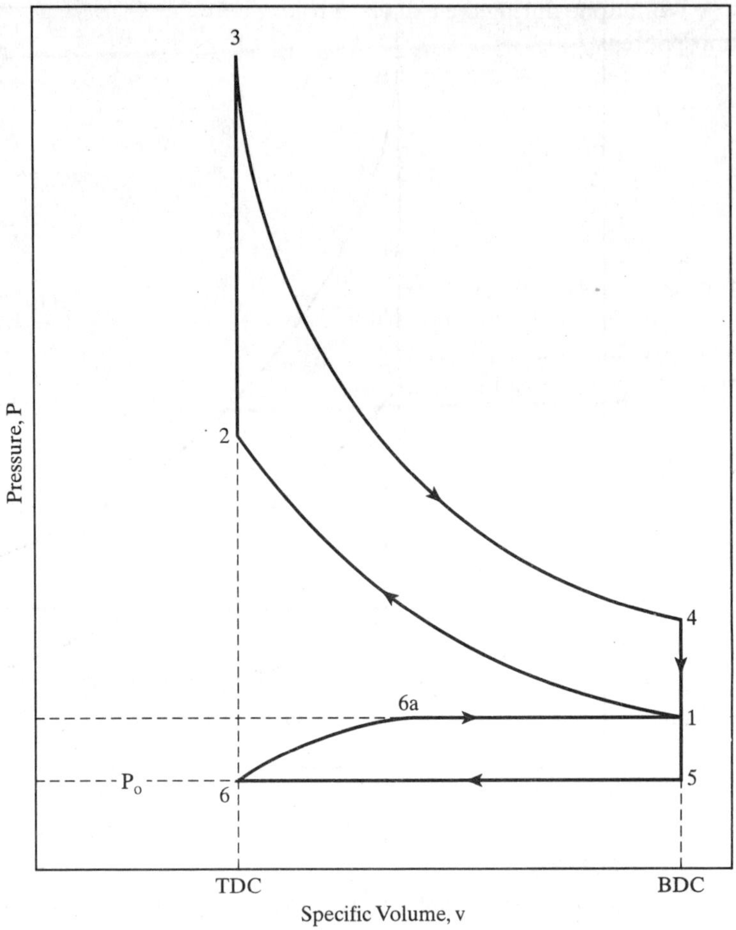

FIGURE 3-5

Four-stroke air-standard Otto cycle, 6-6a-1-2-3-4-5-6, for SI engine equipped with a turbocharger or supercharger.

knocking problems. With a lower compression ratio there will be less compressive heating in the compression stroke, which will compensate for the higher temperature at the start of the stroke.

When an engine without a supercharger or turbocharger is operated at WOT, it can be assumed that the air pressure in the intake manifold is P_o = one atmosphere. At part throttle, the partially closed butterfly valve creates a flow restriction, resulting in a lower inlet pressure P_i in the intake manifold (point 6a in Fig. 3-4). Work done during the intake stroke is, therefore,

$$W_{6-1} = P_i(V_1 - V_6) = P_i V_d \tag{3-33}$$

where V_d is the displacement volume.

Work done during the exhaust stroke where the pressure is about constant at one atmosphere is

$$W_{5-6} = P_{ex}(V_6 - V_5) = -P_{ex}V_d \tag{3-34}$$

The net indicated pumping work for the cycle at part throttle is

$$(W_{pump})_{net} = (P_i - P_{ex})V_d \tag{3-35}$$

The negative value of this pump work means that it lowers the net indicated work of the cycle.

If the engine is equipped with a supercharger or turbocharger, the inlet pressure can be greater than one atmosphere, as shown in Fig. 3-5. Net indicated pump work for this cycle is still given by Eq. (3-35), but now $P_i > P_{ex}$, pump work is positive, and net indicated work is increased.

Using Eqs. (2-29) and (3-35) for pump mean effective pressure, we have

$$\text{pmep} = (W_{pump})_{net}/V_d = (P_i - P_{ex}) \tag{3-36}$$

This can have positive or negative values.

3.5 EXHAUST PROCESS

The exhaust process consists of two steps: blowdown and exhaust stroke. When the exhaust valve opens near the end of the expansion stroke (point 4 in Fig. 3-6), the high-temperature gases are suddenly subjected to a pressure decrease as the resulting blowdown occurs. A large percentage of the gases leaves the combustion chamber during this blowdown process, driven by the pressure differential across the open exhaust valve. When the pressure across the exhaust valve is finally equalized, the cylinder is still filled with exhaust gases at the exhaust manifold pressure of about one atmosphere. These gases are then pushed out of the cylinder through the still open exhaust valve by the piston as it travels from BDC to TDC during the exhaust stroke.

Temperature of the exhaust gases is cooled by expansion cooling when the pressure is suddenly reduced during blowdown. Although this expansion is not reversible, the ideal gas isentropic relationship between pressure and temperature serves as a good model to approximate the exhaust temperature T_7 in the hypothetical process 4-7 of Fig. 3-6. According to that model,

$$\begin{aligned} T_7 &= T_4(P_7/P_4)^{(k-1)/k} = T_3(P_7/P_3)^{(k-1)/k} \\ &= T_4(P_{ex}/P_4)^{(k-1)/k} = T_4(P_o/P_4)^{(k-1)/k} \end{aligned} \tag{3-37}$$

where

$P_7 = P_{ex} = P_o$

P_{ex} = exhaust pressure, which generally can be considered equal to surrounding pressure

P_7 is the pressure in the exhaust system and is almost always very close to one atmosphere in value.

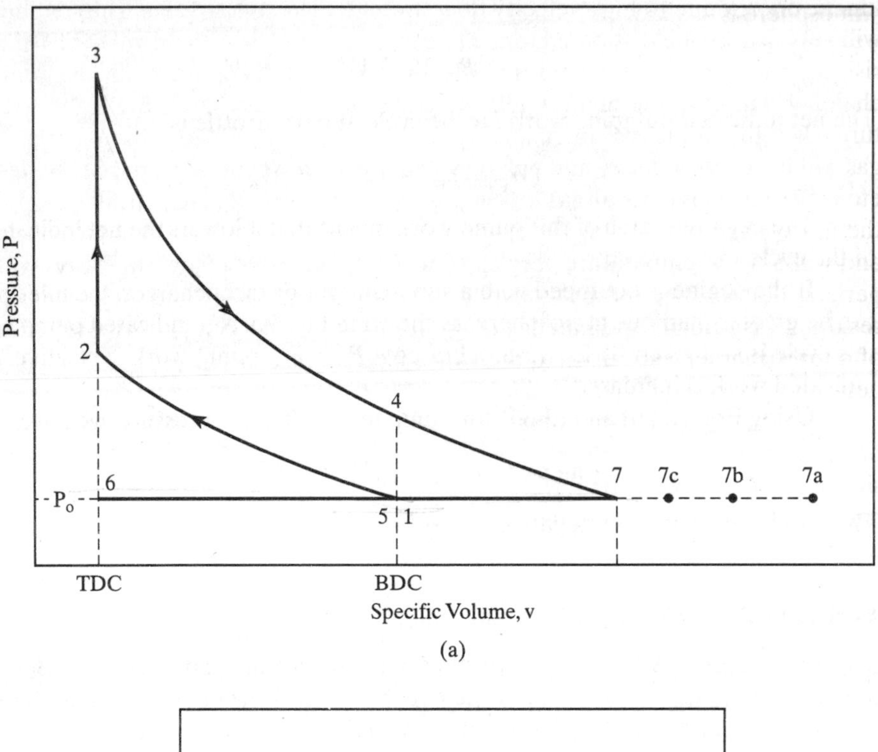

FIGURE 3-6

Air-standard Otto cycle for engine at WOT, showing process 4-7 experienced by exhaust during blowdown.

Gas leaving the combustion chamber during the blowdown process will also have kinetic energy due to high velocity flow through the exhaust valve. This kinetic energy will very quickly be dissipated in the exhaust manifold, and there will be a subsequent rise in enthalpy and temperature. The first elements of gas leaving the combustion chamber will have the highest velocity and will therefore reach the highest temperature when this velocity is dissipated (point 7a in Fig. 3-6). Each subsequent element of gas will have less velocity and will thus experience less temperature rise (points 7b, 7c, etc.). The last elements of gas leaving the combustion chamber during blowdown and the gas pushed out during the exhaust stroke will have relatively low kinetic energy and will have a temperature very close to T_7. Choked flow (sonic velocity) will be experienced across the exhaust valve at the start of blowdown, and this determines the resulting gas velocity and kinetic energy. If possible, it is desirable to mount the turbine of a turbocharger very close to the exhaust manifold. This is done so that exhaust kinetic energy can be utilized in the turbine.

The state of the exhaust gas during the exhaust stroke is best approximated by a pressure of one atmosphere, a temperature of T_7 given in Eq. (3-37), and a specific volume shown at point 7 in Fig. 3-6. It will be noted that this is inconsistent with Fig. 3-6 for the exhaust stroke process 5-6. The figure would suggest that the specific volume v changes during process 5-6. This inconsistency occurs because Fig. 3-6 uses a closed system model to represent an open system process, the exhaust stroke. Also, it should be noted that point 7 is a hypothetical state and corresponds to no actual physical piston position.

At the end of the exhaust stroke, there is still a residual of exhaust gas trapped in the clearance volume of the cylinder. This exhaust residual gets mixed with the new incoming charge of air and fuel and is carried into the new cycle. Exhaust residual is defined as

$$x_r = m_{ex}/m_m \tag{3-38}$$

where m_{ex} is the mass of exhaust gas carried into the next cycle and m_m is the mass of gas mixture within the cylinder for the entire cycle. Values of exhaust residual range from 3% to 7% at full load, increasing to as much as 20% at part-throttle light loads. CI engines generally have less exhaust residual because their higher compression ratios give them smaller relative clearance volumes. In addition to the effect of the clearance volume, the location of the valves and the amount of valve overlap affect the amount of exhaust residual.

In Fig. 3-6, if the blowdown process 4-7 is modeled as an isentropic expansion, then

$$P_4/P_7 = (v_7/v_4)^k = P_4/P_{ex} = P_4/P_o \tag{3-39}$$

$$P_3/P_7 = (v_7/v_3)^k = P_3/P_{ex} = P_3/P_o \tag{3-40}$$

The mass of exhaust in the cylinder after blowdown, but before the exhaust stroke, will be

$$m_7 = V_5/v_{ex} = V_5/v_7 = V_1/v_7 \tag{3-41}$$

The mass of exhaust in the cylinder at the end of the exhaust stroke will be

$$m_{ex} = V_6/v_7 = V_2/v_7 \tag{3-42}$$

where v_7 is calculated using either Eq. (3-39) or (3-40) and represents the constant specific volume of exhaust gas within the cylinder for the entire exhaust stroke 5-6. The mass of gas mixture in Eq. (3-38) can be obtained from

$$m_m = V_1/v_1 = V_2/v_2 = V_3/v_3 = V_4/v_4 = V_7/v_7 \tag{3-43}$$

Combining this with Eqs. (3-38) and (3-42), we get

$$x_r = (V_2/v_7)/(V_7/v_7) = V_2/V_7 \tag{3-44}$$

V_7 is the hypothetical volume of m_m expanded to P_o after combustion. Using Eqs. (3-42) and (3-43), the exhaust residual can also be written as

$$\begin{aligned} x_r &= (V_6/v_7)/(V_4/v_4) = (V_6/V_4)(v_4/v_7) = (1/r_c)(v_4/v_7) \\ &= (1/r_c)[(RT_4/P_4)/(RT_7/P_7)] \end{aligned} \tag{3-45}$$

$$x_r = (1/r_c)(T_4/T_{ex})(P_{ex}/P_4) \tag{3-46}$$

where

r_c = compression ratio

$P_{ex} = P_7 = P_o$ = one atmosphere under most conditions

$T_{ex} = T_7$ from Eq. (3-37)

and T_4 and P_4 are conditions in the cylinder when the exhaust valve opens.

When the intake valve opens, a new charge of inlet air m_a enters the cylinder and mixes with the remaining exhaust residual from the previous cycle. The mixing occurs such that total enthalpy remains constant and

$$m_{ex}h_{ex} + m_a h_a = m_m h_m \tag{3-47}$$

where h_{ex}, h_a, and h_m are the specific enthalpy values of exhaust, air, and mixture, all of which are treated as air in air-standard analysis. If specific enthalpy values are referenced to zero value at an absolute temperature value of zero, then $h = c_pT$ and

$$m_{ex}c_pT_{ex} + m_a c_p T_a = m_m c_p T_m \tag{3-48}$$

Canceling c_p and dividing by m_m, we have

$$(m_{ex}/m_m)T_{ex} + (m_a/m_m)T_a = T_m \tag{3-49}$$

Combining this equation with Eq. (3-38) gives the temperature of the gas mixture in the cylinder at the start of compression in terms of the exhaust residual x_r:

$$(T_m)_1 = x_r T_{ex} + (1 - x_r)T_a \tag{3-50}$$

where $T_{ex} = T_7$ and T_a is the temperature of the incoming air in the intake manifold.

As air enters the cylinder, it mixes with the small charge of hot exhaust residual, heating the air and reducing its density. This, in turn, reduces the volumetric efficiency of the engine. Part of this loss is gained back by the substantial cooling of the small amount of exhaust residual, which increases its density. The partial vacuum this creates in the clearance volume can then be filled with additional intake air.

Example Problem 3-2

The engine operating at the conditions of Example Problem 3-1 has an exhaust pressure of 100 kPa.

Calculate:

1. exhaust temperature
2. exhaust residual
3. temperature of air entering cylinder

(1) Use Fig. 3-6 and Eq. (3-37) to calculate the exhaust temperature:

$$T_{ex} = T_7 = T_3(P_7/P_3)^{(k-1)/k}$$
$$= (3915\ \text{K})(100/10{,}111)^{(1.35-1)/1.35} = \underline{1183\ \text{K} = 910°\text{C}}$$

(2) Use Eq. (3-46) to find exhaust residual:

$$x_r = (1/r_c)(T_4/T_{ex})(P_{ex}/P_4) = (1/8.6)(1844/1183)(100/554) = \underline{0.033}$$

It was assumed that $x_r = 0.04$ when the engine was analyzed in Example Problem 3-1. That analysis should now be redone using this better value of $x_r = 0.033$. When this is done, the following corrected values are obtained:

$$P_3 = 10{,}300\ \text{kPa}$$
$$T_3 = 3988\ \text{K}$$
$$P_4 = 564\ \text{kPa}$$
$$T_4 = 1878\ \text{K}$$
$$T_{ex} = 1199\ \text{K}$$
$$x_r = 0.033$$

The consistent value for the exhaust residual means that an additional iteration is not needed. With a reasonable exhaust residual approximation to start with, two iterations in the analysis will normally be sufficient. Other parameters (e.g., power, mep, etc.) should now be recalculated, with slight changes in their values to be expected.

(3) Equation (3-50) is now used to find the temperature of the air entering the cylinder from the intake manifold:

$$T_1 = x_r T_{ex} + (1 - x_r)T_a$$
$$333 = (0.033)(1199) + (1 - 0.033)T_a$$
$$\underline{T_a = 303\ \text{K} = 30°\text{C}}$$

Example Problem 3-3

The engine in Example Problems 3-1 and 3-2 is now run at part throttle such that the intake pressure is 50 kPa. Calculate the temperature in the cylinder at the start of the compression stroke.

The temperature of the intake air can be assumed to be the same even though it has experienced a pressure reduction expansion when passing the throttle valve. This is because it still flows through the same hot intake manifold after the throttle. However, the temperature of the exhaust residual will be reduced due to the expansion cooling it undergoes when the intake valve opens and the pressure in the cylinder drops to 50 kPa. The temperature of the exhaust residual after expansion can be approximated using Fig. 3-4 and the isentropic expansion model such that

$$T_{6a} = T_{ex}(P_{6a}/P_6)^{(k-1)/k} = (1199\text{ K})(50/100)^{(1.35-1)/1.35} = \underline{1002\text{ K} = 729°\text{C}}$$

Equation (3-50) is now used to find the temperature at the start of compression (point 1):

$$T_1 = x_r T_{6a} + (1 - x_r)T_a$$

$$T_1 = (0.033)(1002\text{ K}) + (1 - 0.033)(303\text{ K}) = \underline{326\text{ K} = 53°\text{C}}$$

This temperature and pressure of 50 kPa should now be used as a starting point and a complete thermodynamic analysis should be done on the cycle with iterations until consistent results are obtained. This is left as an exercise for the student.

3.6 DIESEL CYCLE

Early CI engines injected fuel into the combustion chamber very late in the compression stroke, resulting in the indicator diagram shown in Fig. 3-7. Due to ignition delay and the finite time required to inject the fuel, combustion lasted into the expansion stroke. This kept the pressure at peak levels well past TDC. This combustion process is best approximated as a constant-pressure heat input in an air-standard cycle, resulting in the **Diesel cycle** shown in Fig. 3-8. The rest of the cycle is similar to the air-standard Otto cycle. The Diesel cycle is sometimes called a **Constant-Pressure cycle**.

Thermodynamic Analysis of Air-Standard Diesel Cycle

Process 6-1—constant-pressure intake of air at P_o.
Intake valve open and exhaust valve closed:

$$w_{6-1} = P_o(v_1 - v_6) \tag{3-51}$$

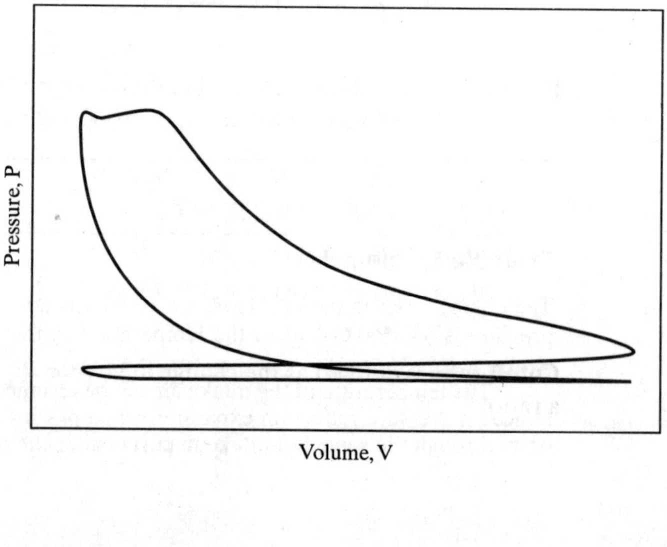

FIGURE 3-7
Indicator diagram of a historic CI engine operating on an early four-stroke cycle.

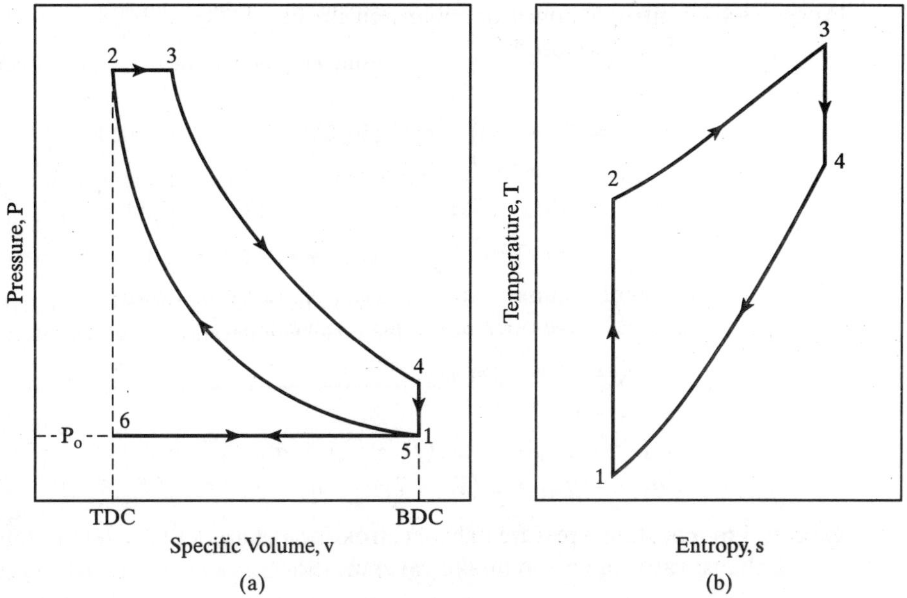

FIGURE 3-8

Air-standard Diesel cycle, 6-1-2-3-4-5-6, which approximates the four-stroke cycle of an early CI engine on **(a)** pressure-specific volume coordinates, and **(b)** temperature-entropy coordinates.

Process 1-2—isentropic compression stroke.
All valves closed:

$$T_2 = T_1(v_1/v_2)^{k-1} = T_1(V_1/V_2)^{k-1} = T_1(r_c)^{k-1} \quad (3\text{-}52)$$
$$P_2 = P_1(v_1/v_2)^k = P_1(V_1/V_2)^k = P_1(r_c)^k \quad (3\text{-}53)$$
$$V_2 = V_{TDC} \quad (3\text{-}54)$$
$$q_{1-2} = 0 \quad (3\text{-}55)$$
$$w_{1-2} = (P_2v_2 - P_1v_1)/(1 - k) = R(T_2 - T_1)/(1 - k)$$
$$= (u_1 - u_2) = c_v(T_1 - T_2) \quad (3\text{-}56)$$

Process 2-3—constant-pressure heat input (combustion).
All valves closed:

$$Q_{2-3} = Q_{in} = m_f Q_{HV}\eta_c = m_m c_p(T_3 - T_2) = (m_a + m_f)c_p(T_3 - T_2) \quad (3\text{-}57)$$
$$Q_{HV}\eta_c = (AF + 1)c_p(T_3 - T_2) \quad (3\text{-}58)$$
$$q_{2-3} = q_{in} = c_p(T_3 - T_2) = (h_3 - h_2) \quad (3\text{-}59)$$
$$w_{2-3} = q_{2-3} - (u_3 - u_2) = P_2(v_3 - v_2) \quad (3\text{-}60)$$
$$T_3 = T_{max} \quad (3\text{-}61)$$

Cutoff ratio is defined as the change in volume that occurs during combustion, given as a ratio:

$$\beta = V_3/V_2 = v_3/v_2 = T_3/T_2 \quad (3\text{-}62)$$

Process 3-4—isentropic power or expansion stroke.
All valves closed:

$$q_{3-4} = 0 \tag{3-63}$$
$$T_4 = T_3(v_3/v_4)^{k-1} = T_3(V_3/V_4)^{k-1} \tag{3-64}$$
$$P_4 = P_3(v_3/v_4)^k = P_3(V_3/V_4)^k \tag{3-65}$$
$$\begin{aligned} w_{3-4} &= (P_4v_4 - P_3v_3)/(1 - k) = R(T_4 - T_3)/(1 - k) \\ &= (u_3 - u_4) = c_v(T_3 - T_4) \end{aligned} \tag{3-66}$$

Process 4-5—constant-volume heat rejection (exhaust blowdown).
Exhaust valve open and intake valve closed:

$$v_5 = v_4 = v_1 = v_{BDC} \tag{3-67}$$
$$w_{4-5} = 0 \tag{3-68}$$
$$Q_{4-5} = Q_{out} = m_m c_v(T_5 - T_4) = m_m c_v(T_1 - T_4) \tag{3-69}$$
$$q_{4-5} = q_{out} = c_v(T_5 - T_4) = (u_5 - u_4) = c_v(T_1 - T_4) \tag{3-70}$$

Process 5-6—constant-pressure exhaust stroke at P_o.
Exhaust valve open and intake valve closed:

$$w_{5-6} = P_o(v_6 - v_5) = P_o(v_6 - v_1) \tag{3-71}$$

Thermal efficiency of Diesel cycle

$$\begin{aligned} (\eta_t)_{DIESEL} &= |w_{net}|/|q_{in}| = 1 - (|q_{out}|/|q_{in}|) \\ &= 1 - [c_v(T_4 - T_1)/c_p(T_3 - T_2)] \\ &= 1 - (T_4 - T_1)/[k(T_3 - T_2)] \end{aligned} \tag{3-72}$$

With rearrangement, this can be shown to equal

$$(\eta_t)_{DIESEL} = 1 - (1/r_c)^{k-1}[(\beta^k - 1)/\{k(\beta - 1)\}] \tag{3-73}$$

where

$$r_c = \text{compression ratio}$$
$$k = c_p/c_v$$
$$\beta = \text{cutoff ratio}$$

If representative numbers are introduced into Eq. (3-73), it is found that the value of the term in brackets is greater than one. When this equation is compared with Eq. (3-31), it can be seen that for a given compression ratio the thermal efficiency of the Otto cycle would be greater than the thermal efficiency of the Diesel cycle. Constant-volume combustion at TDC is more efficient than constant-pressure combustion. However, it must be remembered that CI engines operate with much higher compression ratios than SI engines (12 to 24 versus 8 to 11) and thus have higher thermal efficiencies.

Example Problem 3-4

A large vintage straight six CI truck engine operates on an air-standard Diesel cycle (Fig. 3-8) using heavy diesel fuel with a combustion efficiency of 98%. The engine has a compression ratio

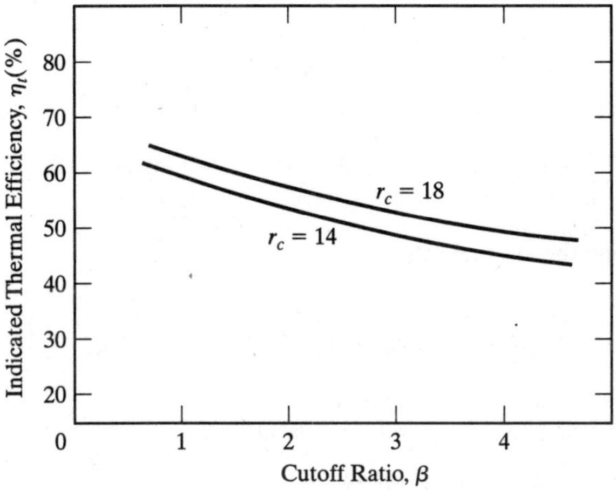

FIGURE 3-9

Indicated thermal efficiency as a function of cutoff ratio for air-standard Diesel cycle (k = 1.35).

of 16.5:1. Temperature and pressure in the cylinders at the start of the compression stroke are 55°C and 102 kPa, and maximum cycle temperature is 2410°C. Calculate:

1. temperature, pressure, and specific volume at each state of the cycle
2. air–fuel ratio of the cylinder gas mixture
3. cylinder temperature when the exhaust valve opens
4. indicated thermal efficiency of the engine

(1) State 1:

$$\underline{T_1 = 55°\text{C} = 328 \text{ K}} \quad \text{given in problem statement}$$
$$\underline{P_1 = 102 \text{ kPa}} \quad \text{given}$$
$$v_1 = RT_1/P_1 = (0.287 \text{ kJ/kg-K})(328 \text{ K})/(102 \text{ kPa}) = \underline{0.9229 \text{ m}^3/\text{kg}}$$

State 2:

Equations (3-52) and (3-53) give temperature and pressure after isentropic compression:

$$T_2 = T_1(r_c)^{k-1} = (328 \text{ K})(16.5)^{1.35-1} \underline{= 875 \text{ K} = 602°\text{C}}$$
$$P_2 = P_1(r_c)^k = (102 \text{ kPa})(16.5)^{1.35} \underline{= 4490 \text{ kPa}}$$
$$v_2 = RT_2/P_2 = (0.287 \text{ kJ/kg-K})(875 \text{ K})/(4490 \text{ kPa}) = \underline{0.0559 \text{ m}^3/\text{kg}}$$

Or, using Eq. (2-12),

$$v_2 = v_1/r_c = (0.9229 \text{ m}^3/\text{kg})/(16.5) = \underline{0.0559 \text{ m}^3/\text{kg}}$$

State 3:

$$\underline{T_3 = T_{max} = 2410°\text{C} = 2683 \text{ K}} \quad \text{given in problem statement}$$
$$\underline{P_3 = P_2 = 4490 \text{ kPa}}$$
$$v_3 = RT_3/P_3 = (0.287 \text{ kJ/kg-K})(2683 \text{ K})/(4490 \text{ kPa}) \underline{= 1715 \text{ m}^3/\text{kg}}$$

Using Eq. (3-62) for cutoff ratio,

$$\beta = v_3/v_2 = (0.1715)/(0.0559) = 3.08$$

State 4:

$$\underline{v_4 = v_1 = 0.9229\ \text{m}^3/\text{kg}}$$

Equations (3-64) and (3-65) give temperature and pressure after isentropic expansion:

$$T_4 = T_3(v_3/v_4)^{k-1} = (2683\ \text{k})(0.1715/0.9229)^{1.35-1} = \underline{1489\ \text{K} = 1216°\text{C}}$$
$$P_4 = P_3(v_3/v_4)^{k} = (4490\ \text{kPa})(0.1715/0.9229)^{1.35} = \underline{463\ \text{kPa}}$$

(2) Eq. (3-58) is used to find the air–fuel ratio:

$$Q_{\text{LHV}}\eta_c = (\text{AF} + 1)c_p(T_3 - T_2)$$
$$= (41{,}400\ \text{kJ/kg})(0.98) = (\text{AF} + 1)(1.108\ \text{kJ/kg-K})(2683 - 875)\text{K}$$
$$\underline{\text{AF} = 19.25}$$

(3) In air-standard Diesel cycle, the exhaust valve opens at state 4:

$$\underline{T_{\text{EVO}} = T_4 = 1489\ \text{K} = 1216°\text{C}}$$

(4) The work produced during the compression stroke is found using Eq. (3-56):

$$w_{1-2} = R(T_2 - T_1)/(1 - k) = (0.287\ \text{kJ/kg-K})(875 - 328)\text{K}/(1 - 1.35)$$
$$= -448.5\ \text{kJ/kg}$$

The work produced during combustion is found using Eq. (3-60):

$$w_{2-3} = P_2(v_3 - v_2) = (4490\ \text{kPa})(0.1715 - 0.0559)\text{m}^3/\text{kg} = +519.0\ \text{kJ/kg}$$

Work during power stroke is calculated using Eq. (3-66):

$$w_{3-4} = R(T_4 - T_3)/(1 - k)$$
$$= (0.287\ \text{kJ/kg-K})(1489 - 2683)\text{K}/(1 - 1.35)$$
$$= +979.1\ \text{kJ/kg}$$

Work during constant-volume blowdown is $w_{4-1} = 0$
Net work per unit mass of gas in cylinder for one cycle

$$w_{\text{net}} = w_{1-2} + w_{2-3} + w_{3-4} = (-448.5) + (+519.0) + (+979.1) = +1049.6\ \text{kJ/kg}$$

Equation (3-59) gives the heat added per unit mass for one cycle:

$$q_{\text{in}} = q_{2-3} = c_p(T_3 - T_2) = (1.108\ \text{kJ/kg-K})(2683 - 875)\text{K} = 2003.3\ \text{kJ/kg}$$

Use Eq. (2-65) with specific quantities to find the indicated thermal efficiency:

$$\eta_t = w/q_{\text{in}} = (1049.6)/(2003.3) = \underline{0.524 = 52.4\%}$$

Thermal efficiency can also be found using either Eq. (3-72) or Eq. (3-73):

$$\eta_t = 1 - (T_4 - T_1)/[k(T_3 - T_2)] = 1 - (1489 - 328)/[1.35(2683 - 875)] = \underline{0.524}$$
$$\eta_t = 1 - (1/r_c)^{1-k}\{(\beta^k - 1)/[k(\beta - 1)]\}$$
$$= 1 - (1/16.5)^{1.35-1}\{[(3.08)^{1.35} - 1]/[1.35(3.08 - 1)]\} = \underline{0.524}$$

3.7 DUAL CYCLE

If Eqs. (3-31) and (3-73) are compared, it can be seen that to have the best of both worlds, an engine ideally would be compression ignition but would operate on the Otto cycle. Compression ignition would operate on the more efficient higher compression ratios, while constant-volume combustion of the Otto cycle would give higher efficiency for a given compression ratio.

The modern high-speed CI engine accomplishes this in part by a simple operating change from early Diesel engines. Instead of injecting the fuel late in the compression stroke near TDC, as was done in early engines, modern CI engines start to inject the fuel much earlier in the cycle, somewhere around 20° bTDC. The first fuel then ignites late in the compression stroke, and some of the combustion occurs almost at constant volume at TDC, much like the Otto cycle. A typical indicator diagram for a modern CI engine is shown in Fig. 3-10. Peak pressure still remains high into the expansion stroke due to the finite time required to inject the fuel. The last of the fuel is still being injected at TDC, and combustion of this fuel keeps the pressure high into the expansion stroke. The resulting cycle shown in Fig. 3-10 is a cross between an SI engine cycle and the early CI cycles. The air-standard cycle used to analyze this modern CI engine cycle is called a **Dual cycle** or sometimes a **Limited Pressure cycle** (Fig. 3-11). It is a Dual cycle because the heat input process of combustion can best be approximated by a dual process of constant volume followed by constant pressure. It can also be considered a modified Otto cycle with a limited upper pressure.

Thermodynamic Analysis of Air-Standard Dual Cycle

The analysis of an air-standard Dual cycle is the same as that of the Diesel cycle, except for the heat input process (combustion) 2-x-3.

Process 2-x—constant-volume heat input (first part of combustion).

All valves closed:

$$V_x = V_2 = V_{TDC} \tag{3-74}$$

$$w_{2-x} = 0 \tag{3-75}$$

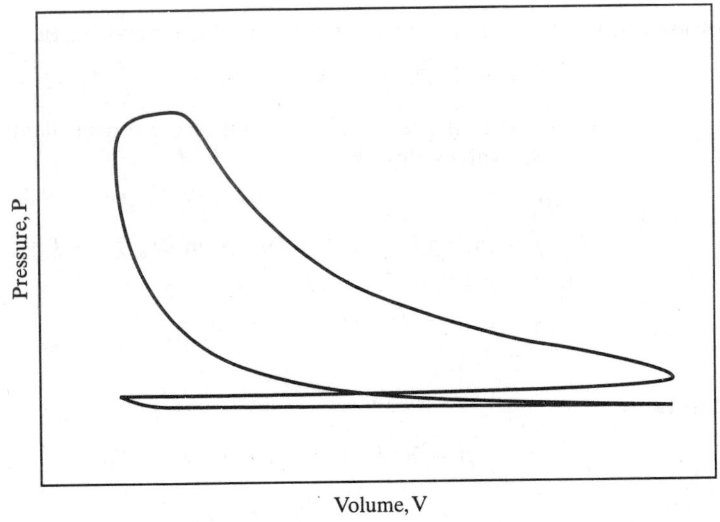

FIGURE 3-10

Indicator diagram of a modern CI engine operating on a four-stroke cycle.

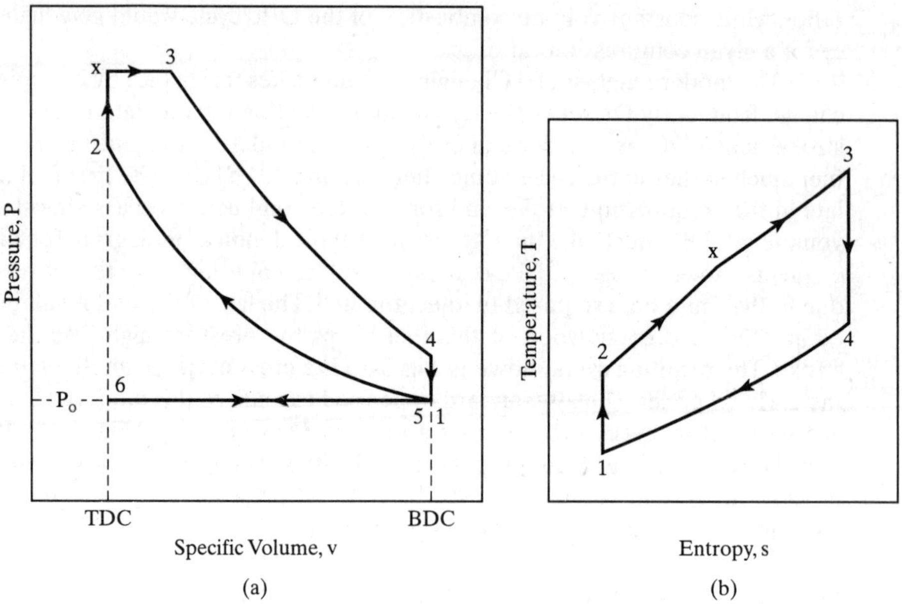

FIGURE 3-11

Air-standard Dual cycle, 6-1-2-x-3-4-5-6, which approximates the four-stroke cycle of a modern CI engine on **(a)** pressure-specific volume coordinates, and **(b)** temperature-entropy coordinates.

$$Q_{2-x} = m_m c_v(T_x - T_2) = (m_a + m_f)c_v(T_x - T_2) \quad (3\text{-}76)$$

$$q_{2-x} = c_v(T_x - T_2) = (u_x - u_2) \quad (3\text{-}77)$$

$$P_x = P_{max} = P_2(T_x/T_2) \quad (3\text{-}78)$$

Pressure ratio is defined as the rise in pressure during combustion, given as a ratio:

$$\alpha = P_x/P_2 = P_3/P_2 = T_x/T_2 = (1/r_c)^k(P_3/P_1) \quad (3\text{-}79)$$

Process x-3—constant-pressure heat input (second part of combustion). All valves closed:

$$P_3 = P_x = P_{max} \quad (3\text{-}80)$$

$$Q_{x-3} = m_m c_p(T_3 - T_x) = (m_a + m_f)c_p(T_3 - T_x) \quad (3\text{-}81)$$

$$q_{x-3} = c_p(T_3 - T_x) = (h_3 - h_x) \quad (3\text{-}82)$$

$$w_{x-3} = q_{x-3} - (u_3 - u_x) = P_x(v_3 - v_x) = P_3(v_3 - v_x) \quad (3\text{-}83)$$

$$T_3 = T_{max} \quad (3\text{-}84)$$

Cutoff ratio:

$$\beta = v_3/v_x = v_3/v_2 = V_3/V_2 = T_3/T_x \quad (3\text{-}85)$$

Heat in:

$$Q_{in} = Q_{2-x} + Q_{x-3} = m_f Q_{HV}\eta_c \quad (3\text{-}86)$$

$$q_{in} = q_{2-x} + q_{x-3} = (u_x - u_2) + (h_3 - h_x) \tag{3-87}$$

Thermal efficiency of Dual cycle:

$$\begin{aligned}(\eta_t)_{DUAL} &= |w_{net}|/|q_{in}| = 1 - (|q_{out}|/|q_{in}|) \\ &= 1 - c_v(T_4 - T_1)/[c_v(T_x - T_2) + c_p(T_3 - T_x)] \\ &= 1 - (T_4 - T_1)/[(T_x - T_2) + k(T_3 - T_x)] \end{aligned} \tag{3-88}$$

This can be rearranged to give

$$(\eta_t)_{DUAL} = 1 - (1/r_c)^{k-1}[\{\alpha\beta^k - 1\}/\{k\alpha(\beta - 1) + \alpha - 1\}] \tag{3-89}$$

where

r_c = compression ratio
$k = c_p/c_v$
α = pressure ratio
β = cutoff ratio

As with the Otto cycle, the air-standard thermal efficiency obtained for CI engines using Eqs. (3-73) or (3-89) is slightly higher than that of real air–fuel cycles [120]. This is because of the same reasons of changing composition, heat losses, valve overlap, and finite time required for cycle processes.

$$(\eta_t)_{actual} \approx 0.85(\eta_t)_{DIESEL} \tag{3-90}$$

$$(\eta_t)_{actual} \approx 0.85(\eta_t)_{DUAL} \tag{3-91}$$

3.8 COMPARISON OF OTTO, DIESEL, AND DUAL CYCLES

Figure 3-12 compares Otto, Diesel, and Dual cycles with the same inlet conditions and the same compression ratios. The thermal efficiency of each cycle can be written as

$$\eta_t = 1 - |q_{out}|/|q_{in}| \tag{3-92}$$

The area under the process lines on T–s coordinates is equal to the heat transfer, so in Fig. 3-12(b) the thermal efficiencies can be compared. For each cycle, q_{out} is the same (process 4-1). q_{in} of each cycle is different, and using Fig. 3-12(b) and Eq. (3-92), it is found that, for these conditions,

$$(\eta_t)_{OTTO} > (\eta_t)_{DUAL} > (\eta_t)_{DIESEL} \tag{3-93}$$

However, this is not the best way to compare these three cycles, because they do not operate on the same compression ratio. Compression ignition engines that operate on the Dual cycle or Diesel cycle have much higher compression ratios than do spark ignition engines operating on the Otto cycle. A more realistic way to compare these three cycles would be to have the same peak pressure—an actual design limitation in engines. This is done in Fig. 3-13. When this figure is compared with Eq. (3-92), it is found that

$$(\eta_t)_{DIESEL} > (\eta_t)_{DUAL} > (\eta_t)_{OTTO} \tag{3-94}$$

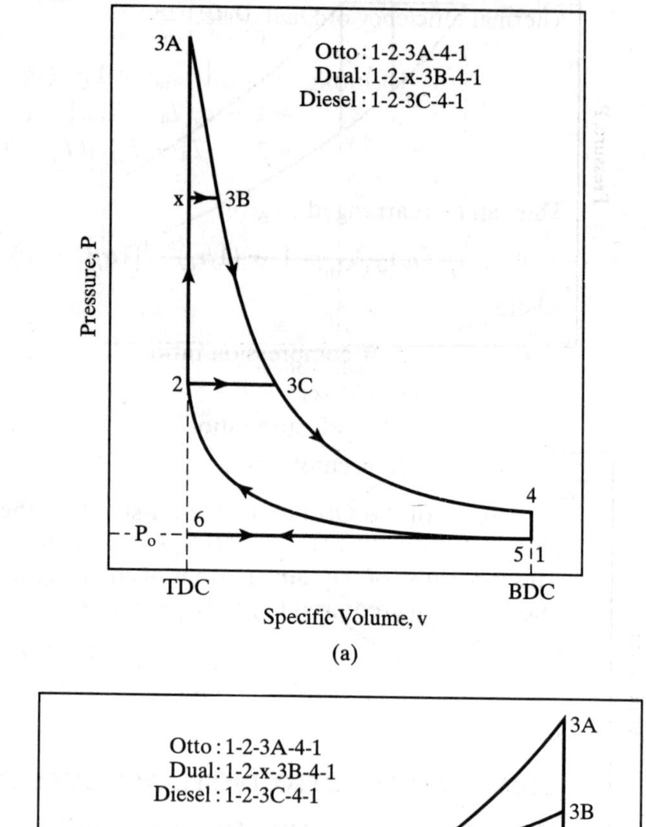

FIGURE 3-12

Comparison of air-standard Otto cycle, Dual cycle, and Diesel cycle. All engines have the same cylinder input conditions and the same compression ratio.

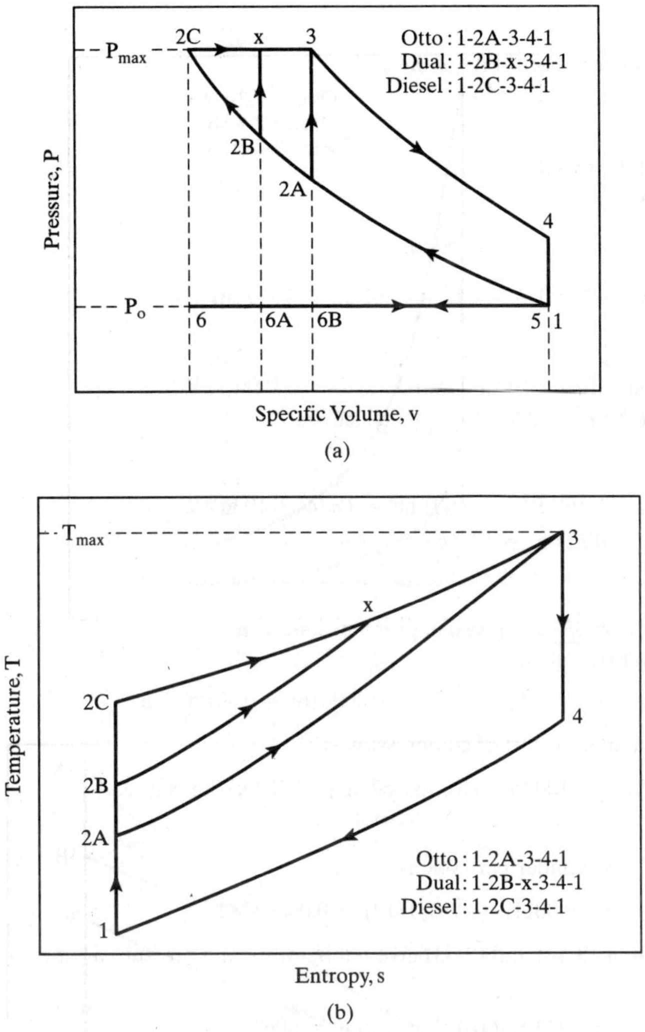

FIGURE 3-13

Comparison of air-standard Otto cycle, Dual cycle, and Diesel cycle. All engines have the same cylinder input conditions and the same maximum temperature and pressure.

Comparing the ideas of Eqs. (3-93) and (3-94) would suggest that the most efficient engine would have combustion as close as possible to constant volume but would be compression ignition and operate at the higher compression ratios which that requires. This is an area where more research and development is needed.

Example Problem 3-5

A small truck has a four-cylinder, four-liter CI engine that operates on the air-standard Dual cycle (Fig. 3-11) using light diesel fuel at an air–fuel ratio of 18. The compression ratio of the engine is 16:1 and the cylinder bore diameter is 10.0 cm. At the start of the compression stroke, conditions in the cylinders are 60°C and 100 kPa with a 2% exhaust residual. It can be assumed that half of the heat input from combustion is added at constant volume and half at constant pressure.

Calculate:

1. temperature and pressure at each state of the cycle
2. indicated thermal efficiency
3. exhaust temperature
4. air temperature in intake manifold
5. engine volumetric efficiency

(1) For one cylinder,

$$V_d = (4\ \text{L})/4 = 1\ \text{L} = 0.001\ \text{m}^3 = 1000\ \text{cm}^3$$

Using Eq. (2-12),

$$r_c = V_{\text{BDC}}/V_{\text{TDC}} = (V_d + V_c)/V_c = 16 = (1000 + V_c)/V_c$$
$$V_c = 66.7\ \text{cm}^3 = 0.0667\ \text{L} = 0.0000667\ \text{m}^3$$

Using Eq. (2-8),

$$V_d = (\pi/4)B^2S = 0.001\ \text{m}^3 = (\pi/4)(0.10\ \text{m})^2S$$
$$S = 0.127\ \text{m} = 12.7\ \text{cm}$$

State 1:

$\underline{T_1 = 60°\text{C} = 333\ \text{K}}$ given in problem statement

$\underline{P_1 = 100\ \text{kPa}}$ given

$V_1 = V_{\text{BDC}} = V_d + V_c = 0.001 + 0.0000667 = \underline{0.0010667\ \text{m}^3}$

Mass of gas in one cylinder at start of compression is

$$m_m = P_1V_1/RT_1 = (100\ \text{kPa})(0.0010667\ \text{m}^3)/(0.287\ \text{kJ/kg-K})(333\ \text{K})$$
$$= 0.00112\ \text{kg}$$

Mass of fuel injected per cylinder per cycle is

$$m_f = (0.00112)(0.98)(1/19) = 0.0000578\ \text{kg}$$

State 2: Equations (3-52) and (3-53) give temperature and pressure after compression:

$$T_2 = T_1(r_c)^{k-1} = (333\ \text{K})(16)^{0.35} = \underline{879\ \text{K} = 606°\text{C}}$$
$$P_2 = P_1(r_c)^k = (100\ \text{kPa})(16)^{1.35} = \underline{4222\ \text{kPa}}$$
$$V_2 = mRT_2/P_2 = (0.00112\ \text{kg})(0.287\ \text{kJ/kg-K})(879\ \text{K})/(4222\ \text{kPa})$$
$$= \underline{0.000067\ \text{m}^3 = V_c}$$

or, using Eq. (2-12),

$$V_2 = V_1/r_c = (0.0010667)/(16) = 0.0000667\ \text{m}^3$$

State x: Heating value of light diesel fuel is obtained from Table A-2 in the Appendix:

$$Q_{\text{in}} = m_fQ_{\text{HV}} = (0.0000578\ \text{kg})(42{,}500\ \text{kJ/kg}) = 2.46\ \text{kJ}$$

If half of Q_{in} occurs at constant volume, then Eq. (3-76) yields:

$$Q_{2-x} = 1.23\ \text{kJ} = m_mc_v(T_x - T_2)$$
$$= (0.00112\ \text{kg})(0.821\ \text{kJ/kg-K})(T_x - 879\ \text{K})$$

$$\underline{T_x = 2217 \text{ K} = 1944°\text{C}}$$
$$\underline{V_x = V_2 = 0.0000667 \text{ m}^3}$$
$$P_x = mRT_x/V_x$$
$$= (0.00112 \text{ kg})(0.287 \text{ kJ/kg-K})(2217 \text{ K})/(0.0000667 \text{ m}^3)$$
$$\underline{= 10{,}650 \text{ kPa} = P_{max}}$$

or

$$P_x = P_2(T_x/T_2) = (4222 \text{ kPa})(2217/879) = 10{,}650 \text{ kPa}$$

State 3:

$$\underline{P_3 = P_x = 10{,}650 \text{ kPa} = P_{max}}$$

Equation (3-81) gives

$$Q_{x-3} = 1.23 \text{ kJ} = m_m c_p(T_3 - T_x)$$
$$= (0.00112 \text{ kg})(1.108 \text{ kJ/kg-K})(T_3 - 2217 \text{ K})$$
$$\underline{T_3 = 3208 \text{ K} = 2935°\text{C} = T_{max}}$$
$$V_3 = mRT_3/P_3 = (0.00112 \text{ kg})(0.287 \text{ kJ/kg-K})(3208 \text{ K})/(10{,}650 \text{ kPa})$$
$$\underline{= 0.000097 \text{ m}^3}$$

State 4:

$$V_4 = V_1 = 0.0010667 \text{ m}^3$$

Equations (3-64) and (3-65) give temperature and pressure after expansion:

$$T_4 = T_3(V_3/V_4)^{k-1} = (3208 \text{ K})(0.000097/0.0010667)^{0.35}$$
$$\underline{= 1386 \text{ K} = 1113°\text{C}}$$
$$P_4 = P_3(V_3/V_4)^k = (10{,}650 \text{ kPa})(0.000097/0.0010667)^{1.35} \underline{= 418 \text{ kPa}}$$

Work out for process x-3 for one cylinder for one cycle, using Eq. (3-83), is

$$W_{x-3} = P(V_3 - V_x) = (10{,}650 \text{ kPa})(0.000097 - 0.0000667)\text{m}^3 \underline{= 0.323 \text{ kJ}}$$

Work out for process 3-4, using Eq. (3-66), is

$$W_{3-4} = mR(T_4 - T_3)/(1 - k)$$
$$= (0.00112 \text{ kg})(0.287 \text{ kJ/kg-K})(1386 - 3208)\text{K}/(1 - 1.35)$$
$$\underline{= 1.673 \text{ kJ}}$$

Work in for process 1-2, using Eq. (3-56), is

$$W_{1-2} = mR(T_2 - T_1)/(1 - k)$$
$$= (0.00112 \text{ kg})(0.287 \text{ kJ/kg-K})(879 - 333)\text{K}/(1 - 1.35)$$
$$= -0.501 \text{ kJ}$$
$$W_{net} = (+0.323) + (+1.673) + (-0.501) = +1.495 \text{ kJ}$$

(2) Equation (3-88) gives indicated thermal efficiency:

$$(\eta_t)_{DUAL} = |W_{net}|/|Q_{in}| = (1.495 \text{ kJ})/(2.46 \text{ kJ}) \underline{= 0.607 = 60.7\%}$$

Pressure ratio is

$$\alpha = P_x/P_2 = 10{,}650/4222 = 2.52$$

Cutoff ratio is

$$\beta = V_3/V_x = 0.000097/0.0000667 = 1.45$$

Use Eq. (3-89) to find thermal efficiency:

$$(\eta_t)_{\text{DUAL}} = 1 - (1/r_c)^{k-1}[\{\alpha\beta^k - 1\}/\{k\alpha(\beta - 1) + \alpha - 1\}]$$
$$= 1 - (1/16)^{0.35}[\{(2.52)(1.45)^{1.35} - 1\}/\{(1.35)(2.52)(1.45 - 1) + 2.52 - 1\}]$$
$$= 0.607$$

(3) Assuming exhaust pressure is the same as intake pressure, and using Eq. (3-37) for exhaust temperature, we get

$$T_{ex} = T_4(P_{ex}/P_4)^{(k-1)/k} = (1386\text{ K})(100/418)^{(1.35-1)/1.35} = \underline{957\text{ K} = 684°\text{C}}$$

Equation (3-46) gives exhaust residual:

$$x_r = (1/r_c)(T_4/T_{ex})(P_{ex}/P_4)$$
$$= (1/16)(1386/957)(100/418) = \underline{0.022 = 2.2\%}$$

(4) Use Eq. (3-50) to find air temperature entering the cylinder:

$$(T_m)_1 = x_r T_{ex} + (1 - x_r)T_a$$
$$(333\text{ K}) = (0.022)(957\text{ K}) + (1 - 0.022)T_a$$
$$\underline{T_a = 319\text{ K} = 46°\text{C}}$$

(5) Mass of air entering one cylinder during intake is

$$m_a = (0.00112\text{ kg})(0.98) = 0.00110\text{ kg}$$

Volumetric efficiency is found using Eq. (2-70):

$$\eta_v = m_a/\rho_a V_d = (0.00110\text{ kg})/(1.181\text{ kg/m}^3)(0.001\text{ m}^3)$$
$$= \underline{0.931 = 93.1\%}$$

HISTORIC—ATKINSON CYCLE

In Otto and Diesel cycles, when the exhaust valve is opened near the end of the expansion stroke, pressure in the cylinder is still on the order of three to five atmospheres. A potential for doing additional work during the power stroke is therefore lost when the exhaust valve is opened and pressure is reduced to atmospheric. If the exhaust valve is not opened until the gas in the cylinder is allowed to expand down to atmospheric pressure, a greater amount of work would be obtained in the expansion stroke, with an increase in engine thermal efficiency. Such an air-standard cycle is called an **Atkinson cycle** or **Overexpanded cycle** (or Complete Expansion cycle) and is shown in Fig. 3-14.

Starting in 1885, a number of crank and valve mechanisms were tried to achieve this cycle, which has a longer expansion stroke than compression stroke. No large number of these engines has ever been marketed, indicating the failure of this development [58].

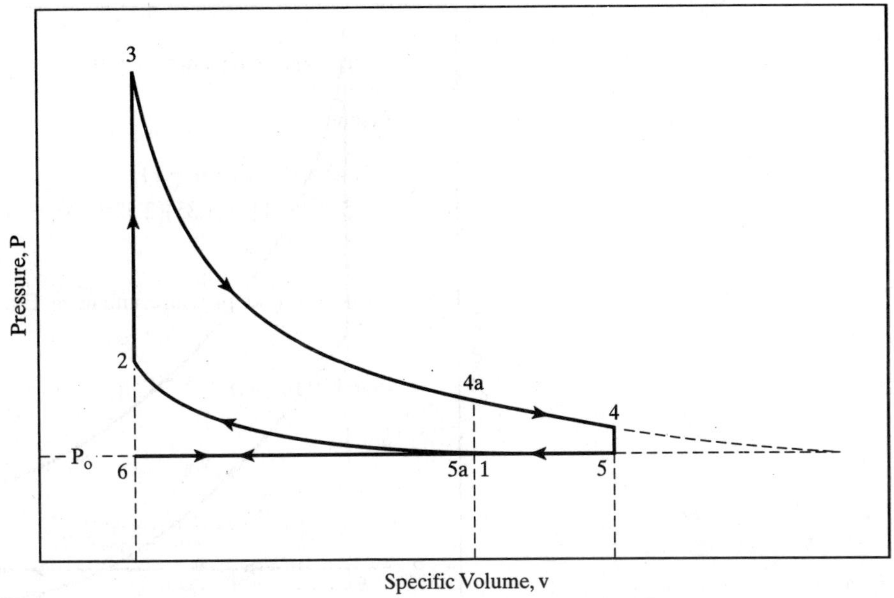

FIGURE 3.14

Air-standard Atkinson cycle, 6-1-2-3-4-5-6, with larger expansion ratio v_4/v_3 than compression ratio v_1/v_3. The same engine operating on an Otto cycle would follow cycle 6-1-2-3-4a-5a-6.

3.9 MILLER CYCLE

The **Miller cycle**, named after R. H. Miller (1890–1967), is a modern modification of the Atkinson cycle and has an expansion ratio greater than the compression ratio. This is accomplished, however, in a much different way. Whereas an engine designed to operate on the Atkinson cycle needed a complicated mechanical linkage system of some kind, a Miller cycle engine uses unique valve timing to obtain the same desired results.

Air intake in a Miller cycle is unthrottled. The amount of air ingested into each cylinder is then controlled by closing the intake valve at the proper time, long before BDC (point 7 in Fig. 3-15). As the piston then continues towards BDC during the latter part of the intake stroke, cylinder pressure is reduced along process 7-1. When the piston reaches BDC and starts back towards TDC, cylinder pressure is again increased during process 1-7. The resulting cycle is 6-7-1-7-2-3-4-5-6. The work produced in the first part of the intake process 6-7 is canceled by part of the exhaust stroke 7-6, process 7-1 is canceled by process 1-7, and the net indicated work is the area within loop 7-2-3-4-5-7. There is essentially no pump work. The compression ratio is

$$r_c = V_7/V_2 \tag{3-95}$$

and the larger expansion ratio is

$$r_e = V_4/V_2 = V_4/V_3 \tag{3-96}$$

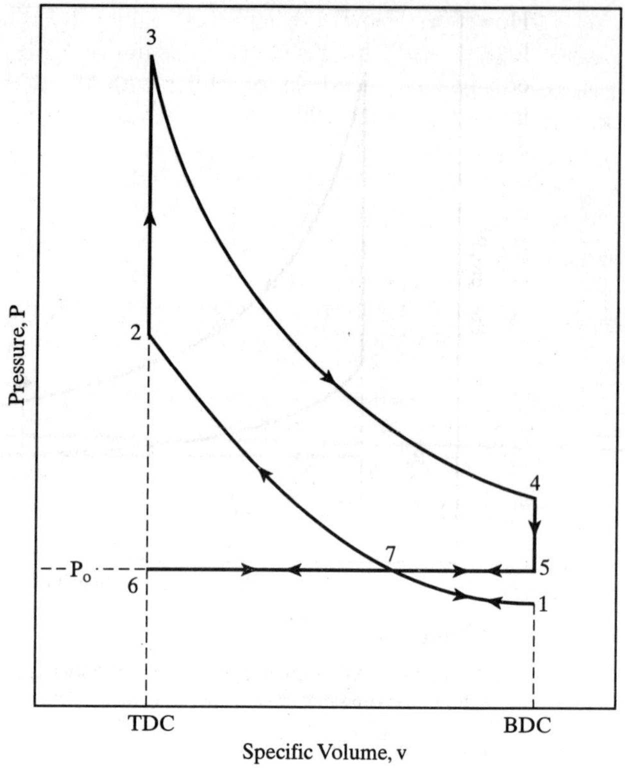

FIGURE 3-15

Air-standard Miller cycle for unthrottled naturally aspirated four-stroke cycle SI engine. If the engine has early intake valve closing, the cycle will be 6-7-1-7-2-3-4-5-7-6. If the engine has late intake valve closing, the cycle will be 6-7-5-7-2-3-4-5-7-6.

The shorter compression stroke, which absorbs work, combined with the longer expansion stroke, which produces work, results in a greater net indicated work per cycle. In addition, by allowing air to flow through the intake system unthrottled, a major loss experienced by most SI engines is eliminated. This is especially true at part throttle, when an Otto cycle engine would experience low pressure in the intake manifold and a corresponding high negative pump work. The Miller cycle engine has essentially no pump work (ideally none), much like a CI engine. This results in higher thermal efficiency.

The mechanical efficiency of a Miller cycle engine would be about the same as that of an Otto cycle engine, which has a similar mechanical linkage system. An Atkinson cycle engine, on the other hand, requires a much more complicated mechanical linkage system, resulting in lower mechanical efficiency.

Another variation of this cycle can be obtained if the intake air is unthrottled and the intake valve is closed after BDC. When this is done, air is ingested during the entire intake stroke, but some of it is then forced back into the intake manifold before the intake valve closes. This results in cycle 6-7-5-7-2-3-4-5-6 in Fig. 3-15. The net indicated work is again the area within loop 7-2-3-4-5-7, with the compression and expansion ratios given by Eqs. (3-95) and (3-96).

For either variation of the cycle to work efficiently, it is extremely important to be able to close the intake valve at the precise correct moment in the cycle (point 7).

However, this point where the intake valve must close changes as the engine speed or load is changed. This control was not possible until variable valve timing was perfected and introduced. Automobiles with Miller cycle engines were first marketed in the latter half of the 1990s. A typical value of the compression ratio is about 8:1, with an expansion ratio of about 10:1.

The first production automobile engines operating on Miller cycles used both early intake valve closing methods and late intake valve closing methods. Several types of variable valve timing systems have been developed and used. This subject is covered in Chapter 5. Opening and closing valves with electronic actuators, without the use of a camshaft, offers the greatest flexibility, both for variable timing and variable lift. This method will become common with the transition to 42-volt electrical systems.

If the intake valve is closed bBDC, less than full displacement volume of the cylinder is available for air ingestion. If the intake valve is closed aBDC, the full displacement volume is filled with air, but some of it is expelled out again before the valve is closed (process 5-7 in Fig. 3-15). In either case, less air and fuel end up in the cylinder at the start of compression, resulting in low output per displacement and low indicated mean effective pressure. To counteract this, Miller cycle engines are usually supercharged or turbocharged with peak intake manifold pressures of 150–200 kPa. Fig. 3-16 shows a supercharged Miller engine cycle.

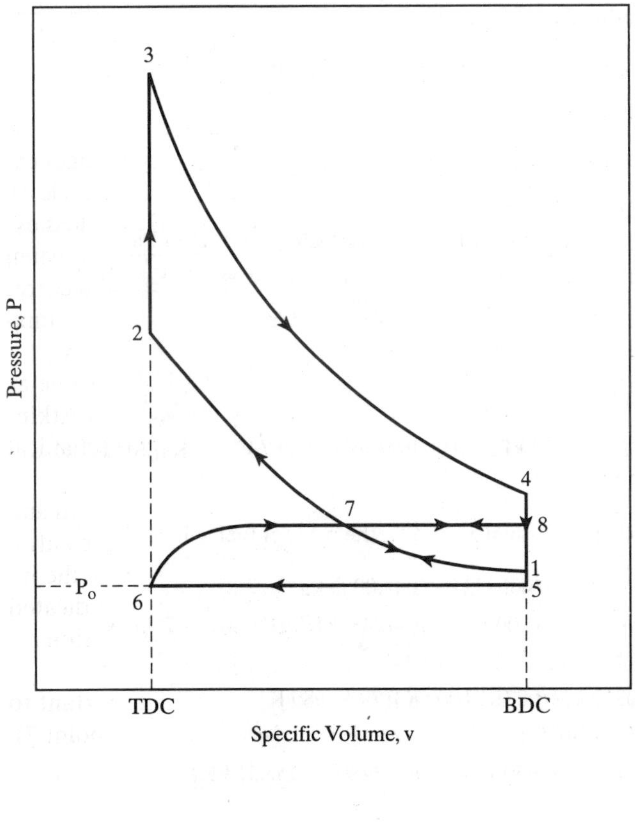

FIGURE 3-16

Air-standard Miller cycle for a four-stroke cycle SI engine equipped with a turbocharger or supercharger. If the engine has early intake valve closing, the cycle will be 6-7-1-7-2-3-4-5-6. If the engine has late intake valve closing, the cycle will be 6-7-8-7-2-3-4-5-6.

Example Problem 3-6

The four-cylinder, 2.5-liter SI automobile engine of Example Problem 3-1 is converted to operate on an air-standard Miller cycle with early valve closing (cycle 6-7-1-7-2-3-4-5-6 in Fig. 3-16). It has a compression ratio of 8:1 and an expansion ratio of 10:1. A supercharger is added that gives a cylinder pressure of 160 kPa when the intake valve closes, as shown in Fig. 3-16. The temperature is again 60°C at this point. The same fuel and AF are used with combustion efficiency $\eta_c = 100\%$.

Calculate:

1. temperature and pressure at all points in the cycle
2. indicated thermal efficiency
3. indicated mean effective pressure
4. exhaust temperature

From Example Problem 3-1, for one cylinder,

$$V_d = 0.000625 \text{ m}^3$$

Expansion ratio is calculated using Eq. (3-96):

$$r_e = V_4/V_3 = (V_d + V_c)/V_c = 10 = (0.000625 + V_c)/V_c$$
$$V_c = 0.000069 \text{ m}^3 = V_2 = V_3 = V_6$$
$$V_1 = V_4 = V_5 = V_d + V_c = 0.000625 + 0.000069 = 0.000694 \text{ m}^3$$

Compression ratio is calculated using Eq. (3-95):

$$V_7 = V_2 r_c = (0.000069)(8) = 0.000552 \text{ m}^3$$

(1) Temperatures and pressures:

$$\underline{T_7 = 60°\text{C} = 333 \text{ K}} \quad \text{given in problem statement}$$
$$\underline{P_7 = P_8 = 160 \text{ kPa}} \quad \text{given}$$
$$P_1 = P_7(V_7/V_1)^k = (160 \text{ kPa})(0.000552/0.000694)^{1.35} = \underline{117 \text{ kPa}}$$
$$T_1 = T_7(V_7/V_1)^{k-1} = (333 \text{ K})(0.000552/0.000694)^{0.35} = \underline{307 \text{ K} = 34°\text{C}}$$
$$T_2 = T_7(r_c)^{k-1} = (333 \text{ K})(8)^{0.35} = \underline{689 \text{ K} = 416°\text{C}}$$
$$P_2 = P_7(r_c)^k = (160 \text{ kPa})(8)^{1.35} = \underline{2650 \text{ kPa}}$$

Mass of gas in cylinder is

$$m_1 = P_1V_1/RT_1 = (117 \text{ kPa})(0.000694 \text{ m}^3)/(0.287 \text{ kJ/kg-K})(307 \text{ K})$$
$$= 0.000922 \text{ kg}$$

If AF = 15 and the exhaust residual $x_r = 4\%$, the mass of fuel will be

$$m_f = (1/16)(0.96)(0.000922) = 0.000055 \text{ kg}$$
$$Q_{in} = m_f Q_{HV}\eta_c = (0.000055 \text{ kg})(44{,}300 \text{ kJ/kg})(1.00) = 2.437 \text{ kJ}$$
$$Q_{in} = m_m c_v(T_3 - T_2) = 2.437 \text{ kJ}$$
$$= (0.000922 \text{ kg})(0821 \text{ kJ/kg-K})(T_3 - 689 \text{ K})$$
$$\underline{T_3 = 3908 \text{ K} = 3635°\text{C}}$$
$$P_3 = P_2(T_3/T_2) = (2650 \text{ kPa})(3908/689) = \underline{15{,}031 \text{ kPa}}$$

$$T_4 = T_3(V_3/V_4)^{k-1} = T_3(1/r_e)^{k-1} = (3908\ \text{K})(1/10)^{0.35}$$
$$= \underline{1746\ \text{K} = 1473°\text{C}}$$
$$P_4 = P_3(1/r_e)^k = (15{,}031\ \text{kPa})(1/10)^{1.35} = \underline{671\ \text{kPa}}$$
$$V_4 = mRT_4/P_4 = (0.000922\ \text{kg})(0.287\ \text{kJ/kg-K})(1746\ \text{K})/(671\ \text{kPa})$$
$$= 0.000694\ \text{m}^3 \quad \text{This agrees with the value for } V_4 \text{ obtained earlier.}$$
$$\underline{P_5 = P_{ex} = 100\ \text{kPa}}$$
$$T_5 = T_4(P_5/P_4) = (1746\ \text{K})(100/671) = \underline{260\ \text{K}}$$

(2) Indicated thermal efficiency is calculated as follows:

$$W_{3-4} = mR(T_4 - T_3)/(1 - k)$$
$$= (0.000922\ \text{kg})(0.287\ \text{kJ/kg-K})(1746 - 3908)\text{K}/(1 - 1.35)$$
$$= 1.635\ \text{kJ}$$
$$W_{7-2} = mR(T_2 - T_7)/(1 - k)$$
$$= (0.000922\ \text{kg})(0.287\ \text{kJ/kg-K})(689 - 333)\text{K}/(1 - 1.35)$$
$$= -0.269\ \text{kJ}$$
$$W_{6-7} = P_7(V_7 - V_6) = (160\ \text{kPa})(0.000552 - 0.000069)\text{m}^3 = 0.077\ \text{kJ}$$
$$W_{5-6} = P_5(V_6 - V_5) = (100\ \text{kPa})(0.000069 - 0.000694)\text{m}^3 = -0.063\ \text{kJ}$$
$$W_{net} = (+1.635) + (-0.269) + (+0.077) + (-0.063) = +1.380\ \text{kJ}$$
$$(\eta_t)_{\text{MILLER}} = |W_{net}|/|Q_{in}| = (1.380\ \text{kJ})/(2.437\ \text{kJ}) = \underline{0.566 = 56.6\%}$$

(3) Indicated mean effective pressure, using Eq. (2-29), is

$$\text{imep} = W_{net}/V_d = (1.380\ \text{kJ})/(0.000625\ \text{m}^3) = \underline{2208\ \text{kPa}}$$

(4) The exhaust temperature is

$$T_{ex} = T_4(P_{ex}/P_4)^{(k-1)/k} = (1746\ \text{K})(100/671)^{(1.35-1)/1.35} = \underline{1066\ \text{K} = 793°\text{C}}$$

3.10 COMPARISON OF MILLER CYCLE AND OTTO CYCLE

When the Otto cycle engine of Example Problems 3-1 and 3-2 is compared with a similar engine operating on a Miller cycle as in Example Problem 3-6, the superiority of the Miller cycle can be seen. Table 3-1 gives such a comparison.

Temperatures in the two cycles are about the same, except for the exhaust temperature. It is important that the temperature at the beginning of combustion for either cycle be low enough so that self-ignition and knock do not become problems. The lower exhaust temperature of the Miller cycle is the result of greater expansion cooling that occurs from the essentially identical maximum cycle temperature. Lower exhaust temperature means less energy is lost in the exhaust, with more of it used as work output in the longer expansion stroke. Pressures throughout the Miller cycle are higher than those of the Otto cycle, mainly because of the supercharged input. The output parameters of imep, thermal efficiency, and work are all higher for the Miller cycle, showing the technical superiority of this cycle. Some of the indicated work and indicated thermal efficiency of the Miller cycle will be lost due to the need to drive the supercharger. Even

TABLE 3-1 Comparison of Otto and Miller Cycles

	Miller Cycle	Otto Cycle
Temperature at start of combustion T_2:	689 K	707 K
Pressure at start of combustion P_2:	2650 kPa	1826 kPa
Maximum temperature T_3:	3908 K	3915 K
Maximum pressure P_3:	15,031 kPa	10,111 kPa
Exhaust temperature:	1066 K	1183 K
Indicated net work per cylinder per cycle for same Q_{in}:	1.380 kJ	1.030 kJ
Indicated thermal efficiency:	56.6 %	52.9 %
Indicated mean effective pressure:	2208 kPa	1649 kPa

with this considered, however, brake work and brake thermal efficiency will be substantially greater than in an Otto cycle engine. If a turbocharger were used instead of a supercharger, brake output parameter values would be even higher.

3.11 TWO-STROKE CYCLES

The first practical two-stroke cycle engines appeared about 1887, and since then many CI and SI engines have been manufactured. The very smallest engines and the largest engines almost always operate on a two-stroke cycle. It is desirable for most small engines (chain saws, leaf blowers, etc.) to be lightweight and inexpensive. Both of these requirements can be met by eliminating engine valves, possible with two-stroke cycles. Very large engines operate at very low speeds, and thus need the two-stroke cycle for smoothness of operation. At very low RPM, a power stroke in every cylinder on every cycle is needed for smoothness.

Two-stroke cycle engines have been used in vehicles off and on throughout the history of the automobile, the last two being made in East Germany until 1990. No modern automobile is now made in high volume with a two-stroke cycle engine because of emission laws of the various countries. Using a two-stroke cycle engine in automobiles is very attractive because of the lower specific weight (engine weight/power) and smoothness of operation (power stroke on every revolution). However, satisfying pollution laws has so far been an insurmountable obstacle. Starting in the late 1980s and going through the 1990s, a large program was instigated by several major world automobile companies to develop a two-stroke cycle automobile engine. This came about after the Orbital Company of Australia developed an air-assisted direct-fuel-injection system for two-stroke cycle engines. Although this greatly reduced hydrocarbon emissions, ever more stringent pollution laws doomed the two-stroke cycle for automobile application, and most development programs were put on the back burner. However, many modern two-stroke cycle engines are being manufactured for applications other than road vehicles (e.g., outboard motors).

With no exhaust stroke and imperfect scavenging (see Chapter 5), large amounts of exhaust residual remain in the cylinder at the start of the next cycle. This dilutes the air–fuel mixture in the cylinder and results in lower combustion temperature. This reduces the generation of NOx emissions, but the lower exhaust temperature creates other problems in the catalytic system.

HISTORIC—TWO-STROKE CYCLE AUTOMOBILES

The last two automobile models manufactured with two-stroke cycle engines were made in the German Democratic Republic until 1990. These were the 0.6 liter Trabant with two-cylinder air-cooled engine, and the 1.0 liter Wartburg with three-cylinder liquid-cooled engine.

Two-Stroke SI Engine Cycle

An air-standard approximation to a typical two-stroke SI engine cycle is shown in Fig. 3-17.

Process 1-2—isentropic power or expansion stroke.

All ports (or valves) closed:

$$T_2 = T_1(V_1/V_2)^{k-1} \tag{3-97}$$

$$P_2 = P_1(V_1/V_2)^k \tag{3-98}$$

$$q_{1-2} = 0 \tag{3-99}$$

$$w_{1-2} = (P_2v_2 - P_1v_1)/(1 - k) = R(T_2 - T_1)/(1 - k) \tag{3-100}$$

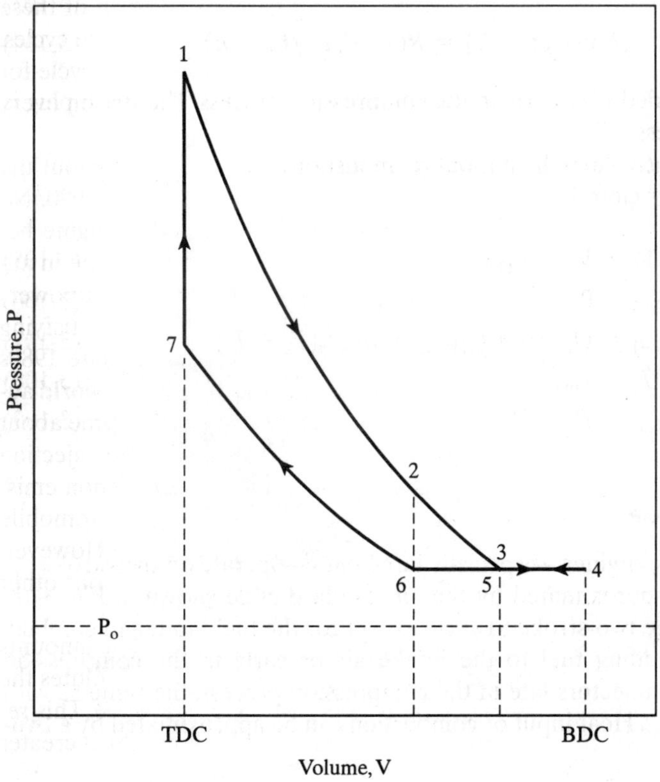

FIGURE 3-17
Air-standard approximation for a two-stroke cycle SI engine, 1-2-3-4-5-6-7-1.

Process 2-3—exhaust blowdown.
Exhaust port open and intake port closed.
Process 3-4-5—intake, and exhaust scavenging.
Exhaust port open and intake port open.

Intake air entering at an absolute pressure on the order of 140–180 kPa fills and scavenges the cylinder. Scavenging is a process in which the air pushes out most of the remaining exhaust residual from the previous cycle through the open exhaust port into the exhaust system, which is at about one atmosphere pressure. The piston uncovers the intake port at point 3, reaches BDC at point 4, reverses direction, and again closes the intake port at point 5. In some engines fuel is mixed with the incoming air. In other engines the fuel is injected later, after the exhaust port is closed.

Process 5-6—exhaust scavenging.
Exhaust port open and intake port closed.
Exhaust scavenging continues until the exhaust port is closed at point 6.
Process 6-7—isentropic compression.
All ports closed:

$$T_7 = T_6(V_6/V_7)^{k-1} \tag{3-101}$$

$$P_7 = P_6(V_6/V_7)^k \tag{3-102}$$

$$q_{6-7} = 0 \tag{3-103}$$

$$w_{6-7} = (P_7v_7 - P_6v_6)/(1 - k) = R(T_7 - T_6)/(1 - k) \tag{3-104}$$

In some engines, fuel is added very early in the compression process. The spark plug is fired near the end of process 6-7.
Process 7-1—constant-volume heat input (combustion).
All ports closed:

$$V_7 = V_1 = V_{TDC} \tag{3-105}$$

$$W_{7-1} = 0 \tag{3-106}$$

$$Q_{7-1} = Q_{in} = m_fQ_{HV}\eta_c = m_mc_v(T_1 - T_7) \tag{3-107}$$

$$T_1 = T_{max} \tag{3-108}$$

$$P_1 = P_{max} = P_7(T_1/T_7) \tag{3-109}$$

Two-Stroke CI Engine Cycle

Many compression ignition engines—especially large ones—operate on two-stroke cycles. These cycles can be approximated by the air-standard cycle shown in Fig. 3-18. This cycle is the same as the two-stroke SI cycle except for the fuel input and combustion process. Instead of adding fuel to the intake air or early in the compression process, fuel is added with injectors late in the compression process, the same as with four-stroke cycle CI engines. Heat input or combustion can be approximated by a two-step (dual) process.

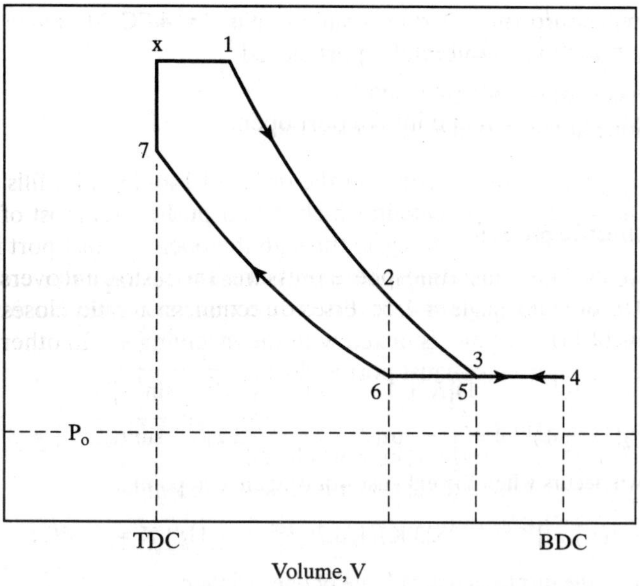

FIGURE 3-18
Air-standard approximation for a two-stroke cycle CI engine, 1-2-3-4-5-6-7-*x*-1.

Process 7-*x*—constant-volume heat input (first part of combustion).
All ports closed:

$$V_7 = V_x = V_{TDC} \tag{3-110}$$

$$W_{7-x} = 0 \tag{3-111}$$

$$Q_{7-x} = m_m c_v (T_x - T_7) \tag{3-112}$$

$$P_x = P_{max} = P_7(T_x/T_7) \tag{3-113}$$

Process *x*-1—constant-pressure heat input (second part of combustion).
All ports closed:

$$P_1 = P_x = P_{max} \tag{3-114}$$

$$W_{x-1} = P_1(V_1 - V_x) \tag{3-115}$$

$$Q_{x-1} = m_m c_p (T_1 - T_x) \tag{3-116}$$

$$T_1 = T_{max} \tag{3-117}$$

Example Problem 3-7

A fishing boat is equipped with an outboard motor that operates on an air-standard two-stroke SI engine cycle at 3100 RPM. The four-cylinder engine has a bore and stroke of B = 5.2 cm and S = 5.8 cm, a mechanical efficiency of η_m = 77%, compression ratio of r_c = 12, and a connecting rod length to crankshaft offset ratio R = r/a = 3.2. The exhaust slot on the side of the cylinder opens at 105° aTDC and the intake slot opens at 50° bBDC. With crankcase compression, the inlet air–fuel mixture enters at a pressure of P = 145 kPa and after mixing with the hot exhaust residual the cylinder gas temperature at the start of compression is T = 48°C. Maximum temperature in the cycle is T_{max} = 2250°C. Calculate:

1. effective compression ratio
2. cylinder temperature at the start of exhaust blowdown
3. indicated power
4. brake power
5. indicated mean effective pressure

(1) Using Fig. 3-17, we see that actual compression starts when the exhaust slot closes at point 6, 105° bTDC or crank angle = 255°. Effective compression ratio can then be obtained using Eq. (2-14):

$$(r_c)_{eff} = V/V_c = 1 + \tfrac{1}{2}(r_c - 1)[R + 1 - \cos\theta - \sqrt{R^2 - \sin^2\theta}]$$
$$= 1 + \tfrac{1}{2}(12 - 1)[3.2 + 1 - \cos(255°) - \sqrt{(3.2)^2 - \sin^2(255°)}] = \underline{8.74}$$

(2) Exhaust blowdown occurs when the exhaust slot is opened at point 2:

$$T_2 = T_1[1/(r_c)_{eff})]^{k-1} = (2523\text{ K})(1/8.74)^{1.35-1} = \underline{1181\text{ K} = 908°\text{C}}$$

(3) Equation (2-8) gives the displacement volume of one cylinder:

$$V_d = (\pi/4)B^2S = (\pi/4)(5.2\text{ cm})^2(5.8\text{ cm}) = 123.2\text{ cm}^3 = 0.0001232\text{ m}^3$$

Clearance volume at TDC, using Eq. (2–12), is

$$r_c = (V_d + V_c)/V_c = 12 = (123.2 + V_c)/V_c, \quad V_c = 11.20\text{ cm}^3 = 0.0000112\text{ m}^3$$

Temperature and pressure at end of compression:

$$T_7 = T_6[(r_c)_{eff}]^{k-1} = (321\text{ K})(8.74)^{1.35-1} = 686\text{ K} = 413°\text{C}$$
$$P_7 = P_6[(r_c)_{eff}]^{k} = (145\text{ kPa})(8.74)^{1.35} = 2707\text{ kPa}$$

Mass of gas in the cylinder during cycle, calculated at end of compression, is

$$m = P_7V_c/RT_7$$
$$= (2707\text{ kPa})(0.0000112\text{ m}^3)/(0.287\text{ kJ/kg-K})(686\text{ K}) = 0.000154\text{ kg}$$

Work produced during the power stroke, using Eq. (3-14), is

$$W_{1-2} = mR(T_2 - T_1)/(1 - k)$$
$$= (0.000154\text{ kg})(0.287\text{ kJ/kg-K})(1181\text{ K} - 2523\text{ K})/(1 - 1.35) = 0.1695\text{ kJ}$$

Work during compression stroke is

$$W_{6-7} = mR(T_7 - T_6)/(1 - k) = (0.000154)(0.287)(686 - 321)/(1 - 1.35)$$
$$= -0.0461\text{ kJ}$$

Use Eq. (2-14) to find cylinder volume at point 3 when the intake slot opens, 50° bBDC or a crank angle of 130°:

$$V_3/V_1 = 1 + \tfrac{1}{2}(12 - 1)[3.2 + 1 - \cos(130°) - \sqrt{(3.2)^2 - \sin^2(130°)} = 10.547$$
$$V_5 = V_3 = (10.547)V_1 = (10.547)(0.0000112\text{ m}^3) = 0.000118\text{ m}^3$$

Use effective compression ratio to find cylinder volume at point 6:

$$V_6 = V_2 = V_1(r_c)_{\text{eff}} = (0.0000112\ \text{m}^3)(8.74) = 0.0000979\ \text{m}^3$$

Work for process 5-6 is

$$W_{5-6} = P(V_6 - V_5) = (145\ \text{kPa})(0.0000979 - 0.000118)\text{m}^3 = -0.0029\ \text{kJ}$$

W_{3-4} is canceled by W_{4-5}
Net work for one cylinder for one cycle is

$$W_{\text{net}} = (0.1695) + (-0.0461) + (-0.0029) = 0.1205\ \text{kJ/cylinder-cycle}$$

Use Eq. (2-42) to find indicated power:

$$\dot{W}_i = WN/n = [(0.1205\ \text{kJ/cyl-cycle})(3100/60\ \text{rev/sec})/(1\ \text{rev/cycle})](4\ \text{cyl})$$
$$= \underline{24.9\ \text{kW} = 33.4\ \text{hp}}$$

(4) Brake power using Eq. (2-47) is

$$\dot{W}_b = \eta_m \dot{W}_i = (0.77)(24.9\ \text{kW}) = \underline{19.17\ \text{kW} = 25.7\ \text{hp}}$$

(5) Equation (2-88) gives indicated mean effective pressure:

$$\text{imep} = [(1000)(24.9)(1)]/[(4\ \text{cylinders})(0.1232)(3100/60)] = \underline{978\ \text{kPa} = 141.8\ \text{psia}}$$

3.12 STIRLING CYCLE

In recent years, a number of experimental engines that operate on the **Stirling cycle** shown in Fig. 3-19 have been tested. The concept of the Stirling engine has been around since 1816, and while it is not a true internal combustion engine, it is included here briefly because it is a heat engine used to propel vehicles as one of its applications. The basic engine uses a free-floating, double-acting piston with a gas chamber on both ends of the cylinder. Combustion does not occur within the cylinder, but the working gas is heated with an external combustion process. Heat input can also come from solar or nuclear sources. Engine output is usually a rotating shaft [8].

A Stirling engine has an internal regeneration process that uses a heat exchanger. Ideally, the heat exchanger uses the rejected heat in process 4-1 to preheat the internal working fluid in the heat addition process 2-3. The only heat transfers with the surroundings then occur with a heat addition process 3-4 at one maximum temperature T_{high}, and a heat rejection process 1-2 at one minimum temperature T_{low}. If the processes in the air-standard cycle in Fig. 3-19 can be considered reversible, the thermal efficiency of the cycle will be

$$(\eta_t)_{\text{STIRLING}} = 1 - (T_{\text{low}}/T_{\text{high}}) \tag{3-118}$$

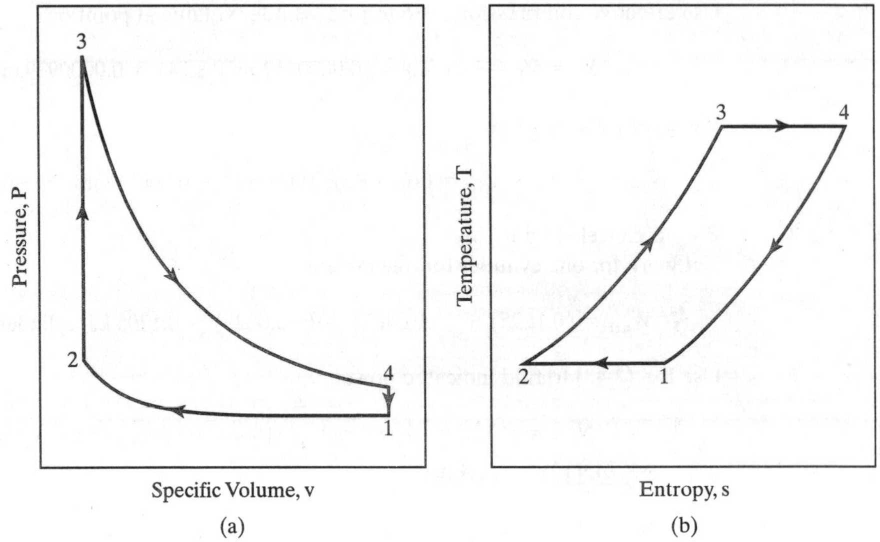

FIGURE 3-19

Ideal air-standard Stirling cycle, 1-2-3-4-1, on **(a)** pressure-specific volume coordinates, and **(b)** temperature-entropy coordinates.

This is the same thermal efficiency as a Carnot cycle and is the theoretical maximum possible. Although a real engine cannot operate reversibly, a well-designed Stirling engine can have a very high thermal efficiency. This is one of the attractions that is generating interest in this type of engine. Other attractions include low emissions with no catalytic converter and the flexibility of many possible fuels that can be used. This is because heat input is from a continuous steady-state combustion in an external chamber at a relatively low temperature around 1000 K. Fuels used have included gasoline, diesel fuel, jet fuel, alcohol, and natural gas. In some engines, the fuel can be changed with no adjustments needed.

Problems with Stirling engines include sealing, warm-up time needed, and high cost. Other possible applications include refrigeration, stationary power, and heating of buildings.

HISTORIC—LENOIR ENGINE

One of the first successful engines developed during the second half of the 1800s was the Lenoir engine (Fig. 3-20). Several hundred of these were built in the 1860s. They operated on a two-stroke cycle and had mechanical efficiencies up to 5% and power output up to 4.5 kW (6 hp). The engines were double acting, with combustion occurring on both ends of the piston. This gave two power strokes per revolution from a single cylinder [29].

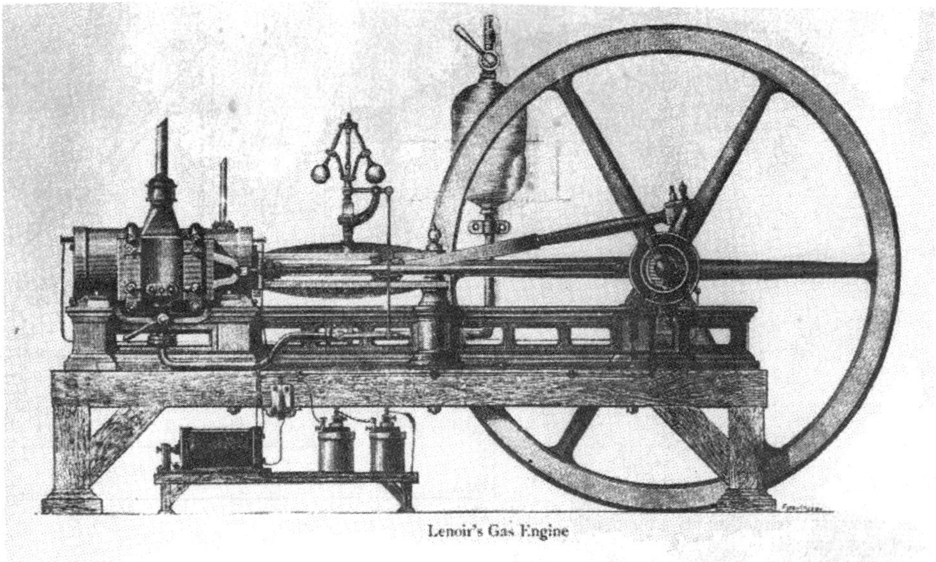

FIGURE 3.20

Lenoir noncompression engine of 1861. Reprinted with permission from Ref. [29], "Internal Fire" by Lyle Cummins.

3.13 LENOIR CYCLE

The **Lenoir cycle** is approximated by the air-standard cycle shown in Fig. 3-21. The first half of the first stroke was intake, with air–fuel entering the cylinder at atmospheric pressure (process 1-2 in Fig. 3-21). At about halfway through the first stroke, the intake valve was closed and the air–fuel mixture was ignited without any compression. Combustion raised the temperature and pressure in the cylinder almost at constant volume in the slow-moving engine (process 2-3). The second half of the first stroke then became the power or expansion process 3-4. Near BDC, the exhaust valve opened and blowdown occurred (4-5). This was followed by the exhaust stroke 5-1, completing the two-stroke cycle. There was essentially no clearance volume.

Thermodynamic Analysis of Air-Standard Lenoir Cycle

The intake process 1-2 and the latter half of the exhaust stroke process 2-1 cancel each other thermodynamically on P–V coordinates and can be left out of the analysis of the Lenoir cycle. The cycle then becomes 2-3-4-5-2.

Process 2-3—constant volume heat input (combustion).

All valves closed:

$$P_2 = P_1 = P_o \tag{3-119}$$

$$v_3 = v_2 \tag{3-120}$$

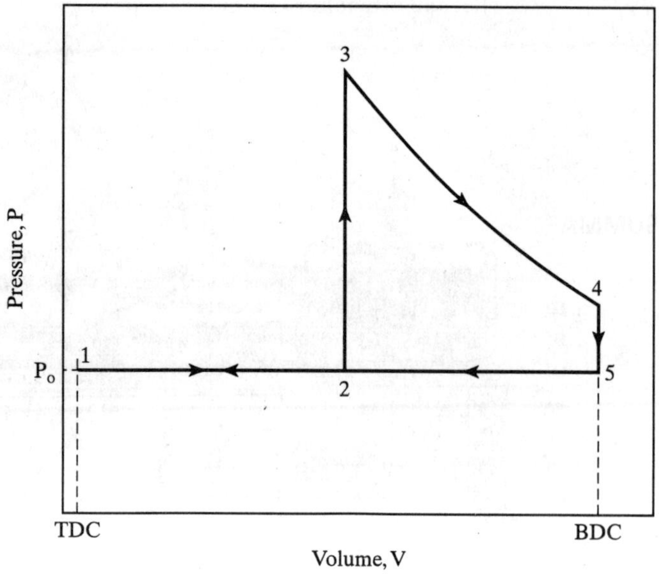

FIGURE 3-21

Air-standard approximation for a historic Lenoir engine cycle, 1-2-3-4-5-1.

$$w_{2\text{-}3} = 0 \tag{3-121}$$

$$q_{2\text{-}3} = q_{in} = c_v(T_3 - T_2) = (u_3 - u_2) \tag{3-122}$$

Process 3-4—isentropic power or expansion stroke.
All valves closed:

$$q_{3\text{-}4} = 0 \tag{3-123}$$

$$T_4 = T_3(v_3/v_4)^{k-1} \tag{3-124}$$

$$P_4 = P_3(v_3/v_4)^k \tag{3-125}$$

$$w_{3\text{-}4} = (P_4v_4 - P_3v_3)/(1 - k) = R(T_4 - T_3)/(1 - k) \tag{3-126}$$

$$= (u_3 - u_4) = c_v(T_3 - T_4)$$

Process 4-5—constant-volume heat rejection (exhaust blowdown).
Exhaust valve open and intake valve closed:

$$v_5 = v_4 = v_{BDC} \tag{3-127}$$

$$w_{4\text{-}5} = 0 \tag{3-128}$$

$$q_{4\text{-}5} = q_{out} = c_v(T_5 - T_4) = (u_5 - u_4) \tag{3-129}$$

Process 5-2—constant-pressure exhaust stroke at P_o.
Exhaust valve open and intake valve closed:

$$P_5 = P_2 = P_1 = P_o \tag{3-130}$$

$$w_{5\text{-}2} = P_o(v_2 - v_5) \tag{3-131}$$

$$q_{5\text{-}2} = q_{out} = (h_2 - h_5) = c_p(T_2 - T_5) \tag{3-132}$$

Thermal efficiency of Lenoir cycle:

$$(\eta_t)_{\text{LENOIR}} = |w_{\text{net}}|/|q_{\text{in}}| = 1 - (|q_{\text{out}}|/|q_{\text{in}}|)$$
$$= 1 - [c_v(T_4 - T_5) + c_p(T_5 - T_2)]/[c_v(T_3 - T_2)] \quad (3\text{-}133)$$
$$= 1 - [(T_4 - T_5) + k(T_5 - T_2)]/(T_3 - T_2)$$

SUMMARY

This chapter reviewed the basic cycles used in internal combustion engines. Although many engine cycles have been developed, for over a century most automobile engines have operated on the basic SI four-stroke cycle developed in the 1870s by Otto and others. This can be approximated and analyzed using the ideal air-standard Otto cycle. During the 1990s a major effort was made by several car manufacturers to develop a two-stroke cycle engine for automobiles. This effort failed because of pollution regulations. Many small SI engines do operate on two-stroke cycles, sometimes (erroneously) called a two-stroke Otto cycle.

Early four-stroke CI engines operated on a cycle that can be approximated by the air-standard Diesel cycle. This cycle was improved in modern CI engines of the type used in automobiles and trucks. Changing the injection timing resulted in a more efficient engine operating on a cycle best approximated by an air-standard Dual cycle. Due to the greater thermal efficiency of these engines, there is an ever-increasing percentage of vehicles being manufactured with four-stroke cycle CI engines, especially in Europe. Most small CI engines and very large CI engines operate on a two-stroke cycle.

At present, most automobiles operate on an SI four-stroke cycle, approximated either by the air-standard Otto cycle, or the more modern Miller cycle. The Miller cycle is an improvement on the Otto cycle brought about by several technology advancements, mainly variable valve timing control. Valve control allows for a more efficient cycle by reducing pumping losses and giving an expansion ratio that is greater than the effective compression ratio.

PROBLEMS

3.1 Cylinder conditions at the start of compression in an SI engine operating at WOT on an air-standard Otto cycle are 60°C and 98 kPa. The engine has a compression ratio of 9.5:1 and uses gasoline with AF = 15.5. Combustion efficiency is 96%, and it can be assumed that there is no exhaust residual.

Calculate:

(a) Temperature at all states in the cycle. [°C]
(b) Pressure at all states in the cycle. [kPa]
(c) Specific work done during power stroke. [kJ/kg]
(d) Heat added during combustion. [kJ/kg]
(e) Net specific work done. [kJ/kg]
(f) Indicated thermal efficiency. [%]

3.2 The engine in Problem 3-1 is a three-liter V6 engine operating at 2400 RPM. At this speed the mechanical efficiency is 84%.
Calculate:

(a) Brake power. [kW]
(b) Torque. [N-m]
(c) Brake mean effective pressure. [kPa]
(d) Friction power lost. [kW]
(e) Brake specific fuel consumption. [gm/kW-hr]
(f) Volumetric efficiency. [%]
(g) Output per displacement. [kW/L]

3.3 The exhaust pressure of the engine in Problem 3-2 is 100 kPa.
Calculate:

(a) Exhaust temperature. [°C]
(b) Actual exhaust residual. [%]
(c) Temperature of air entering cylinders from intake manifold. [°C]

3.4 The engine of Problems 3-2 and 3-3 is operated at part throttle with intake pressure of 75 kPa. Intake manifold temperature, mechanical efficiency, exhaust residual, and air–fuel ratio all remain the same.
Calculate:

(a) Temperature in cylinder at start of compression stroke. [°C]
(b) Temperature in cylinder at start of combustion. [°C]

3.5 An SI engine operating at WOT on a four-stroke air-standard cycle has cylinder conditions at the start of compression of 100°F and 14.7 psia. Compression ratio is $r_c = 10$, and the heat added during combustion is $q_{in} = 800$ BTU/lbm. During compression the temperature range is such that a value for the ratio of specific heats $k = 1.4$ would be correct. During the power stroke the temperature range is such that a value of $k = 1.3$ would be correct. Use these values for compression and expansion, respectively, when analyzing the cycle. Use a value for specific heat of $c_v = 0.216$ BTU/lbm-°R, which best corresponds to the temperature range during combustion.
Calculate:

(a) Temperature at all states in cycle. [°F]
(b) Pressure at all states in cycle. [psia]
(c) Average value of k which would give the same indicated thermal efficiency value as the analysis in parts (a) and (b).

3.6 A CI engine operating on the air-standard Diesel cycle has cylinder conditions at the start of compression of 65°C and 130 kPa. Light diesel fuel is used at an equivalence ratio of $\phi = 0.8$ with a combustion efficiency $\eta_c = 0.98$. Compression ratio is $r_c = 19$.
Calculate:

(a) Temperature at each state of the cycle. [°C]
(b) Pressure at each state of the cycle. [kPa]
(c) Cutoff ratio.

(d) Indicated thermal efficiency. [%]
(e) Heat lost in exhaust. [kJ/kg]

3.7 A compression ignition engine for a small truck is to operate on an air-standard Dual cycle with a compression ratio of $r_c = 18$. Due to structural limitations, maximum allowable pressure in the cycle will be 9000 kPa. Light diesel fuel is used at a fuel–air ratio of FA = 0.054. Combustion efficiency can be considered 100%. Cylinder conditions at the start of compression are 50°C and 98 kPa.
Calculate:

(a) Maximum indicated thermal efficiency possible with these conditions. [%]
(b) Peak cycle temperature under conditions of part (a). [°C]
(c) Minimum indicated thermal efficiency possible with these conditions. [%]
(d) Peak cycle temperature under conditions of part (c). [°C]

3.8 An in-line six, 3.3-liter CI engine using light diesel fuel at an air–fuel ratio of AF = 20 operates on an air-standard Dual cycle. Half the fuel can be considered burned at constant volume, and half at constant pressure with combustion efficiency $\eta_c = 100\%$. Cylinder conditions at the start of compression are 60°C and 101 kPa. Compression ratio $r_c = 14:1$.
Calculate:

(a) Temperature at each state of the cycle. [K]
(b) Pressure at each state of the cycle. [kPa]
(c) Cutoff ratio.
(d) Pressure ratio.
(e) Indicated thermal efficiency. [%]
(f) Heat added during combustion. [kJ/kg]
(g) Net indicated work. [kJ/kg]

3.9 The engine in Problem 3-8 produces 57 kW of brake power at 2000 RPM.
Calculate:

(a) Torque. [N-m]
(b) Mechanical efficiency. [%]
(c) Brake mean effective pressure. [kPa]
(d) Brake specific fuel consumption. [gm/kW-hr]

3.10 An Otto cycle SI engine with a compression ratio of $r_c = 9$ has peak cycle temperature and pressure of 2800 K and 9000 kPa. Cylinder pressure when the exhaust valve opens is 460 kPa and exhaust manifold pressure is 100 kPa.
Calculate:

(a) Exhaust temperature during exhaust stroke. [°C]
(b) Exhaust residual after each cycle. [%]
(c) Velocity out of the exhaust valve when the valve first opens. [m/sec]
(d) Theoretical momentary maximum temperature in the exhaust. [°C]

3.11 An SI engine operates on an air-standard four-stroke Otto cycle with turbocharging. Air–fuel enters the cylinders at 70°C and 140 kPa, and heat in by combustion equals $q_{in} = 1800$ kJ/kg. Compression ratio $r_c = 8$ and exhaust pressure $P_{ex} = 100$ kPa.

Calculate:

(a) Temperature at each state of the cycle. [°C]
(b) Pressure at each state of the cycle. [kPa]
(c) Work produced during expansion stroke. [kJ/kg]
(d) Work of compression stroke. [kJ/kg]
(e) Net pumping work. [kJ/kg]
(f) Indicated thermal efficiency. [%]
(g) Compare with Problems 3-12 and 3-13.

3.12 An SI engine operates on an air-standard four-stroke Miller cycle with turbocharging. The intake valves close late, resulting in cycle 6-7-8-7-2-3-4-5-6 in Fig. 3-16. Air–fuel enters the cylinders at 70°C and 140 kPa, and heat in by combustion equals q_{in} = 1800 kJ/kg. Compression ratio r_c = 8, expansion ratio r_e = 10, and exhaust pressure P_{ex} = 100 kPa.
Calculate:

(a) Temperature at each state of the cycle. [°C]
(b) Pressure at each state of the cycle. [kPa]
(c) Work produced during expansion stroke. [kJ/kg]
(d) Work of compression stroke. [kJ/kg]
(e) Net pumping work. [kJ/kg]
(f) Indicated thermal efficiency. [%]
(g) Compare with Problems 3-11 and 3-13.

3.13 An SI engine operates on an air-standard four-stroke Miller cycle with turbocharging. The intake valves close early, resulting in cycle 6-7-1-7-2-3-4-5-6 in Fig. 3-16. Air–fuel enters the cylinders at 70°C and 140 kPa, and heat in by combustion equals q_{in} = 1800 kJ/kg. Compression ratio r_c = 8, expansion ratio r_e = 10, and exhaust pressure P_{ex} = 100 kPa.
Calculate:

(a) Temperature at each state of the cycle. [°C]
(b) Pressure at each state of the cycle. [kPa]
(c) Work produced during expansion stroke. [kJ/kg]
(d) Work of compression stroke. [kJ/kg]
(e) Net pumping work. [kJ/kg]
(f) Indicated thermal efficiency. [%]
(g) Compare with Problems 3-11 and 3-12.

3.14 A six-cylinder SI engine with a compression ratio of 10.5:1 operates on a four-stroke cycle at 3000 RPM. Air enters the cylinders at T_1 = 110°F and P_1 = 12.8 psia. The intake valve closes at 20° aBDC and combustion starts when the spark plug fires at 15° bTDC. Connecting rod length is 6.64 in. and crank offset is 1.66 in.
Calculate:

(a) Cylinder temperature at start of combustion using air-standard Otto cycle analysis. [°F]
(b) Cylinder temperature when spark plug fires assuming compression doesn't start until intake valve closes. [°F]

3.15 A spark ignition four-stroke cycle engine operates on an air-standard Miller cycle with no supercharger, and early intake valve closing (cycle 6-7-1-7-2-3-4-5-6 in Fig 3-15). Compression ratio is 8.2 and expansion ratio is 10.2. Cylinder conditions when the intake valve closes are $T_7 = 57°C$ and $P_7 = 100$ kPa. Maximum temperature and pressure in the cycle are $T_{max} = 3427°C$ and $P_{max} = 9197$ kPa.
Calculate:

(a) Minimum cylinder pressure during cycle. [kPa]
(b) Pump work for one cylinder for one cycle. [kJ]
(c) Pressure in cylinder when exhaust valve opens. [kPa]

3.16 An experimental two-stroke cycle SI automobile engine operates on the air-standard cycle shown in Fig. 3-17, with a compression ratio of 10.5:1. When the intake valve opens at 52° bBDC a supercharger supplies air-fuel to the cylinders at a pressure of $P_{intake} = 17.8$ psia. The exhaust port opens at 70° bBDC. Maximum temperature and pressure in the cycle are $T_{max} = 4200°F$ and $P_{max} = 1137$ psia. Connecting rod length is 9.5 inches and crank offset is 2.5 inches.
Calculate:

(a) Cylinder temperature when exhaust port opens. [°F]
(b) Effective compression ratio.
(c) Temperature and pressure at the end of the compression stroke. [°F, psia]

3.17 A six cylinder, two-stroke cycle CI ship engine with bore $B = 35$ cm and stroke $S = 105$ cm produces 3600 kW of brake power at 210 RPM.
Calculate:

(a) Torque at this speed. [kN-m]
(b) Total displacement. [L]
(c) Brake mean effective pressure. [kPa]
(d) Average piston speed. [m/sec]

3.18 A single-cylinder, two-stroke cycle model airplane engine with a 7.54-cm^3 displacement produces 1.42 kW of brake power at 23,000 RPM using glow plug ignition. The square engine (bore = stroke) uses 31.7 gm/min of castor oil–methanol–nitromethane fuel at an air–fuel ratio AF = 4.5. During intake scavenging, 65% of the incoming air–fuel mixture gets trapped in the cylinder, while 35% of it is lost with the exhaust before the exhaust port closes. Combustion efficiency $\eta_c = 0.94$.
Calculate:

(a) Brake specific fuel consumption. [gm/kW-hr]
(b) Average piston speed. [m/sec]
(c) Unburned fuel exhausted to atmosphere. [gm/min]
(d) Torque. [N-m]

3.19 A historic single-cylinder engine with a mechanical efficiency $\eta_m = 5\%$ operates at 140 RPM on the Lenoir cycle shown in Fig. 3-21. The cylinder has a double acting piston with a 12-in. bore and a 36-in. stroke. The fuel has a heating value $Q_{LHV} = 12{,}000$ BTU/lbm and is used at an air–fuel ratio AF = 18. Combustion occurs at constant volume half way through the intake-power stroke when cylinder conditions equal 70°F and 14.7 psia.

Calculate:

(a) Temperature at each state of cycle. [°F]

(b) Pressure at each state of cycle. [psia]

(c) Indicated thermal efficiency. [%]

(d) Brake power. [hp]

(e) Average piston speed. [ft/sec]

3.20 Cylinder conditions at the start of compression of a four-stroke cycle SI engine are 27°C and 100 kPa. The engine has a compression ratio of $r_c = 8{:}1$, and heat addition from combustion is $q_{in} = 2000$ kJ/kg.
Calculate:

(a) Temperature and pressure at each state of the cycle, using air-standard Otto cycle analysis with constant specific heats. [°C, kPa]

(b) Indicated thermal efficiency in part (a). [%]

(c) Temperature and pressure at each state of the cycle, using any standard air tables that are based on variable specific heats as functions of temperature (e.g., reference [73]). [°C, kPa]

(d) Indicated thermal efficiency in part (c). [%]

DESIGN PROBLEMS

3.1D Design an SI engine to operate on a six-stroke cycle. The first four strokes of the cycle are the same as a four-stroke Otto cycle. This is followed with an additional air-only intake stroke and an air-only exhaust stroke. Draw simple schematics, and explain the speed and operation of the camshafts when the valves open and close. Also, explain the control of the ignition process.

3.2D Design a mechanical linkage system for a four-stroke cycle, reciprocating SI engine to operate on the Atkinson cycle (i.e., normal compression stroke and a power stroke which expands until cylinder pressure equals ambient pressure). Explain using simple schematic drawings.

3.3D An SI engine operating on a four-stroke air-standard cycle using stoichiometric gasoline is to have a maximum cylinder pressure of 11,000 kPa at WOT. Inlet pressure can be 100 kPa without supercharging, or it can be as high as 150 kPa with a supercharger. Pick a compression ratio and inlet pressure combination to give maximum indicated thermal efficiency. Pick a compression ratio and inlet pressure to give maximum imep.

CHAPTER 4

Thermochemistry and Fuels

"The supply of . . . petrol fuels . . . will soon become unequal to the demand, with the result that a critical situation is bound to arise in the not very distant future. Civilization is now so deeply committed to the use of internal-combustion engines for all road transport and for many other purposes, that it is a matter of absolute necessity to find an alternative fuel. Fortunately such a fuel is in sight in the form of alcohol; this is a vegetable product whose consumption involves no drain on the world's storage and which, in tropical countries at all events, can ultimately be produced in quantities, sufficient to meet the world's demand, at all events at the present rate of consumption."

The High-Speed Internal-Combustion Engine

by *Harry R. Ricardo* (1923)

This chapter reviews basic thermochemistry principles as applied to IC engines. It studies ignition characteristics and combustion in engines, the *octane number* of SI fuels, and the *cetane number* of CI fuels. Gasoline and other possible *alternate fuels* are examined.

4.1 THERMOCHEMISTRY

Combustion Reactions

Most IC engines obtain their energy from the combustion of a hydrocarbon fuel with air, which converts chemical energy of the fuel to internal energy in the gases within the engine. This internal energy is then converted to the rotating crankshaft output by the mechanical linkages of the engine. There are many thousands of different hydrocarbon fuel components, which consist mainly of hydrogen and carbon but may also contain oxygen (alcohols), nitrogen, sulfur, etc. The maximum possible amount of chemical energy is released (heat) from the fuel when it reacts (combusts) with a *stoichiometric* amount of oxygen. Stoichiometric oxygen (sometimes called theoretical oxygen) is just enough to convert all carbon in the fuel to CO_2 and all hydrogen to H_2O, with no oxygen left over. The balanced chemical equation of the simplest hydrocarbon fuel, methane CH_4, burning with stoichiometric oxygen is

$$CH_4 + 2\,O_2 \rightarrow CO_2 + 2\,H_2O$$

It takes two moles of oxygen to react with one mole of fuel, and this gives one mole of carbon dioxide and two moles of water vapor. If isooctane is the fuel component, the balanced stoichiometric combustion with oxygen would be

$$C_8H_{18} + 12.5\,O_2 \rightarrow 8\,CO_2 + 9\,H_2O$$

Molecules react with molecules, so in balancing a chemical equation, molar quantities (fixed number of molecules) are used and not mass quantities. One kgmole of a substance has a mass in kilograms equal in number to the molecular weight (molar mass) of that substance. In English units, the lbmmole is used

$$m = NM \tag{4-1}$$

where

m = mass
N = number of moles
M = molecular weight

In SI units:

$$\begin{aligned} 1 \text{ kgmole of } CH_4 &= 16.04 \text{ kg} \\ 1 \text{ kgmole of } O_2 &= 32.00 \text{ kg} \\ 1 \text{ kgmole} &= 6.02 \times 10^{26} \text{ molecules} \end{aligned}$$

In English units:

$$\begin{aligned} 1 \text{ lbmmole of } CH_4 &= 16.04 \text{ lbm} \\ 1 \text{ lbmmole of } O_2 &= 32.00 \text{ lbm} \\ 1 \text{ lbmmole} &= 2.73 \times 10^{26} \text{ molecules} \end{aligned}$$

The components on the left side of a chemical reaction equation, which are present before the reaction, are called *reactants*, while the components on the right side of the equation, which are present after the reaction, are called *products* or exhaust.

Very small powerful engines could be built if fuel was burned with pure oxygen. However, the cost of using pure oxygen would be prohibitive, and thus is not done. Air is used as the source of oxygen to react with fuel. Atmospheric air is made up of about:

78% nitrogen by mole
21% oxygen
1% argon
traces of CO_2, Ne, CH_4, He, H_2O, etc.

Nitrogen and argon are essentially chemically neutral and do not react in the combustion process. Their presence, however, does affect the temperature and pressure in the combustion chamber. To simplify calculations without causing any large error, the neutral argon in air is assumed to be combined with the neutral nitrogen, and atmospheric air then can be modeled as being made up of 21% oxygen and 79% nitrogen. For every 0.21 moles of oxygen there is also 0.79 moles of nitrogen, or for one mole of oxygen there are 0.79/0.21 moles of nitrogen. For every mole of oxygen needed for combustion, 4.76 moles of air must be supplied: one mole of oxygen plus 3.76 moles of nitrogen.

Stoichiometric combustion of methane with *air* is then

$$CH_4 + 2\,O_2 + 2(3.76)\,N_2 \rightarrow CO_2 + 2\,H_2O + 2(3.76)\,N_2$$

and the combustion of isooctane with air is

$$C_8H_{18} + 12.5\,O_2 + 12.5(3.76)\,N_2 \rightarrow 8\,CO_2 + 9\,H_2O + 12.5(3.76)\,N_2$$

It is convenient to balance combustion reaction equations for one kgmole of fuel. The energy released by the reaction will thus have units of energy per kgmole of fuel, which is easily transformed to total energy when the flow rate of fuel is known. This convention will be followed in this textbook. Molecular weights can be found in Table 4-1 and Table A-2 in the Appendix. The molecular weight of 29 will be used for air. Combustion can occur, within limits, when more than stoichiometric air is present (lean) or when less than stoichiometric air is present (rich) for a given amount of fuel. If methane is burned with 150% stoichiometric air, the excess oxygen ends up in the products:

$$CH_4 + 3\,O_2 + 3(3.76)\,N_2 \rightarrow CO_2 + 2\,H_2O + 3(3.76)\,N_2 + O_2$$

If isooctane is burned with 80% stoichiometric air, there is not enough oxygen to convert all the carbon to CO_2, and carbon monoxide CO is found in the products:

$$C_8H_{18} + 10\,O_2 + 10(3.76)\,N_2 \rightarrow 3\,CO_2 + 9\,H_2O + 5\,CO + 10(3.76)\,N_2$$

Carbon monoxide is a colorless, odorless, poisonous gas which can be further burned to form CO_2. It is formed in any combustion process when there is a deficiency of

TABLE 4-1 Molecular Weights

Substance		Molecular Weight (kg/kgmole) or (lbm/lbmmole)
Air		28.97
Argon	Ar	39.95
Carbon	C	12.01
Carbon Monoxide	CO	28.01
Carbon Dioxide	CO_2	44.01
Hydrogen	H_2	2.02
Water Vapor	H_2O	18.02
Helium	He	4.00
Nitrogen	N_2	28.01

oxygen. It is also very likely that some of the fuel will not get burned when there is a deficiency of oxygen. This unburned fuel ends up as pollution in the exhaust of the engine.

Various terminology is used for the amount of air or oxygen used in combustion:

$$80\% \text{ stoichiometric air} = 80\% \text{ theoretical air} = 80\% \text{ air}$$
$$= 20\% \text{ deficiency of air}$$

$$133\% \text{ stoichiometric oxygen} = 133\% \text{ theoretical oxygen}$$
$$= 133\% \text{ oxygen} = 33\% \text{ excess oxygen}$$

For actual combustion in an engine, the equivalence ratio is a measure of the fuel–air mixture relative to stoichiometric conditions. It is defined as follows:

$$\phi = (\text{FA})_{\text{act}}/(\text{FA})_{\text{stoich}} = (\text{AF})_{\text{stoich}}/(\text{AF})_{\text{act}} \tag{4-2}$$

where

$$\text{FA} = m_f/m_a = \text{fuel–air ratio}$$
$$\text{AF} = m_a/m_f = \text{air–fuel ratio}$$
$$m_a = \text{mass of air}$$
$$m_f = \text{mass of fuel}$$

when

$\phi < 1$ running lean, oxygen in exhaust
$\phi > 1$ running rich, CO and fuel in exhaust
$\phi = 1$ stoichiometric, maximum energy released from fuel

SI engines normally operate with an equivalence ratio in the range of 0.9 to 1.2, depending on the type of operation.

Example Problem 4-1

Isooctane is burned with 120% theoretical air in a small three-cylinder turbocharged automobile engine.
Calculate:

1. air–fuel ratio
2. fuel–air ratio
3. equivalence ratio

Stoichiometric reaction is

$$C_8H_{18} + 12.5\,O_2 + 12.5(3.76)\,N_2 \rightarrow 8\,CO_2 + 9\,H_2O + 12.5(3.76)\,N_2$$

With 20% excess air, the reaction is

$$C_8H_{18} + 15\,O_2 + 15(3.76)\,N_2 \rightarrow 8\,CO_2 + 9\,H_2O + 15(3.76)\,N_2 + 2.5\,O_2$$

With 20% excess air, all the fuel gets burned, and the same amount of CO_2 and H_2O is found in the products. In addition, there is some oxygen and additional nitrogen in the products (the excess air).

(1) Equations (2-55) and (4-1) are used to find the air–fuel ratio:

$$AF = m_a/m_f = N_a M_a/N_f M_f = [(15)(4.76)(29)]/[(1)(114)]$$
$$= \underline{18.16}$$

(2) Equation (2-56) is used to find the fuel–air ratio:

$$FA = m_f/m_a = 1/AF = 1/18.16 = \underline{0.055}$$

(3) Fuel–air ratio of stoichiometric combustion is

$$(FA)_{stoich} = [(1)(114)]/[(12.5)(4.76)(29)] = 0.066$$

Equivalence ratio is obtained using Eq. (4-2):

$$\phi = (FA)_{act}/(FA)_{stoich} = (0.055)/(0.066) = \underline{0.833}$$

Even when the flow of air and fuel into an engine is controlled exactly at stoichiometric conditions, combustion will not be "perfect," and components other than CO_2, H_2O, and N_2 are found in the exhaust products. One major reason for this is the extremely short time available for each engine cycle, which often means that less than complete mixing of the air and fuel is obtained. Some fuel molecules do not find an oxygen molecule to react with, and small quantities of both fuel and oxygen end up in the exhaust. Chapter 7 goes into more detail on this and other reasons that ideal combustion is not obtained. SI engines have a combustion efficiency in the range of 95% to 98% for lean mixtures and lower values for rich mixtures, where there is not enough air to react with all the fuel (see Fig. 4-1). CI engines operate lean overall and typically have combustion efficiencies of about 98%.

Chemical Equilibrium

If a general chemical reaction is represented by

$$\nu_A A + \nu_B B \rightarrow \nu_C C + \nu_D D \qquad (4\text{-}3)$$

where A and B represent all reactant species, whether one or two or more, and C and D represent all products, regardless of number, then ν_A, ν_B, ν_C, and ν_D represent the stoichiometric coefficients of A, B, C, and D.

Equilibrium composition for this reaction can be found if one knows the *chemical equilibrium constant*

$$K_e = [(N_C^{\nu_C} N_D^{\nu_D})/(N_A^{\nu_A} N_B^{\nu_B})](P/N_t)^{\Delta\nu} \qquad (4\text{-}4)$$

where

$\Delta\nu = \nu_C + \nu_D - \nu_A - \nu_B$
N_i = number of moles of component i at equilibrium
N_t = total number of moles at equilibrium
P = total absolute pressure in units of atmospheres

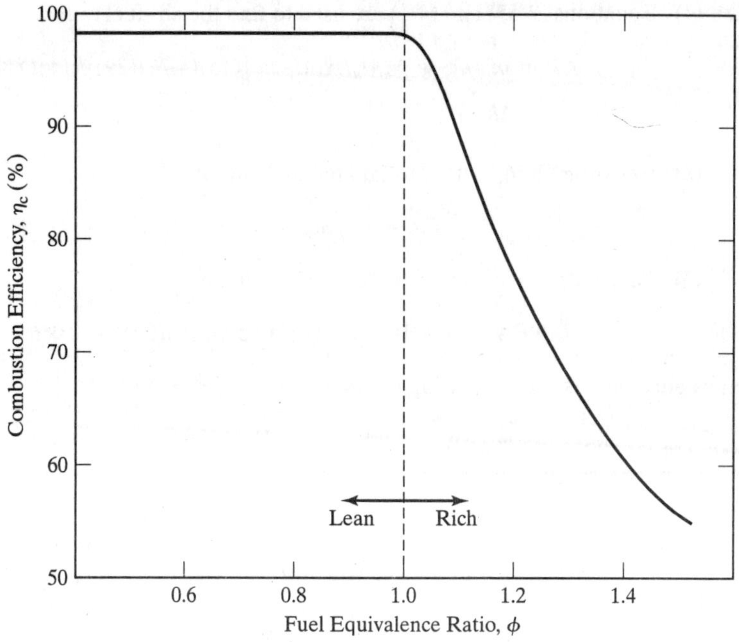

FIGURE 4-1

Combustion efficiency as a function of fuel equivalence ratio. Efficiency for engines operating lean is generally on the order of 98%. When an engine operates fuel rich, there is not enough oxygen to react with all the fuel, and combustion efficiency decreases. CI engines operate lean and typically have high combustion efficiency. Adapted from [58].

Equilibrium constants for many reactions can be found in thermodynamic textbooks or chemical handbooks, tabulated in logarithmic form (ln or $\log_{10}$). An abbreviated table can be found in the Appendix of this book (Table A-3).

K_e is very dependent on temperature, changing many orders of magnitude over the temperature range experienced in an IC engine. As K_e gets larger, equilibrium is more towards the right (products). This is the maximizing of entropy. For hydrocarbon fuels reacting with oxygen (air) at high engine temperatures, the equilibrium constant is very large, meaning that there are very few reactants (fuel and air) left at final equilibrium. However, at these high temperatures another chemical phenomenon takes place that affects the overall combustion process in the engine and what ends up in the engine exhaust.

Examination of the equilibrium constants in Table A-3 shows that dissociation of normally stable components will occur at these high engine temperatures. CO_2 dissociates to CO and O, O_2 dissociates to monatomic O, N_2 dissociates to monatomic N, etc. This not only affects chemical combustion, but is a cause of one of the major emission problems of IC engines. Nitrogen as diatomic N_2 does not react with other substances, but when it dissociates to monatomic nitrogen at high temperature it readily reacts with oxygen to form nitrogen oxides, NO and NO_2, a major pollutant from automobiles. To avoid generating large amounts of nitrogen oxides, combustion temperatures in automobile engines are lowered, which reduces the dissociation of N_2. Unfortunately, this also lowers the thermal efficiency of the engine.

Example Problem 4-2

One major reason for nitrogen oxides NOx in automobile engine exhaust is that at the high temperatures experienced in the engine a small amount of the normally stable diatomic nitrogen N_2 in the air dissociates to monatomic nitrogen N, which is very reactive. Near the end of combustion in a modern diesel engine operating on a Dual cycle the temperature and pressure in the cylinders are 3500 K and 10,500 kPa. Use Equation (4-4) to get an order-of-magnitude approximation of the percent of diatomic nitrogen that dissociates to monatomic nitrogen in the reaction: $N_2 \rightarrow 2\,N$

In the reaction there is only one reactant (N_2), so this will be component A in Eq. (4-4); there is no component B. Component C represents the only product (N); there is no component D.

The total pressure in atmosphere units is

$$P = (10{,}500\ \text{kPa})/(101\ \text{kPa/atm}) = 104\ \text{atm}$$

The actual dissociation reaction for 1 mole of N_2 is

$$N_2 \rightarrow 2x\text{N} + (1 - x)\,N_2$$

where x = extent of reaction

The chemical equilibrium constant K_e is obtained from Table A-3 in the Appendix at $T = 3500$ K:

$$\log_{10} K_e = -7.346 \qquad K_e = 4.508 \times 10^{-8}$$

Eq. (4-4) is then

$$K_e = 4.508 \times 10^{-8} = [(2x)^2/(1 - x)][(104)/(2x + (1 - x))]^{2-1}$$

or extent of reaction $\underline{x = 0.00001041 = 0.001041\%}$

This number should be considered only as a crude order-of-magnitude approximation for predicting the formation of NO and NO_2 in the engine cylinders. Chemical kinetics and the time–temperature–pressure relationship in the cylinders must be considered, time being on the order of milliseconds and chemical equilibrium never being obtained. This very low number for the production of N and the consequent formation of NOx underlines the seriousness of NOx emissions, which is considered one of the major air pollutants generated in automobile engines.

Exhaust Dew Point Temperature

When exhaust gases of an IC engine are cooled below their dew point temperature, water vapor in the exhaust starts to condense to liquid. It is common to see water droplets come out of an automobile exhaust pipe when the engine is first started and the pipe is cold. Very quickly the pipe is heated above the dew point temperature, and condensing water is then seen only as vapor when the hot exhaust is cooled by the surrounding air, much more noticeable in the cold wintertime.

Example Problem 4-3

At what temperature will water start to condense out of exhaust gases of the engine in Example Problem 4-1? Exhaust pressure is one atmosphere.
Calculate this for:

1. dry inlet air
2. inlet air with relative humidity of 55%

(1) The reaction equation from Example Problem 4-1 is

$$C_8H_{18} + 15\,O_2 + 15(3.76)\,N_2 \rightarrow 8\,CO_2 + 9\,H_2O + 2.5\,O_2 + 15(3.76)\,N_2$$

Mole fraction of water vapor in the exhaust products is

$$x_v = N_v/N_{\text{total}} = (9)/[8 + 9 + 2.5 + 15(3.76)] = 0.1186$$

Partial pressure of water vapor is

$$P_v = x_v P_{\text{total}} = (0.1186)(101\text{ kPa}) = 11.98\text{ kPa}$$

The dew point is the temperature at which this water vapor pressure becomes saturated. From steam tables [90], we have

$$\underline{T_{DP} = 49°C}$$

(2) If the relative humidity (rh) of the inlet air is 55% at $T = 25°C$, then the vapor pressure of water at the inlet will be

$$P_v = (rh)P_{\text{Sat at 25C}} = (0.55)(3.169\text{ kPa}) = 1.743\text{ kPa}$$

Using psychrometric equations and steam tables from any thermodynamics textbook (e.g. [90]), we find that the specific humidity is

$$\omega_v = m_v/m_a = 0.622[P_v/(P - P_v)] = (0.622)[(1.743)/(101 - 1.743)]$$
$$= 0.0109\text{ kg}_v/\text{kg}_a$$

Changing this mass ratio to a molar ratio using the molecular weights of air (29) and water vapor (18) gives the number of moles of water carried in with the air for one mole of fuel:

$$N_v = N_{\text{air}}\omega_v(M_{\text{air}}/M_v) = [(15)(4.76)](0.0109)[(29)/(18)] = 1.25$$

The reaction equation then becomes

$$C_8H_{18} + 15\,O_2 + 15(3.76)\,N_2 + 1.25\,H_2O \rightarrow$$
$$8\,CO_2 + 10.25\,H_2O + 2.5\,O_2 + 15(3.76)\,N_2$$

The mole fraction of water vapor in the exhaust is

$$x_v = (10.25)/[8 + 10.25 + 2.5 + 15(3.76)] = 0.1329$$

The partial pressure of water vapor is

$$P_v = x_v P_{\text{total}} = (0.1329)(101\text{ kPa}) = 13.42\text{ kPa}$$

The dew point temperature is

$$\underline{T_{DP} = 52°C}$$

Combustion Temperature

Heat liberated by the combustion reaction of a hydrocarbon fuel with air is the difference between the total enthalpy of the products and the total enthalpy of the reactants. This is called *heat of reaction, heat of combustion*, or *enthalpy of reaction* and is given by

$$Q = \sum_{\text{PROD}} N_i h_i - \sum_{\text{REACT}} N_i h_i \tag{4-5}$$

where

N_i = number of moles of component i

$h_i = (h_f^o)_i + \Delta h_i$

h_f^o = enthalpy of formation, the enthalpy needed to form one mole of that component at standard conditions of 25°C and 1 atm

Δh_i = change of enthalpy from standard temperature for component i

Q will be negative, meaning that heat is given up by the reacting gases. Values of h_f^o and Δh are molar-specific quantities and can be found in most thermodynamic textbooks.

Table A-2 gives *heating values* for a number of fuels. Heating value Q_{HV} is the negative of the heat of reaction for one unit of fuel, and thus is a positive number. It is calculated assuming both the reactants and the products are at 25°C. Care must be used when using heating values, which almost always are given in mass units (kJ/kg), whereas heats of reaction are obtained using molar quantities as in Eq. (4-5). Two values of heating value are given in the table; *higher heating value* is used when water in the exhaust products is in the liquid state, and *lower heating value* is used when water in the products is vapor. The difference is the heat of vaporization of the water:

$$Q_{HHV} = Q_{LHV} + \Delta h_{vap} \tag{4-6}$$

Higher heating value is usually listed on fuel containers. For engine analysis, lower heating value is the logical value to use. All energy exchange in the combustion chamber occurs at high temperature, and only somewhere down the exhaust process, where it can no longer affect engine operation, does the exhaust gas get cooled to the dew point temperature so that some of the water vapor condenses to water. Heat into the engine for conversion to output work can be given as

$$Q_{in} = \eta_c m_f Q_{LHV} \tag{4-7}$$

where

n_c = combustion efficiency

m_f = mass of fuel

An estimation of the maximum possible temperature that could be reached in an IC engine can be obtained by calculating the *adiabatic flame temperature* of the input air–fuel mixture. This is done by using Eq. (4-5) and setting Q = 0, yielding

$$\sum_{\text{PROD}} N_i h_i = \sum_{\text{REACT}} N_i h_i \tag{4-8}$$

Assuming that inlet conditions of the reactants are known, it is necessary to find the temperature of the products such that this equation will be satisfied. This is the adiabatic flame temperature.

Adiabatic flame temperature is the ideal theoretical maximum temperature that can be obtained for a given fuel and air mixture. The actual peak temperature in an engine cycle will be several hundred degrees less than this. There is some heat loss even in the very short time of one cycle, combustion efficiency is less than 100% so a small amount of fuel does not get burned, and some components dissociate at the high engine temperatures. All these factors contribute to making the actual peak engine temperature somewhat less than adiabatic flame temperature.

Example Problem 4-4

An SI engine operating on stoichiometric propane fuel burns 0.00005 kg of fuel in each cylinder during each cycle with a combustion efficiency of 95%. When combustion starts at the end of compression, the temperature and pressure in the cylinder are T = 700 K and P = 2000 kPa. Exhaust temperature leaving the cylinder after combustion is T_{ex} = 1200 K.
Calculate:

1. combustion heat input into each cylinder during each cycle using Eq. (4-5)
2. heat input using Eq. (4-7)

(1) The combustion reaction equation:

$$C_3H_8 + 5\,O_2 + 5(3.76)\,N_2 \rightarrow 3\,CO_2 + 4\,H_2O + 5(3.76)\,N_2$$

Equation (4-5) is used to find heat input.

Enthalpy of reactants at T = 700 K before combustion, using data from ref. [90], is

$$\begin{aligned} H_{\text{react}} &= 1[(-103{,}850) + (29{,}771)]_{\text{C3H8}} + 5[(0) + (12{,}499)]_{\text{O2}} \\ &\quad + 5(3.76)[(0) + (11{,}937)]_{\text{N2}} \\ &= +212{,}832 \text{ KJ/kgmole} \end{aligned}$$

Enthalpy of products at T_{ex} = 1200 K after combustion is

$$\begin{aligned} H_{\text{prod}} &= 3[(-393{,}522) + (44{,}473)]_{\text{CO2}} + 4[(-241{,}826) + (34{,}506)]_{\text{H2O}} \\ &\quad + 5(3.76)[(0) + (28{,}109)]_{\text{N2}} \\ &= -1{,}347{,}978 \text{ KJ/kgmole} \end{aligned}$$

Equation (4-5),

$$\begin{aligned} Q_{\text{in}} &= \sum H_{\text{prod}} - \sum H_{\text{react}} \\ &= \{[(-1{,}347{,}978 \text{ KJ/kgmole}) - (212{,}832)]/(44.097 \text{ kg/kgmole})\}(0.00005 \text{ kg}) \\ &= \underline{1.770 \text{ kJ}} \end{aligned}$$

(2) Equation 4-7 is used to find the heat input for one cylinder during one cycle:

$$Q_{\text{in}} = \eta_c m_f Q_{\text{LHV}} = (0.95)(0.00005 \text{ kg})(46{,}190 \text{ kJ/kg}) = \underline{2.194 \text{ kJ}}$$

The solution (2.194 kJ) given by Eq. (4-7) is consistent with air-standard analysis, while the solution (1.770 kJ) given by Eq. (4-5) is closer to the actual heat input. This is one reason for the difference in the actual indicated thermal efficiency of a cycle and the thermal efficiency obtained by air-standard Otto cycle analysis given in Eq. (3-32):

$$(\eta_t)_{\text{actual}} \approx 0.85\,(\eta_t)_{\text{OTTO}} \tag{3-32}$$

Example Problem 4-5

Find the adiabatic flame temperature of isooctane burned with an equivalence ratio of 0.833 in dry air as in Example Problem 4-1. It can be assumed that the reactants are at a temperature of 427°C (700 K) after the compression stroke.

From Example Problem 4-1, we have

$$C_8H_{18} + 15\,O_2 + 15(3.76)\,N_2 \rightarrow 8\,CO_2 + 9\,H_2O + 2.5\,O_2 + 15(3.76)\,N_2$$

Equations (4-5) and (4-8) are used for adiabatic combustion. Enthalpy values can be obtained from most thermodynamic textbooks. The values used here are from [90]. The calculation is as follows:

$$\sum_{\text{PROD}} N_i(h_f^o + \Delta h)_i = \sum_{\text{REACT}} N_i(h_f^o + \Delta h)_i$$

$$8[(-393{,}522) + \Delta h_{CO_2}] + 9[(-241{,}826) + \Delta h_{H_2O}] + 2.5[0 + \Delta h_{O_2}] + 15(3.76)[0 + \Delta h_{N_2}]$$
$$= [(-259{,}280) + (73{,}473)] + 15[0 + (12{,}499)] + 15(3.76)[0 + (11{,}937)]$$

This can be simplified to

$$8\Delta h_{CO_2} + 9\Delta h_{H_2O} + 2.5\Delta h_{O_2} + 56.4\Delta h_{N_2} = 5{,}999{,}535$$

By trial and error, find the temperature that satisfies this equation (try $T = 2400$ K):

$$8(115{,}779) + 9(93{,}741) + 2.5(74{,}453) + 56.4(70{,}640) = 5{,}940{,}130$$

This is too low, so try $T = 2600$ K:

$$8(128{,}074) + 9(104{,}520) + 2.5(82{,}225) + 56.4(77{,}963) = 6{,}567{,}948$$

This is too high, so the adiabatic flame temperature is found by interpolation:

$$\underline{T_{max} = 2419\text{ K} = 2146°\text{C}}$$

Engine Exhaust Analysis

It is common practice to analyze the exhaust of an IC engine. The control system of a modern *smart* automobile engine includes sensors that continuously monitor the exhaust leaving the engine. These sensors determine the chemical composition of the hot exhaust by various chemical, electronic, and thermal methods. This information, along with information from other sensors, is used by the engine management system (EMS) to regulate the operation of the engine by controlling the air–fuel ratio, ignition timing, inlet tuning, valve timing, etc.

Repair shops and highway check stations also routinely analyze automobile exhaust to determine operating conditions or emissions. This is done by taking a sample of the exhaust gases and running it through an external analyzer. When this is done, there is a high probability that the exhaust gas will cool below its dew point temperature before it is fully analyzed, and the condensing water will change the composition of the exhaust. To compensate for this, a *dry analysis* can be performed by first removing all water vapor from the exhaust, usually by some thermo-chemical means.

Example Problem 4-6

The four-cylinder engine of a light truck owned by a utility company has been converted to run on propane fuel. A dry analysis of the engine exhaust gives the following volumetric percentages:

$$\begin{array}{ll} CO_2 & 4.90\% \\ CO & 9.79\% \\ O_2 & 2.45\% \end{array}$$

Calculate the equivalence ratio at which the engine is operating. The three components identified sum up to 4.90 + 9.79 + 2.45 = 17.14% of the total, which means that the remaining gas (nitrogen) accounts for 82.86% of the total. Volume percent equals molar percent, so if an unknown amount of fuel is burned with an unknown amount of air, the resulting reaction is

$$xC_3H_8 + y\,O_2 + y(3.76)\,N_2 \rightarrow 4.90\,CO_2 + 9.79\,CO + 2.45\,O_2 + 82.86\,N_2 + z\,H_2O$$

where z = number of moles of water vapor removed before dry analysis.

Conservation of nitrogen during reaction gives

$$y(3.76) = 82.86 \quad \text{or} \quad y = 22.037$$

Conservation of carbon yields

$$3x = 4.90 + 9.79 \quad \text{or} \quad x = 4.897$$

Conservation of hydrogen gives

$$8x = 8(4.897) = 2z \quad \text{or} \quad z = 19.588$$

The reaction is

$$4.90\,C_3H_8 + 22.037\,O_2 + 22.037(3.76)\,N_2 \rightarrow$$
$$4.90\,CO_2 + 9.79\,CO + 2.45\,O_2 + 82.86\,N_2 + 19.588\,H_2O$$

Dividing by 4.90 yields

$$C_3H_8 + 4.50\,O_2 + 4.50(3.76)\,N_2 \rightarrow CO_2 + 2\,CO + 0.50\,O_2 + 16.92\,N_2 + 4\,H_2O$$

Actual air–fuel ratio is

$$AF_{act} = m_a/m_f = [(4.50)(4.76)(29)]/[(1)(44)] = 14.12$$

Stoichiometric combustion reaction is

$$C_3H_8 + 5\,O_2 + 5(3.76)\,N_2 \rightarrow 3\,CO_2 + 4\,H_2O + 5(3.76)\,N_2$$

Stoichiometric air–fuel ratio is

$$AF_{stoich} = m_a/m_f = [(5)(4.76)(29)]/[(1)(44)] = 15.69$$

Equivalence ratio, using Eq. (4-2), is

$$\phi = (AF)_{stoich}/(AF)_{act} = 15.69/14.12 = \underline{1.11}$$

4.2 HYDROCARBON FUELS—GASOLINE

The main fuel for SI engines is gasoline, which is a mixture of many hydrocarbon components and is manufactured from crude petroleum. Crude oil was first discovered in Pennsylvania in 1859, and the fuel product line generated from it developed along with the development of the IC engine. Crude oil is made up almost entirely of carbon and hydrogen with some traces of other species. It varies from 83% to 87% carbon and 11% to 14% hydrogen by weight. The carbon and hydrogen can combine in many ways and form many different molecular compounds. One test of a crude oil sample identified over 25,000 different hydrocarbon components [93].

The crude oil mixture that is taken from the ground is separated into component products by *cracking* and/or distillation using thermal or catalytic methods at an oil refinery. Cracking is the process of breaking large molecular components into more useful components of smaller molecular weight. Preferential distillation is used to separate the mixtures into single components or smaller ranges of components. Generally, the larger the molecular weight of a component, the higher is its boiling temperature. Low boiling temperature components (smaller molecular weights) are used for solvents and fuels (gasoline), while high boiling temperature components with their large molecular weights are used for tar and asphalt or returned to the refining process for further cracking. The component mixture of the refining process is used for many products, including:

automobile gasoline
diesel fuel
aircraft gasoline
jet fuel
home heating fuel
industrial heating fuel
natural gas
lubrication oil
asphalt
alcohol
rubber
paint
plastics
explosives

The availability and cost of gasoline fuel, then, is a result of a market competition with many other products. This becomes more critical with the depletion of the earth's crude oil reserves, which looms on the horizon.

Crude oil obtained from different parts of the world contains different amounts and combinations of hydrocarbon species. In the United States, two general classifications are identified: Pennsylvania crude and western crude. Pennsylvania crude has a high concentration of paraffins with little or no asphalt, while western crude has an asphalt base with little paraffin. The crude oil from some petroleum fields in the Mideast is made up of component mixtures that could be used immediately for IC engine fuel with little or no refining.

Figure 4-2 shows a temperature–vaporization curve for a typical gasoline mixture. The various components of different molecular weights will vaporize at different temperatures, small molecular weights boiling at low temperature and larger molecular weights at higher temperature. This makes a very desirable fuel. A small percentage of components that vaporize (boil) at low temperature is needed to assure the starting of a cold engine; fuel must vaporize before it can burn. However, too much of this *front-end volatility* can cause problems when the fuel vaporizes too quickly. Volumetric efficiency of the engine will be reduced if fuel vapor replaces air too early in the intake system. Another serious problem this can cause is *vapor lock*, which occurs when fuel vaporizes in the fuel supply lines or in the carburetor in the hot engine compartment. When this happens, the supply of fuel is cut off and the engine stops. A large percent of the fuel should be vaporized at the normal intake system temperature during the short time of the intake process. To maximize volumetric efficiency, some of the fuel should not vaporize until late into the compression stroke and even into the start of combustion. This is why some high-molecular-weight components are included in gasoline mixtures. If too much of this *high-end volatility* is included in the gasoline, however, some of the fuel never gets vaporized and ends up as exhaust pollution or condenses on the cylinder walls and dilutes the lubricating oil.

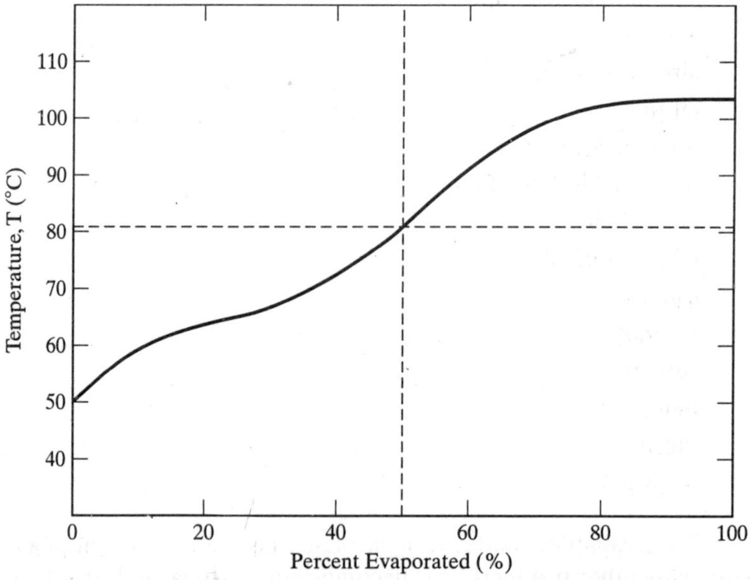

FIGURE 4-2

Temperature–vaporization curve for a typical gasoline mixture. The components that vaporize at low temperatures are called *front-end volatility* components and are useful for starting a cold engine. Those components which vaporize at the highest temperatures are called *high-end volatility* components and increase engine volumetric efficiency. This gasoline could be classified as 57-81-103°C. Fifty percent of the gasoline would be vaporized at 81°C.

One way that is sometimes used to describe a gasoline is to use three temperatures: the temperature at which 10% is vaporized, at which 50% is vaporized, and at which 90% is vaporized. The gasoline in Fig. 4-2 could therefore be classified as 57-81-103°C.

If different commercial brands of gasoline are compared, there is found to be little difference in the volatility curves for a given season and location in the country. There is usually about a 5°C shift down in temperature on the vaporization curve for winter gasoline compared with summer.

If gasoline is approximated as a single-component hydrocarbon fuel, it would have a molecular structure of about C_8H_{15} and a corresponding molecular weight of 111. These are the values that will be used in this textbook. Sometimes, gasoline is approximated by the real hydrocarbon component isooctane C_8H_{18}, which best matches its component structure and thermodynamic properties. Table A-2 lists properties of gasoline, isooctane, and some other common fuels.

4.3 SOME COMMON HYDROCARBON COMPONENTS

Carbon atoms form four bonds in molecular structures, while hydrogen has one bond. A *saturated* hydrocarbon molecule will have no double or triple carbon-to-carbon bonds and will have a maximum number of hydrogen atoms. An *unsaturated* molecule will have double or triple carbon-to-carbon bonds.

A number of different *families* of hydrocarbon molecules have been identified; a few of the more common ones are described.

Paraffins

The paraffin family (sometimes called alkanes) are chain molecules with a carbon-hydrogen combination of C_nH_{2n+2}, n being any number. The simplest member of this family, and the simplest of all stable hydrocarbon molecules is methane (CH_4), which is the main component of natural gas. It can be pictured as:

methane (CH_4)

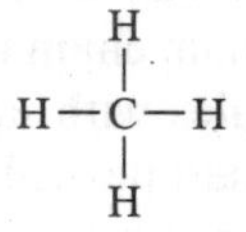

Other species of this family include:

propane (C_3H_8)

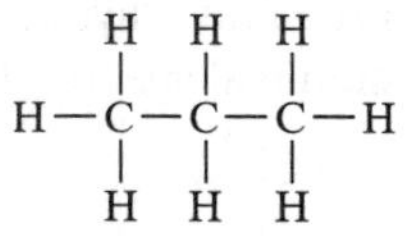

butane (C_4H_{10})

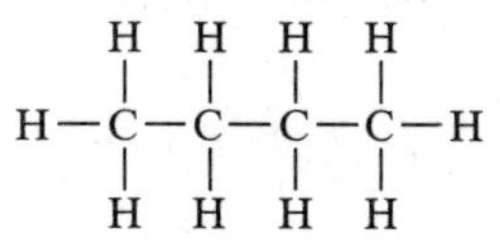

TABLE 4-2 Prefixes for Hydrocarbon Fuel Components

Number of Carbon Atoms in Main Chain or Ring	Prefix
1	meth
2	eth
3	prop
4	but
5	pent
6	hex
7	hept
8	oct
9	non
10	dec
11	undec
12	dodec

Table 4-2 gives the prefixes used to identify paraffins and other hydrocarbon families according to the number of carbon atoms in the main molecular structure. The paraffin family, then, uses the suffix ane—thus methane, propane, etc.

Sometimes the chains in the molecule are branched, and other molecular structures are obtained with the same number of carbon and hydrogen atoms. One such *isomer* is isobutane, which has the same chemical formula as butane (C_4H_{10}) but has a different structure:

isobutane (C_4H_{10})

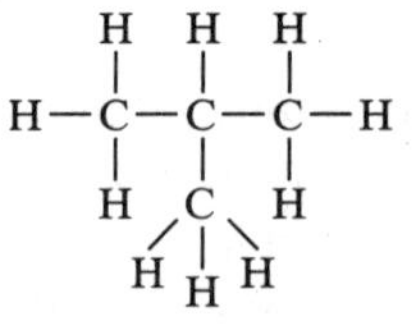

Isobutane can also be called methylpropane—propane because it has three carbon atoms in the main chain and one methyl *radical*, CH_3, replacing one of the hydrogen atoms. Molecules with no branches in their chain are sometimes called normal; thus butane is sometimes called normal butane or n-butane. Even though isobutane and n-butane have the same chemical formula, C_4H_{10}, and almost identical molecular weights, they have different thermal and physical properties. This is true for any two chemical species that have different molecular structures, even if they have the same chemical formula.

There are many ways chemical chains can be branched, giving a very large number of possible chemical species. Isooctane (C_8H_{18}) has the following molecular structure:

isooctane (C_8H_{18})

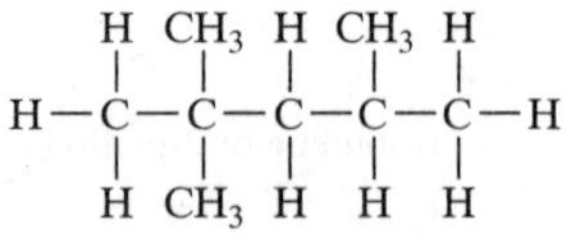

It can also be called 2,2,4-trimethylpentane, pentane because it has five carbon atoms in the main chain, trimethyl because it has three methyl radicals (CH_3), and 2,2,4 because the three radicals are on the second, second, and fourth carbon atoms in the chain. Note that 2,4,4 trimethylpentane would be the same molecule. Other examples of isomers are:

2-ethylpentane (C_7H_{16})

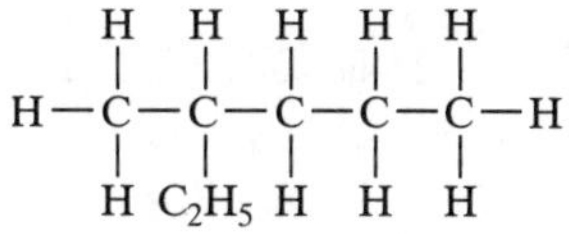

2-methyl-3-ethylhexane (C_9H_{20})

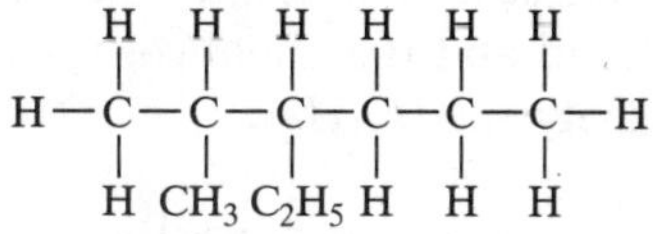

Olefins

The olefin family consists of chain molecules that contain one double carbon–carbon bond, and are therefore unsaturated. The prefixes from Table 4-2 are used with the suffix "ene." The chemical makeup is C_nH_{2n}. The following are examples of olefins:

ethene (C_2H_4)

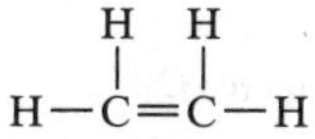

butene-1 (C_4H_8)

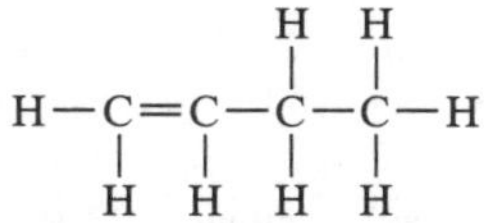

butene-2 (C_4H_8)

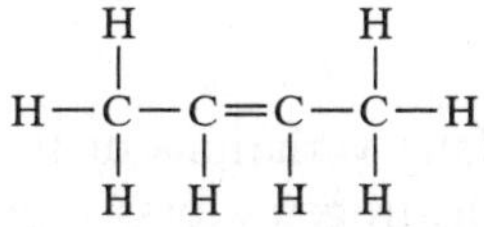

isobutene or 2-methylpropene (C_4H_8)

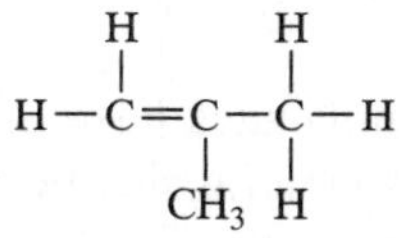

Diolefins

Diolefins are chain molecules similar to olefins, except that they have two double carbon–carbon bonds. These unsaturated compounds have the chemical formula C_nH_{2n-2} and use the suffix "diene."

2,5-heptadiene (C_7H_{12})

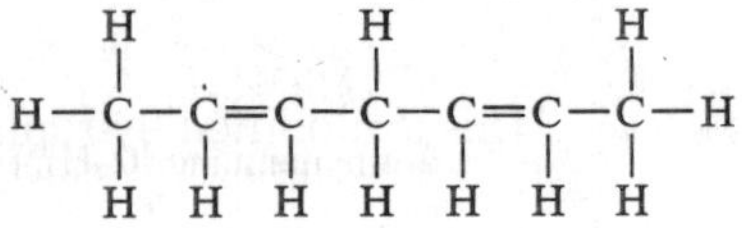

Acetylene

The acetylene family has unsaturated chain molecules with a triple carbon–carbon bond and the chemical formula C_nH_{2n-2}. The best known member of the family is acetylene (C_2H_2).

acetylene (C_2H_2) $H-C\equiv C-H$

Cycloparaffins

Cycloparaffins have unsaturated molecules with a single-bond ring and a chemical formula of C_nH_{2n}.

cyclobutane (C_4H_8)

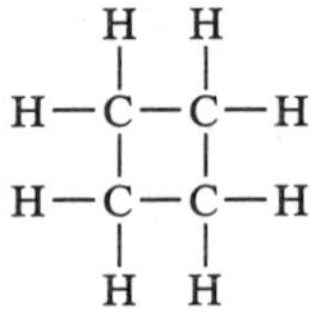

cyclopentane (C_5H_{10})

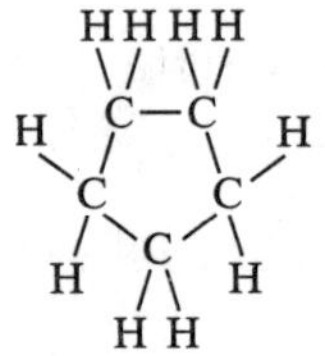

Many variations of these molecules are possible, with one or more of the attached hydrogen atoms replaced with various side radicals or chains. Cycloparaffins make good automobile gasoline components.

Aromatics

Aromatic molecules have an unsaturated ring structure with double carbon–carbon bonds and a general chemical formula of C_nH_{2n-6}. The basic molecule in this family is the benzene ring:

benzene (C_6H_6)

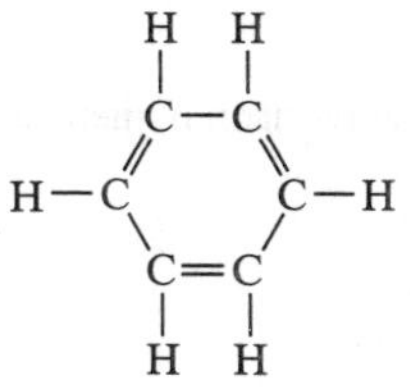

This is modified by replacing the hydrogen atoms with various groups:

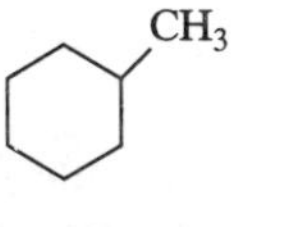

toluene (C_7H_8)

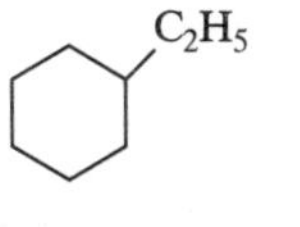

ethylbenzene (C_8H_{10})

When more than one hydrogen atom is replaced, many isomers are possible:

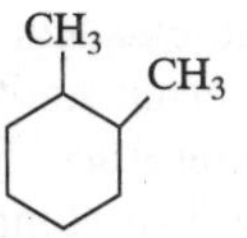

orthoxylene (C_8H_{10})

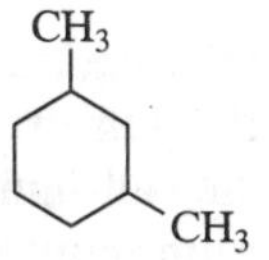

metaxylene (C_8H_{10})

paraxylene (C_8H_{10})

When more than one ring combine in a single large molecule, many additional species are possible:

$C_{10}H_8$

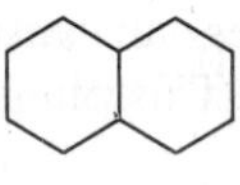

$C_{11}H_{10}$

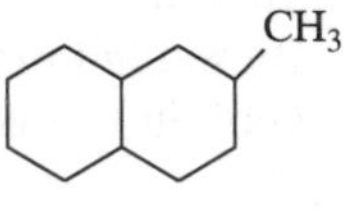

$C_{14}H_{10}$

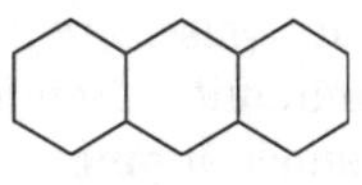

Aromatics generally make good gasoline fuel components, with some exceptions due to exhaust pollution. They have high densities in the liquid state and thus have high energy content per unit volume. Aromatics have high solvency characteristics, and care must be used in material selection for the fuel delivery system (e.g., they will dissolve or swell some gasket materials). Aromatics will dissolve a greater amount of water than some other hydrocarbons and thus can create fuel line freezing problems when the temperature is lowered and some of the water comes out of solution. Aromatics make poor CI engine fuel.

Alcohol

Alcohols are similar to paraffins with one of the hydrogen atoms replaced with the hydroxyl radical OH. The most common alcohols are:

methyl alcohol (methanol), (CH_3OH)

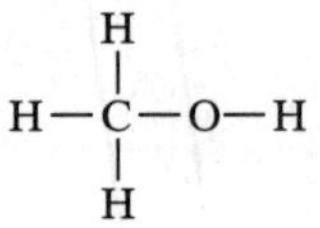

ethyl alcohol (ethanol), (C_2H_5OH)

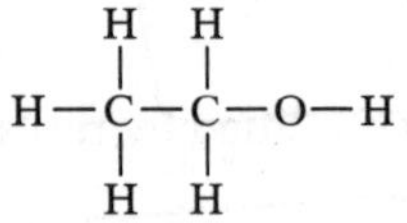

propyl alcohol (propanol), (C_3H_7OH)

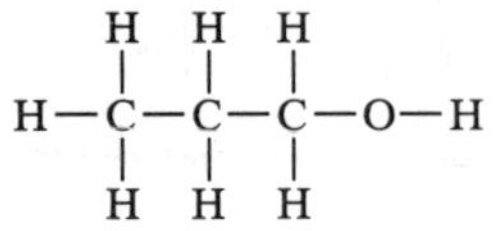

4.4 SELF-IGNITION AND OCTANE NUMBER

Self-Ignition Characteristics of Fuels

If the temperature of an air–fuel mixture is raised high enough, the mixture will self-ignite without the need of a spark plug or other external igniter. The temperature above which this occurs is called the **self-ignition temperature** (SIT). This is the basic principle of ignition in a compression ignition engine. The compression ratio is high enough that the temperature rises above SIT during the compression stroke. Self-ignition then occurs when fuel is injected into the combustion chamber. On the other hand, self-ignition (or preignition, or autoignition) is not desirable in an SI engine, where a spark plug is used to ignite the air–fuel at the proper time in the cycle. The compression ratios of gasoline-fueled SI engines are limited to about 11:1 to avoid self-ignition. When self-ignition does occur in an SI engine, higher than desirable pressure pulses are generated. These high pressure pulses can cause damage to the engine and quite often are in the audible frequency range. This phenomenon is often called **knock** or **ping**.

Figure 4-3 shows the basic process that takes place when self-ignition occurs. If a combustible air–fuel mixture is heated to a temperature less than SIT, no ignition will occur and the mixture will cool off. If the mixture is heated to a temperature above SIT, self-ignition will occur after a short time delay called **ignition delay** (ID). The higher the initial temperature rise above SIT, the shorter will be ID. The values for SIT and ID for a given air–fuel mixture are ambiguous, depending on many variables which include temperature, pressure, density, turbulence, swirl, fuel–air ratio, presence of inert gases, etc. [93].

Ignition delay is generally a very small fraction of a second. During this time, preignition reactions occur, including oxidation of some fuel components and even cracking of some large hydrocarbon components into smaller HC molecules. These preignition reactions raise the temperature at local spots, which then promotes additional reactions until, finally, the actual combustion reaction occurs.

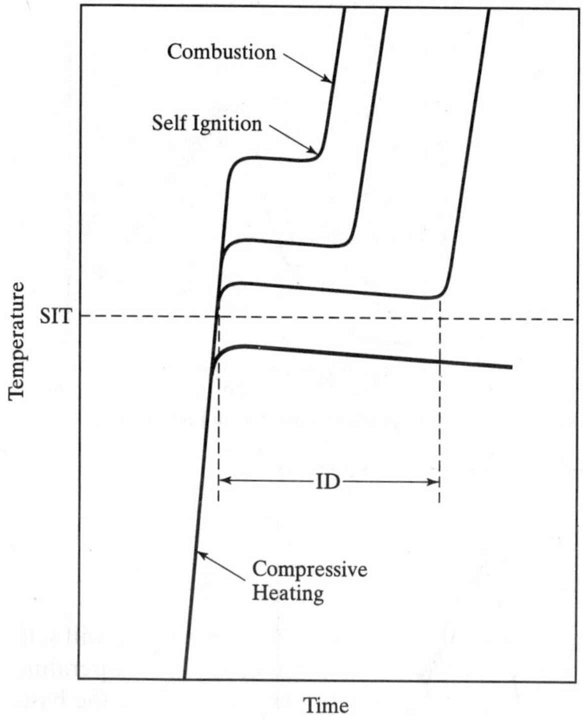

FIGURE 4-3

Self-ignition characteristics of fuels. If the temperature of a fuel is raised above the self-ignition temperature (SIT), the fuel will spontaneously ignite after a short ignition delay (ID) time. The higher above SIT the fuel is heated, the shorter will be the ID. Ignition delay is generally on the order of thousandths of a second. Adapted from [126].

Figure 4-4 shows the pressure–time history within a cylinder of a typical SI engine. With no self-ignition, the pressure force on the piston follows a smooth curve, resulting in smooth engine operation. When self-ignition does occur, pressure forces on the piston are not smooth and engine knock occurs.

For illustrative reasons, a combustion chamber can be visualized schematically as a long hollow tube, as shown in Fig. 4-5. Obviously, this is not the shape of a real engine combustion chamber, but it allows visualization of what happens during combustion. These ideas can then be extrapolated to real combustion engine shapes. Before combustion, the chamber is divided into four equal mass units, each occupying an equal volume. Combustion starts at the spark plug on the left side, and the flame front travels from left to right. As combustion occurs, the temperature of the burned gases is increased to a high value. This, in turn, raises the pressure of the burned gases and expands the volume of that mass as shown in Fig. 4-5(b). The unburned gases in front of the flame front are compressed by this higher pressure, and compressive heating raises the temperature of the gas. The temperature of the unburned gas is further raised by radiation heating from the flame, and this then raises the pressure even higher. Heat transfer by conduction and convection are not important during this process due to the very short time interval involved.

The flame front moving through the second mass of air–fuel does so at an accelerated rate because of the higher temperature and pressure, which increase the reaction rate. This, in turn, further compresses and heats the unburned gases in front of the

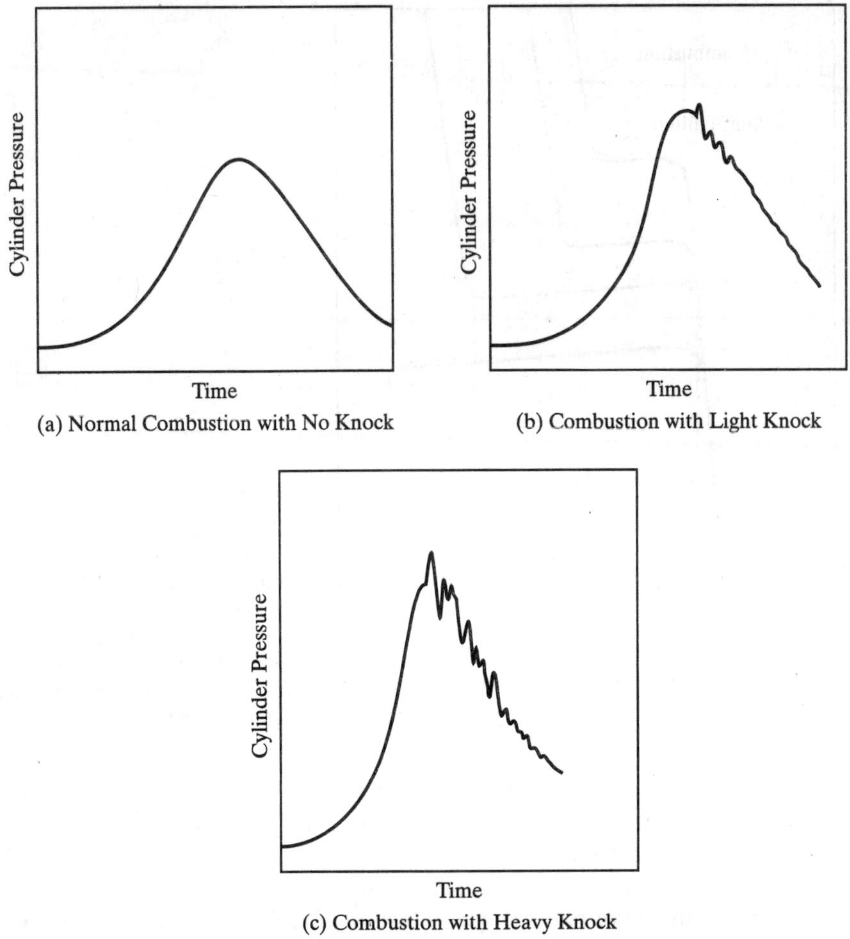

FIGURE 4-4

Cylinder pressure as a function of time in a typical SI engine combustion chamber showing **(a)** normal combustion, **(b)** combustion with light knock, and **(c)** combustion with heavy knock. Adapted from [33].

flame as shown in Fig. 4-5(c). In addition, the energy release in the combustion process further raises the temperature and pressure of the burned gases behind the flame front. This occurs both by compressive heating and radiation. Thus, the flame front continues its travel through an unburned mixture that is progressively higher in temperature and pressure. By the time the flame reaches the last portion of unburned gas, this gas is at a very high temperature and pressure. In this *end gas* near the end of the combustion process is where self-ignition and knock occur. To avoid knock, it is necessary for the flame to pass through and consume all unburned gases which have risen above self-ignition temperature before the ignition delay time elapses. This is done by a combination of fuel property control and design of combustion chamber geometry.

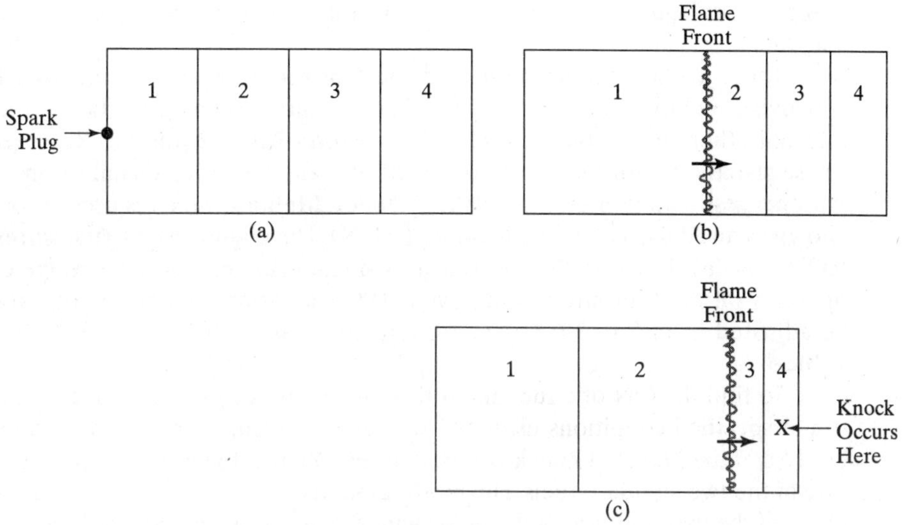

FIGURE 4-5

SI engine combustion chamber schematically visualized as a long hollow cylinder with the spark plug located at the left end. **(a)** Mass of air–fuel is equally distributed as spark plug is fired to start combustion. **(b)** As flame front moves across chamber, unburned mixture in front of flame is compressed into smaller volume. **(c)** Flame front continues to compress unburned mixture into smaller volume, which increases its temperature and pressure. If compression raises temperature of end gas above SIT, self-ignition and knock can occur.

At the end of the combustion process, the hottest region in the cylinder is near the spark plug where combustion was initiated. This region became hot at the start of combustion and then continued to increase in temperature due to compressive heating and radiation as the flame front passed through the rest of the combustion chamber.

By limiting the compression ratio in an SI engine, the temperature at the end of the compression stroke where combustion starts is limited. The reduced temperature at the start of combustion then reduces the temperature throughout the entire combustion process, and knock is avoided. On the other hand, a high compression ratio will result in a higher temperature at the start of combustion. This will cause all temperatures for the rest of the cycle to be higher. The higher temperature of the end gas will create a short ID time, and knock will occur.

Octane Number and Engine Knock

The fuel property that describes how well a fuel will or will not self-ignite is called the **octane number** or just **octane**. This is a numerical scale generated by comparing the self-ignition characteristics of the fuel to that of standard fuels in a specific test engine at specific operating conditions. The two standard reference fuels used are isooctane (2,2,4 trimethylpentane), which is given the octane number (ON) of 100, and n-heptane, which is given the ON of 0. The higher the octane number of a fuel, the less likely it will self-ignite. Engines with low compression ratios can use fuels with lower octane numbers, but high-compression engines must use high-octane fuel to avoid self-ignition and knock.

There are several different tests used for rating octane numbers, each of which will give a slightly different ON value. The two most common methods of rating gasoline and other automobile SI fuels are the **Motor Method** and the **Research Method**. These give the Motor Octane Number (MON) and Research Octane Number (RON). Another less common method is the **Aviation Method**, which is used for aircraft fuel and gives an Aviation Octane Number (AON). The engine used to measure MON and RON was developed in the 1930s. It is a single-cylinder, overhead valve engine that operates on the four-stroke Otto cycle. It has a variable compression ratio that can be adjusted from 3 to 30. Test conditions to measure MON and RON are given in Table 4-3.

To find the ON of a fuel, the following test procedure is used. The test engine is run at specified conditions using the fuel being tested. Compression ratio is adjusted until a standard level of knock is experienced. The test fuel is then replaced with a mixture of the two standard fuels. The intake system of the engine is designed such that the blend of the two standard fuels can be varied to any percent from all isooctane to all n-heptane. The blend of fuels is varied until the same knock characteristics are observed as with the test fuel. The percent of isooctane in the fuel blend is the ON given to the test fuel. For instance, a fuel that has the same knock characteristics as a blend of 87% isooctane and 13% n-heptane would have an ON of 87.

On the fuel pumps at an automobile service station is found the **anti-knock index**:

$$AKI = (MON + RON)/2 \tag{4-9}$$

This is often referred to as the octane number of the fuel.

Because the test engine has a combustion chamber designed in the 1930s and because the tests are conducted at low speed, the octane number obtained will not always totally correlate with operation in modern high-speed engines. Octane numbers should not be taken as absolute in predicting knock characteristics for a given engine. If there are two engines with the same compression ratio, but with different combustion chamber geometries, one may not knock using a given fuel while the other may experience serious knock problems with the same fuel.

TABLE 4-3 Test Conditions for Octane Number Measurement

	RON	MON
Engine Speed (RPM):	600	900
Inlet Air Temperature (°C):	52 (125°F)	149 (300°F)
Coolant Temperature (°C):	100 (212°F)	100
Oil Temperature (°C):	57 (135°F)	57
Ignition Timing:	13° bTDC	19°–26° bTDC
Spark Plug Gap (mm):	0.508 (0.020 in.)	0.508
Inlet Air Pressure:	atmospheric pressure	
Air–Fuel Ratio:	adjusted for maximum knock	
Compression Ratio:	adjusted to get standard knock	

Adapted from [58].

Operating conditions used to measure MON are more severe than those used to measure RON. Some fuels, therefore, will have a RON greater than MON (see Table A-2). The difference between these is called **fuel sensitivity**:

$$FS = RON - MON \tag{4-10}$$

Fuel sensitivity is a good measure of how sensitive knock characteristics of a fuel will be to engine geometry. A low FS number will usually mean that knock characteristics of that fuel are insensitive to engine geometry. FS numbers generally range from 0 to 10.

For measuring octane numbers above 100, fuel additives are mixed with isooctane and other standard points are established. A common additive used for many years to raise the octane number of a fuel was tetraethyl lead (TEL).

Common octane numbers (anti-knock index) for gasoline fuels used in automobiles range from 87 to 95, with higher values available for special high-performance and racing engines. Reciprocating SI aircraft engines usually use low-lead fuels with octane numbers in the 85 to 100 range.

The octane number of a fuel depends on a number of variables, some of which are not fully understood. Things that affect ON are combustion chamber geometry, turbulence, swirl, temperature, inert gases, etc. This can be seen by the difference in RON and MON for some fuels, brought about by different operating characteristics of the test engine. Other fuels will have identical RON and MON. The higher the flame speed in an air–fuel mixture, the higher the octane number. This is because, with a higher flame speed, the air–fuel mixture that is heated above SIT will be consumed during ignition delay time, and knock will be avoided.

If several fuels of known ON are mixed, a good approximation of the mixture octane number is

$$ON_{mix} = (\%\text{ of A})(ON_A) + (\%\text{ of B})(ON_B) + (\%\text{ of C})(ON_c) \tag{4-11}$$

where % = mass percent.

Early crude fuels for automobiles had very low octane numbers that required low compression ratios. This was not a serious handicap to early engines, which needed low compression ratios because of the technology and materials of that day. High compression ratios generate higher pressures and forces that could not be tolerated in early engines.

Fuel components with long chain molecules generally have lower octane numbers: The longer the chain, the lower is the ON. Components with more side chains have higher octane numbers. For a compound with a given number of carbon and hydrogen atoms, the more these atoms are combined in side chains and not in a few long chains, the higher will be the octane number. Fuel components with ring molecules have higher octane numbers. Alcohols have high octane numbers because of their high flame speeds.

There are a number of gasoline additives that are used to raise octane number (e.g., hydrotreated aliphatics and methylcyclopentadienyl manganese tricarbonyl (MMT)). For many years, the standard additive was tetraethyl lead TEL, $(C_2H_5)_4Pb$. A few milliliters of TEL in several liters of gasoline could raise the ON several points in a very predictable manner (Fig. 4-6).

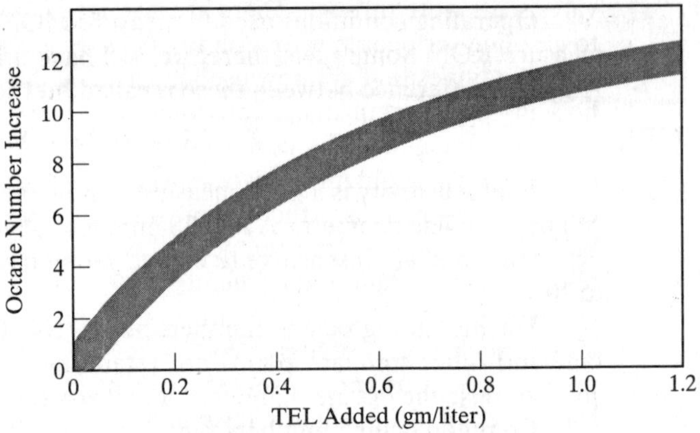

FIGURE 4-6

Octane number increase as a function of TEL added to gasoline; varies somewhat with gasoline blend. Octane number is MON, RON, or AKI. Adapted from [58].

When TEL was first used, it was mixed with the gasoline at the local fuel service station. The process was to pour liquid TEL into the fuel tank and then add gasoline, which would mix with the TEL due to the natural turbulence of the pouring. This was not a safe way of handling TEL, which has toxic vapors and is even harmful in contact with human skin. Soon after this, TEL was blended into the gasoline at the refineries, which made it much safer to handle. However, this created a need for additional storage tanks and gasoline pumps at the service station. High-octane and low-octane fuels were now two different gasolines and could not be blended at the service station from a common gasoline base.

Figure 2-5 shows how the compression ratios of automobile engines increased after the introduction of TEL in the 1920s.

The major problem with TEL is the lead that ends up in the engine exhaust. Lead is a very toxic engine emission. For many years, the lead emissions problem was not considered serious just due to the lower numbers of automobiles. However, by the late 1940s and in the 1950s, the pollution problem of automobile exhaust was recognized, first in the Los Angeles basin area of California. The reason that awareness of the problem started there was a combination of a high density of automobiles and the unique weather conditions in the basin. In the 1960s and 1970s, as the number of automobiles proliferated both in the United States and in the rest of the world, it was recognized that lead could no longer be tolerated in gasoline fuel. In the 1970s low-lead and no-lead gasolines were being marketed, and by the early 1990s lead in fuel was unlawful in the United States for most vehicles.

The elimination of lead from gasoline created a problem for older automobiles and other older engines. When TEL is consumed in the combustion process in the cylinder of an engine, one of the results is lead being deposited on the walls of the combustion chamber. This lead reacts with the hot walls and forms a very hard surface. When older engines were manufactured, softer steels were used in the cylinder walls, heads, and valve seats. It was then expected that when these engines were operated using leaded fuel, these parts would become heat treated and hardened during use. Now, when these engines are operated using unleaded fuel, they do not experience this hardening treatment with possible long-term wear problems. The wear that occurs on valve seats is the most critical, and there have been catastrophic engine failures when valve seats wore through. There are now lead substitutes available which can be added to gasoline for people who wish to operate older automobiles for extended lengths of time. Additives that are now used in gasoline to raise the octane number include alcohols and organomanganese compounds.

As an engine ages, deposits build up on the combustion chamber walls. This increases knock problems in two ways. First, it makes the clearance volume smaller and consequently increases the compression ratio. Second, the deposits act as a thermal barrier and increase the temperatures throughout the engine cycle, including peak temperature. Octane requirements can go up as an engine ages, with an average increase needed of about three or four for older engines.

Knock usually occurs at WOT when the engine is loaded (e.g., fast startup or going up a hill). Serious knock problems can be reduced by retarding the ignition spark and starting combustion slightly later in the compression stroke. Many modern smart engines have knock detection to help determine optimum operating conditions. These are usually transducers that detect knock pressure pulses. Some spark plugs are equipped with pressure transducers for this purpose. The human ear is a good knock detector.

Engine knock can also be caused by surface ignition. If any local hot spot exists on the combustion chamber wall, this can ignite the air–fuel mixture and cause the same kind of loss of cycle combustion control. This can occur on surface deposits of older engines, with hot exhaust valves, on hot spark plug electrodes, or on any sharp corner in the combustion chamber. The worst kind of surface ignition is preignition, which starts combustion too early in the cycle. This causes the engine to run hotter, which causes more surface hot spots, which causes more surface ignition. In extreme surface ignition problems, when the combustion chamber walls are too hot, run-on will occur. This means the engine will continue to run after the spark ignition has been turned off.

Example Problem 4-7

A gasoline-type fuel is generated by blending 15% by weight of butene-1, 70% triptane, and 15% isodecane.
Determine:

1. the anti-knock index
2. the anti-knock index if 0.4 gm of TEL is added per liter of fuel
3. if the fuel is sensitive to combustion chamber geometry

(1) Using data from Table A-2 and Eq. (4-11) gives MON and RON for the mixture:
Research:

$$\text{RON} = (0.15)(99) + (0.70)(112) + (0.15)(113) = 110.2$$

Motor:

$$\text{MON} = (0.15)(80) + (0.70)(101) + (0.15)(92) = 96.5$$

Anti-knock index is obtained from Eq. (4-9):

$$\text{AKI} = (\text{MON} + \text{RON})/2 = (96.5 + 110.2)/2 = \underline{103}$$

(2) Using Fig. 4-6 gives an increase of about 7 for 0.4 gm/L:

$$\underline{(\text{AKI})_{\text{with TEL}} = 110}$$

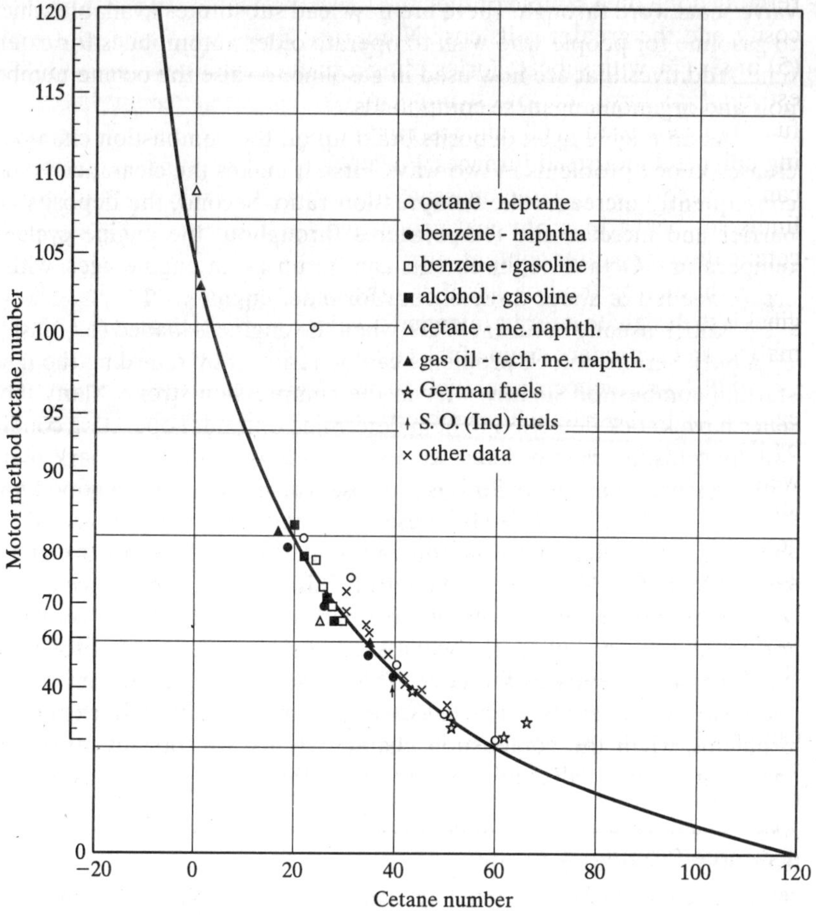

FIGURE 4-7

Relationship of cetane number (CN) and Motor Octane Number (MON) for several fuels. From "The Internal Combustion Engine in Theory and Practice," by C. F. Taylor, MIT Press Publishers, copyright MIT Press, [120].

(3) Fuel sensitivity is obtained from Eq. (4-10):

$$FS = RON - MON = 110.2 - 96.5 = 13.7$$

This is a fairly large number, which suggests that knock characteristics of this fuel are very sensitive to combustion chamber geometry.

4.5 DIESEL FUEL

Diesel fuel (diesel oil, fuel oil) is obtainable over a large range of molecular weights and physical properties. Various methods are used to classify it, some using numerical scales and some designating it for various uses. Generally speaking, the greater the refining done on a sample of fuel, the lower is its molecular weight, the lower is its viscosity, and the greater is its cost. Numerical scales usually range from one (1) to five (5) or six (6), with subcategories using alphabetical letters (e.g., A1, 2D, etc). The lowest numbers have the lowest molecular weights and lowest viscosity. These are the fuels typically used in CI engines. Higher numbered fuels are used in residential heating units and industrial furnaces. Fuels with the largest numbers are very viscous and can be used only in large, massive heating units. Each classification has acceptable limits set on various physical properties, such as viscosity, flash point, pour point, cetane number, sulfur content, etc.

Another method of classifying diesel fuel to be used in internal combustion engines is to designate it for its intended use. These designations include bus, truck, railroad, marine, and stationary engine fuel, going from lowest molecular weight to highest.

For convenience, diesel fuels for IC engines can be divided into two extreme categories. Light diesel fuel has a molecular weight of about 170 and can be approximated by the chemical formula $C_{12.3}H_{22.2}$ (see Table A-2). Heavy diesel fuel has a molecular weight of about 200 and can be approximated as $C_{14.6}H_{24.8}$. Most diesel fuel used in engines will fit in this range. Light diesel fuel will be less viscous and easier to pump, will generally inject into smaller droplets, and will be more costly. Heavy diesel fuel can generally be used in larger engines with higher injection pressures and heated intake systems. Often an automobile or light truck can use a less costly heavier fuel in the summer, but must change to a lighter, less viscous fuel in cold weather because of cold starting and fuel line pumping problems.

One environmental problem that occurs with the combustion of diesel fuel is caused by the sulfur contained in the fuel. The sulfur ends up in the exhaust products where it combines with water vapor to form acid. Because of this, the allowable amount of sulfur in diesel fuel is continuously being lowered by emission law standards. In mid-2006 the allowable amount of sulfur in diesel fuel in the United States is being lowered from 500 ppm to 15 ppm. This will raise the cost of fuel for the consumer by about ten cents per gallon [174].

Cetane Number

In a compression ignition engine, self-ignition of the air–fuel mixture is a necessity. The correct fuel must be chosen which will self-ignite at the precise proper time in the engine cycle. It is therefore necessary to have knowledge and control of the ignition delay time of the fuel. The property that quantifies this is called the **cetane number**. The larger the cetane number, the shorter is the ID and the quicker the fuel will self-ignite in the combustion chamber environment. A low cetane number means the fuel will have a long ID.

Like octane number rating, cetane numbers are established by comparing the test fuel to two standard reference fuels. The fuel component n-cetane (hexadecane), $C_{16}H_{34}$, is given the cetane number value of 100, while heptamethylnonane (HMN), $C_{12}H_{34}$, is given the value of 15. The cetane number (CN) of other fuels is then obtained by comparing the ID of that fuel to the ID of a mixture blend of the two reference fuels with

$$\text{CN of fuel} = (\text{percent of n-cetane}) + (0.15)(\text{percent of HMN}) \tag{4-12}$$

A special CI test engine is used which has the capability of having the compression ratio changed as it operates. Fuel being rated is injected into the engine cylinder late in the compression stroke at 13° bTDC. The compression ratio is then varied until combustion starts at TDC, giving an ID of 13° of engine rotation. Without changing the compression ratio, the test fuel is replaced with a blend of the two reference fuels. Using two fuel tanks and two flow controls, the blend of the fuels is varied until combustion is again obtained at TDC, an ID of 13°.

The difficulty of this method, in addition to requiring a costly test engine, is to be able to recognize the precise moment when combustion starts. The very slow rise in pressure at the start of combustion is very difficult to detect.

The normal cetane number range for vehicle fuel is about 40 to 60. For a given engine injection timing and rate, if the cetane number of the fuel is low, the ignition delay will be too long. When this occurs, more fuel than desirable will be injected into the cylinder before the first fuel particles ignite, causing a very large, fast pressure rise at the start of combustion. This results in low thermal efficiency and a rough-running engine. Long ignition delay of fuels with cetane numbers below 40 results in a very rich fuel–air mixture in the cylinder when ignition finally occurs. This results in unacceptable levels of exhaust smoke, and these fuels are illegal by many emission laws. If the CN of the fuel is high, combustion will start too soon in the cycle. Pressure will rise before TDC and more work will be required in the compression stroke.

The cetane number of a fuel can be raised with certain additives which include nitrates and nitrites. There is a strong inverse correlation between the cetane number of a fuel and its octane number (Fig. 4-7).

Because of the cost and difficulty of measuring the CN of a fuel, several empirical approximation methods using physical properties of the fuel have been developed. One such approximation is the **cetane index** [40], given by

$$\begin{aligned} \mathrm{CI} &= -420.34 + 0.016\,\mathrm{G}^2 + 0.192\,\mathrm{G}(\log_{10}T_{\mathrm{mp}}) \\ &\quad + 65.01(\log_{10}T_{\mathrm{mp}})^2 - 0.0001809\,T_{\mathrm{mp}}^2 \end{aligned} \tag{4-13}$$

where

$G = (141.5/S_g) - 131.5$
S_g = specific gravity
T_{mp} = midpoint boiling temperature in°F

The following equation, given in reference [183], is a semiempirical equation that predicts the ignition delay as a function of cetane number and other operating parameters:

$$\begin{aligned} \mathrm{ID(ca)} &= (0.36 + 0.22\,\overline{U}_p)\exp\{E_A[(1/R_uT_ir_c^{\,k-1}) \\ &\quad - (1/17{,}190)][(21.2)/(P_ir_c^{\,k} - 12.4)]^{0.63}\} \end{aligned} \tag{4-14}$$

where

ID(ca) = ignition delay in crank angle degrees
$E_A = (618{,}840)/(\mathrm{CN} + 25)$ = activation energy

CN = cetane number
$\overline{U}_p$ = average piston speed in m/sec
R_u = universal gas constant = 8.314 kJ/kgmole-K
T_i, P_i = temperature (K) and pressure (bars) at start of compression stroke
r_c = compression ratio
$k = c_p/c_v$ = 1.35 in air-standard analysis

For an engine speed of N (RPM), ignition delay in milliseconds is

$$\text{ID(ms)} = \text{ID(ca)}/(0.006\,N) \tag{4-15}$$

The accuracy of Eq. (4-14) and (4-15) is shown in Fig. 4-9.

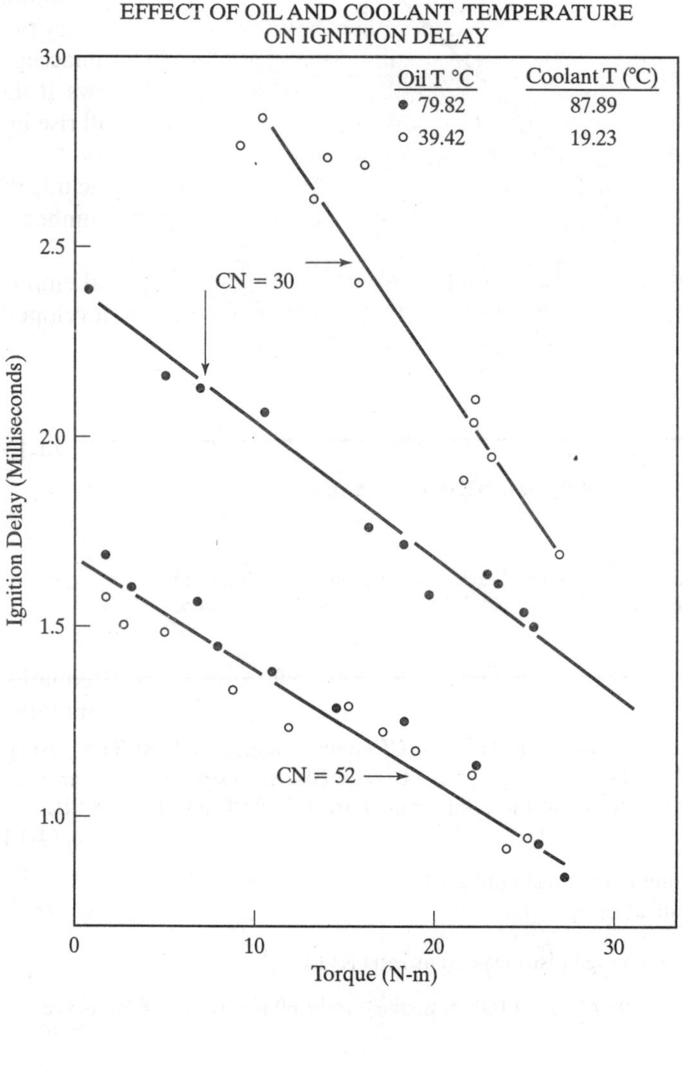

FIGURE 4-8

Ignition delay as a function of engine load (torque) and oil and coolant temperatures for two CI fuels. Reprinted with permission from SAE Paper No. 821231 © 1982, SAE International, [237].

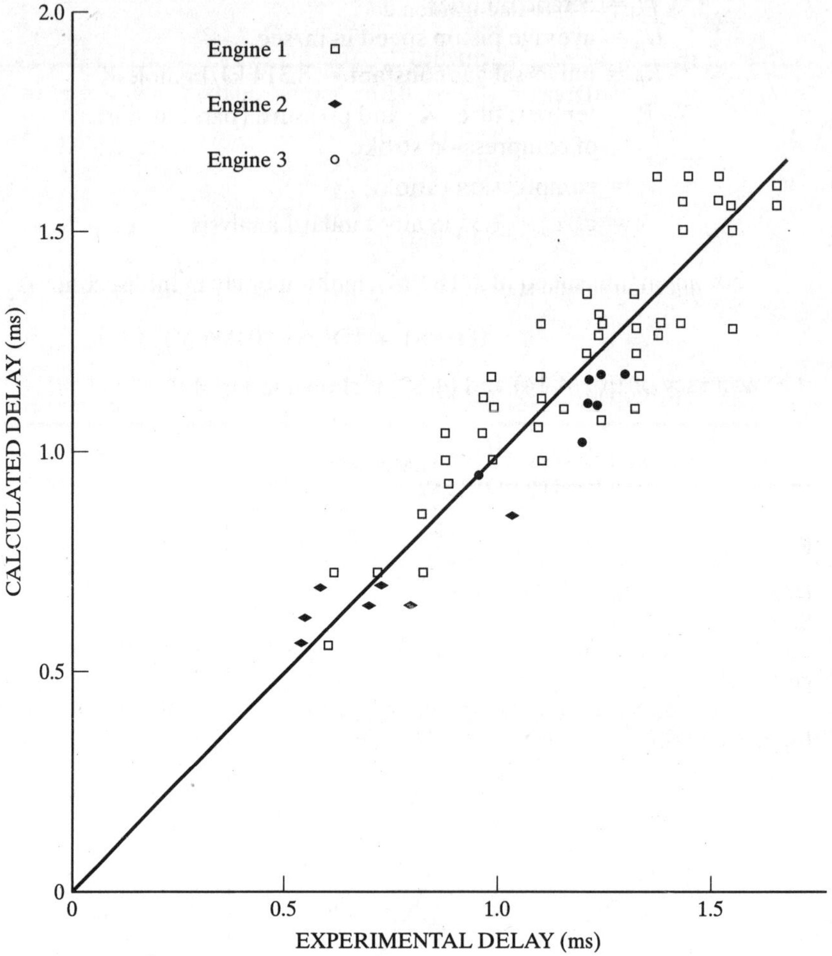

FIGURE 4-9

Experimental data for ignition delay of diesel fuel vs values predicted by Eq. (4-15). Reprinted with permission from SAE Paper No. 811235 © 1981, SAE International, [170].

Example Problem 4-8

It is desired for combustion to start at 15° bTDC in a CI engine running at 1600 RPM, using diesel fuel with a cetane number of 45. The engine has a bore of 10.4 cm, a stroke of 16.0 cm, and a compression ratio of 14. Air conditions at the cylinder inlet are $T_i = 33°\text{C}$ and $P_i = 98$ kPa. Calculate:

1. crank angle when fuel injection should start
2. ignition delay in milliseconds

(1) Equation (2-2) gives average piston speed at 1600 RPM:

$$\overline{U}_p = 2SN = (2\ \text{strokes/rev})(0.160\ \text{m/stroke})(1600/60\ \text{rev/sec}) = 8.53\ \text{m/sec}$$

Use Eq. (4-14) to find ignition delay:

$$E_A = (618{,}840)/(\text{CN} + 25) = (618{,}840)/(45 + 25) = 8841$$

$$\begin{aligned}\text{ID(ca)} &= (0.36 + 0.22\bar{U}_p)\exp\{E_A[(1/R_u T_i r_c^{k-1}) \\ &\quad - (1/17{,}190)][(21.2)/(P_i r_c^k - 12.4)]^{0.63}\} \\ &= [0.36 + (0.22)(8.53)]\exp\{8841[(1/(8.314)(306)(14)^{0.35}) \\ &\quad - (1/17{,}190)][(21.2)/((0.98)(14)^{1.35} - 12.4)]^{0.63}\} \\ &= 5.19^\circ \text{ of crank angle}\end{aligned}$$

Injection should start at

$$(15^\circ \text{ bTDC}) + (5.19^\circ) = \underline{20.19^\circ \text{ bTDC}}$$

(2) Equation (4-15) gives ignition delay in milliseconds:

$$\text{ID(ms)} = \text{ID(ca)}/(0.006\ N) = (5.19)/[(0.006)(1600)] = \underline{0.541 \text{ ms}}$$

Example Problem 4-9

Hexadecane (n-cetane, $C_{16}H_{34}$) is one of the standard fuels used to establish the Cetane Number Scale for fuels used in CI engines. It is given the cetane number of 100. Calculate the percent error obtained when the cetane number of hexadecane is approximated by the cetane index given in Eq. (4-13).

From reference [93], the specific gravity (0.773) and boiling temperature (548°F) are found for hexadecane. Using Eq. (4-13),

$$G = (141.5/S_g) - 131.5 = (141.5/0.773) - 131.5 = 51.55$$

$$\begin{aligned}\text{CI} &= -420.34 + 0.016\,G^2 + 0.192\,G(\log_{10}T_{mp}) + 65.01(\log_{10}T_{mp})^2 - 0.0001809\,T_{mp}^{\ 2} \\ &= (-420.34) + (0.016)(51.55)^2 + (0.192)(51.55)[\log_{10}(548)] \\ &\quad + (65.01)[\log_{10}(548)]^2 - (0.0001809)(548)^2 = 83\end{aligned}$$

Percent error is

$$\Delta\% = [(83 - 100)/(100)](100) = \underline{-17\%}$$

4.6 ALTERNATE FUELS

Sometime during the 21st century, crude oil and petroleum products will become very scarce and costly to find and produce. At the same time, there will likely be an increase in the number of automobiles and other IC engines. Although fuel economy of engines is greatly improved from the past and will probably continue to be improved, numbers alone dictate that there will be a great demand for fuel in the coming decades. Gasoline will become scarce and costly. Alternate fuel technology, availability, and use must and will become more common in the coming decades.

Although there have always been some IC engines fueled with non-gasoline or diesel oil fuels, their numbers have been relatively small. Because of the high cost of petroleum products, some third-world countries have for many years been using manufactured alcohol as their main vehicle fuel.

Many pumping stations on natural gas pipelines use the pipeline gas to fuel the engines driving the pumps. This solves an otherwise complicated problem of delivering fuel to the pumping stations, many of which are in very isolated regions. Some large displacement engines have been manufactured especially for pipeline work. These consist of a bank of engine cylinders and a bank of compressor cylinders connected to the same crankshaft and contained in a single engine block similar to a V-style engine.

Another reason motivating the development of alternate fuels for the IC engine is concern over the emission problems of gasoline engines. Combined with other air-polluting systems, the large number of automobiles is a major contributor to the air quality problem of the world. Vast improvements have been made in reducing emissions given off by an automobile engine. If a 30% improvement is made over a period of years and during the same time the number of automobiles in the world increases by 30%, there is no net gain. Actually, the net improvement in cleaning up automobile exhaust since the 1950s, when the problem became apparent, is over 95%. However, additional improvement is needed due to the ever-increasing number of automobiles.

A third reason for alternate fuel development in the United States and other industrialized countries is the fact that a large percentage of crude oil must be imported from other countries which control the larger oil fields. In recent years, up to a third of the United States foreign trade deficit has been from the purchase of crude oil, tens of billions of dollars.

Listed next are the major alternate fuels that have been and are being considered and tested for possible high-volume use in automobile and other kinds of IC engines. These fuels have been used in limited quantities in automobiles and small trucks and vans. Quite often, fleet vehicles have been used for testing (e.g., taxies, delivery vans, utility company trucks). This allows for comparison testing with similar gasoline-fueled vehicles, and simplifies fueling of these vehicles.

It must be remembered that, in just about all alternate fuel testing, the engines used are modified engines that were originally designed for gasoline fueling. They are, therefore, not the optimum design for the other fuels. Only when extensive research and development is done over a period of years will maximum performance and efficiency be realized from these engines. However, the research and development is difficult to justify until the fuels are accepted as viable for large numbers of engines (the chicken-and-egg problem).

Some diesel engines are starting to appear on the market which use dual fuel. They use methanol or natural gas and a small amount of diesel fuel that is injected at the proper time to ignite both fuels.

Most alternate fuels are very costly at present. This is often because of the quantity used. Many of these fuels will cost much less if the amount of their usage gets to the same order of magnitude as gasoline. The cost of manufacturing, distribution, and marketing all would be less.

Another problem with alternate fuels is the lack of distribution points (service stations) where the fuel is available to the public. The public will be reluctant to purchase an automobile unless there is a large-scale network of service stations available where fuel for that automobile can be purchased. On the other hand, it is difficult to justify building a network of these service stations until there are enough automobiles to make them profitable. Some cities are starting to make available a few distribution points for some of these fuels, such as propane, natural gas, and methanol. The transfer from one major fuel type to another will be a slow, costly, and sometimes painful process.

In the fuels discussed in the next several sections, some of the drawbacks for a particular fuel may become less of a problem if large quantities of that fuel are used (i.e., cost, distribution, etc.).

Alcohol

Alcohols are an attractive alternate fuel because they can be obtained from a number of sources, both natural and manufactured. Methanol (methyl alcohol) and ethanol (ethyl alcohol) are two kinds of alcohol that seem most promising and have had the most development as engine fuel.

The advantages of alcohol as a fuel include:

1. Can be obtained from a number of sources, both natural and manufactured.
2. Is a high octane fuel with anti-knock index numbers (octane number on fuel pump) of over 100. High octane numbers result, at least in part, from the high flame speed of alcohol. Engines using high-octane fuel can run more efficiently by using higher compression ratios.
3. Generally lower overall emissions compared with those of gasoline.
4. When burned, it forms more moles of exhaust, which gives higher pressure and more power in the expansion stroke.
5. Has high evaporative cooling (h_{fg}), which results in a cooler intake process and compression stroke. This raises the volumetric efficiency of the engine and reduces the required work input in the compression stroke.
6. Low sulfur content in the fuel.

The disadvantages of alcohol fuels include:

1. Low energy content of the fuel, as can be seen in Table A-2. This means that almost twice as much alcohol as gasoline must be burned to give the same energy input to the engine. With equal thermal efficiency and similar engine output usage, twice as much fuel would have to be purchased, and the distance which could be driven with a given fuel tank volume would be cut in half. The same amount of automobile use would require twice as much storage capacity in the distribution system, twice the number of storage facilities, twice the volume of storage at the service station, twice as many tank trucks and pipelines, etc. Even with the lower energy content of alcohol, engine power for a given displacement would be about the same. This is because of the lower air–fuel ratio needed by alcohol. Alcohol contains oxygen and thus requires less air for stoichiometric combustion. More fuel can be burned with the same amount of air.
2. More aldehydes in the exhaust. If as much alcohol fuel was consumed as gasoline, aldehyde emissions would be a serious exhaust pollution problem.
3. Much more corrosive than gasoline on copper, brass, aluminum, rubber, and many plastics. This puts some restrictions on the design and manufacturing of engines to

be used with this fuel. This should also be considered when alcohol fuels are used in engine systems designed to be used with gasoline. Fuel lines and tanks, gaskets, and even metal engine parts can deteriorate with long-term alcohol use (resulting in cracked fuel lines, the need for a special fuel tank, etc). Methanol is very corrosive on metals. This is because residual methanol reacts with the exhaust CO_2 and H_2O and forms formic acid (CH_2O_2) and carbonic acid (H_2CO_3). Methanol is considered a health hazard if it comes in contact with the skin.

4. Poor cold weather starting characteristics due to low vapor pressure and evaporation. Alcohol-fueled engines generally have difficulty starting at temperatures below 10°C. Often a small amount of gasoline is added to alcohol fuel, which greatly improves cold-weather starting. The need to do this, however, greatly reduces the attractiveness of any alternate fuel.
5. Poor ignition characteristics in general.
6. Almost invisible flames, which is considered dangerous when handling fuel. Again, a small amount of gasoline removes this danger.
7. Danger of storage tank flammability due to low vapor pressure. Air can leak into storage tanks and create a combustible mixture.
8. Low flame temperatures, which generate less NOx, but the resulting lower exhaust temperatures take longer to heat the catalytic converter to an efficient operating temperature.
9. Strong, possibly offensive odor. Headaches and dizziness have been experienced when refueling an automobile.
10. Vapor lock in fuel delivery systems.
11. Mixes with water, so any leakage can contaminate groundwater.

Methanol

Of all the fuels being considered as an alternate to gasoline, methanol is one of the more promising and has experienced major research and development. Pure methanol and mixtures of methanol and gasoline in various percentages have been extensively tested in engines and vehicles for a number of years [88, 130]. The most common mixtures are M85 (85% methanol and 15% gasoline) and M10 (10% methanol and 90% gasoline). The data from these tests, which include performance and emission levels, are compared with those obtained from tests with pure gasoline (M0) and pure methanol (M100). Some smart **flexible-fuel** (or **variable-fuel**) engines are capable of using any random mixture combination of methanol and gasoline ranging from pure methanol to pure gasoline. Two fuel tanks are used and various flow rates of the two fuels can be pumped to the engine, passing through a mixing chamber. Using information from sensors in the intake and exhaust, the EMS adjusts to the proper air–fuel ratio, ignition timing, injection timing, and valve timing (where possible) for the fuel mixture being used. Fast, abrupt changes in fuel mixture combinations must be avoided to allow for these adjustments to occur smoothly.

One problem with gasoline–alcohol mixtures as a fuel is the tendency for alcohol to combine with any water present. When this happens the alcohol separates locally from the gasoline, resulting in a nonhomogeneous mixture. This causes the engine to run erratically due to the large AF differences between the two fuels.

At least one automobile company has been experimenting with a three-fuel vehicle that can use any combination of gasoline–methanol–ethanol [11].

Methanol can be obtained from many sources, both fossil and renewable. These include coal, petroleum, natural gas, biomass, wood, landfills, and even the ocean. However, any source that requires extensive manufacturing or processing raises the price of the fuel and requires an energy input back into the overall environmental picture, both unattractive.

In some parts of the country, M10 fuel (10% methanol and 90% gasoline) is sold at some service stations in place of pure gasoline. It is advisable to read the (sometimes) small print on the fuel pump to determine the type of fuel that is being used in your automobile.

Emissions from an engine using M10 fuel are about the same as those using gasoline. The advantage (and disadvantage) of using this fuel is mainly the 10% decrease in gasoline use. With M85 fuel there is a measurable decrease in HC and CO exhaust emissions. However, there is an increase in NOx and a large (≈500%) increase in formaldehyde formation.

Methanol is used in some dual-fuel CI engines. Methanol by itself is not a good CI fuel because of its high octane number, but if a small amount of diesel oil is used for ignition, it can be used with good results. This is very attractive for third-world countries, where methanol can often be obtained much cheaper than diesel oil. Older CI bus engines have been converted to operate on methanol in tests conducted in California. This resulted in an overall reduction of harmful emissions compared with worn engines operating with diesel fuel [115].

Ethanol

Ethanol has been used as automobile fuel for many years in various regions of the world. Brazil is probably the leading user, where in the 1990s, 4.5 million vehicles operated on fuels that were 93% ethanol. For a number of years, **gasohol** has been available at service stations in the United States, mostly in the Midwest corn-producing states. Gasohol is a mixture of 90% gasoline and 10% ethanol. As with methanol, the development of systems using mixtures of gasoline and ethanol continues. Two mixture combinations that are important are E85 (85% ethanol) and E10 (gasohol). E85 is basically an alcohol fuel with 15% gasoline added to eliminate some of the problems of pure alcohol (i.e., cold starting, tank flammability, etc.). E10 reduces the use of gasoline with no modification needed to the automobile engine. Flexible-fuel engines that can operate on any ratio of ethanol–gasoline are being tested [122].

Ethanol can be made from ethylene or from fermentation of grains and sugar. Much of it is made from corn, sugar beets, sugar cane, and even cellulose (wood and paper). In the United States, corn is the major source. The present cost of ethanol is high due to the manufacturing and processing required. This would be reduced if larger amounts of this fuel were used. However, very high production would create a food–fuel competition, with resulting higher costs for both. Some studies show that, at present in the United States, crops grown for the production of ethanol consume more energy in plowing, planting, harvesting, fermenting, and delivery than what is in the final product. This defeats one major reason for using an alternate fuel [95].

Ethanol has less HC emissions than gasoline but more than methanol.

Example Problem 4-10

A taxicab is equipped with a flexible-fuel four-cylinder SI engine running on a mixture of methanol and gasoline at an equivalence ratio of 0.95. How must the air–fuel ratio change as the fuel flow to the engine shifts from 10% methanol (M10) to 85% methanol (M85)?

Change masses of fuel to moles at M10:

Fuel	Mass, m (kg)	Molecular Weight, M	Moles $N = m/M$ (kgmoles)	Mole Fraction
CH_3OH	0.10	32	0.003125	0.278
C_8H_{15}	0.90	111	0.008108	0.722
	1.00		0.011233	1.000

For 1 kgmole of fuel reacting with stoichiometric air,

$$0.278\ CH_3OH + 0.722\ C_8H_{15} + 8.9005\ O_2 + 8.9005(3.76)\ N_2 \rightarrow$$
$$6.054\ CO_2 + 5.971\ H_2O + 8.9005(3.76)\ N_2$$

For 1 kgmole of fuel reacting with an equivalence ratio $\phi = 0.95$

$$0.278\ CH_3OH + 0.722\ C_8H_{15} + (8.9005/0.95)\ O_2 + (8.9005/0.95)(3.76)\ N_2 \rightarrow$$
$$6.054\ CO_2 + 5.971\ H_2O + (8.9005/0.95)(3.76)\ N_2 + 0.468\ O_2$$

Air–fuel ratio:

$$AF = m_a/m_f$$
$$= [(8.9005/0.95)(1 + 3.76)(29)]/[(0.278)(32) + (0.722)(111)] = \underline{14.53}$$

Repeating calculations for M85.

Fuel	Mass, m (kg)	Molecular Weight, M	Moles $N = m/M$ (kgmoles)	Mole Fraction
CH_3OH	0.85	32	0.026563	0.952
C_8H_{15}	0.15	111	0.001351	0.048
	1.00		0.027914	1.000

The stoichiometric reaction is

$$0.952\ CH_3OH + 0.048\ C_8H_{15} + 1.992\ O_2 + 1.992(3.76)\ N_2 \rightarrow$$
$$1.336\ CO_2 + 2.264\ H_2O + 1.992(3.76)\ N_2$$

Reaction with an equivalence ratio $\phi = 0.95$ is

$$0.952\ CH_3OH + 0.048\ C_8H_{15} + (1.992/0.95)\ O_2 + (1.992/0.95)(3.76)\ N_2$$
$$\rightarrow 1.336\ CO_2 + 2.264\ H_2O + (1.992/0.95)(3.76)\ N_2 + 0.105\ O_2$$

Air–fuel ratio:

$$AF = m_a/m_f$$
$$= [(1.992/0.95)(1 + 3.76)(29)]/[(0.952)(32) + (0.048)(111)] = \underline{8.09}$$

As the fuel flow composition is changed, the engine management system (EMS) must adjust the AF from 14.53 to 8.09.

Example Problem 4-11

As an old-car collector enthusiast is cruising along in his 1950 chrome-covered Buick "dream car," he realizes he is running low on fuel, so he pulls up to the fuel pump at the local convenience store and fills the fuel tank. The car has a large-displacement straight-eight engine with a carburetor adjusted to supply stoichiometric air at normal operating conditions using gasoline fuel. The man does not notice that the fuel he is putting into his automobile contains 10% methanol (M10), but he does notice a slight loss in power as the fuel is consumed in the following days. Calculate the actual equivalence ratio the carburetor is supplying to the engine when it is adjusted to give $\phi = 1$ with gasoline but is actually operating with M10 fuel.

Stoichiometric combustion of gasoline:

$$C_8H_{15} + 11.75\ O_2 + 11.75(3.76)\ N_2 \rightarrow 8\ CO_2 + 7.5\ H_2O + 11.75(3.76)\ N_2$$

Air–fuel ratio:

$$AF = m_a/m_f = [(11.75)(1 + 3.76)(29)]/[(1)(111)] = 14.61$$

The carburetor is adjusted to give stoichiometric combustion with gasoline, so this is the actual $(AF)_{act}$ that it supplies.

When M10 fuel is used, the stoichiometric requirements would be (using result obtained in Example Problem 4-10)

$$0.278\ CH_3OH + 0.722\ C_8H_{15} + 8.9005\ O_2 + (8.9005)(3.76)\ N_2 \rightarrow 6.054\ O_2 + 5.971\ H_2O + (8.9005)(3.76)\ N_2$$

$$(AF)_{stoich} = [(8.9005)(1 + 3.76)(29)]/[(0.278)(32) + (0.722)(111)] = 13.80$$

Equivalence ratio:

$$\phi = (AF)_{stoich}/(AF)_{act} = (13.80)/(14.61) = \underline{0.945}$$

This assumes that the mass densities of gasoline and M10 fuel are the same when M10 actually has a slightly higher density. If M10 fuel were going to be used extensively, the carburetor should be readjusted for stoichiometric operation. Using alcohol fuel in older engines designed for gasoline might also create material compatibility problems between the fuel and engine components.

Hydrogen

Hydrogen is projected as a possible major fuel of the future to replace the eventual dwindling gasoline supplies. This could be as a fuel for internal combustion engines, or for fuel cells if that technology replaces IC engines for vehicle propulsion. As fuel for a combustion engine, it offers a high octane number and no CO, CO_2, or HC emissions. Fuel storage and refueling for an automobile are the two greatest problems that must be solved to make this a viable vehicle fuel.

A number of companies have built automobiles with prototype or modified engines that operate on hydrogen fuel [64, 154, 186]. A few of these companies include hydrogen-fueled, IC-engine-powered automobiles in their medium-range projections of commercially available products. At least one company (BMW) is developing a dual-fuel gasoline–hydrogen automobile. Mazda has adapted an experimental rotary Wankel engine to run on hydrogen fuel, reasoning that this was a good type of engine for this fuel. The fuel intake is on the opposite side of the engine from where combustion occurs, lowering the chance of preignition from a hot engine block; hydrogen fuel ignites very easily [86].

Fuel storage in an automobile can be done in any of the three phases: liquid, gaseous, or solid. Several of the present experimental automobiles store the onboard fuel as a cryogenic liquid. At atmospheric pressure this must be done at a temperature of −250°C (23 K, −420°F) which requires a bulky, superinsulated fuel tank. Storage temperature could be raised somewhat if the fuel tank was pressurized. It requires 140 liters of liquid hydrogen fuel to have the same energy content as 40 liters of gasoline. Refueling a liquid hydrogen tank in an automobile would be difficult and possibly dangerous; the general public could not be allowed to handle cryogenic liquids. At the start of the 20th century the only liquid hydrogen refueling station in the world available to the public was at the Munich Air Port in Bavaria.

Hydrogen can be stored as a compressed gas with pressures up to 35 MPa (5000 psia). However, enough fuel for a reasonable vehicle range would require an unacceptable tank volume. Refueling a compressed gas fuel would be difficult for the average public.

Hydrogen can be stored as a component in a metal hydride solid (e.g., lanthanumnickle hydride, or titanium–iron hydride). When the hydride solid is cooled, it chemically absorbs hydrogen; when it is slightly heated with an electric generator or battery, it releases hydrogen as a gas, which can fuel either an IC engine or a fuel cell. Any vehicle using this method of fuel storage would experience a vehicle weight penalty. At present, a hydride can absorb about 3.5% of its weight as hydrogen. Experimental cars can get about a 300 mile range with an 86 kg (190 lbm) storage unit. Storage will probably be improved to 7% to 15% of weight with additional development. Many cooling (absorbing H_2) and heating (releasing) cycles will reduce the efficiency of a hydride with time. Liquid hydrides (boron, sodium, calcium) might also be used to store hydrogen in the future.

Hydrogen as a fuel has the following advantages:

1. High octane number, and, when used at a fuel equivalence ratio of 0.6 (67% excess air), compression ratios of about 14:1 can be utilized without serious knock problems. This makes for high thermal efficiency.
2. Low emissions. Essentially no CO, CO_2, or HC in the exhaust as there is no carbon in the fuel. NOx emissions can be kept low if the engine is operated at low fuel equivalence ratio. Need for a catalytic converter could be eliminated. Most exhaust would be H_2O and N_2.
3. Fuel availability. There are a number of different ways of obtaining hydrogen, including natural gas and electrolysis of water. However, when the total energy and environmental picture is considered in converting natural gas to hydrogen, it can be argued that it would be better to use the natural gas as the engine fuel directly. Also, the fuel burned to generate electricity for electrolysis and the impact this

has on the environment must be considered. Solar energy or improved fuel cell technology could diminish this problem.

4. Fuel leakage to the environment is not a pollutant.

Hydrogen as a fuel has the following disadvantages:

1. Heavy, bulky fuel storage, both in vehicle and at the service station. This would be much less of a problem for a stationary engine.
2. Difficult to refuel.
3. Poor engine volumetric efficiency. Any time a gaseous fuel is used in an engine, fuel will displace some of the inlet air and poorer volumetric efficiency will result. This is especially true for hydrogen fuel, which is generally burned with a large amount of excess air.
4. High fuel cost, given present-day technology and availability.
5. Fuel can detonate.

HISTORIC—HYDROGEN TRAGEDIES

Hydrogen was a factor in two major tragedies of transportation vehicles during the 20th century. On January 28, 1986 the space shuttle *Challenger* blew up shortly after takeoff from Kennedy Space Center in Florida, killing all seven astronauts on board. A faulty o-ring allowed liquid hydrogen fuel to leak, causing the vehicle to explode. On May 6, 1937 the German dirigible *Hindenburg* exploded as it was landing in Lakehurst, New Jersey, killing 36 of the 97 people on board. In this tragedy the exploding hydrogen was not a fuel, but was being used to provide lift for the lighter-than-air ship. Several possible causes of the explosion have been suggested, including static electricity, lightning, or sabotage. At 804 ft long, the *Hindenburg* was the largest rigid airship ever built. The disaster ended the use of lighter-than-air ships for carrying passengers.

Natural Gas—Methane

Natural gas is a mixture of components, consisting mainly of methane (60–98%) with small amounts of other hydrocarbon fuel components. In addition, it contains various amounts of N_2, CO_2, He, and traces of other gases. Its sulfur content ranges from very little (sweet) to larger amounts (sour). It is stored as compressed natural gas (CNG) at pressures of 16 to 25 MPa, or as liquid natural gas (LNG) at pressures of 70 to 210 kPa and a temperature around −160°C. As a fuel, it works best in an engine system with a single-throttle body fuel injector. This gives a longer mixing time, which is needed by this fuel. Tests using CNG in various-sized vehicles continue to be conducted by government agencies and private industry [12, 94, 101, 176].

The advantages of natural gas as a fuel include the following:

1. Octane number of 120, which makes it a very good SI engine fuel. One reason for this high octane number is a fast flame speed. Engines can operate with a high compression ratio.

2. Low engine emissions. Less aldehydes than with methanol, and less CO_2.
3. Fuel is fairly abundant worldwide with much available in the United States. It can be made from coal but this would make it more costly.

The disadvantages of natural gas as an engine fuel include the following:

1. Low energy density resulting in low engine performance.
2. Low engine volumetric efficiency because it is a gaseous fuel.
3. Need for large pressurized fuel storage tank. Most test vehicles have a range of only about 120 miles. There is some safety concern with a pressurized fuel tank.
4. Inconsistent fuel properties.
5. Refueling is slow process.

Dual-fuel CI engines using natural gas and diesel oil are being developed for trucks and stationary applications. These engines use low-sulfur natural gas (or pure methane in some cases) as the main fuel, and a small amount of high-grade, low-sulfur diesel fuel for ignition. This is done both for economic reasons (natural gas being much cheaper per energy unit than diesel fuel) and environmental considerations. Natural gas has lower combustion temperatures than diesel fuel, and by using late injection the combustion temperature can be reduced still further. This greatly reduces the production of NOx. In addition, because of the lower amount of carbon in the fuel, less CO_2 is generated, and very little solid particulate emissions. These engines use single-unit injectors, which inject both fuels simultaneously. The natural gas is injected at almost sonic velocity which produces high turbulence and consequent high flame speed. Tests show that ignition delay with the dual fuel is almost the same as for diesel fuel alone, or possibly very slightly longer [213]. Flame speed is higher, possibly because of higher turbulence. The flame can be classified as a combination of premixed (as in SI combustion) and diffusion (CI combustion) (see Chapter 7).

A unique application of using natural gas/methane as a fuel is in landfill gas engines. These are large stationary engines used to drive electric generators, which get their fuel from the gases released from landfills. In the year 2000, two percent of the electricity produced in England came from landfill gas engines [174]. Landfill gas as a fuel is "extremely dirty and variable in quality" [174], generally consisting of 45–65% methane. Contaminates include silicon, chlorine, fluorine, solid particles, etc. High fuel filtration is required, with input and operating temperatures kept much higher than normal to avoid condensation of contaminates. Engine components, such as pistons and valves, are made of special materials because of the corrosiveness of the fuel. Special lubricating oils are needed because of the acidity and higher operating temperatures.

HISTORIC—BUSES OPERATING ON NATURAL GAS

In some countries in eastern and southern Asia, buses using natural gas as fuel have a unique fuel reservoir system. The gas is stored at about one atmosphere pressure in a large inflatable rubber diaphragm on the roof of the bus. With a full load of fuel, the bus is about twice the height as it is when it has no fuel. No fuel gauge is needed on these buses.

Propane–Butane

Propane has been tested in fleet vehicles for a number of years; after gasoline and diesel oil it is the third most used vehicle fuel in the United States. It is a good high-octane fuel (AKI $>$ 100) and produces lower emissions than gasoline: about 60% less CO, 30% less HC, and 20% less NOx. The fact that it is a single-component fuel allows for high optimization of engine and catalytic systems. Propane is stored at normal temperatures as a liquid under pressure and is delivered through a high-pressure line to the engine, where it is vaporized.

Propane is produced as a by-product in natural gas production and petroleum refinery processing. This limits its availability, as it is very unlikely these processes would be increased just to produce propane. Use of the supply of propane as a vehicle fuel would have to compete with home heating, where propane has a very large market. There exists a large countrywide distribution system for propane fuel, although most of these fuel stations are not designed for vehicle fueling.

Recent testing has been done with propane–butane mixtures as a vehicle fuel (20% butane/80% propane, 30/70, and 50/50) [160, 172]. The greater availability of butane makes this a possible fuel in the future. Butane has a slightly lower octane number than propane ($\approx$92), but still higher than most gasolines. It has a greater energy content per volume when stored as a liquid, which increases vehicle range.

Reformulated Gasoline

Reformulated gasoline is normal gasoline with a slightly modified formulation and additives to help reduce engine emissions. Included in the fuel are oxidation inhibitors, corrosion inhibitors, metal deactivators, detergents, and deposit control additives. Oxygenates such as methyl tertiary-butyl ether $CH_3OC_4H_9$ (MTBE) and alcohols are added, such that there is 1–3% oxygen by weight. This is to help reduce CO in the exhaust. Levels of benzene, aromatics, and high-boiling temperature components are reduced, as is the vapor pressure. Recognizing that engine deposits contribute to emissions, cleaning additives are included. Some additives clean carburetors, some clean fuel injectors, and some clean intake valves, each of which often does not clean other components. The use of MTBE has recently been questioned, critics claiming that it contaminates groundwater when leaks occur in fuel storage tanks.

On the plus side is the fact that all gasoline-fueled engines, old and new, can use this fuel without modification. On the negative side is that only moderate emission reduction is realized, cost is increased, and the use of petroleum products is not reduced. [121].

Coal–Water Slurry

In the latter half of the 1800s, before petroleum-based fuels were perfected, many other fuels were tested and used in IC engines. When Rudolf Diesel was developing his engine, one of the fuels he used was a coal dust–water slurry. Fine particles of coal (carbon) were dispersed in water and injected and burned in early diesel engines. Although this never became a common fuel, a number of experimental engines using this fuel have been built over the last hundred years. Even today, some work continues on this fuel technology. The major improvement in this type of fuel has been the reduction of the average coal particle size. In 1894 the average particle size was on the order of

100 μ (1 μ = 1 micron = 10^{-6} m). This was reduced to about 75 μ in the 1940–1970 period and further reduced to about 10 μ by the early 1980s. The typical slurry is about 50% coal and 50% water by mass. One major problem with this fuel is the abrasiveness of the solid particles, which manifests itself in worn injectors and piston rings [27].

Coal is an attractive fuel because of the large supply that is available. It is not so attractive from an environmental viewpoint, in that when burned, it produces large amounts of CO_2, a major greenhouse gas. When coal is used as an engine fuel, other methods than a slurry seem more feasible. These include liquefaction or gasification of the coal.

Example Problem 4-12

A 50% coal–water slurry (50% coal and 50% water by mass) is burned in stoichiometric air. Calculate:

1. air–fuel ratio
2. heating value of fuel

(1) Assuming coal is carbon, the fuel mixture in molar quantities will consist of one mole of carbon to (12/18) moles of water.
The stoichiometric combustion reaction is given by

$$C(s) + (12/18)\,H_2O + O_2 + (3.76)\,N_2 \rightarrow CO_2 + \tfrac{2}{3}H_2O + (3.76)\,N_2$$

$$AF = [(1)(4.76)(29)]/[(1)(12) + (12/18)(18)] = \underline{5.75}$$

(2) Heating value of carbon from Table A-2 is

$$(Q_{\text{LHV}})_c = 33{,}800 \text{ kJ/kg}$$

Heating value of water is

$$(Q_{\text{LHV}})_w = 0$$

Therefore, the heating value of a fuel mixture made of 50% carbon and 50% water by mass is

$$(Q_{\text{LHV}})_{\text{mix}} = [(33{,}800 + 0)/2] = \underline{16{,}900 \text{ kJ/kg}}$$

This is a very low heating value compared with the heating values of normal hydrocarbon fuels, which generally are in the 40,000 to 50,000 kJ/kg range. However, the low stoichiometric AF means that a greater amount of fuel can be used with the same amount of inlet air, and the power output developed by the engine can equal that of engines using other fuels. Specific fuel consumption would be very high.

Other Fuels

Attempts to use many other types of fuel have been tried throughout the history of IC engines. Often, this was done out of necessity or to promote financial gain by some group. At present, a number of biomass fuels are being evaluated, mainly in Europe. These include CI fuel made from wood, barley, soybeans, rapeseed, cottonseed, sunflower seeds, corn oil, and even beef tallow and pig manure. Advantages of these fuels generally include availability and low cost, low sulfur, and low emissions. Disadvantages include low energy content (heating value) and corresponding high specific fuel consumption. These fuels also create a market fuel–food competition and the corresponding higher cost of both.

HISTORIC—WHEN AUTOMOBILES RAN ON CHARCOAL

In the late 1930s and early 1940s, petroleum products became very scarce, especially in Europe, due to World War II. Just about all gasoline products were claimed by the German army, leaving no fuel for civilian automobile use. Although this was an inconvenience for the civilian population, it did not stop them from using their beloved automobiles [44].

Enterprising people in several countries, mainly Sweden and Germany, developed a way to operate their automobiles using solid fuels like charcoal, wood, or coal. Using technology first researched 20 years earlier, they converted their vehicles by building a combustion chamber in the trunk of the car or on a small trailer pulled by the car. In this combustion chamber, the coal, wood, or other solid or waste fuel was burned with a restricted supply of oxygen (air). This generated a supply of carbon monoxide, which was then piped to the engine and used to fuel the engine. With coal being basically carbon, the ideal reaction in the combustion chamber (CO generator) would be

$$C + \tfrac{1}{2}O_2 + \tfrac{1}{2}(3.76)\,N_2 \rightarrow CO + \tfrac{1}{2}(3.76)\,N_2$$

This CO was then piped to the automobile engine, where, ideally, the reaction would be

$$\underbrace{CO + \tfrac{1}{2}(3.76)N_2}_{\text{fuel}} + \tfrac{1}{2}O_2 + \tfrac{1}{2}(3.76)\,N_2 \rightarrow CO_2 + 3.76\,N_2$$

Obviously the carburetor on the engine had to be modified and adjusted to supply the gaseous fuel and to give the correct AF. When this was done, the automobile could be operated, but with much less power and longevity. Impurities from the CO generator very quickly dirtied up the combustion chambers of the engine, even when extensive filtering was done. Another problem was with leakage of CO, which is a poisonous, odorless, colorless gas. When this occurred and CO got into the passenger compartment, the driver and other occupants were in danger of sickness and death. Drivers subjected to CO would react as if intoxicated in much the same way as a drunk driver, with the same results of erratic driving, accidents, and even death.

As late as the 1970s, Sweden was still working on the development of this method of using solid fuel to power automobiles [44].

Example Problem 4-13

During World War II a Swedish car lover found himself unable to obtain gasoline to operate his vehicle. Like many of his neighbors, he built himself a trailer on which he mounted a charcoal-burning carbon monoxide generator, a combustion chamber with restricted air input. The CO generated was then piped to the engine and used as fuel for the automobile pulling the trailer. Estimate the loss of power that was experienced when CO was used in this manner as fuel for the carbureted engine originally designed to operate on stoichiometric gasoline.

With gasoline, the reaction is

$$C_8H_{15} + 11.75\,O_2 + 11.75(3.76)\,N_2 \rightarrow 8\,CO_2 + 7.5\,H_2O + 11.75(3.76)\,N_2$$

Thus, for 11.75(4.76) kgmoles of air, heat in is

$$Q_{in} = m_f Q_{HV} = (1 \text{ kgmole of fuel})(111 \text{ kg/kgmole})(43.0 \text{ MJ/kg}) = 4773 \text{ MJ}$$

For 1 kgmole of input oxygen (4.76 kgmoles of air), heat in is

$$Q_{in} = (4773 \text{ MJ})/(11.75) = 406.2 \text{ MJ}$$

When CO (and the N_2 that comes with it from the generator) is burned stoichiometrically with one kgmole of oxygen, the reaction is

$$2\,CO + 2[\tfrac{1}{2}(3.76)]\,N_2 + O_2 + (3.76)\,N_2 \rightarrow 2\,CO_2 + 2(3.76)\,N_2$$

Therefore,

$$Q_{in} = (2 \text{ kgmoles of CO})(28 \text{ kg/kgmole})(10.1 \text{ MJ/kg}) = 565.6 \text{ MJ}$$

This, however, is reduced because CO is a gaseous fuel which displaces some of the air in the intake system. For each kgmole of oxygen (4.76 kgmoles of air) in the intake system, there will be $[2 + 2(\frac{1}{2})(3.76)] = 5.76$ kgmoles of gaseous fuel. For the same total gas volumetric flow rate, only $(4.76)/[(4.76) + (5.76)] = 0.452$ will be the fraction that is new inlet air. Therefore,

$$Q_{in} = (565.6 \text{ MJ})(0.452) = 255.7 \text{ MJ}$$

The percent loss in heat is then

$$\% \text{ loss of } Q_{in} = \{[(406.2) - (255.7)]/(406.2)\}(100) = 37.1\% \text{ loss}$$

Assuming the same thermal efficiency and same engine mechanical efficiency with the two fuels gives brake power output as

$$(\dot{W}_b)_{CO} = 62.9\%(\dot{W}_b)_{gasoline}$$

These calculations are all based on ideal reactions. An actual engine–CO generator system operated under these conditions would experience many additional losses, including less than ideal reactions, solid impurities and filtering problems, and fuel delivery complications. All of these would significantly lower the actual engine output.

4.7 CONCLUSIONS

For most of the 20th century, the two main fuels that have been used in internal combustion engines have been gasoline (SI engines) and fuel oil (diesel oil for CI engines). During this time, these fuels have experienced an evolution of composition and additives according to the contemporary needs of the engines and environment. In the latter part of the century, alcohol fuels made from various farm products and other sources have become increasingly more important, both in the United States and in other countries. With increasingly air pollution problems and a petroleum shortage looming on the horizon, major research and development programs are being conducted throughout the world to find suitable alternate fuels to supply engine needs for the coming decades.

PROBLEMS

4.1 C_4H_8 is burned in an engine with a fuel-rich air–fuel ratio. Dry analysis of the exhaust gives the following volume percents: $CO_2 = 14.95\%$, $C_4H_8 = 0.75\%$, $CO = 0\%$, $H_2 = 0\%$, $O_2 = 0\%$, with the rest being N_2. Higher heating value of this fuel is $Q_{HHV} = 46.9$ MJ/kg. Write the balanced chemical equation for one mole of this fuel at these conditions.

Calculate:

(a) Air–fuel ratio.

(b) Equivalence ratio.

(c) Lower heating value of fuel. [MJ/kg]

(d) Energy released when one kg of this fuel is burned in the engine with a combustion efficiency of 98%. [MJ]

4.2 Draw the chemical structural formula of 2-methyl-2,3-ethylbutane. This is an isomer of what chemical family? Write the balanced chemical reaction equation for one mole of this fuel burning with an equivalence ratio of $\phi = 0.7$. Calculate the stoichiometric AF for this fuel.

4.3 Draw the chemical structural formula of **(a)** 3,4-dimethylhexane, **(b)** 2,4-diethylpentane, **(c)** 3-methyl-3-ethylpentane. These are isomers of what other molecules?

4.4 Hydrogen is used as a fuel in an experimental engine and is burned with stoichiometric *oxygen*. Reactants enter at a temperature of 25°C and complete combustion occurs at constant pressure. Write the balanced chemical reaction equation.

Calculate:

(a) Fuel–air (fuel–oxygen) ratio.

(b) Equivalence ratio.

(c) Theoretical maximum temperature from this combustion. (Use enthalpy values from a thermodynamics textbook.) [K]

(d) Dew point temperature of exhaust if exhaust pressure is 101 kPa. [°C]

4.5 Isooctane is burned with air in an engine at an equivalence ratio of 0.8333. Assuming complete combustion, write the balanced chemical reaction equation.

Calculate:

(a) Air–fuel ratio.

(b) How much excess air is used. [%]

(c) AKI and FS of this fuel.

4.6 A race car burns nitromethane with air at an equivalence ratio of 1.25. Except for unburned fuel, all nitrogen ends up as N_2 and all carbon ends up as CO_2 (no CO). Write the balanced chemical reaction equation.

Calculate:

(a) Percent stoichiometric air. [%]

(b) Air–fuel ratio.

4.7 Methanol is burned in an engine with air at an equivalence ratio of $\phi = 0.75$. Exhaust pressure and inlet pressure are 101 kPa. Write the balanced chemical equation for this reaction.

Calculate:

(a) Air–fuel ratio.

(b) Dew point temperature of the exhaust if the inlet air is dry. [°C]

(c) Dew point temperature of the exhaust if the inlet air has a relative humidity of 40% at 25°C. [°C]

(d) Anti-knock index of methanol.

4.8 Compute the indicated power generated at WOT by a three-liter, four-cylinder, four-stroke cycle SI engine operating at 4800 RPM using either gasoline or methanol. For each case, the intake manifold is heated such that all fuel is evaporated before the intake ports, and the air–fuel mixture enters the cylinders at 60°C and 100 kPa. Compression ratio $r_c = 8.5$, fuel equivalence ratio $\phi = 1.0$, combustion efficiency $\eta_c = 98\%$, and volumetric efficiency $\eta_v = 100\%$. Calculate the indicated specific fuel consumption for each fuel. [gm/kW-hr]

4.9 A four-cylinder SI engine with a compression ratio $r_c = 10$ operates on an air-standard Otto cycle at 3000 RPM using ethyl alcohol as fuel. Conditions in the cylinders at the start of the compression stroke are 60°C and 101 kPa. Combustion efficiency $\eta_c = 97\%$. Write the balanced stoichiometric chemical equation for this fuel.

Calculate:

(a) AF if the engine operates at an equivalence ratio $\phi = 1.10$.

(b) Peak temperature in cycle of part (a). [°C]

(c) Peak pressure in cycle of part (a). [kPa]

4.10 Tim's 1993 Buick has a six-cylinder, four-stroke cycle SI engine with multipoint port fuel injectors operating on an Otto cycle at WOT. The fuel injectors are set to deliver an AF such that gasoline would burn at stoichiometric conditions. (Approximate gasoline using isooctane properties.)

Calculate:

(a) Equivalence ratio of air–gasoline mixture.

(b) Equivalence ratio if gasoline is replaced with ethanol without readjusting the AF delivered by the fuel injectors.

(c) Increase or decrease in brake power using alcohol instead of gasoline under these conditions, with the same air flow rate and same thermal efficiency. Assume ethanol would burn under these conditions with the same combustion efficiency. [%]

4.11 For the same air flow rate, what would be the percentage increase in engine power if stoichiometric gasoline is replaced with stoichiometric nitromethane? Assume the same thermal efficiency and the same combustion efficiency. [%]

4.12 Compare the indicated power generated in an engine using stoichiometric gasoline, stoichiometric methanol, or stoichiometric nitromethane. Assume the same combustion efficiency, thermal efficiency, and air flow rate for all fuels.

4.13 Isodecane is used as a fuel.

Calculate:

(a) Anti-knock index.

(b) MON if 0.2 gm/L of TEL is added to the fuel.

(c) How many gallons of butene-1 should be added to 10 gallons of isodecane to give a mixture MON of 87. Density of isodecane $\rho_{isod} = 768\ kg/m^3$, and density of butene-1 $\rho_{but} = 595\ kg/m^3$.

4.14 A six-liter, eight-cylinder, four-stroke cycle SI race car engine operates at 6000 RPM using stoichiometric nitromethane as fuel. Combustion efficiency is 99%, and the fuel input rate is 0.198 kg/sec.

Calculate:

(a) Volumetric efficiency of the engine. [%]
(b) Flow rate of air into the engine. [kg/sec]
(c) Heat generated in each cylinder per cycle. [kJ]
(d) How much chemical energy there is in unburned fuel in the exhaust. [kW]

4.15 (a) Give three reasons why methanol is a good alternate fuel for automobiles. (b) Give three reasons why it is not a good alternate fuel.

4.16 When one-half mole of oxygen and one-half mole of nitrogen are heated to 3000 K at a pressure of 5000 kPa, some of the mixture will react to form NO by the reaction equation $\frac{1}{2}O_2 + \frac{1}{2}N_2 \rightarrow NO$. Assume these are the only components that react.

Calculate:

(a) Chemical equilibrium constant for this reaction at these conditions using Table A-3.
(b) Number of moles of NO at equilibrium.
(c) Number of moles of O_2 at equilibrium.
(d) Number of moles of NO at equilibrium if the total pressure is doubled.
(e) Number of moles of NO at equilibrium if there was originally one-half mole of oxygen, one-half mole of nitrogen, and one mole of argon at 5000 kPa total pressure.

4.17 An experimental truck engine uses hydrogen fuel (H_2) with an air–fuel ratio AF = 30. For one kgmole of H_2 burned at this condition, with reactants and products at 25°C, and water in the products considered all vapor:

Calculate:

(a) Equivalence ratio.
(b) Lambda value.
(c) Heat released using Eq (4-5). [kJ]

4.18 A new fuel being developed for use in internal combustion engines consists of $^1/_2$ methanol and $^1/_2$ butene-1 by mole.

Calculate:

(a) Stoichiometric AF. [kg_a/kg_f]
(b) AKI.

4.19 When one kgmole of diatomic hydrogen (H_2) is heated at a pressure of 101 KPa, some of it dissociates into monatomic hydrogen (H) in the reaction: $H_2 \rightarrow 2\,H$. At a certain temperature there will be x moles of H for every 2x moles of H_2.

Calculate:

(a) Number of moles of monatomic H at this temperature (x = ?).
(b) Chemical equilibrium constant at this temperature.
(c) Temperature. [K]

4.20 A fuel mixture consists of 20% isooctane, 20% triptane, 20% isodecane, and 40% toluene by moles. Write the chemical reaction formula for the stoichiometric combustion of one mole of this fuel.

Calculate:

(a) Air–fuel ratio.

(b) Research octane number.

(c) Lower heating value of fuel mixture. [kJ/kg]

4.21 A flexible-fuel vehicle operates with a stoichiometric fuel mixture of one-third isooctane, one-third ethanol, and one-third methanol, by mass.

Calculate:

(a) Air–fuel ratio.

(b) MON, RON, FS, and AKI.

4.22 It is desired to find the cetane number of a fuel oil that has a density of 860 kg/m^3 and a midpoint boiling temperature of 229°C. When tested in the standard test engine, the fuel is found to have the same ignition characteristics as a mixture of 23% hexadecane and 77% heptamethylnonane.

Calculate:

(a) Cetane number of fuel.

(b) Percent error if cetane index is used to approximate the cetane number. [%]

4.23 A CI engine running at 2400 RPM has an ignition delay of 15° of crankshaft rotation. What is the ID in seconds?

4.24 Hexane C_6H_{14} is used for fuel in a CI engine at an air–fuel ratio AF = 25. Hexane has a specific gravity of 0.659 and a midpoint boiling temperature of 69°C.

Calculate:

(a) Equivalence ratio.

(b) Lambda value.

(c) Cetane index of hexane.

4.25 A ten-cylinder CI engine running at 1295 RPM powers an electric generator. The engine has a compression ratio of 16:1 and an average piston speed of 9.50 m/sec. Inlet temperature and pressure at the start of compression are 47°C and 110 kPa. It is desired that combustion start 12° bTDC using fuel with a cetane number of 51.

Calculate:

(a) Crank angle when fuel injection should start. [° bTDC]

(b) Ignition delay of fuel in milliseconds.

4.26 A fuel blend has a density of 720 kg/m^3 and a midpoint boiling temperature (temperature at which 50% will be evaporated) of 91°C. Calculate the cetane index.

4.27 The fuel which a coal-burning carbon monoxide generator supplies to an automobile engine consists of $CO + \frac{1}{2}(3.76)N_2$.

Calculate:

(a) Q_{HHV} and Q_{LHV} of fuel. [kJ/kg]

(b) Stoichiometric air–fuel ratio.

(c) Dew point temperature of exhaust. [°C]

DESIGN PROBLEMS

4.1D Using data from Table A-2 and boiling point data from chemistry handbooks, design a three-component gasoline blend. Give a three-temperature classification of your blend and draw a vaporization curve similar to Fig. 4-2. What is RON, MON, and AKI of your blend?

4.2D An automobile will use hydrogen as fuel. Design a fuel *tank* (i.e., a fuel storage system for the vehicle) and a method to deliver the fuel from the tank to the engine.

4.3D An automobile will use propane as fuel. Design a fuel *tank* (i.e., a fuel storage system for the vehicle) and a method to deliver the fuel from the tank to the engine.

CHAPTER 5

Air and Fuel Induction

"Our old troublesome enemy the carburetor is evidently being pushed to the scrap heap by the atomizer and direct-feed systems of supply."

Resume of the New York Automobile Show

The Automobile Magazine (January 1902)

This chapter describes intake systems of engines—how air and fuel are delivered into the cylinders. The object of the intake system is to deliver the proper amount of air and fuel accurately and equally to all cylinders at the proper time in the engine cycle. Flow into an engine is pulsed as the intake valves open and close, but can generally be modeled as quasi-steady state flow.

The intake system consists of an intake manifold, a throttle, intake valves, and either fuel injectors or a carburetor to add fuel. Fuel injectors can be mounted by the intake valves of each cylinder (multipoint port injection), at the inlet of the manifold (throttle body injection), or direct injection in the cylinder head (CI engines, modern two-stroke cycle SI engines, and some four-stroke cycle SI automobile engines).

5.1 INTAKE MANIFOLD

The **intake manifold** is a system designed to deliver air to the engine through pipes to each cylinder, called *runners*. The inside diameter of the runners must be large enough that a high flow resistance and the resulting low volumetric efficiency do not occur. On the other hand, the diameter must be small enough to assure high air velocity and turbulence, which enhances its capability of carrying fuel droplets and increases evaporation and air–fuel mixing.

The length of a runner and its diameter should be sized together to equalize, as much as possible, the amount of air and fuel that is delivered to each separate cylinder. Some engines have active intake manifolds with the capability of changing runner length and diameter for different engine speeds. At low speeds, the air is directed through longer, smaller diameter runners to keep the velocity high and to assure proper mixing of air and fuel. At high engine speeds, shorter, larger diameter runners are used, which minimizes flow resistance but still enhances proper mixing. The amount of air and fuel in one runner length is about the amount that gets delivered to one cylinder each cycle.

To minimize flow resistance, runners should have no sharp bends, and the interior wall surface should be smooth with no protrusions such as the edge of a gasket.

Some intake manifolds are heated to accelerate the evaporation of the fuel droplets in the air–fuel mixture flow. This is done by heating the walls with hot engine coolant flow, by designing the intake manifold to be in close thermal contact with the hot exhaust manifold, or sometimes with electrical heating.

On SI engines, air flow rate through the intake manifold is controlled by a throttle plate (butterfly valve) usually located at the upstream end. The throttle is incorporated into the carburetor for those engines so equipped.

Fuel is added to inlet air somewhere in the intake system—before the manifold, in the manifold, or directly into each cylinder. The further upstream the fuel is added, the more time there is to evaporate the fuel droplets and to get proper mixing of the air and fuel vapor. However, this also reduces engine volumetric efficiency by displacing incoming air with fuel vapor. Early fuel addition also makes it more difficult to get good cylinder-to-cylinder AF consistency because of the asymmetry of the manifold and different lengths of the runners.

It is found that when fuel is added early in the intake system, the flow of fuel through the manifold occurs in three different manners. Fuel vapor mixes with the air and flows with it. Very small liquid fuel droplets are carried by the air flow, smaller droplets following the streamlines better than larger droplets. With a higher mass inertia, liquid particles will not always flow at the same velocity as the air and will not flow around corners as readily, larger droplets deviating more than smaller ones. The third way fuel flows through the manifold is in a thin liquid film along the walls. This film occurs because gravity separates some droplets from the flow, and when other droplets strike the wall where the runner executes a corner. These latter two types of liquid fuel flow make it difficult to deliver the same air–fuel ratio to each of the cylinders. The length of a runner to a given cylinder and the bends in it will influence the amount of fuel that gets carried by a given air flow rate. The liquid film on the manifold walls also makes it difficult to have precise throttle control. When the throttle position is changed quickly and the air flow rate changes, the time rate of change of fuel flow will be slower due to this liquid wall film.

Gasoline components evaporate at different temperatures and at different rates. Because of this, the composition of vapor in the air flow will not be exactly the same as that of the fuel droplets carried by the air or the liquid film on the manifold walls. The air–fuel mixture that is then delivered to each cylinder can be quite different, both in composition and in air–fuel ratio. One result of this is that the possibility of knock problems will be different in each cylinder. The minimum fuel octane number that can be used in the engine is dictated by the worst cylinder (i.e., the cylinder with the greatest knock problem). This problem is further complicated by the fact that the engine is operated over a range of throttle settings. At part throttle, there is a lower total pressure in the intake manifold, and this changes the evaporation rate of the various fuel components. Most of these problems are reduced or eliminated by using multipoint port fuel injection, with each cylinder receiving its own individual fuel input.

5.2 VOLUMETRIC EFFICIENCY OF SI ENGINES

It is desirable to have maximum volumetric efficiency in the intake of any engine. This will vary with engine speed, as shown in Fig. 5-1, which represents the efficiency curve of reciprocating engines. There will be a certain engine speed at which volumetric efficiency will be maximum, decreasing at both higher and lower speeds. There are many physical and operating variables that shape this curve. These will be examined.

Fuel

In a naturally aspirated engine, volumetric efficiency will always be less than 100% because fuel is also being added and the volume of fuel vapor will displace some incoming air. The type of fuel, as well as how and when it is added, will determine how much the volumetric efficiency is affected. Systems with carburetors or throttle body injection add fuel early in the intake flow and generally have lower overall volumetric efficiency. This is because the fuel will immediately start to evaporate and fuel vapor will

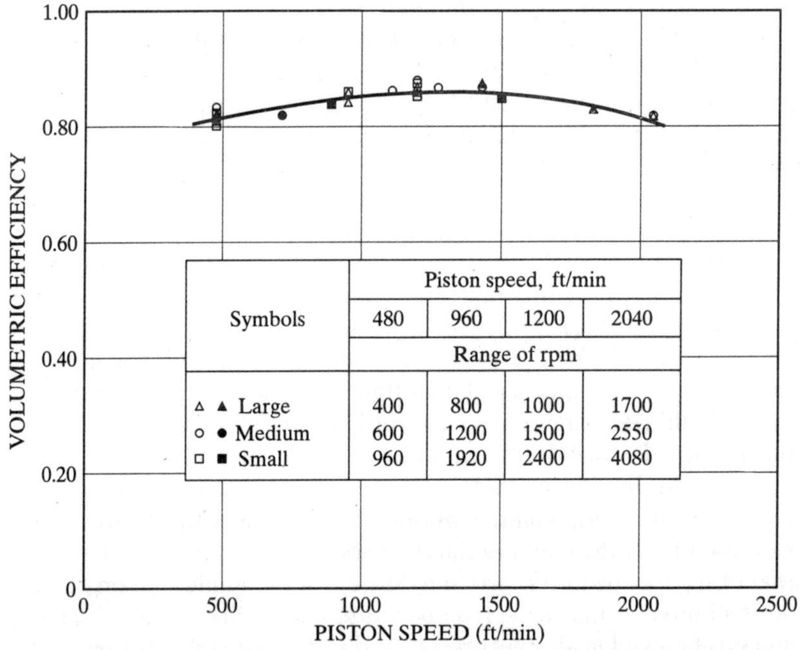

FIGURE 5-1

Volumetric Efficiency of three reciprocating internal combustion engines as a function of piston speed or engine speed. Reprinted with permission from *The Internal Combustion Engine in Theory and Practice* by C. F. Taylor, © MIT Press, [120].

displace incoming air. Multipoint injectors, which add fuel at the intake valve ports, will have better efficiency because no air is displaced until after the intake manifold. Fuel evaporation does not occur until the flow is entering the cylinder at the intake valve. Those engines that inject fuel directly into the cylinders after the intake valve is closed will experience no volumetric efficiency loss due to fuel evaporation. Manifolds with late fuel addition can be designed to further increase volumetric efficiency by having larger diameter runners. High velocity and turbulence to promote evaporation are not needed. They can also be operated cooler, which results in a denser inlet air flow.

Those fuels with a smaller air–fuel ratio, such as alcohol, will experience a greater loss in volumetric efficiency. Fuels with high heat of vaporization will regain some of this lost efficiency due to the greater evaporation cooling that will occur with these fuels. This cooling will create a denser air–fuel flow for a given pressure, allowing for more air to enter the system. Alcohol has high heat of vaporization, so some efficiency lost due to AF is gained back again.

Gaseous fuels, such as hydrogen and methane, displace more incoming air than liquid fuels, which are only partially evaporated in the intake system. This must be considered when trying to modify engines made for gasoline fuel to operate on these gaseous fuels. It can be assumed that fuel vapor pressure in the intake system is between 1% and 10% of total pressure when gasoline-type liquid fuel is being used. When gaseous fuels or alcohol is being used, the fuel vapor pressure is often greater than 10% of the total. Intake manifolds can be operated much cooler when gaseous fuel is used, as no vaporization is required. This will gain back some lost volumetric efficiency.

The later that fuel vaporizes in the intake system, the better is the volumetric efficiency. On the other hand, the earlier that fuel vaporizes, the better are the mixing process and cylinder-to-cylinder distribution consistency.

In older carbureted automobile engines, somewhere around 60% evaporation of the fuel in the intake manifold was considered desirable, with the rest of the evaporation taking place during the compression stroke and combustion process. If fuel is evaporated too late in the cycle, a small percent of the high-molecular-weight components may not vaporize. Some of this unvaporized fuel ends up on the cylinder walls, where it gets by the piston rings and dilutes the lubricating oil in the crankcase.

Heat Transfer—High Temperature

All intake systems are hotter than the surrounding air temperature and will consequently heat the incoming air. This lowers the density of the air, which reduces volumetric efficiency. Intake manifolds for carbureted systems or throttle body injection systems are purposely heated to enhance fuel evaporation. At lower engine speeds, the air flow rate is slower and the air remains in the intake system for a longer time. It thus gets heated to higher temperatures at low speeds, which lowers the volumetric efficiency curve in Fig. 5-1 at the low-speed end.

Some systems that have been tried inject small amounts of water into the intake manifold. This is to improve the volumetric efficiency by increasing the resulting evaporative cooling that occurs. Probably the most successful use of this principle was with large high-performance aircraft engines of World War II. Power was increased by substantial amounts when water injection was added to some of these engines.

Valve Overlap

At TDC at the end of the exhaust stroke and the beginning of the intake stroke, both intake and exhaust valves are open simultaneously for a brief moment. When this happens, some exhaust gas can get pushed through the open intake valve back into the intake system. The exhaust then gets carried back into the cylinder with the intake air–fuel charge, displacing some of the incoming air and lowering volumetric efficiency. This problem is greatest at low engine speeds, when the real time of valve overlap is greater and there is low pressure in the intake manifold. This effect lowers the efficiency curve in Fig. 5-1 at the low engine speed end. Other factors that affect this problem are the intake and exhaust valve location and engine compression ratio.

Fluid Friction Losses

Air moving through any flow passage or past any flow restriction undergoes a pressure drop. For this reason, the pressure of the air entering the cylinders is less than the surrounding atmospheric air pressure, and the amount of air entering the cylinder is subsequently reduced. The viscous flow friction that affects the air as it passes through the air filter, carburetor, throttle plate, intake manifold, and intake valve reduces the volumetric efficiency of the engine intake system. Viscous drag, which causes the pressure loss, increases with the square of flow velocity. This results in decreasing the efficiency on the high-speed end of the curve in Fig. 5-1. Much development work has been done to reduce pressure losses in air intake systems. Smooth walls in the intake manifold, the avoidance of sharp corners and bends, elimination of the carburetor, and close-fitting parts alignment with no gasket protrusions all contribute to decreasing intake pressure loss. One of the greatest flow restrictions is the flow through the intake valve. To reduce this restriction, the intake valve flow area has been increased by building multivalve engines having two or even three intake valves per cylinder.

Air–fuel flow into the cylinders is usually diverted into a rotational flow pattern within the cylinder. This is done to enhance evaporation, mixing, and flame speed and will be explained in the next chapter. This flow pattern is accomplished by shaping the intake runners and contouring the surface of the valves and valve ports. This increases inlet flow restriction and decreases volumetric efficiency.

If the diameter of the intake manifold runners is increased, flow velocity will be decreased and pressure losses will be decreased. However, a decrease in velocity will result in poorer mixing of the air and fuel and less accurate cylinder-to-cylinder distribution. Compromises in design must be made.

In some low-performance, high fuel-efficient engines, the walls of the intake manifold are made rough to enhance turbulence to get better air–fuel mixing. High volumetric efficiency is not as important in these engines.

Choked Flow

The extreme case of flow restriction is when choked flow occurs at some location in the intake system. As air flow is increased to higher velocities, it eventually reaches sonic velocity at some point in the system. This *choked flow* condition is the maximum flow rate that can be produced in the intake system regardless of how controlling conditions are changed. The result of this is a lowering of the efficiency curve on the high-speed end in Fig. 5-1. Choked flow occurs in the most restricted passage of the system, usually at the intake valve or in the carburetor throat on those engines with carburetors.

Closing Intake Valve After BDC

The timing of the closure of the intake valve affects how much air ends up in the cylinder. Near the end of the intake stroke, the intake valve is open and the piston is moving from TDC towards BDC. Air is pushed into the cylinder through the open intake valve due to the vacuum created by the additional volume being displaced by the piston. There is a pressure drop in the air as it passes through the intake valve, and the pressure inside the cylinder is less than the pressure outside the cylinder in the intake manifold. This pressure differential still exists the instant the piston reaches BDC and air is still entering the cylinder. This is why the closing of the intake valve is timed to occur aBDC. When the piston reaches BDC, it starts back towards TDC and in so doing starts to compress the air in the cylinder. Until the air is compressed to a pressure equal to the pressure in the intake manifold, air continues to enter the cylinder. The ideal time for the intake valve to close is when this pressure equalization occurs between the air inside the cylinder and the air in the manifold. If it closes before this point, air that was still entering the cylinder is stopped and a loss of volumetric efficiency is experienced. If the valve is closed after this point, air being compressed by the piston will force some air back out of the cylinder, again with a loss in volumetric efficiency.

This valve-closing point in the engine cycle, at which the pressure inside the cylinder is the same as the pressure in the intake manifold, is highly dependent on engine speed. At high engine speeds, there is a greater pressure drop across the intake valve due to the higher flow rate of air. This, plus the reduced real cycle time at high speed, would indicate that ideally, the intake valve should close at a later cycle position. On the other hand, at low engine speeds the pressure differential across the intake valve is less, and pressure equalization would occur earlier after BDC. Ideally, the valve should close at an earlier position in the cycle at low engine speeds.

The position where the intake valve closes on many engines is controlled by a camshaft and cannot change with engine speed. On these engines, the closing cycle position is designed for one engine speed, depending on the use for which the engine is designed. This is no problem for a single-speed industrial engine but is a compromise for an automobile engine that operates over a large speed range. The result of this single-position valve timing is to reduce volumetric efficiency of the engine at both high and low speeds. Some automobile engines have limited adjustment on their camshafts and can change valve closing somewhat to match engine speed. When 42-volt electrical systems become standard, camshafts will eventually be replaced with electronic valve actuators. This will allow more complete control of valve timing and thus improve volumetric efficiency.

Intake Tuning

When gas flows in a pulsed manner, as in the intake manifold of an engine, pressure waves are created that travel down the length of the flow passage. The wavelength of these waves is dependent on pulse frequency and air flow rate or velocity. When these waves reach the end of the runner or an obstruction in the runner, they create a reflected pressure wave back along the runner. The pressure pulses of the primary waves and the reflected waves can reinforce or cancel each other, depending on whether they are in or out of phase.

If the length of the intake manifold runner and the flow rate are such that the pressure waves reinforce at the point where the air enters the cylinder through the intake valve, the pressure pushing the air in will be slightly higher, and slightly more air will enter the cylinder. When this happens, the system is **tuned** and volumetric efficiency is increased. However, when the flow rate of air is such that the reflected pressure pulses are out of phase with the primary pulses, the pressure pushing air into the cylinder is slightly reduced and volumetric efficiency is lower. All older engines and many modern engines have passive constant-length intake runner systems that can be tuned for one engine speed (i.e., length of runner designed for one certain air flow rate and pulse timing). At other speeds the system will be out of tune, and volumetric efficiency will be less at both higher and lower engine speeds.

Some modern engines have active intake systems that can tune the manifold over a range of engine speeds. This is done by changing the length of the intake runners to match the air flow rate at various engine operating conditions. Various methods are used to accomplish this. Some systems have single-path runners that can be changed in length during operation by various mechanical methods. Other systems have dual-path runners with controlling valves and/or secondary throttle plates. As the engine speed changes, the air is directed through runners of various lengths, which optimally tune the flow for the current speed. All active systems are controlled by the EMS.

Exhaust Residual

During the exhaust stroke, not all of the exhaust gases get pushed out of the cylinder by the piston, a small residual being trapped in the clearance volume. The amount of this residual depends on the compression ratio, and somewhat on the location of the valves and valve overlap. In addition to displacing some incoming air, this exhaust gas residual interacts with the air in two other ways. When the very hot gas mixes with the incoming air, it heats the air, lowers the air density, and decreases volumetric efficiency. This is counteracted slightly, however, by the partial vacuum created in the clearance volume when the hot exhaust gas is in turn cooled by the incoming air.

EGR

In all modern automobile engines and in many other engines, some exhaust gas is recycled (EGR) into the intake system to dilute the incoming air. This reduces combustion temperatures in the engine, which results in less nitrogen oxides in the exhaust. Up to about 20% of exhaust gases will be diverted back into the intake manifold, depending on how the engine is being operated. Not only does this exhaust gas displace some incoming air, but it also heats the incoming air and lowers its density. Both of these interactions lower the volumetric efficiency of the engine.

In addition, engine crankcases are vented into the intake systems, displacing some of the incoming air and lowering the volumetric efficiency. Gases forced through the crankcase can amount to about 1% of the total gas flow through the engine.

Example Problem 5-1

A 5.6-liter V8 engine with a compression ratio of 10.2:1 is equipped with cylinder cutout, which converts it to a 2.8 liter four-cylinder engine at low load requirements. The engine operates on an Otto cycle using gasoline, and with eight cylinders at 1800 RPM it has an AF = 14.9, a volumetric efficiency of 57%, a combustion efficiency of 91%, and a mechanical efficiency of 92%. If cylinder cutout occurs, the engine speeds up to produce the same brake power output. Using only four cylinders at this condition the engine has an AF = 14.2, a volumetric efficiency of 66%, and a combustion efficiency of 99%, but a mechanical efficiency of only 90%.
Calculate:

1. percent reduction in fuel consumption operating on four cylinders to produce same brake power output
2. engine speed needed to produce same power output using only four cylinders

(1) Equation (3-31) gives indicated thermal efficiency of Otto cycle:

$$\eta_t = 1 - (1/r_c)^{k-1} = 1 - (1/10.2)^{1.35-1} = 0.556 = 55.6\%$$

Use Eq. (2-71) for air flow into an eight cylinder engine:

$$\begin{aligned}\dot{m}_a &= \eta_v \rho_a V_d N/n \\ &= (0.57)(1.181\ \text{kg/m}^3)(0.0056\ \text{m}^3/\text{cycle})(1800/60\ \text{rev/sec})/(2\ \text{rev/cycle}) \\ &= 0.0565\ \text{kg/sec}\end{aligned}$$

Use Eq. (2-55) for fuel flow into an eight cylinder engine:

$$\dot{m}_f = \dot{m}_a/\text{AF} = (0.0565\ \text{kg/sec})/(14.9) = 0.00379\ \text{kg/sec}$$

Equations (2-65) and (2-47) give brake power output of eight cylinder engine:

$$\begin{aligned}\dot{W}_b &= \eta_m \dot{W}_i = \eta_m \eta_t \dot{m}_f Q_{HV} \eta_c \\ &= (0.92)(0.556)(0.00379\ \text{kg/sec})(43{,}000\ \text{kJ/kg})(0.91) \\ &= 75.9\ \text{kW}\end{aligned}$$

Equations (2-65) and (2-47) are now used to find fuel flow rate needed for the same brake power output, using only four cylinders:

$$(75.9\ \text{kW}) = (0.90)(0.556)\dot{m}_f(43{,}000\ \text{kJ/kg})(0.99) \quad \dot{m}_f = 0.00356\ \text{kg/sec}$$

The percent decrease in fuel usage is

$$\Delta\% = [(0.00356 - 0.00379)/(0.00379)](100) = \underline{-6.1\%}$$

(2) Use Eq. (2-55) to find air flow rate needed with four cylinders:

$$\dot{m}_a = (\text{AF})\dot{m}_f = (14.2)(0.00356\ \text{kg/sec}) = 0.0506\ \text{kg/sec}$$

Equation (2-71) gives the engine speed needed to produce same brake power output:

$$\begin{aligned}N &= n\dot{m}_a/\eta_v \rho_a V_d \\ &= [(2\ \text{rev/cycle})(0.0506\ \text{kg/sec})]/[(0.66)(1.181\ \text{kg/m}^3)(0.0028\ \text{m}^3/\text{cycle})] \\ &= \underline{46.4\ \text{rev/sec} = 2782\ \text{RPM}}\end{aligned}$$

5.3 INTAKE VALVES

Intake valves of most IC engines are poppet valves that are spring loaded closed and pushed open at the proper cycle time by the engine camshaft, shown schematically in Fig. 1-13. Much rarer are rotary valves or sleeve valves, found on some engines.

Most valves and valve seats against which they close are made of hard alloy steel or, in some rarer cases, ceramic. They are connected by hydromechanical or mechanical linkage to the camshaft. Ideally, they would open and close almost instantaneously at the proper times. This is impossible in a mechanical system, and slower openings and closings are necessary to avoid wear, noise, and chatter. The lobes on a camshaft are designed to give quick but smooth opening and closing without bounce at the mechanical interface. This requires some compromise in the speed of valve actuation. When camshafts are replaced with electronic actuators, valves will be able to open and close much faster, with a resulting improvement in engine operation.

Earlier engines had camshafts mounted close to the crankshaft and the valves mounted in the engine block. As combustion chamber technology progressed, valves were moved to the cylinder head (overhead valves), and a long mechanical linkage system (push rods, rocker arms, tappets) was required. This was improved by also mounting the camshaft in the engine head (i.e., overhead cam engines). Most modern automobile engines have one or two camshafts mounted in the head of each bank of cylinders. The closer the camshaft is mounted to the stems of the valves, the greater is the mechanical efficiency of the system.

The distance that a valve opens (dimension 1 in Fig. 5-2) is called **valve lift** and is generally on the order of a few millimeters to more than a centimeter, depending on the engine size. The valve lift is usually about 5 to 10 mm for automobile engines. Generally,

$$l_{max} < d_v/4 \tag{5-1}$$

where

l_{max} = valve lift when valve is fully open
d_v = diameter of valve

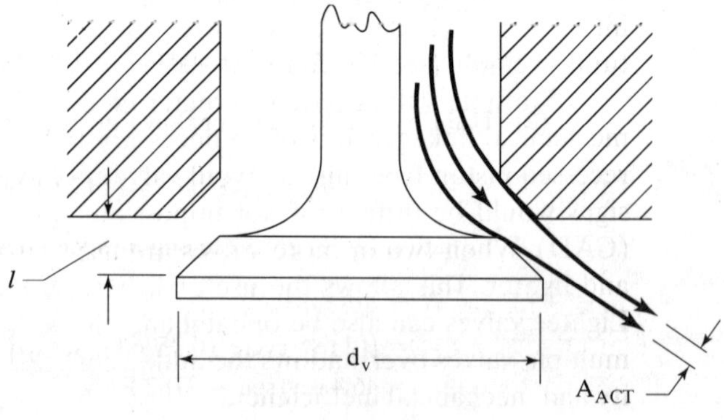

FIGURE 5-2

Flow through a poppet valve. When flow separates from the surface at the corners, the actual flow area is less than the geometric passage area of the valve. The ratio of these areas is called the discharge coefficient. Valve diameter is d_v and valve lift is 1.

The angle of the valve surface at the interface with the valve seat is generally designed to give minimum flow restriction. As air flows around corners, the streamlines separate from the surface, and the actual cross-sectional area of flow is less than the flow passage area, as shown in Fig. 5-2. The ratio of the actual flow area to the flow passage area is called the *valve discharge coefficient*:

$$C_{Dv} = A_{act}/A_{pass} \tag{5-2}$$

The passage area of flow is

$$A_{pass} = \pi d_v l \tag{5-3}$$

Shape and angle of valve surfaces are sometimes designed to give special mass flow patterns to improve overall engine efficiency.

Intake valves offer the greatest restriction to incoming air in most engines. This is especially true at higher engine speeds. Various empirical formulas can be found in technical engine literature for sizing intake valves. Equations giving the minimum valve intake area necessary for a modern engine can be given [40] in the form

$$A_i = CB^2[(\overline{U}_p)_{max}/c_i] = (\pi/4)d_v^2 \tag{5-4}$$

where

C = constant having a value of about 1.3
B = bore
$(\overline{U}_p)_{max}$ = average piston speed at maximum engine speed
c_i = speed of sound at inlet conditions
d_v = diameter of valve

A_i is the total inlet valve area for one cylinder, whether it has one, two, or three intake valves.

On many newer engines with overhead valves and small fast-burn combustion chambers, there is often not enough wall space in the combustion chambers to fit the spark plug and exhaust valve and still have room for an intake valve large enough to satisfy Eq. (5-4). For this reason, most engines are now built with more than one intake valve per cylinder. Two or three smaller intake valves give more flow area and less flow resistance than one larger valve, as was used in older engines. At the same time, these two or three intake valves, along with usually two exhaust valves, can be more easily fit into a given cylinder head size with enough clearance to maintain the required structural strength. (see Fig. 5-3.)

Multiple valves require greater complexity of design with more camshafts and mechanical linkages. It is often necessary to have specially-shaped cylinder heads and recessed piston faces just to avoid valve-to-valve or valve-to-piston contact. These designs would be difficult if not impossible without the use of computer-aided design (CAD). When two or more valves are used instead of one, the valves will be smaller and lighter. This allows the use of lighter springs and reduces forces in the linkage. Lighter valves can also be opened and closed faster. Greater volumetric efficiency of multiple valves overshadows the added cost of manufacturing and the added complexity and mechanical inefficiency.

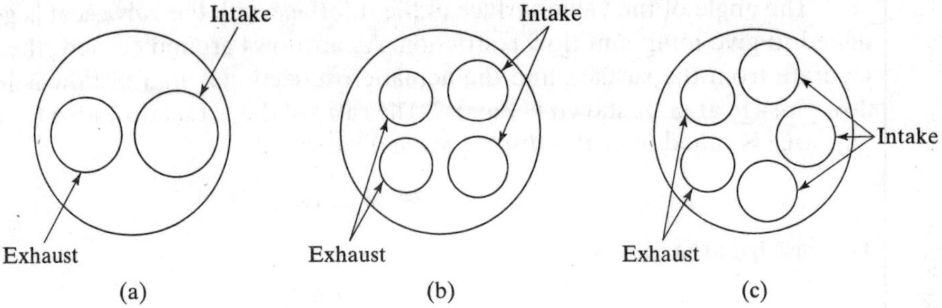

FIGURE 5-3

Possible valve arrangements for a modern overhead valve engine. For a given combustion chamber size, two or three smaller intake valves will give greater flow area than one larger valve. For each cylinder, the flow area of the intake valve(s) is generally about 10 percent greater than the flow area of the exhaust valve(s). (**a**) Most early overhead valve engines (1950s–1980s) and a few modern engines. (**b**) Most present-day automobile engines. (**c**) Some modern high-performance automobile engines.

Some engines with multiple intake valves are designed so that only one intake valve per cylinder operates at low speed. As speed is increased, less time per cycle is available for air intake, and the second (and sometimes third) valve actuates, giving additional inlet flow area. This allows for increased control of the flow of air within the cylinder at various speeds, which results in more efficient combustion. In some of these systems, the valves will have different timing. The low-speed valve will close at a relatively early point aBDC. When operating, the high-speed valve(s) will then close at a later position (up to 20° later) to avoid lowering the volumetric efficiency, as explained in Section 5.2.

Mass flow through the intake valve into a cylinder is shown in Fig. 5-4. Reverse flow can result when valve overlap occurs near TDC. Reverse flow out of the cylinder will also occur at lower engine speeds as the intake valve is closing aBDC, as previously explained.

Intake valves normally start to open somewhere between 10° and 25° bTDC and should be totally open by TDC to get maximum flow during the intake stroke. The higher the speed for which the engine is designed, the earlier in the cycle the intake valve will be opened. In most engines valve timing is set for one engine speed, with losses occurring at any lower speed or higher speed. At lower than design speed the intake valve opens too early, creating valve overlap that is larger than necessary. This problem is made worse because low engine speeds generally have low intake manifold pressures. At higher than design speeds, the intake valve opens too late and intake flow has not been fully established at TDC, with a loss in volumetric efficiency. Automobile engines operate at many different speeds, with valve timing set for optimization at only one speed. Industrial engines which operate at only one speed can obviously have their valve timing set for that speed. Modern automobile engines have longer valve overlap because of their higher operating speeds.

Intake valves normally finish closing about 40°–50° aBDC for engines operating on an Otto cycle. Again, the correct point of closing can be designed for only one engine speed, with increased losses at either higher or lower speeds.

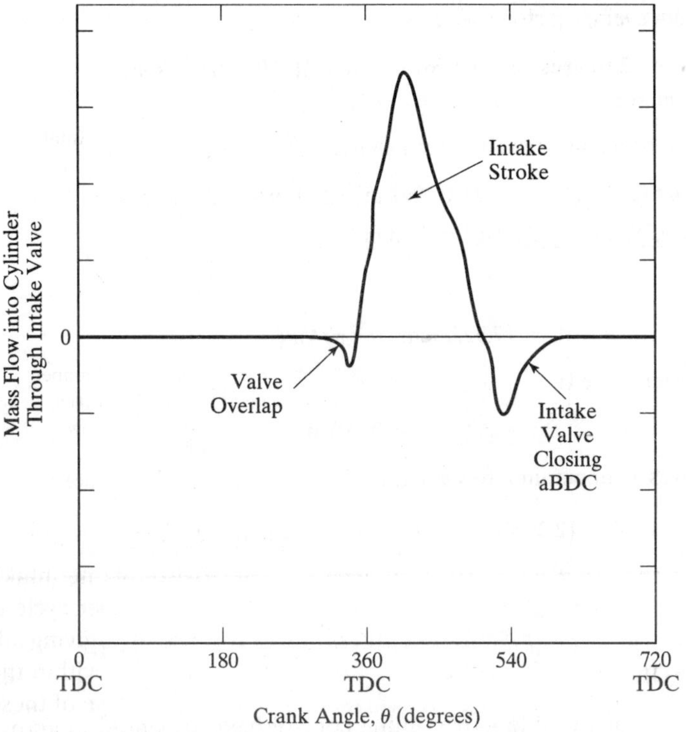

FIGURE 5-4
Flow of air–fuel mixture through the intake valve(s) into an engine cylinder. Possible backflow can occur during valve overlap and when the intake valve closes after BDC. Adapted from [28].

Example Problem 5-2

A 2.8-liter four-cylinder square engine (bore = stroke) with two intake valves per cylinder is designed to have a maximum speed of 7500 RPM. Intake temperature is 60°C. Calculate:

1. intake valve area
2. diameter of intake valves
3. maximum valve lift

Use Eq. (3-1j) to find the speed of sound at inlet conditions:

$$c_i = \sqrt{kRT} = \sqrt{(1.40)(287\ \text{J/kg-K})(333\ \text{K})} = 366\ \text{m/sec}$$

Here,

R = gas constant
T = temperature
$k = c_p/c_v = 1.40$ The value of 1.40 is used here because this corresponds to the inlet temperature of 333 K.

For one cylinder,

$$V_d = (2.8\ \text{L})/4 = 0.7\ \text{L} = 0.0007\ \text{m}^3$$

Use Eq. (2-8) to get stroke with $B = S$:

$$V_d = (\pi/4)B^2S = (\pi/4)S^3 = 0.0007\ \text{m}^3$$
$$S = 0.0962\ \text{m} = 9.62\ \text{cm} = B$$

Use Eq. (2-2) to get maximum average piston speed:

$$(\overline{U}_p)_{max} = 2SN = (2\ \text{strokes/rev})(0.0962\ \text{m/stroke})(7500/60\ \text{rev/sec}) = 24.1\ \text{m/sec}$$

(1) Equation (5-4) gives total intake valve area needed:

$$A_i = 1.3B^2(\overline{U}_p)_{max}/c_i = (1.3)(0.0962\ \text{m})^2(24.1\ \text{m/sec})/(366\ \text{m/sec}) = \underline{0.000792\ \text{m}^2 = 7.92\ \text{cm}^2 = 1.23\ \text{in.}^2}$$

(2) For each valve,

$$A_i = (7.92/2)\text{cm}^2 = (\pi/4)d_v^2$$

The diameter of each valve is

$$\underline{d_v = 2.25\ \text{cm} = 0.886\ \text{in.}}$$

(3) Equation (5-1) gives an upper limit to valve lift:

$$l_{max} = d_v/4 = (2.25\ \text{cm})/4 = \underline{0.56\ \text{cm} = 5.6\ \text{mm} = 0.22\ \text{in.}}$$

5.4 VARIABLE VALVE CONTROL

In recent years several forms of variable valve timing control have appeared in automobile engines. These systems allow for more efficient engine operation by varying the times at which valves open, duration of opening, and valve overlap period (Fig. 5-5). The systems are called by various names in automobile advertising and technical literature: variable valve control (VVC), variable valve actuation (VVA), variable valve event (VVE), variable valve system (VVS), variable timing control (VTC), and others. This book will use VVC. In addition to variable timing, the most modern systems also allow for variable valve lift.

For over a hundred years the vast majority of automobiles and trucks used engines that operated either on the basic Otto cycle (SI engines) or on the Diesel or Dual cycles (CI engines). The valves in these engines were operated with camshafts, which were generally designed for average operation and had no method of varying either timing or lift. This gave adequate operation at medium speeds and loads, but was not optimized for high speed or for low speed and idle.

At high engine speed, real time of a cycle is less and more air–fuel per cycle is generally required. Optimization for this situation requires the intake valve to open earlier in the cycle, have a longer open duration, and if possible have a greater lift. The exhaust valve should also have greater lift and open earlier to allow more real time for blowdown; it should also close a little later at the end of the exhaust stroke. The longer valve overlap in cycle time could be tolerated because of the higher intake manifold pressure and the shorter real time. At low engine speed and idle, real time of a cycle is greater and less air–fuel input is needed. Both the intake valve and the exhaust valve should open later and close earlier. At low speed, there is very low pressure in the intake system so less valve overlap is desirable. If overlap is too great, a large exhaust residual

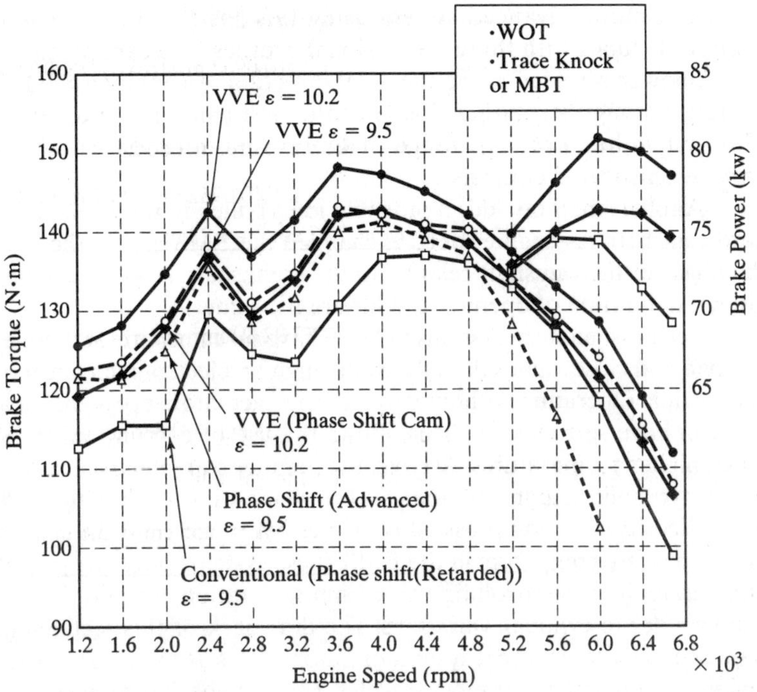

FIGURE 5-5
Torque as a function of engine speed for engine using variable valve control (VVC). The upper curves, which are torques using various valve timing controls, show improvement over conventional set control data of lower curve. Reprinted with permission from SAE Paper No. 011224 © 2001 SAE International, [226].

backflows into the intake manifold and displaces some of the incoming air–fuel. This is why a fuel-rich mixture is desirable for good combustion at low speed. The intake valve lift should be reduced at low engine speed so that inlet flow velocity remains high enough to provide a proper flow pattern and mixing.

Some very early systems used engine lubrication oil and bleed holes in the mechanical components connecting the camshaft with the valves. At different engine speeds, real time of a cycle would change, and the flow of oil would slightly change the hydromechanical cam–valve connection, and thus change valve timing. Since the late 1990s, engines with VVC have used several methods to obtain this goal. Most of these earlier systems worked only on the intake valves and had no lift variation. The most modern of systems can now control both intake and exhaust valves, timing and lift.

One method uses a camshaft with dual lobes for each valve, one for high speed and one for low speed. The camshaft is mounted such that it can be moved fore and aft along its axis of rotation. At low engine speeds the camshaft is positioned such that the low-speed lobes are in contact with the valve mechanisms. At some preset higher speed the camshaft moves and brings the high-speed lobes in contact with the valves. Various systems use mechanical, hydraulic, or electrical means to move the camshaft. This system is an improvement, but only gives optimization at two engine speeds. A more advanced system using this basic method incorporates a camshaft using wide lobes with three dimensional profiles. The cam profile which contacts the cam follower is varied along the axial direction. As engine speed is changed the cam is moved along its rotating axis until the best cam profile for that speed is utilized. This system has a limited range of timing and duration variability, and a need for more sophisticated controls.

Another method adds a movable idler pulley in the chain–belt drive of the standard camshaft. As engine speed is changed the EMS moves the idler pulley, changing the phase of the camshaft relative to the crankshaft. A valve can be opened earlier or later, but duration of opening and lift are the same.

The most modern systems of VVC have no camshafts and use electrical solenoids to open and close valves directly, or through an electromechanical or electrohydraulic connection. Operating without valve springs, actuators open and close valves very fast, and can have soft closings. This allows the use of ceramic valves which can tolerate much higher temperatures. A typical system uses an electronically controlled double-acting hydraulic actuator to open and close the valves. When this is done, the temperature, viscosity, and compressibility of the hydraulic fluid must be considered. Using more powerful computers in the EMS, these systems have the potential of almost infinite variability in controlling valve timing, duration, and lift, including cycle-to-cycle and cylinder-to-cylinder variations. The drawback of this technology, when using the standard 12-volt electrical system of automobiles, is the large size of the components, making it impractical for most vehicles. This drawback disappears when a 42-volt electrical system is used and much smaller components can be used. The elimination of camshafts was a major factor in promoting the change to a 42-volt standard. Elimination of the camshaft reduces engine friction and increases its mechanical efficiency.

With total variability possible for valve timing, duration, and lift, several additional facets of engine operation can be improved, including elimination of the throttle valve, better low-speed torque, greater power, lower emissions, and better fuel economy. With timing and lift control of the intake valve(s), the throttle can be eliminated, and inlet air flow can be regulated with valve control (i.e., Miller cycle instead of Otto cycle). This eliminates, or greatly reduces, negative pumping work, shortens the compression stroke relative to a longer expansion stroke, and allows control of cylinder inlet flow velocity at all speeds by varying lift. With two or three intake valves, each controlled separately, better overall control of the engine cycle is possible. At high speed, all valves are open with maximum lift, giving maximum volumetric efficiency and power. At low speed, some valves might not be opened, and lift is controlled to give better inlet velocity and mixing. Higher valve lift at some lower speeds can improve low-end torque. Having different timing on multiple intake valves can assist in creating a stratified air–fuel charge in the combustion chamber, very desirable in modern combustion philosophy. When a light power output is required from a large engine, the EMS can change the normal four-stroke cycle to a more efficient six-stroke cycle. With proper valve and fuel control two additional dummy strokes can be added after the normal exhaust stroke. With no fuel added and all valves fully open, these two strokes add nothing to the engine cycle, except there is now a power stroke from each cylinder only on every third revolution. For the same power, engine RPM would have to be increased, and would run at a more efficient higher speed.

By controlling each valve separately on engines equipped with multiple intake valves, major improvements can be made in fuel consumption and emissions. With proper programming of ignition timing, valve timing, and valve lift, optimum combustion can be obtained at all engine speeds. Turbulence, swirl, tumble, air–fuel ratio, and charge stratification can each be adjusted to give minimum fuel consumption and/or minimum emissions. Unfortunately, these often can't be obtained simultaneously.

Example Problem 5-3

A small five-cylinder CI automobile engine operates with a 42 volt electrical system and variable valve timing control without a camshaft. At operational speed of 3500 RPM the intake valve opens 32° bTDC and closes 57° aBDC. The exhaust valve opens 52° bBDC and closes 21° aTDC. When the engine is idling at 400 RPM the intake valve opens 12° bTDC and closes 18° aBDC. The exhaust valve opens 21° bBDC and closes 8° aTDC.
Calculate:

1. valve overlap in crankcase degrees for operational speed and for idle speed
2. valve overlap in real time for operational speed and for idle speed

(1) Valve overlap at operational speed:

$$VO = (\angle \text{ when IVO}) + (\angle \text{ when EVC}) = (32° \text{ bTDC}) + (21° \text{ aTDC}) = \underline{53°}$$

Valve overlap at idle condition:

$$VO = (12° \text{ bTDC}) + (8° \text{ aTDC}) = \underline{20°}$$

(2) Time of overlap at 3500 RPM:

$$t = [(53°)/(360°/\text{rev})]/[(3500 \text{ rev/min})/(60 \text{ sec/min})] = \underline{0.0025 \text{ sec}}$$

Time of overlap at idle speed:

$$t = (20/360 \text{ rev})/(400/60 \text{ rev/sec}) = \underline{0.0083 \text{ sec}}$$

5.5 FUEL INJECTION

Fuel Injectors

Fuel injectors are nozzles that inject a spray of fuel into the intake air. They are normally controlled electronically, but mechanically controlled injectors, which are cam operated, also exist (Fig. 5-6). A metered amount of fuel is trapped in the nozzle end of the injector, and a high pressure is applied to it, usually by a mechanical compression process of some kind. At the proper time, the nozzle is opened and fuel is sprayed into the surrounding air. The amount of fuel injected each cycle is controlled by injector pressure and time duration of injection.

An electronic fuel injector consists of the following basic components: valve housing, magnetic plunger, solenoid coil, helical spring, fuel manifold, and pintle (needle valve). When not activated, the coil spring holds the plunger against its seat, which blocks the inlet flow of fuel. When activated, the electric solenoid coil is excited, which moves the plunger and connected pintle. This opens the needle valve and allows fluid

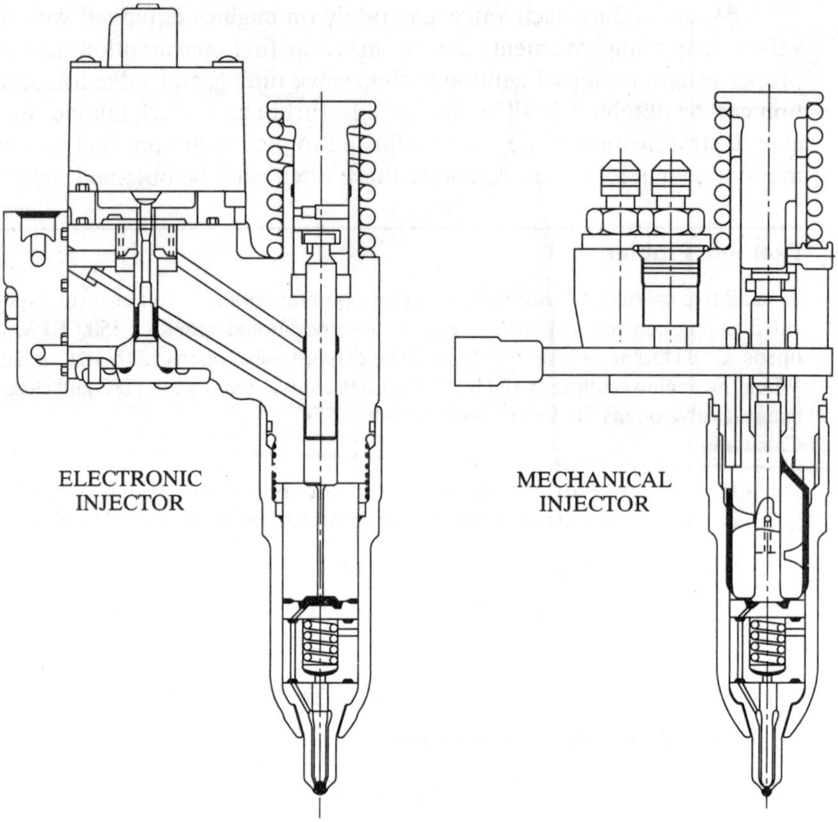

FIGURE 5-6

Electronic and mechanical fuel injectors. Reprinted with permission from SAE Paper No. 850542 © 1985 SAE International, [181].

from the manifold to be injected out the valve orifice. The valve can either be pushed opened by added pressure from the plunger, or it can be opened by being connected to the plunger, which then releases the pressurized fuel. Each valve can have one or several orifice openings, each having a diameter of about 0.2 to 1.0 mm. The fuel exits the injectors at velocities greater than 100 m/sec, and flow rates of 3 to 4 gm/sec. In mechanically controlled injectors there is no solenoid coil, and the plunger is moved by the action of a camshaft.

Some systems have a single fuel pump (**common rail**) supplying the injectors of all cylinders or a bank of cylinders. The fuel can be supplied at high pressure with the injectors acting only as a metering device. Other systems supply the fuel to the injectors at lower pressure, and the injector must then increase the pressure and meter the flow. There is generally a return line from each injector for excess fuel. Some systems have a fuel pump for each cylinder, with the pump and injector sometimes built as a single unit. Engine operating conditions and information from sensors in the engine and exhaust system are used to continuously adjust AF, timing, injection pressure, and

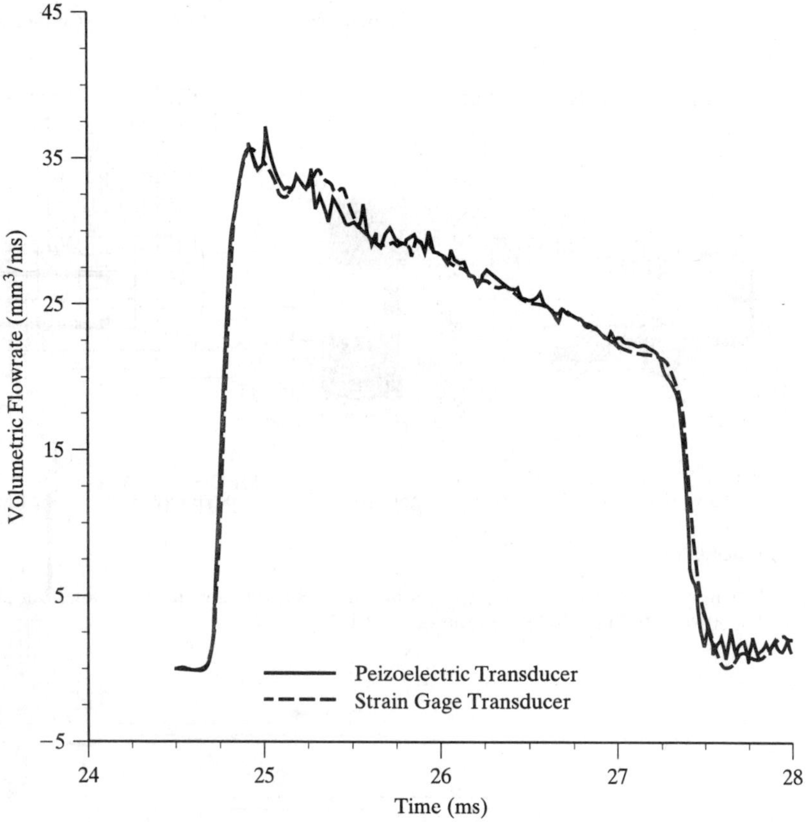

FIGURE 5-7

Flow rate of fuel as a function of time through fuel injector with peak injector pressure of 140 MPa. Reprinted with permission from SAE Paper No. 910724 © 1991 SAE International, [155].

injection duration. Engines with pump systems and controls on each cylinder can be more finely adjusted than those with single-pump systems for the entire engine or bank of cylinders.

Only moderate injector pressure is required for SI engines using port injectors or throttle body injectors (e.g., 200–300 kPa absolute), which inject into the lower pressure engine intake system. Injectors used for injecting directly into the combustion chamber (DI) operate at much higher pressures, up to 10 MPa. Higher pressure is required because injection operates against high cylinder pressure, and because finer droplet size is needed due to the very short allotted evaporation time. Many SI engine direct injection systems that use gasoline (GDI) inject a combination of fuel and air (Figs. 5-9, 5-10, 5-11). With these injectors, air is injected through a separate orifice during and immediately after the fuel injection. This type of injection greatly speeds the atomization, vaporization, and mixing of the fuel droplets, which is needed because of the extremely short time period (less than 0.008 sec at 3000 RPM).

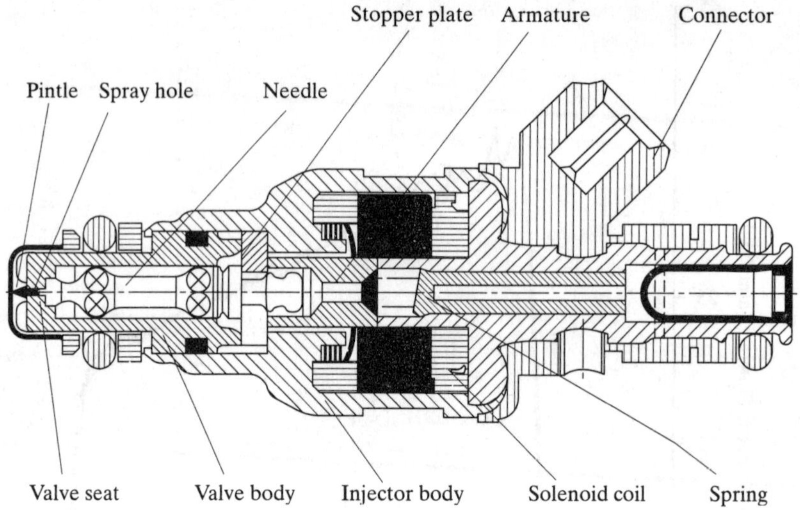

FIGURE 5-8

Schematic of electronic fuel injector for SI engine. Reprinted with permission from SAE Paper No. 870124 © 1987 SAE International, [179].

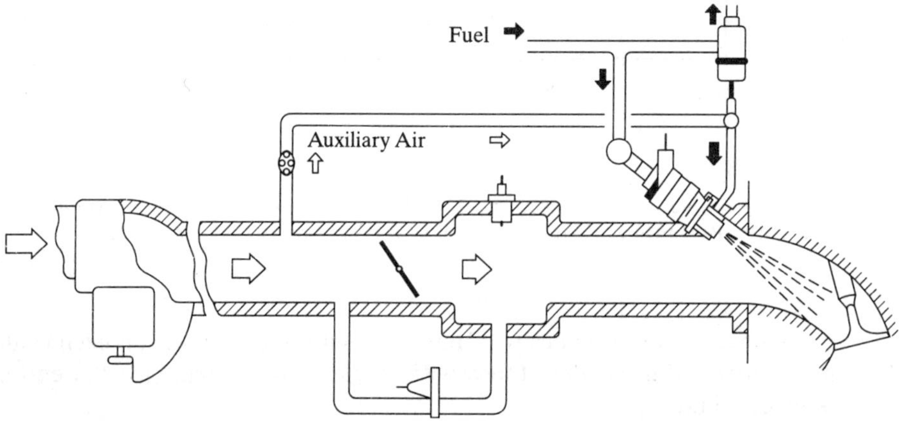

FIGURE 5-9

Schematic of port injection system using injection of air–fuel mixture. By injecting air with the fuel, evaporation and mixing are accelerated. Reprinted with permission from SAE Paper No. 870124 © 1987 SAE International, [179].

Injection pumps deliver fuel volumetrically and must be controlled to compensate for thermal expansion of the fuel at different temperatures, and for compressibility at different pressures. Time average of fuel flow into an engine at different operating conditions can vary by as much as a factor of 50.

The amount of fuel injected for each cycle can be adjusted by injection duration time, which is on the order of 1.5 to 10 ms. This corresponds to an engine crank angle rotation of

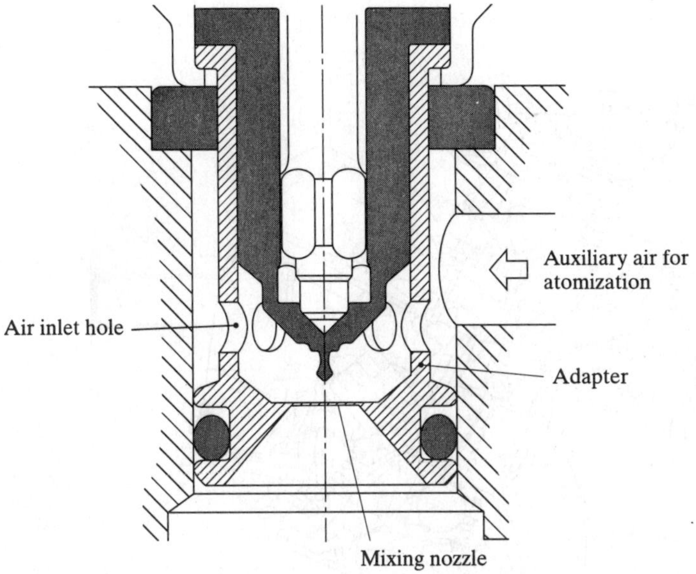

FIGURE 5-10
Schematic of fuel injector nozzle for injecting air–fuel mixture in multipoint port injection system. Reprinted with permission from SAE Paper No. 870124 © 1987 SAE International, [179].

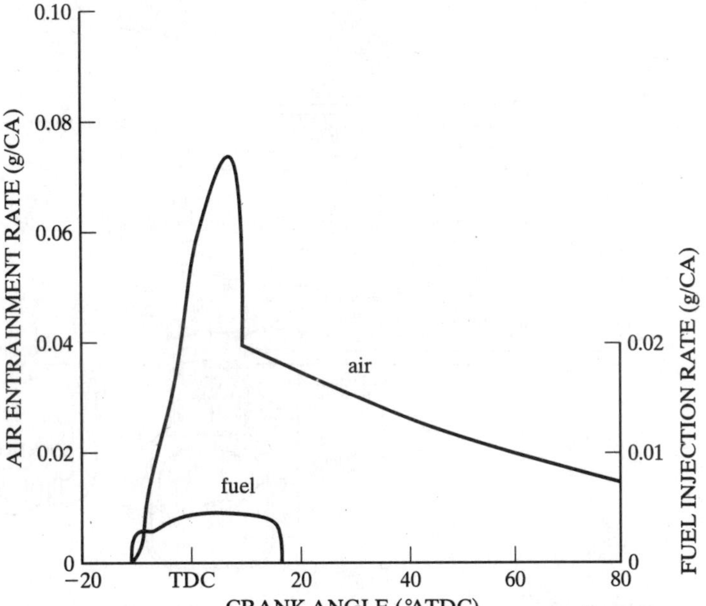

FIGURE 5-11
Mass flow rate of fuel and air injected into cylinder in typical system using air–fuel injection. Mass flow in units of (grams/crank angle degree). Reprinted with permission from SAE Paper No. 811235 © 1981 SAE International, [170].

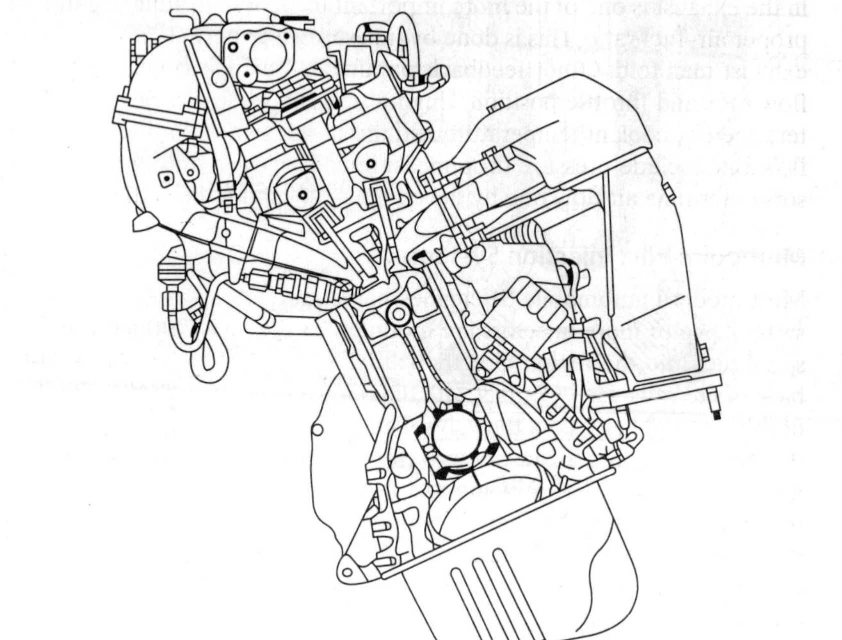

FIGURE 5-12

Engine cutaway and system layout of Toyota engine with gasoline direct injection [GDI]. Reprinted with permission from SAE Paper No. 970627 © 1997 SAE International, [241].

10° to 300°, depending on immediate operating conditions. The duration of injection is determined by feedback from engine and exhaust sensors. Sensing the amount of oxygen in the exhaust is one of the more important feedbacks in adjusting injection duration for proper air–fuel ratio. This is done by measuring the partial pressure of the oxygen in the exhaust manifold. Other feedback parameters include engine speed, temperatures, air flow rate, and throttle position. Engine start-up, when a richer mixture is needed, is determined by coolant temperature and the starter switch. Various ways of determining air flow rate include pressure drop measurement and hot-wire flow sensors. Hot-wire sensors determine air flow rate by the cooling affect on hot electrical resistors.

Multipoint Port Injection Systems

Most modern automobile SI engines have multipoint port fuel injectors. In this type of system, one or more injectors are mounted by the intake valve(s) of each cylinder. They spray fuel into the region directly behind the intake valve, sometimes directly onto the back of the valve face. Contact with the relatively hot valve surface enhances evaporation of the fuel and helps cool the valve. The injectors are usually timed to spray the fuel into the quasi-stationary air just before the intake valve opens. High liquid spray velocity is necessary to assure evaporation and mixing with the air. Because injection starts before the intake valve is open, there is a momentary pause in the air flow, and the air velocity does not promote the needed mixing and evaporation enhancement. When the valve then opens, the fuel vapor and liquid droplets are carried into the cylinder by the onrush of air, often with the injector continuing to spray. Any backflow of hot residual exhaust gas that occurs when the intake valve opens also enhances the evaporation of fuel droplets. Each cylinder has its own injector or set of injectors which give a fairly constant fuel input cycle-to-cycle and cylinder-to-cylinder, depending on the manufactured quality of the injector parts. Even with perfect control of the fuel flow, there would still be variations in AF due to the imperfect air flow cycle-to-cycle and cylinder-to-cylinder. Multipoint port injector systems are better than carburetors or throttle body injector systems at giving consistent AF delivery. Some multipoint systems have an additional auxiliary injector or injectors mounted upstream in the intake manifold to give added fuel when rich mixtures are needed for startup, idling, WOT acceleration, or high RPM operation.

Because there is such a short duration (time and length) after fuel injection for evaporation and mixing to occur, it is essential that port injectors spray very tiny droplets of fuel. Ideally, droplet size could be varied with engine speed, smaller at higher speeds, when real time is shorter.

Intake systems with multipoint injectors can be built with improved volumetric efficiency. There is no venturi throat to create a pressure drop as with a carburetor. Because little or no air–fuel mixing occurs in most of the intake manifold, high velocity is not as important, and larger diameter runners with less pressure loss can be used. There is also no displacement of incoming air with fuel vapor in the manifold.

HISTORIC—FUEL INJECTION

The first production automobile in the United States to be equipped with fuel injection was the 1957 Chevrolet Corvette. Out of a total of 6339 Corvettes built that year, 240 were equipped with Rochester Ramjet, continuous flow, throttle body fuel injectors [78].

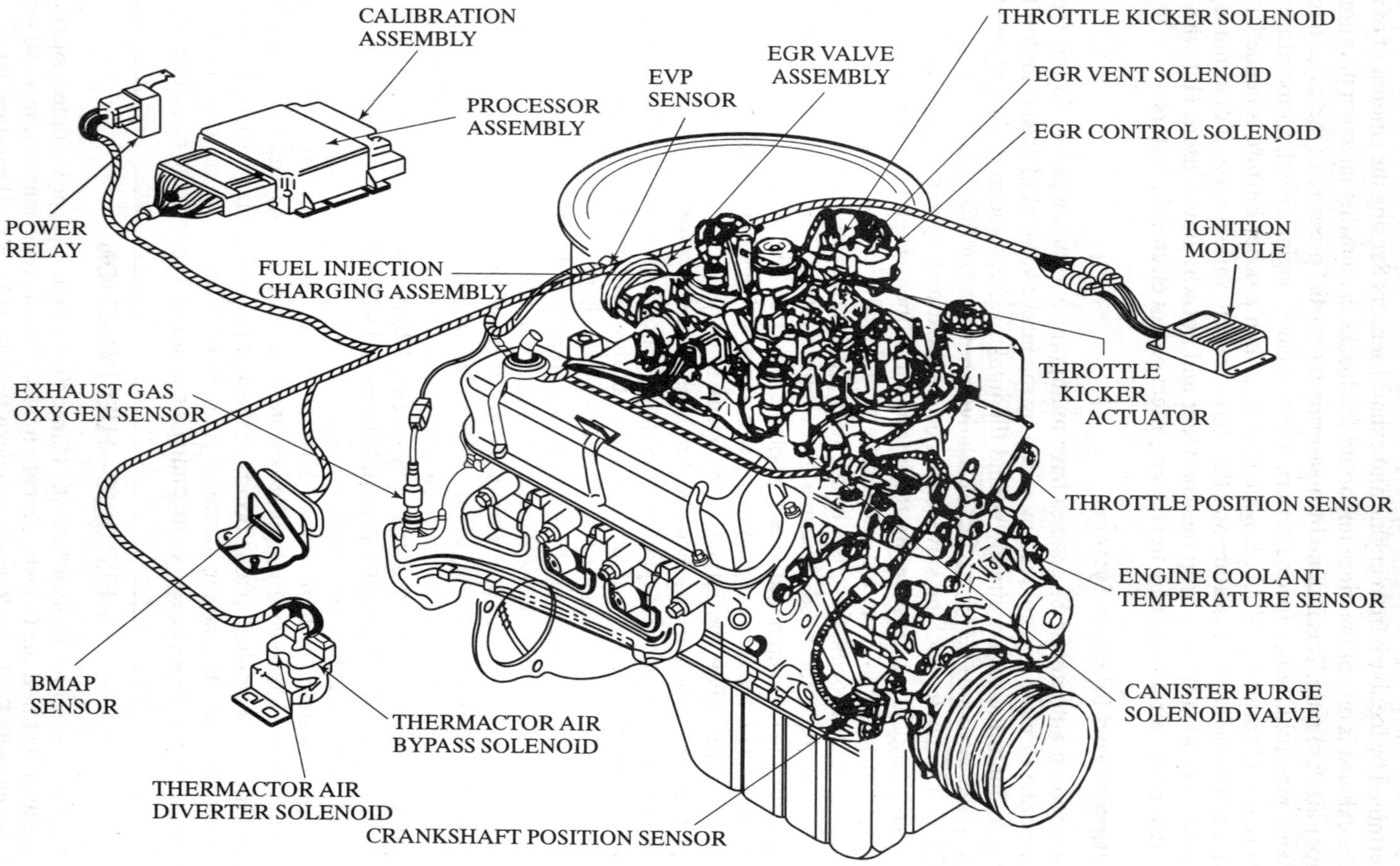

FIGURE 5-13

Cutaway view of electronic two-injector throttle-body fuel injection system used in 1980 Ford automobiles. Reprinted with permission from SAE Paper No. 790742 © 1979 SAE International, [167].

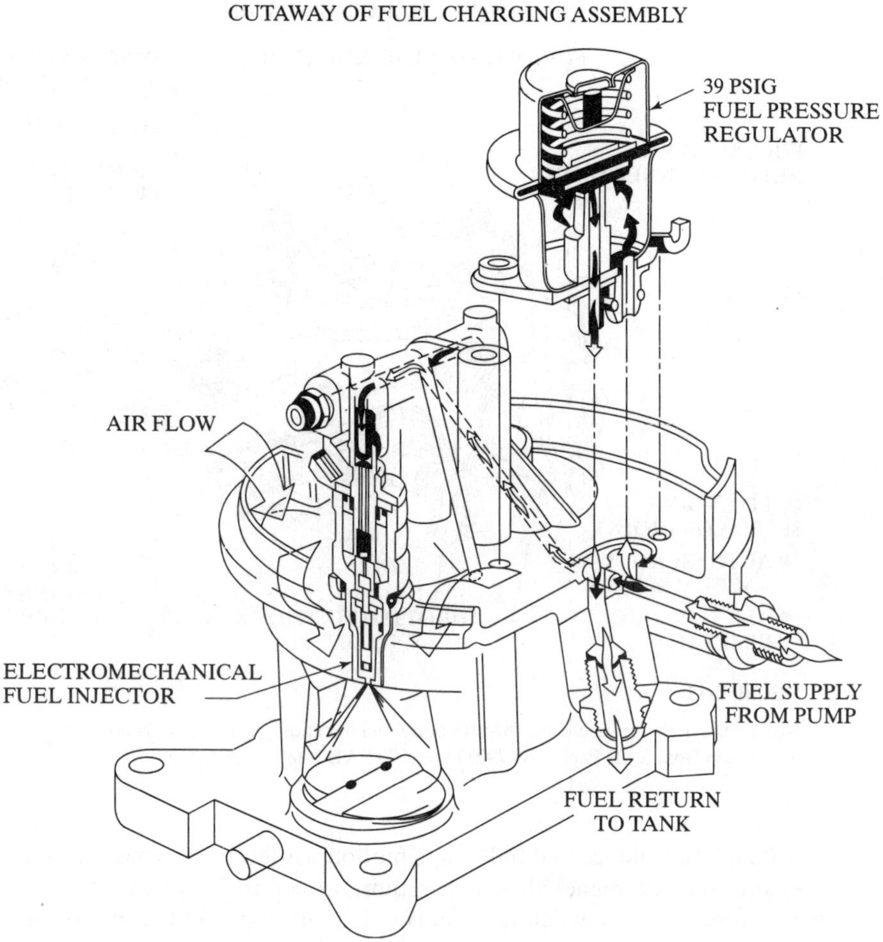

FIGURE 5-14

Fuel charging assembly for electronic throttle-body fuel injection system. Reprinted with permission from SAE Paper No. 790742 © 1979 SAE International, [167].

Direct Injection Systems for SI Engines

Major work continues to develop direct injection for SI engines and some models of automobiles now commercially available are equipped with such systems (Fig. 5-12). These systems inject the fuel directly into the combustion chamber, either during the intake stroke or during the compression stroke. There are two basic types of **gasoline direct injection** (GDI), either injection of gasoline alone, or a dual injection of gasoline and air together. Injection of fuel alone is usually done during the compression stroke and is somewhat similar to fuel injection in a CI engine. Because of the very short time available for vaporization and mixing with air, very fine droplets of fuel are required,

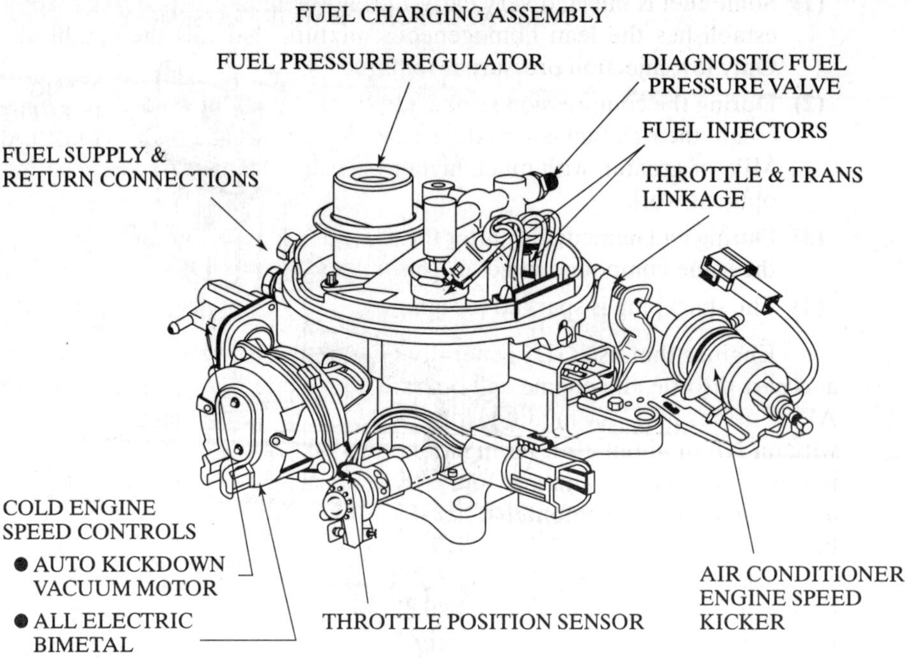

FIGURE 5-15

System assembly for electronic throttle-body fuel injection system on 1980 Ford. Reprinted with permission from SAE Paper No. 790742 © 1979 SAE International, [167].

as are high turbulence and bulk mass motion within the combustion chamber. Injector pressure must be higher than that required for port injection. This is because of the higher pressure into which the fuel is being injected, and because of the requirement for finer droplets of fuel. Injection will sometimes be in two stages, a pilot injection for ignition, followed by the main injection.

The most modern of GDI systems for automobile and other engines inject a combination of air and fuel (Figs. 5-9, 5-10, 5-11). By injecting air with the fuel, vaporization and mixing times can be greatly reduced. This method also makes possible the stratification of the air–fuel mixture, which is the way these engines are generally operated. In stratified charge combustion, a rich mixture is established around the sparkplug, while a very lean mixture fills the rest of the combustion chamber. The overall air–fuel ratio can be as high as 50:1, a mixture which would not ignite if it were homogeneous. By operating with a very lean mixture combustion temperature is reduced, and this reduces heat loss (giving higher thermal efficiency), knock problems, and the generation of harmful emissions.

The fuel-rich mixture near the sparkplug readily ignites and burns with a fast flame speed. This then ignites the lean mixture in the rest of the combustion chamber. To establish this stratified air–fuel distribution, a sequence of injections is required:

(1) Some fuel is injected very early, during the intake stroke (e.g., 120° aTDC). This establishes the lean homogeneous mixture that fills the combustion chamber. Only low injection pressure is required.

(2) During the compression stroke additional fuel is injected at very high pressure to create the rich fuel–air mixture near the sparkplug. Pressures can be as high as 10 MPa or greater, with much higher pressures being tested in experimental development work.

(3) During and immediately after the second fuel injection, air is injected, usually with the same compound injector. This enhances evaporation of the fuel just injected.

(4) The sparkplug is sparked for ignition.

Engines that use GDI generally operate in three different modes. At light load and part throttle, the engine will operate with a stratified charge mode and an overall AF of about 50:1. At medium load, it will still operate with stratified charge, but with an AF of about 20:1. At high load or WOT operation, fuel is injected only during the intake stroke and maximum thermal efficiency is obtained by operating with a homogeneous stoichiometric air–fuel mixture. High levels of EGR are used at high load.

To be able to operate in these different modes requires precise control of valve timing, lift, and duration, and control of bulk motion of the air–fuel mixture. These parameters are controlled by intake adjustments for different loads by the engine management system (EMS). Engines will often have more than one injector per cylinder.

Throttle Body Fuel Injection Systems

Some fuel injection systems, including most very early ones, consist of throttle body injection (Figs. 5-13, 5-14, 5-15). Engines with these systems have one or more injectors mounted directly before the butterfly throttle valve. The throttle valve is mounted in a throttle body located at the air inlet to the intake manifold much like a carburetor, and in vehicles is usually controlled in a similar manner (i.e., by a pedal–cable arrangement [167]). The fuel injectors can fire simultaneously, alternately, or continuously, depending on operating conditions and control logic.

Injectors for throttle body systems are generally constant-pressure steady-state devices, which control fuel input amounts by time length of injection. In a typical system fuel is supplied to the injectors from a fuel pump under pressure of about 250 to 300 kPa (40–50 psia). Injection is controlled by exciting the solenoid, which lifts the injector plunger and opens the pintle, allowing fuel to flow.

Often the system will operate under several different modes, these being determined by sensing engine RPM, throttle position, coolant temperature, etc. Four such modes might be the following:

(1) Cranking, with engine RPM under some set minimum

(2) Closed throttle

(3) Part throttle

(4) Wide open throttle

Example Problem 5-4

A 4.8-liter V8 SI engine operates on a four-stroke GDI cycle at 4200 RPM. The engine operates with two direct injections of gasoline in each cylinder during each cycle with an overall air–fuel ratio of 28:1. The first injection in each cylinder consists of one-fourth of the total fuel input and occurs late in the intake stroke and early part of the compression stroke, 10° bBDC to 80° aBDC. The second injection inputs the rest of the fuel near the sparkplug shortly before ignition, 70° bTDC to 30° bTDC. A supercharger gives the engine a volumetric efficiency of 98% at this speed.
Calculate:

1. pseudo-steady-state mass flow rate of fuel into the engine
2. length of time of one first fuel injection
3. mass flow rate through injector during first fuel injection
4. mass flow rate through injector during second fuel injection

(1) Equation (2-71) gives mass flow rate of air into engine:

$$\dot{m}_a = \eta_v \rho_a V_d N/n$$
$$= (0.98)(1.181\ \text{kg/m}^3)(0.0048\ \text{m}^3/\text{cycle})(4200/60\ \text{rev/sec})/(2\ \text{rev/cycle})$$
$$= 0.1944\ \text{kg/sec}$$

Equation (2-55) gives mass flow rate of fuel into engine:

$$\dot{m}_f = \dot{m}_a/\text{AF} = (0.1944\ \text{kg/sec})/(28) = \underline{0.00694\ \text{kg/sec}}$$

(2) Crank angle rotation for one first injection is

$$\angle = 10^\circ\ \text{bBDC to } 80^\circ\ \text{aBDC} = (90^\circ)/(360^\circ/\text{rev}) = 0.25\ \text{rev}$$

Time of one first injection is

$$t = (0.25\ \text{rev})/(4200/60\ \text{rev/sec}) = \underline{0.00357\ \text{sec} = 3.57\ \text{msec}}$$

(3) Use Eqs. (2-55) and (2-70) to find mass of fuel for first injection in one cylinder for one cycle:

$$m_f = m_a/\text{AF} = \{[(\eta_v \rho_a V_d)/(\text{AF})]/(8\ \text{cyl})\}(0.25\ \text{of total fuel})$$
$$= \{[(0.98)(1.181\ \text{kg/m}^3)(0.0048\ \text{m}^3/\text{cycle})/(28)]/(8)\}(0.25) = 0.0000062\ \text{kg}$$

Mass flow rate of fuel through the injector during first injection is

$$\dot{m}_f = (0.0000062\ \text{kg})/(0.00357\ \text{sec}) = \underline{0.00174\ \text{kg/sec}}$$

(4) Crank angle rotation for one second injection is

$$\angle = (40^\circ)/(360^\circ/\text{rev}) = 0.1111\ \text{rev}$$

Time of one second revolution is

$$t = (0.1111\ \text{rev})/(4200/60\ \text{rev/sec}) = 0.00159\ \text{sec}$$

Mass of fuel of one second injection in one cylinder is

$$m_f = \{[(.98)(1.181\ \text{kg/m}^3)(0.0048\ \text{m}^3/\text{cycle})/(28)]/(8\ \text{cyl})\}(0.75\ \text{of total})$$
$$= 0.0000186\ \text{kg}$$

Mass flow rate through fuel injector during second injection is

$$\dot{m}_f = (0.0000186\ \text{kg})/(0.00159\ \text{sec}) = \underline{0.0117\ \text{kg/sec}}$$

5.6 CARBURETORS

For several decades, carburetors were used on most SI engines as the means of adding fuel to the intake air. The basic principle on which the carburetor works is extremely simple, but by the 1980s, when fuel injectors finally replaced it as the main fuel input system, it had evolved into a complicated, sophisticated, expensive system. Carburetors are still found on a few automobiles, but the vast majority of car engines use simpler, better controlled, more flexible fuel injector systems. Many small engines like those on lawn mowers and model airplanes still use carburetors, although much simpler ones than those found on the automobile engines of the 1960s and 1970s. This is to keep the cost of these engines down, simple carburetors being cheap to manufacture while fuel injectors require more costly control systems. Even on some of these small engines, carburetors are being replaced with fuel injectors as pollution laws become more stringent.

Figure 5-16 shows that the basic carburetor is a venturi tube (A) mounted with a throttle plate (B) (butterfly valve) and a capillary tube to input fuel (C). It is usually mounted on the upstream end of the intake manifold, with all air entering the engine passing first through this venturi tube. Most of the time, there will be an air filter mounted directly on the upstream side of the carburetor. Other main parts of the carburetor are the fuel reservoir (D), main metering needle valve (E), idle speed adjustment (F), idle valve (G), and choke (H).

As air enters the engine due to the pressure differential between the surrounding atmospheric air and the partial vacuum in the cylinders during intake strokes, it is accelerated to high velocity in the throat of the venturi of the carburetor. By Bernoulli's Principle, this causes the pressure in the throat P_2 to be reduced to a value less than the surrounding pressure P_1, which is about one atmosphere. The pressure above the fuel in the fuel reservoir is equal to atmospheric pressure as the reservoir is vented to the surroundings ($P_3 = P_1 > P_2$). There is, therefore, a pressure differential through the fuel supply capillary tube, and this forces fuel flow into the venturi throat. As the fuel flows out of the end of the capillary tube, it breaks into very small droplets which are carried away by the high-velocity air. These droplets then evaporate and mix with the air in the following intake manifold. As engine speed is increased, the higher flow rate of air will create an even lower pressure in the venturi throat. This creates a greater pressure differential through the fuel capillary tube, which increases the fuel flow rate to keep up with the greater air flow rate and engine demand. A properly designed carburetor can supply the correct AF at all engine speeds, from idle to WOT. There is a main metering valve (E) in the fuel capillary tube for flow rate adjustment.

The level in the fuel reservoir is controlled by a float shutoff. Fuel comes from a fuel tank supplied by an electric fuel pump on most modern automobiles, by a mechanically driven fuel pump on older automobiles, or even by gravity on some small engines (lawn mowers) and historic automobiles.

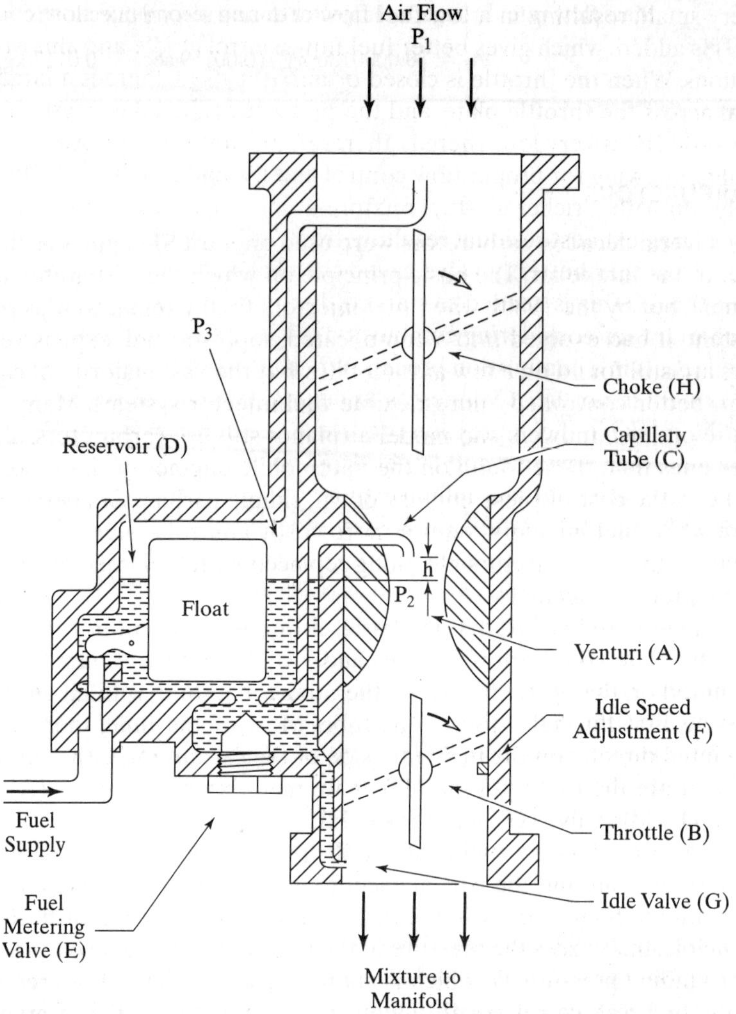

FIGURE 5-16

Basic automobile carburetor showing (A) venturi, (B) throttle valve, (C) fuel capillary tube, (D) fuel reservoir, (E) main metering valve, (F) idle speed adjustment, (G) idle valve, and (H) choke.

The throttle controls the air flow rate and thus the engine speed. There is an idle speed adjustment (throttle stop) which sets the closed throttle position such that some air can flow even at fully closed throttle. This adjustment, which is usually about 5°–15° of throttle plate rotation, controls how fast the engine will run at idle conditions. Because the air flow rate through the venturi throat will be minimal at idle conditions when the throttle is closed, the pressure in the throat will be only slightly less than atmospheric pressure. The pressure differential through the fuel capillary tube will be very small, resulting in a low fuel flow rate and very poor flow control. An idle valve (G) is added, which gives better fuel flow control at idle and almost-closed throttle positions. When the throttle is closed or almost closed, there is a large pressure differential across the throttle plate, and the pressure in the intake system downstream of the throttle (B) is very low. There is, therefore, a substantial pressure drop through the idle valve, allowing for proper flow control and a greater flow rate of fuel. Engines are usually run with a richer air–fuel mixture at low and idle speeds to avoid misfires caused by a large exhaust residual resulting from valve overlap.

Another butterfly valve called the **choke** (H) is positioned upstream of the venturi throat. This is needed to start cold engines. It is not really the air–fuel ratio that is important for considering combustion, but the air–vapor ratio; only fuel that is vaporized reacts in a combustion process. When an engine is cold (as in an automobile sitting overnight in northern Minnesota in January), a very small percent of fuel will vaporize in the intake and compression processes. The fuel is cold and much more viscous, creating a lower flow rate and larger droplets which vaporize more slowly. The metal engine parts are cold and inhibit vaporization. Even in the compression stroke, which heats the air–fuel mixture, the cold cylinder walls absorb heat and reduce vaporization. Engine lubrication is cold and more viscous, making the engine turn more slowly in the starting process. As the engine turns over slowly with the starter, only a very small air flow is generated through the carburetor. This creates only a very small pressure differential through the fuel capillary tube and a very low flow rate. At starting conditions, the throttle is wide open, so no substantial pressure differential is established through the idle valve. All of this creates very little fuel evaporation, and if normal carburetor action were used, there would not be enough fuel vapor in the cylinder to create combustion and get the engine started. For this reason, the choke was added to carburetors. When starting a cold engine, the first step is to close the choke. This restricts air flow and creates a vacuum in the entire intake system downstream of the choke, even at the very low air flow rates encountered in starting. There is, therefore, a large pressure differential across both the fuel capillary tube and the idle valve, causing a large fuel flow to mix with the low air flow. This gives a very rich air–fuel mixture entering the cylinders, up to $AF = 1:1$ for very cold starts. With only a small percent of fuel evaporating, a combustible air–vapor mixture is created, combustion occurs, and the engine starts. Only a few engine cycles are required before everything starts to heat up and more normal operation occurs. As the engine heats up, the choke is opened and has no effect on final steady-state operation.

It does not require frigid winter temperatures to create the need for a choke for starting an engine. Anyone who has ever tried to start a chokeless lawn mower at 10°C will agree with this.

Most later automobile carburetors were equipped with automatic chokes. These would be closed by the vehicle operator before cold starting was attempted, usually by stepping the throttle pedal to the floor. After the engine was started, the choke would slowly open automatically as the engine temperature increased, controlled by thermal or vacuum means. Small lawn-mower-type engines and older automobile engines have manually controlled chokes.

Many small low-cost engines have no choke. Some constant-speed engines, such as those for model airplanes or some industrial applications, have no throttle.

Another added feature of modern automobile carburetors is an accelerator pump. When fast acceleration is desired, the throttle is quickly opened to WOT, and the air and fuel flowing into the engine are quickly increased. The gaseous air and fuel vapor react to this acceleration very quickly due to their low mass inertia. The fuel that is still liquid in larger droplets and in the film along the walls of the intake manifold has much higher density and mass inertia and, therefore, accelerates much slower. The engine experiences a momentary deficiency of fuel and a major reduction in the fuel–air ratio. This causes an undesirable hesitation in the acceleration of the engine speed to the possible extreme of stalling the engine. To avoid this, an acceleration pump is added, which injects an added quantity of fuel into the air flow when the throttle is opened quickly. Instead of experiencing a momentary lean air–fuel mixture, the engine experiences a momentary rich mixture that assists in the acceleration process.

Air and Fuel Flow in Carburetors

From gas dynamics [58], the air flow through a venturi throat can be written

$$\dot{m}_a = (C_{Dt}A_tP_0/\sqrt{RT_0})(P_t/P_0)^{1/k}\{[2k/(k-1)][1-(P_t/P_0)^{(k-1)/k}]\}^{\frac{1}{2}} \quad (5\text{-}5)$$

where

C_{Dt} = discharge coefficient of venturi throat
A_t = flow area of venturi throat
P_0, T_0 = ambient pressure and temperature
P_t = throat pressure
R = gas constant

The pressure differential in the air will be

$$\Delta P_a = P_0 - P_t = P_1 - P_2 \quad (5\text{-}6)$$

where P_1 and P_2 are as shown in Fig. 5-16.

The pressure differential through the fuel capillary tube will be

$$\Delta P_f = \Delta P_a - \rho_f g h \quad (5\text{-}7)$$

where

ρ_f = density of fuel
g = acceleration due to gravity
h = height differential in the fuel capillary tube

The second term in Eq. (5-7) is the hydraulic head between the fuel reservoir and throat. The elevation h is built into a carburetor to avoid fuel leaking out when the vehicle is parked on a slope. Values of h are typically 1 to 2 cm.

Liquid fuel flow through a capillary tube is

$$\dot{m}_f = C_{Dc}A_c\sqrt{2\rho_f\Delta P_f} \quad (5\text{-}8)$$

where

C_{Dc} = discharge coefficient of the capillary tube
A_c = cross-sectional flow area of the capillary tube

Following the method of [58] and using Eqs. (5-5)–(5-8), the air–fuel ratio supplied by the carburetor can be obtained as

$$AF = \dot{m}_a/\dot{m}_f = (C_{Dt}/C_{Dc})(A_t/A_c)(\rho_a/\rho_f)^{\frac{1}{2}}\,\Omega\,\Pi \tag{5-9}$$

with

$$\Omega = [\Delta P_a/(\Delta P_a - \rho_f g h)]^{\frac{1}{2}}$$

$$\Pi = \{[k/(k-1)][(P_t/P_0)^{2/k} - (P_t/P_0)^{(k+1)/k}]/[1 - (P_t/P_0)]\}^{\frac{1}{2}}$$

If the air velocity through the carburetor throat is increased by increasing the engine speed, a maximum flow rate will be reached when sonic velocity occurs. This will happen [112] when

$$(P_t/P_0) = [2/(k+1)]^{k/(k-1)} \tag{5-10}$$

Using the value of $k = 1.4$ because of the lower temperature of the air through the carburetor, Eq. (5-10) reduces to

$$P_t = 0.5283\ P_0 = 53.4 \text{ kPa at standard conditions}$$

Maximum air flow through the carburetor will then be

$$(\dot{m}_a)_{max} = \rho_0 c_0 C_{Dt} A_t \sqrt{[2/(k+1)]^{(k+1)/(k-1)}} \tag{5-11}$$

where $c_0 = \sqrt{kRT_0}$ = ambient speed of sound.

At standard conditions,

$$\rho_0 = 1.181 \text{ kg/m}^3$$

$$c_0 = [(1.4)(287 \text{ J/kg-K})(298 \text{ K})]^{\frac{1}{2}} = 346 \text{ m/sec}$$

Using $k = 1.4$, the square root term in Eq. (5-11) is equal to 0.5787, and the equation can be reduced to

$$[(\dot{m}_a)_{max} \text{ in kg/sec}] = 236.5\ C_{Dt}[A_t \text{ in m}^2] \tag{5-12}$$

This equation can be used to size the carburetor throat needed for an engine. Equation (5-8) can then be used to establish the cross-sectional area of the fuel capillary tube A_c relative to other parameters.

Fortunately, it is found that, once the diameters of the carburetor throat and fuel capillary tube are established, carburetors can be built that will give correct air–fuel mixtures over a large range of operating conditions. These include starting, WOT, cruise, and sudden deceleration. Cold-engine starting characteristics of a carburetor have already been explained.

WOT is used for high-speed operation or when accelerating. Here the carburetor delivers a rich mixture that gives maximum power at a cost of poorer fuel economy.

At steady-state cruising, a carburetor delivers a lean mixture ($AF \approx 16$) which gives less power but good fuel economy. A modern midsize automobile requires only about 5 to 6 kW (7–8 hp) to cruise at 55 MPH on a level highway.

When an engine is operating at high speed and the throttle is suddenly closed to decelerate the automobile, a slightly rich mixture will be delivered by the carburetor. The combination of a closed throttle and high engine speed will create a high vacuum in the intake system downstream from the throttle plate. This will cause very little fuel flow into the carburetor throat but will cause a flow through the idle valve. This fuel, mixed with the low air flow rate, will give the rich mixture needed to keep good combustion. With the high vacuum in the intake system, a very large exhaust residual will occur during valve overlap, and a rich mixture is needed to sustain combustion. Misfires are common during this type of deceleration. Fuel injectors give much better AF control in fast deceleration.

When air flows through a venturi nozzle, the pressure drops as the air is accelerated through the throat and then rises again as the air velocity is reduced past the throat. There is always a net pressure loss through a venturi, with downstream pressure never equaling upstream pressure. For a given flow rate, the smaller the throat diameter, the greater will be this net pressure loss. This loss directly reduces the volumetric efficiency of the engine. This would suggest that the throat diameter of the carburetor should be made large. However, a large throat area would have low air velocity and a small pressure differential through the fuel capillary tube, causing poorer AF control, larger fuel droplets, and poorer mixing of air and fuel. This would be especially true at low engine speeds and corresponding low air flow rates. Generally, it would be desirable to have large-throat carburetors on high-performance engines, which usually operate at high speeds and where fuel economy is a secondary priority. Small economy engines that do not need high power would have small-throat carburetors.

One way to avoid compromising on the throat diameter is to use a *two-barrel carburetor* (i.e., two separate, smaller diameter, parallel venturi nozzles mounted in a single carburetor body). At low engine speeds, only one carburetor barrel is used. This gives a higher pressure differential to better control fuel flow and mixing without causing a large pressure loss through the carburetor. At higher engine speeds and higher air flow rates, both barrels are used, giving the same better control without a large pressure loss.

Another type of carburetor uses a secondary venturi mounted inside the primary larger venturi, as shown in Fig. 5-17. The large diameter of the primary venturi avoids a large pressure loss, while the small diameter of the secondary venturi gives a higher pressure differential for good fuel flow control and mixing. Still another type of carburetor changes the air flow area in the throat, increasing it at high speed and decreasing it at low speed. Several methods of doing this on various carburetors have been tried with mostly less-than-ideal results. There has even been some success in making a variable-diameter capillary tube for the fuel flow. When modern electronic controls are used for the various operations of a carburetor, a more dependable, accurate, and flexible system is realized. However, along with the emergence of electronic controls came an even better fuel input system: fuel injectors.

When a four-stroke cycle engine is operating, each cylinder has intake occurring about one-fourth of the time. A single carburetor can, therefore, supply an air–fuel

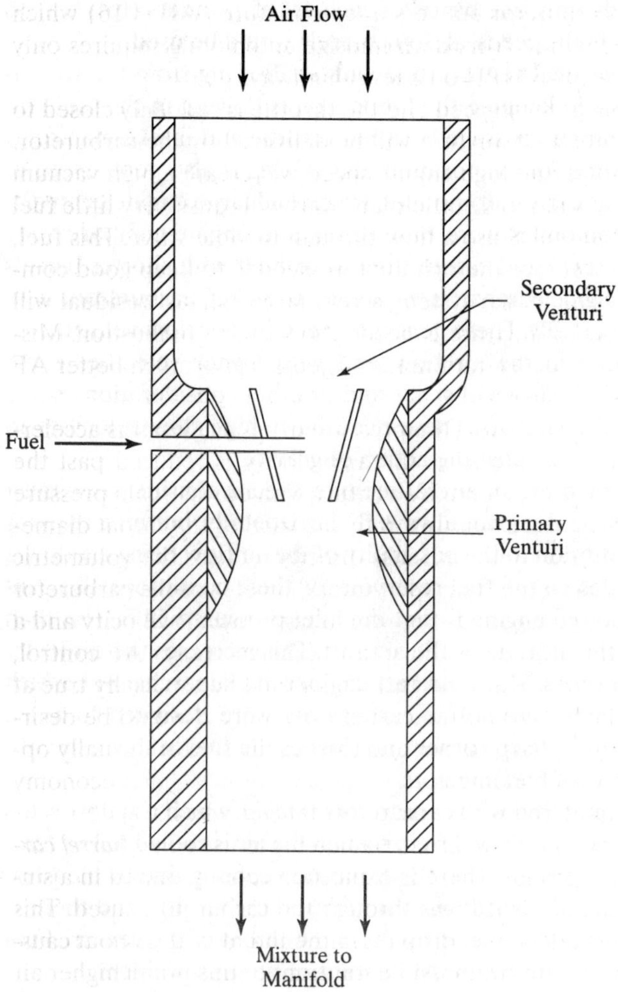

FIGURE 5-17

Throat of carburetor with secondary venturi. The small secondary venturi gives a large pressure drop and good fuel flow control, while the larger primary throat offers less resistance to the main air flow.

mixture to as many as four cylinders without the need for enlarging the throat area. Instead of supplying a given flow rate intermittently one-fourth of the time to one cylinder, the same carburetor can supply the same flow rate to four cylinders at almost steady-state flow if the cylinder cycles are dispersed evenly about the engine rotation, the normal way of operating an engine. The same size carburetor would be correct for two or three cylinders also, with flow occurring on and off. If five or more cylinders are connected to a single carburetor, the throat area would have to be larger to accommodate the higher flow rates when more than one cylinder is taking in air and fuel, which would occur at least some of the time.

When eight-cylinder automobile engines were popular during the 1950s through the 1980s, two- and four-barrel carburetors were quite common. Each barrel of a two-barrel carburetor would be used to supply four cylinders at about steady-state air flow. Four-barrel carburetors were also used, with each of four cylinders being supplied by two barrels. At low engine speeds, only one barrel of each set of cylinders (two barrels total) would be in operation. At high speeds, all four barrels would be used.

A *downdraft* carburetor (vertical venturi tube with air flowing from top to bottom) is best in that gravity assists in keeping the fuel droplets flowing in the same direction as the air flow. A long runner (passage between throttle and intake manifold) that allows more distance and time for evaporation and mixing is also good. Both of these concepts were acceptable in early automobiles, which had large engine compartments and a high hood. As automobiles were built lower and engine compartments smaller, compromising was necessary and carburetors were built with shorter barrels and runners. To further reduce engine compartment height, *side-draft* carburetors were developed with air flowing horizontally. These generally need higher flow velocities to keep the fuel droplets suspended in the air flow, and with higher velocities come greater pressure losses. For special reasons of space and/or other considerations, some engines are fitted with *updraft* carburetors. These need fairly high flow velocities to carry the fuel droplets in suspension against the action of gravity.

When a carburetor is designed for an aircraft engine, special consideration must be given to the fact that the vehicle does not always fly horizontally, but may bank or even go into inverted flight. In addition to the possibility of the air flow being up, down, or horizontal, it is necessary to design the fuel reservoir for these conditions also. Another difference from an earthbound engine is that the inlet pressure will be less than one atmosphere, depending on the altitude of the aircraft. This increases the difficulty of keeping the correct AF at all times. Many aircraft engines are supercharged, which minimizes this problem. Even later automobile carburetors were designed to avoid fuel starvation as the vehicle turns a sharp corner and sloshes the fuel in the reservoir, a problem which does not occur with fuel injectors.

A problem sometimes encountered with carburetors is *icing*, which usually occurs on the throttle plate. Water vapor in the air will freeze when the air is cooled to low temperatures. Cooling occurs for two reasons: There is expansion cooling due to the pressure reduction experienced by the air as it flows through the carburetor, and there is evaporative cooling due to the just-added fuel droplets in the throat of the venturi. Fuel additives and heating the carburetor are two possible solutions to this problem.

Another problem of carburetors is the splitting of the air flow around the throttle plate immediately after the fuel has been added. This makes it very difficult to get homogeneous mixing and is a major reason why the air–fuel mixture delivered to the cylinders is often nonuniform. This problem is more serious with later short-barrel, short-runner carburetors.

At conditions other than WOT, the major pressure drop in an intake system will be at the throttle plate of the carburetor. This may be as much as 90% of the total pressure drop, or greater. The flow may become choked (sonic velocity) at a partially closed throttle. When the throttle position is suddenly changed, it takes several engine revolutions to reestablish steady-state flow through the carburetor.

Example Problem 5-5

A six-cylinder, 3.6-liter SI engine is designed to have a maximum speed of 6000 RPM. At this speed the volumetric efficiency of the engine is 0.92. The engine will be equipped with a two-barrel

carburetor, one barrel for low speeds and both barrels for high speed. Gasoline density can be considered to be 750 kg/m^3.
Calculate:

1. throat diameters for the carburetor (assume discharge coefficient $C_{Dt} = 0.94$)
2. fuel capillary tube diameter (assume discharge coefficient $C_{Dc} = 0.74$)

Use Eq. (2-71) to get air flow at maximum speed:

$$(\dot{m}_a)_{max} = \eta_v \rho_a V_d N/n$$
$$= (0.92)(1.181 \text{ kg/m}^3)(0.0036 \text{ m}^3\text{/cycle})(6000/60 \text{ rev/sec})/(2 \text{ rev/cycle})$$
$$= 0.1956 \text{ kg/sec}$$

(1) Equation (5-12) is used to find the throat area needed at maximum engine speed:

$$(\dot{m}_a)_{max} = 236.5\, C_{Dt} A_t = (236.5)(0.94) A_t = 0.1956$$
$$A_t = 0.00088 \text{ m}^2 = 8.8 \text{ cm}^2$$

For each barrel,

$$A_t = (\pi/4) d_t^2 = 0.00044 \text{ m}^2$$
$$\underline{d_t = 0.0237 \text{ m} = 2.37 \text{ cm} = 0.93 \text{ in.}}$$

Throttle of the second barrel would remain closed until engine speed exceeded about 3000 RPM.

(2) Using Eq. (5-10) at maximum flow rate in one barrel gives a pressure in the throat of

$$P_t = 53.4 \text{ kPa}$$

Equation (5-6) gives the pressure drop of the air:

$$\Delta P_a = P_o - P_t = 101 - 53.4 = 47.6 \text{ kPa}$$

Assuming a fuel capillary tube height differential of 1.5 cm, Eq. (5-7) gives the pressure drop across the tube:

$$\Delta P_f = \Delta P_a - \rho_f g h$$
$$= (47.6 \text{ kPa}) - (750 \text{ kg/m}^3)(9.81 \text{ m/sec}^2)(0.015 \text{ m})/(1 \text{ kg-m/N-sec}^2)$$
$$= 47.49 \text{ kPa}$$

Designing for an air–fuel ratio of AF = 15.2, Eq. (5-9) is used to determine the cross-sectional area of the fuel capillary tube A_c:

$$\Omega = [\Delta P_a/(\Delta P_a - \rho_f g h)]^{\frac{1}{2}} = [(47.6 \text{ kPa})/(47.49 \text{ kPa})]^{\frac{1}{2}} = 1.0012$$

$$\Pi = \{[k/(k-1)][(P_t/P_o)^{2/k} - (P_t/P_o)^{(k+1)/k}]/[1 - (P_t/P_o)]\}^{\frac{1}{2}}$$
$$= \{[1.4/0.4][(53.4/101)^{2/1.4} - (53.4/101)^{2.4/1.4}]/[1 - (53.4/101)]\}^{\frac{1}{2}}$$
$$= 0.7053$$

$$\text{AF} = (C_{Dt}/C_{Dc})(A_t/A_c)(\rho_a/\rho_f)^{\frac{1}{2}}\, \Omega\, \Pi$$
$$15.2 = (0.94/0.74)(0.00044/A_c)(1.181/750)^{\frac{1}{2}}(1.0012)(0.7053)$$

Solving this for the flow area A_c in the capillary tube yields

$$A_c = 1.03 \times 10^{-6} \text{ m}^2 = (\pi/4) d_c^2$$

This gives a fuel capillary tube diameter of

$$\underline{d_c = 0.00115 \text{ m} = 1.15 \text{ mm} = 0.045 \text{ in.}}$$

5.7 SUPERCHARGING AND TURBOCHARGING

Superchargers

Superchargers and turbochargers are compressors mounted in the intake system and used to raise the pressure of the incoming air. This results in more air and fuel entering each cylinder during each cycle. This added air and fuel creates more power during combustion, and the net power output of the engine is increased (Fig. 5-18). The pressure increase for most engines is generally somewhere between 20 and 50 kPa, with most engines on the lower end of this scale. At least one automobile manufacturer (Saab) has developed a high-pressure supercharger ($\Delta P = 280$ kPa) to use with its variable-compression engine (see Section 7-7)[148,151].

Superchargers are mechanically driven directly off the engine crankshaft. They are generally positive displacement compressors running at speeds about the same as engine speed (Fig. 1-8). The power to drive the compressor is a parasitic load on the engine output, and this is one of the major disadvantages compared with a turbocharger. Other disadvantages include higher cost, greater weight, and noise. A major advantage of a supercharger is very quick response to throttle changes. Being mechanically linked to the crankshaft, any engine speed change is immediately transferred to the compressor.

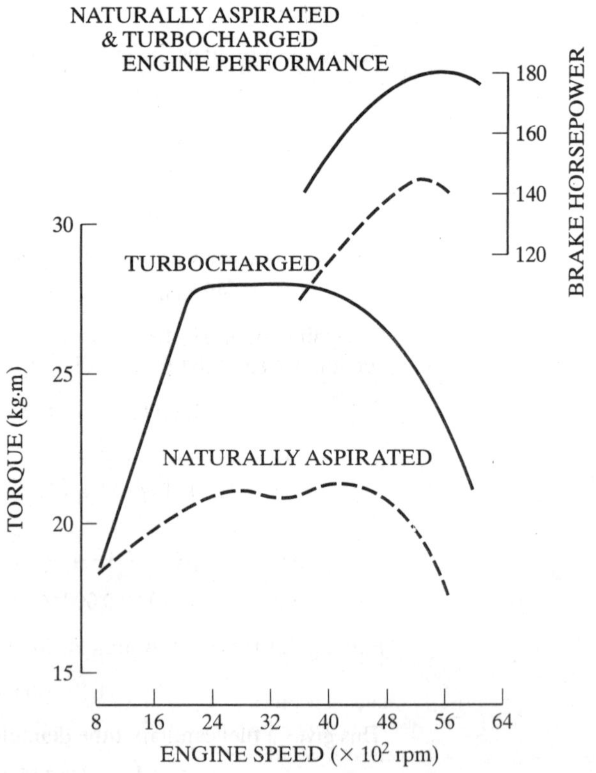

FIGURE 5-18

Power and torque curves of 1982 Datsun 280ZX engines with and without turbocharger. The six-cylinder, 2.75 L, SI engine had bore of 8.61 cm and stroke of 7.87 cm. The turbocharged model had a compression ratio of 7.4, while the naturally aspirated model had a compression ratio of 8.8. Reprinted with permission from SAE Paper No. 820442 © 1982 SAE International, [203].

Some high-performance automobile engines and just about all large CI engines are supercharged. All two-stroke cycle engines that are not crankcase compressed (a form of supercharging) must be either supercharged or turbocharged.

When the first law of thermodynamics is applied to the air flowing through a supercharger compressor,

$$\dot{W}_{sc} = \dot{m}_a(h_{out} - h_{in}) = \dot{m}_a c_p(T_{out} - T_{in}) \tag{5-13}$$

where

$\dot{W}_{sc}$ = power needed to drive the supercharger
$\dot{m}_a$ = mass flow rate of air into the engine
c_p = specific heat of air
h = specific enthalpy
T = temperature

This assumes that the compressor heat transfer, kinetic energy terms, and potential energy terms are negligibly small, true for most compressors. All compressors have isentropic efficiencies less than 100%, so the actual power needed will be greater than the ideal. In Fig. 5-19, process $1-2_s$ represents ideal isentropic compression, while process $1-2_A$ is the actual process with an increase in entropy. The isentropic efficiency η_s of the supercharger compressor is

$$\begin{aligned}(\eta_s)_{sc} &= \dot{W}_{isen}/\dot{W}_{act} = [\dot{m}_a(h_{2s} - h_1)]/[\dot{m}_a(h_{2A} - h_1)] \\ &= [\dot{m}_a c_p(T_{2s} - T_1)]/[\dot{m}_a c_p(T_{2A} - T_1)] = (T_{2s} - T_1)/(T_{2A} - T_1)\end{aligned} \tag{5-14}$$

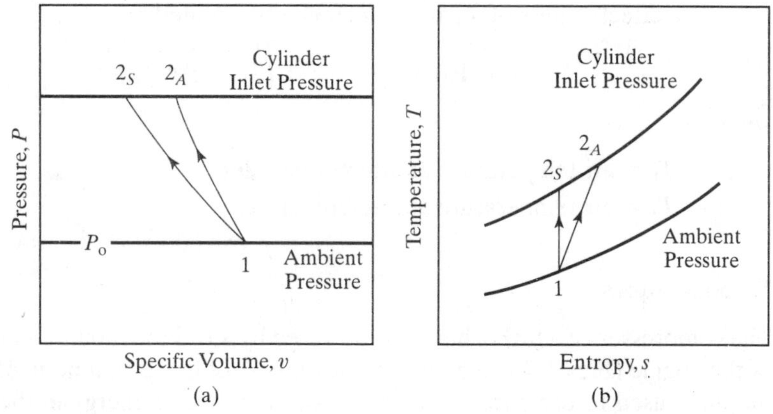

FIGURE 5-19

Ideal flow process $(1-2_s)$ and actual flow process $(1-2_A)$ through a supercharger or a turbocharger compressor in (a) pressure–volume coordinates, and (b) temperature–entropy coordinates.

If the inlet temperature and pressure, as well as the designed output pressure, are known, the ideal gas isentropic relationship can be used to find T_{2s}:

$$T_{2s} = T_1(P_2/P_1)^{(k-1)/k} \tag{5-15}$$

The actual outlet temperature T_{2A} can then be calculated from Eq. (5-14) if the isentropic efficiency is known. When using Eq. (5-15), a value of $k = 1.40$ should be used because of the lower temperature at this point.

There is also a mechanical efficiency of less than 100% between the power taken from the engine and that delivered to the compressor:

$$\eta_m = (\dot{W}_{\text{act}})_{\text{sc}}/\dot{W}_{\text{from engine}} \tag{5-16}$$

For added engine output power, it is desirable to have the higher input air pressure supplied by the supercharger. However, the supercharger also raises the inlet air temperature by compressive heating, as can be seen in Eq. (5-15). This is undesirable in SI engines. If the temperature at the start of the compression stroke is higher, all temperatures in the rest of the cycle will also be higher. Often, this will cause self-ignition and knocking problems during combustion. To avoid this, many superchargers are equipped with an aftercooler that cools the compressed air back to a lower temperature. The aftercooler can be either an air-to-air heat exchanger or an air-to-liquid heat exchanger. The cooling fluid can be air flowing through the engine compartment, or it can be engine liquid coolant in a more complex system. Some superchargers are made up of two or more compressor stages with an aftercooler following each stage. Aftercoolers are not needed on superchargers used on CI engines because there is no concern about knock problems. Aftercoolers are costly and take up space in the engine compartment. For these reasons, the superchargers on some automobiles do not have aftercoolers. These engines generally have reduced compression ratios to avoid problems of self-ignition and knock.

The effectiveness of an aftercooler can be defined as

$$\text{Eff} = (T_1 - T_2)/(T_1 - T_{\text{coolant}}) \tag{5-17}$$

where

T_1 = air temperature at aftercooler inlet

T_2 = air temperature at aftercooler exit

Turbochargers

The compressor of a turbocharger is powered by a turbine mounted in the exhaust flow of the engine (Figs. 1-9 and 5-20). The advantage of this is that none of the engine shaft output is used to drive the compressor, and only waste energy in the exhaust is used. However, the turbine in the exhaust flow causes a more restricted flow, resulting in a slightly higher pressure at the cylinder exhaust port. This reduces the engine power output very slightly. Turbocharged engines generally have lower specific fuel consumption rates. They produce more power, while the friction power lost remains about the same.

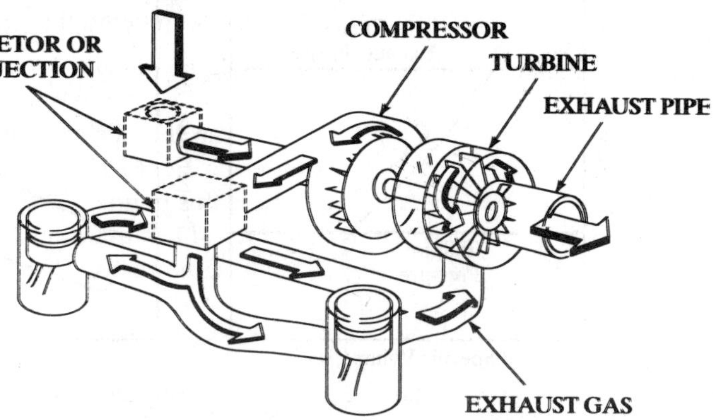

FIGURE 5-20

Schematic showing operation of turbocharger for an SI engine. Reprinted with permission from SAE Paper No. 780413 © 1978 SAE International, [136].

Maximum pressure in an engine exhaust system is only slightly above atmospheric, so there can be only a very small pressure drop through the turbine. Because of this, it is necessary to run the turbine at very high speeds so that enough power is produced to run the compressor. Speeds of 100,000 to 130,000 RPM are common. These high speeds, and the fact that exhaust gas is a hot, corrosive environment, demand special materials and concern for long-term reliability.

A disadvantage of turbochargers is **turbo lag**, which occurs with a sudden throttle change. When the throttle is quickly opened to accelerate an automobile, the turbocharger will not respond quite as quickly as a supercharger. It takes several engine revolutions to change the exhaust flow rate and to speed up the rotor of the turbine. Turbo lag has been greatly reduced by using lightweight ceramic rotors that can withstand the high temperatures and that have very little mass inertia. Turbo lag can also be reduced by using a smaller intake manifold.

Most turbochargers, like superchargers, are equipped with an aftercooler to again lower the compressed air temperature. Many also have a bypass that allows the exhaust gases to be routed around the turbocharger when an inlet air pressure boost is not needed. Some modern turbines have variable blade angle, which can be adjusted to give maximum efficiency for any air flow rate when engine speed or load is changed.

Radial flow centrifugal compressors, turning at high speed, are generally used on automobile-size engines. On very large engines, axial flow compressors are used because of their greater efficiency at the higher air flow rates. The isentropic efficiency of a compressor is defined as

$$(\eta_s)_{comp} = (\dot{W}_c)_{isen}/(\dot{W}_c)_{act} \tag{5-18}$$

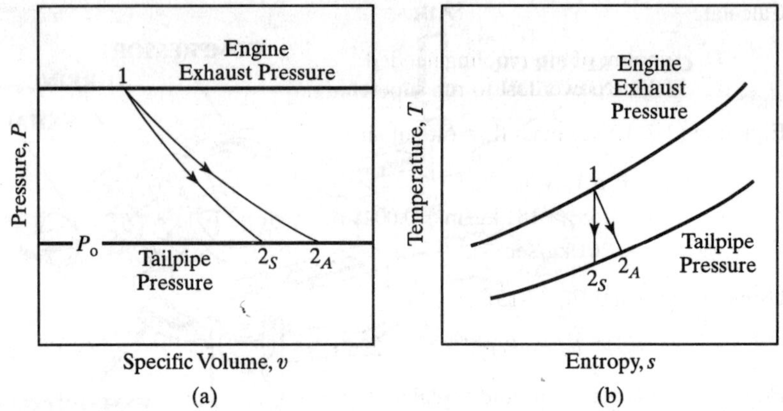

FIGURE 5-21

Ideal flow process ($1\text{–}2_s$) and actual exhaust flow process ($1\text{–}2_A$) through the turbine of a turbocharger in (a) pressure–volume coordinates, and (b) temperature–entropy coordinates. The inlet air flow through the compressor of the turbocharger is the same as shown in Fig. 5-19 for a supercharger.

Using Fig. 5-21 the turbine driving the compressor has an isentropic efficiency defined as

$$(\eta_s)_{\text{turb}} = (\dot{W}_t)_{\text{act}}/(\dot{W}_t)_{\text{isen}}$$
$$= [\dot{m}_a(h_1 - h_{2A})]/[\dot{m}_a(h_1 - h_{2S})] = (T_1 - T_{2A})/(T_1 - T_{2S}) \quad (5\text{-}19)$$

where

η_s = isentropic efficiency
$\dot{W}_c$ = power to drive compressor
$\dot{W}_t$ = turbine power

The pulsing nature of the exhaust flow reduces this efficiency to less than steady-state flow values. There is a mechanical efficiency between the turbine and compressor:

$$\eta_m = (\dot{W}_c)_{\text{act}}/(\dot{W}_t)_{\text{act}} \quad (5\text{-}20)$$

The overall efficiency of the turbocharger can then be considered:

$$\eta_{\text{turbo}} = (\eta_s)_{\text{comp}}\,(\eta_s)_{\text{turb}}\,\eta_m \quad (5\text{-}21)$$

Values of overall efficiency range from 70% to 90%.

Example Problem 5-6

A six-cylinder, 4.8-liter, supercharged engine operating at 3500 RPM has an overall volumetric efficiency of 158%. The supercharger has an isentropic efficiency of 92% and a mechanical efficiency in its link with the engine of 87%. It is desired that air be delivered to the cylinders at 65°C and 180 kPa, while ambient conditions are 23°C and 98 kPa.

Calculate:

1. amount of aftercooling needed
2. engine power lost to run supercharger

Equation (2-71) gives mass flow rate of air to the engine:

$$\begin{aligned}\dot{m}_a &= \eta_v \rho_a V_d N/n \\ &= (1.58)(1.181\ \text{kg/m}^3)(0.0048\ \text{m}^3/\text{cycle})(3500/60\ \text{rev/sec})/(2\ \text{cycles/rev}) \\ &= 0.261\ \text{kg/sec}\end{aligned}$$

Using Fig. 5-19 and Eq. (5-15)

$$T_{2s} = T_1(P_2/P_1)^{(k-1)/k} = (296\ \text{K})(180/98)^{(1.4-1)/1.4} = 352\ \text{K} = 79°\text{C}$$

Equation (5-14) is used to find actual air temperature at compressor exit:

$$\begin{aligned}(\eta_s)_{\text{sc}} &= (T_{2s} - T_1)/(T_{2A} - T_1) = 0.92 = (352 - 296)/(T_{2A} - 296) \\ T_{2A} &= 357\ \text{K} = 84°\text{C}\end{aligned}$$

(1) Amount of aftercooling needed to reduce air temperature back to 65°C is

$$\begin{aligned}\dot{Q} &= \dot{m}_a c_p (T_{2A} - T_{\text{in}}) \\ &= (0.261\ \text{kg/sec})(1.005\ \text{kJ/kg-K})(357 - 338)\text{K} = \underline{5.0\ \text{kW}}\end{aligned}$$

(2) Equations (5-13) and (5-16) are combined to find the engine power needed to drive the supercharger:

$$\begin{aligned}\dot{W} &= \dot{m}_a c_p (T_{\text{out}} - T_{\text{in}})/\eta_m \\ &= (0.261\ \text{kg/sec})(1.005\ \text{kJ/kg-K})(357 - 296)\text{K}/(0.87) \\ &= \underline{18.4\ \text{kW} = 24.7\ \text{hp}}\end{aligned}$$

5.8 DUAL-FUEL ENGINES

For various technical and financial reasons, some engines are designed to operate using a combination of two fuels. For instance, in some third-world countries dual-fuel engines are used because of the high cost of diesel fuel. Large CI engines are run on a combination of natural gas and diesel oil. Natural gas is the main fuel because it is more cheaply available. However, natural gas is not a good CI fuel by itself because it does not readily self-ignite (due to its high octane number). A small amount of diesel oil is injected at the proper cycle time. This ignites in a normal manner and initiates combustion in the natural gas–air mixture filling the cylinder. Combinations of fuel input systems are needed on these types of engines.

Example Problem 5-7

A large nine-cylinder, two-stroke cycle CI engine operating at 257 RPM uses a dual fuel combination of natural gas as the main fuel and a small amount of light diesel fuel for ignition. Ninety-five percent of the air flow into the engine is used for stoichiometric combustion with natural gas. The other 5% of air flow is used with the diesel fuel at an equivalence ratio of 1.3, a rich mixture being better for self-ignition. The engine has a bore of 32 cm, a stroke of 61 cm, a volumetric efficiency of 98%, an indicated thermal efficiency of 61%, a mechanical efficiency of 91%, and a combustion efficiency of 99%. In the country where the engine is being used the cost of diesel fuel is \$0.52/kg, and the cost of natural gas is \$0.09/kg.

Calculate:

1. mass flow rate of natural gas into the engine (use methane to approximate natural gas)
2. mass flow rate of diesel fuel into the engine
3. brake power output of the engine
4. cost savings by not using all diesel fuel

(1) Total displacement of engine using Eq. (2-9) is:

$$V_d = N_c(\pi/4)B^2S$$
$$= (9)(\pi/4)(32\ \text{cm})^2(61\ \text{cm}) = 441{,}500\ \text{cm}^3 = 441.5\ \text{L} = 0.4415\ \text{m}^3$$

Eq. (2-71) gives air flow into engine:

$$\dot{m}_a = \eta_v \rho_a V_d N/n$$
$$= (0.98)(1.181\ \text{kg/m}^3)(0.4415\ \text{m}^3/\text{cycle})(257/60\ \text{rev/sec})/(1\ \text{rev/cycle})$$
$$= 2.189\ \text{kg/sec}$$

Mass flow of natural gas (methane) into engine using Eq. (2-55) is:

$$\dot{m}_{ng} = \dot{m}_a/\text{AF} = (2.189\ \text{kg/sec})(0.95\ \text{of total})/(17.2) = \underline{0.121\ \text{kg/sec}}$$

(2) Mass flow rate of light diesel fuel into engine is:

$$\dot{m}_{df} = [(2.189\ \text{kg/sec})(0.05)/(14.5)](1.3) = \underline{0.00981\ \text{kg/sec}}$$

(3) Eq. (2-65) is used to find indicated power output:

$$\dot{W}_i = \eta_t \eta_c \dot{m}_f Q_{\text{LHV}} = (0.61)(0.99)[(0.121\ \text{kg/sec})(49770\ \text{kJ/kg}) + (0.00981)(42500)]$$
$$= 3889\ \text{kW}$$

Brake power using Eq. (2-27) is

$$\dot{W}_b = \eta_m \dot{W}_i = (0.91)(3889\ \text{kW}) = \underline{3539\ \text{kW}}$$

(4) Cost of engine operation using dual fuel is:

$$\text{cost} = [(0.121\ \text{kg/sec})(\$0.09/\text{kg}) + (0.00981\ \text{kg/sec})(\$0.52/\text{kg})]$$
$$\times (3600\ \text{sec/hr})/(3539\ \text{kW})$$
$$= \$0.0163/\text{kW-hr}$$

To get same indicated power using all diesel fuel, use Eq. (2-65) to get mass flow rate of diesel fuel needed:

$$\dot{m}_{df} = \dot{W}_i/\eta_t \eta_c Q_{\text{LHV}} = (3889\ \text{kJ/sec})/[(0.61)(0.99)(42500\ \text{kJ/kg})] = 0.1515\ \text{kg/sec}$$
$$\text{cost} = [(0.1515\ \text{kg/sec})(\$0.52/\text{kg})][(3600\ \text{sec/hr})/(3539\ \text{kW})] = \$0.0801/\text{kW-hr}$$
$$\text{savings} = \$0.0801 - \$0.0163 = \underline{\$0.0638/\text{kW-hr}}$$
$$\text{percent savings} = [(0.0801 - 0.0163)/(0.0801)](100) = \underline{79.7\%\ \text{savings}}$$

5.9 INTAKE FOR TWO-STROKE CYCLE ENGINES

Inlet air in two-stroke cycle engines must be input at a pressure greater than atmospheric. Following blowdown, at the start of the intake process, the cylinder is still filled with exhaust gas at atmospheric pressure. There is no exhaust stroke or intake stroke. Air under pressure enters the cylinder and pushes most of the remaining exhaust residual out the still-open exhaust port. This is called **scavenging**. When most of the exhaust gas is out, the exhaust port closes and the cylinder is left filled with mostly air and fuel.

There are two general methods of putting air into the cylinders: through normal intake valves, or through intake slots in the cylinder walls. The intake air is pressurized using a supercharger, turbocharger, or crankcase compression.

Two-stroke cycle engines have open combustion chambers. It would be extremely difficult to get proper scavenging in a cylinder with a divided chamber.

Some modern experimental two-stroke cycle automobile engines use standard-type superchargers and input the air through intake valves with no fuel added. The compressed air scavenges the cylinder and leaves it filled with air and a small amount of exhaust residual. After the intake and exhaust valves are closed, fuel is injected directly into the combustion chamber by injectors mounted in the cylinder head. This is done to avoid HC pollution from fuel passing into the exhaust system when both exhaust and intake valves are open. In some automobile engines, air is injected with the fuel. This speeds evaporation and mixing, which is required because of the very short time of the compression stroke. Fuel injection pressure is on the order of 500 to 600 kPa, while air injection pressure is slightly less at about 500 kPa. For SI engines fuel injection occurs early in the compression stroke, immediately after the exhaust valve closes. In CI engines the injection occurs late in the compression stroke, a short time before combustion starts.

Other modern experimental automobile engines, and just about all small two-stroke cycle engines, due to cost, use crankcase compression to force air into and scavenge the cylinders. In these engines, air is introduced at atmospheric pressure into the cylinder below the piston through a one-way valve when the piston is near TDC. The power stroke pushes the piston down and compresses the air in the crankcase, which has been designed for this dual purpose. The compressed air then passes through an input channel into the combustion chambers. In larger engines, the fuel is added with injectors, as with supercharged engines. In small engines, the fuel is usually added with a carburetor to the air as it enters the crankcase. This is done to keep the cost down on small engines, simple carburetors being cheap to build. However, as pollution laws become more stringent, fuel injectors will probably become more common in small engines.

Exhaust blowdown occurs at about 100° to 110° aTDC when the exhaust valve is opened or when the exhaust slots in the cylinder walls are uncovered. Slightly later, at about 50° bBDC, intake occurs by means of valves or intake slots located a short distance below the exhaust slots in the cylinder walls. Either air or an air–fuel mixture enters the cylinder at a pressure of 1.2 to 1.8 atmospheres, as explained previously. The pressurized air pushes out most of the remaining exhaust gas through the still-open exhaust valves or slots. Ideally, the incoming air will force most of the exhaust gas out of the cylinder without mixing with it and without too much air–fuel going out the open exhaust valve. Some mixing will occur and some fuel will be lost out the exhaust valve. This will cause lower fuel economy and HC pollution in the exhaust. To avoid this, only air is input and used for scavenging in modern experimental two-stroke cycle automobile engines. Fuel is added with injectors after the intake valve is closed.

Lubricating oil must be added to the inlet air on those engines that use crankcase compression. The crankcase on these engines cannot be used as the oil reservoir as with most other engines. Instead, the surfaces of the engine components are lubricated by oil vapor carried by the intake air. On some engines, lubricating oil is mixed directly with the fuel and is vaporized in the carburetor along with the fuel. Other engines have a separate oil reservoir and feed lubricant directly into the intake air flow. Two negative results occur because of this method of lubrication. First, some oil vapor gets into the exhaust flow during valve overlap and contributes directly to HC exhaust emissions. Second, combustion is less efficient due to the poorer fuel quality of the oil. Engines that use superchargers or turbochargers generally use standard pressurized lubrication systems, with the crankcase serving as the oil reservoir.

To avoid an excess of exhaust residual, no pockets of stagnant flow or dead zones can be allowed in the scavenging process. This is controlled by the size and position of the intake and exhaust slots or valves, by the geometry of the slots in the wall, and by contoured flow deflectors on the piston face. Figure 5-22 shows several geometric configurations of scavenging that are used.

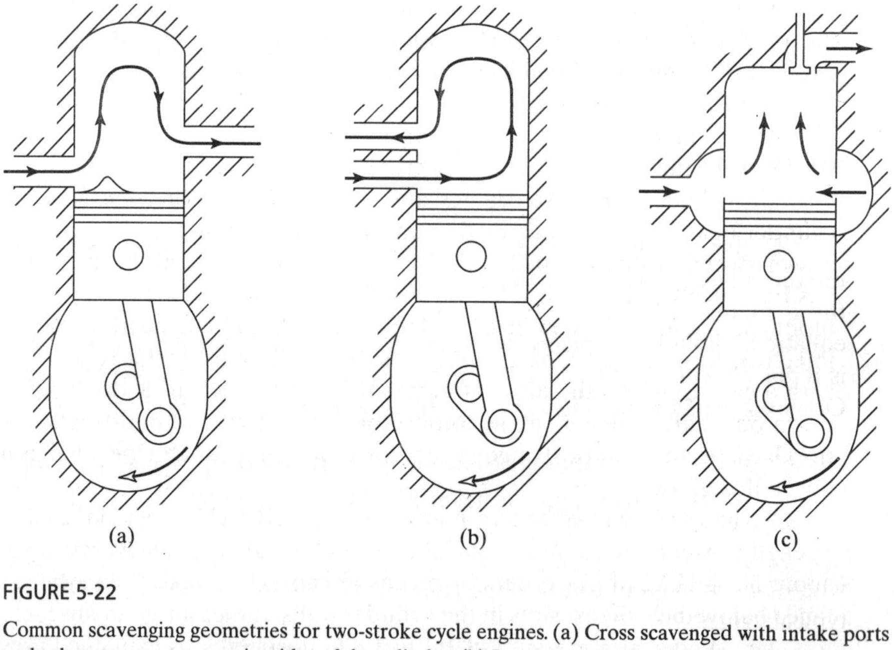

FIGURE 5-22

Common scavenging geometries for two-stroke cycle engines. (a) Cross scavenged with intake ports and exhaust ports on opposite sides of the cylinder. (b) Loop scavenged with intake ports and exhaust ports on the same side of the cylinder. (c) Uniflow scavenged (or through-flow scavenged) with intake ports in cylinder walls and exhaust valve in head. Other variations and combinations of these types exist, depending on the placement of slots and/or valves.

Cross Scavenged Intake slots and exhaust slots are located on opposite sides of the cylinder wall. Proper design is required to assure that the intake air deflects up without *short-circuiting* and leaving a stagnant pocket of exhaust gas at the head end of the cylinder.

Loop Scavenged Intake and exhaust ports are on the same side of the cylinder wall, and incoming air flows in a loop.

Uniflow Scavenged or Through-Flow Scavenged Intake ports are in the cylinder walls and exhaust valves in the head (or intake valves are in the head and exhaust ports are in the wall, which is less common). This is the most efficient system of scavenging but requires the added cost of valves.

For the same power generation, more air input is required in a two-stroke cycle engine than in a four-stroke cycle engine. This is because some of the air is lost in the overlap period of the scavenging process. A number of different intake and performance efficiencies are defined for the intake process of a two-stroke cycle engine. Volumetric efficiency of a four-stroke cycle engine can be replaced by either delivery ratio or charging efficiency,

$$\text{delivery ratio} = \lambda_{dr} = m_{mi}/V_d\rho_a \tag{5-22}$$

$$\text{charging efficiency} = \lambda_{ce} = m_{mt}/V_d\rho_a \tag{5-23}$$

where

m_{mi} = mass of air–fuel mixture ingested into cylinder
m_{mt} = mass of air–fuel trapped in cylinder after all valves are closed
V_d = displacement volume (swept volume)
ρ_a = density of air at ambient conditions

with typical values $0.65 < \lambda_{dr} < 0.95$
$0.50 < \lambda_{ce} < 0.75$

Delivery ratio is greater than charging efficiency because some of the air–fuel mixture ingested into the cylinder is lost out of the exhaust port before it is closed. For those engines that inject fuel after the valves are closed, the mass of mixture in these equations should be replaced with mass of ingested air. Sometimes, ambient air density is replaced with the density of air in the inlet runner downstream of the supercharger. Other efficiencies include:

$$\text{trapping efficiency} = \lambda_{te} = m_{mt}/m_{mi} = \lambda_{ce}/\lambda_{dr} \tag{5-24}$$

$$\text{scavenging efficiency} = \lambda_{se} = m_{mt}/m_{tc} \tag{5-25}$$

$$\text{relative charge} = \lambda_{rc} = m_{tc}/V_d\rho_a = \lambda_{ce}/\lambda_{se} \tag{5-26}$$

where m_{tc} = mass of total charge trapped in cylinder, including exhaust residual with typical values:

$$0.65 < \lambda_{te} < 0.80$$
$$0.85 < \lambda_{se} < 0.95$$
$$0.50 < \lambda_{rc} < 0.90$$

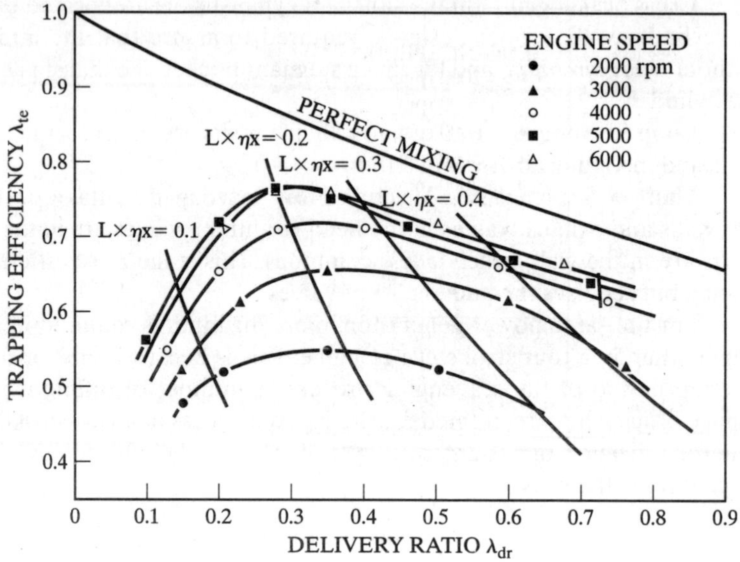

FIGURE 5-23

Trapping efficiency as a function of delivery ratio for a two-cylinder, 0.347 L, two-stroke cycle SI motorcycle engine. Reprinted with permission from SAE Paper No. 750908 © 1975 SAE International, [229].

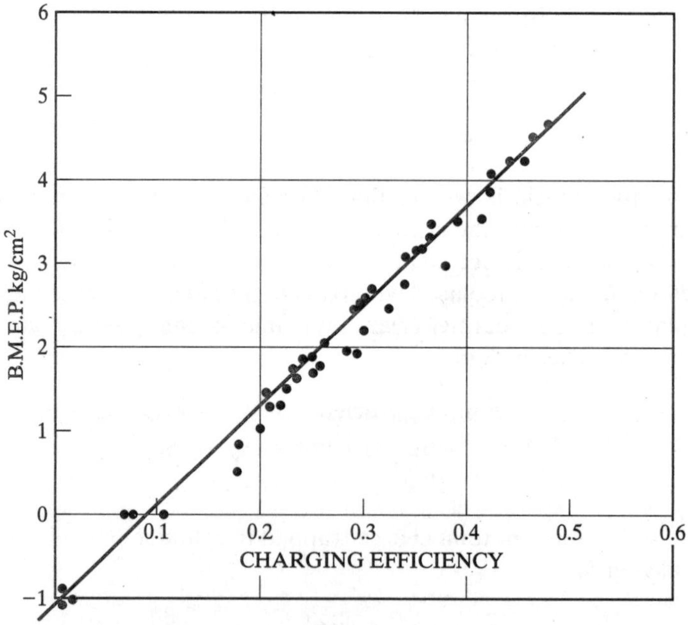

FIGURE 5-24

Brake mean effective pressure as a function of charging efficiency for two-stroke cycle SI motorcycle engine of 0.347 L displacement. Reprinted with permission from SAE Paper No. 750908 © 1975 SAE International, [229].

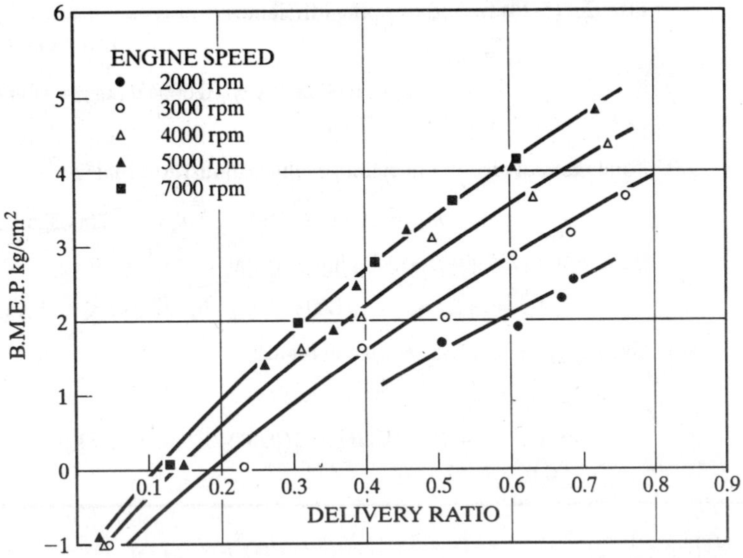

FIGURE 5-25

Brake mean effective pressure as a function of delivery ratio for two-stroke cycle SI motorcycle engine of 0.347 L displacement. Reprinted with permission from SAE Paper No. 750908 © 1975 SAE International, [229].

Example Problem 5-8

An experimental six-cylinder, two-stroke cycle, SI automobile engine has a delivery ratio of 0.88 when operating at 3700 RPM. At this speed, when the exhaust slots close during the cycle, there is an air–fuel mass of 0.000310 kg in each cylinder, plus a 5.3% exhaust residual from the preceding cycle. The engine has a bore of 7.62 cm and a stroke of 8.98 cm.
Calculate:

1. charging efficiency
2. trapping efficiency
3. scavenging efficiency
4. relative charge

(1) Use Eq. (2-8) for displacement volume of one cylinder:

$$V_d = (\pi/4)B^2S$$
$$= (\pi/4)(7.62\text{ cm})^2(8.98\text{ cm}) = 409.5\text{ cm}^3 = 0.4095\ L = 0.0004095\text{ m}^3$$

Equation (5-23) gives charging efficiency:

$$\lambda_{ce} = m_{mt}/V_d\rho_a = (0.000310\text{ kg})/[(0.0004095\text{ m}^3)(1.181\text{ kg/m}^3)] = \underline{0.641 = 64.1\%}$$

(2) Equation (5-22) is used to find mass of air–fuel ingested into cylinder:

$$m_{mi} = \lambda_{dr}V_d\rho_a = (0.88)(0.0004095\text{ m}^3)(1.181\text{ kg/m}^3) = 0.000426\text{ kg}$$

Use Eq. (5-24) to find trapping efficiency:

$$\lambda_{te} = m_{mt}/m_{mi}$$
$$= \lambda_{ce}/\lambda_{dr} = (0.000310 \text{ kg})/(0.000426 \text{ kg}) = (0.641)/(0.88)$$
$$= \underline{0.728 = 72.8\%}$$

(3) Total mass of air–fuel in cylinder, plus exhaust residual is

$$m_{tc} = (0.000310)(1 + 0.053) = 0.000326 \text{ kg}$$

Equation (5-25) gives scavenging efficiency:

$$\lambda_{se} = m_{mt}/m_{tc} = (0.000310 \text{ kg})/(0.000326 \text{ kg}) = \underline{0.951 = 95.1\%}$$

(4) Use Eq. (5-26) to find relative charge:

$$\lambda_{rc} = m_{tc}/V_d\rho_a$$
$$= \lambda_{ce}/\lambda_{se} = (0.000326 \text{ kg})/[(0.0004095 \text{ m}^3)(1.181 \text{ m}^3)] = (0.641)/(0.951)$$
$$= \underline{0.674 = 67.4\%}$$

5.10 INTAKE FOR CI ENGINES

CI engines are operated unthrottled, with engine speed and power controlled by the amount of fuel injected during each cycle. This allows for high volumetric efficiency at all speeds, with the intake system designed for very little flow restriction of the incoming air. Further raising the volumetric efficiency is the fact that no fuel is added until late in the compression stroke, after air intake is fully completed. In addition, many CI engines are turbocharged, which enhances air intake even more.

Fuel is added late in the compression stroke, starting somewhere around 20° bTDC. Injectors mounted in the cylinder head inject directly into the combustion chamber, where self-ignition occurs due to the high temperature of the air caused by compressive heating. It takes a short period of time for the fuel to evaporate, mix with the air, and then self-ignite, so combustion starts shortly bTDC. At this time fuel is still being injected, which keeps combustion occurring well into the power stroke. It is important that fuel with the correct cetane number be used in an engine so that self-ignition initiates the start of combustion at the proper cycle position. A distribution of fuel droplet sizes is desirable so that the start of combustion of all fuel particles is not simultaneous, but is spread over a short period of cycle time. This slows the start of the pressure pulse on the piston and gives smoother engine operation.

Injection pressure for CI engines must be much higher than that required for SI engines. The cylinder pressure into which the fuel is first injected is very high near the end of the compression stroke, due to the high compression ratio of CI engines. By the time the final fuel is injected, peak pressure during combustion is being experienced. Pressure must be high enough that fuel spray will penetrate across the entire combustion chamber. Injection pressures of 20 to 200 MPa are common (3000 to 30,000 psia). A typical engine could have injection pressure of 25 MPa at idle, 135 MPa at nominal speed, and 160 MPa at high speed [156]. Average fuel droplet size generally decreases with increasing pressure. Orifice hole size of injectors is typically in the range of of 0.2 to 1.0 mm diameter. Often the injection process is divided into two to five short injections

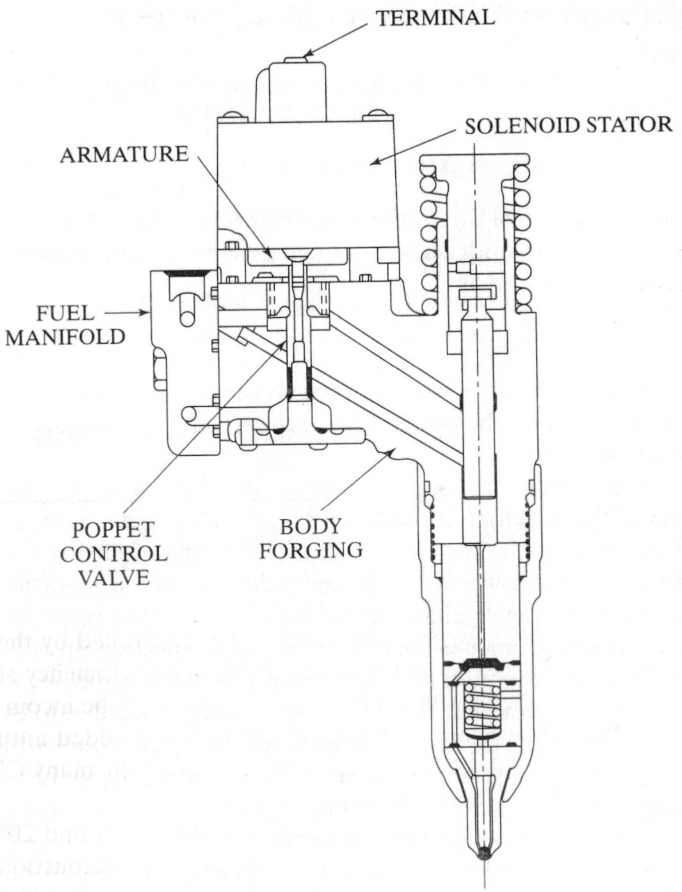

FIGURE 5-26
Electronic fuel injector for CI engine. Reprinted with permission from SAE Paper No. 850542 © 1985 SAE International, [181].

per cycle. There is a short preinjection at about 4 ms before the main injection. This is to shorten the ignition delay of the main injection. This can be followed with two or three short injections which help to continue combustion of the solid carbon in the combustion chamber and into the exhaust system.

During injection, the mass flow rate of fuel through an injector is

$$\dot{m}_f = C_D A_n \sqrt{2\rho_f \Delta P} \tag{5-27}$$

The total mass of fuel injected into one cylinder during one cycle is

$$m_f = C_D A_n \sqrt{2\rho_f \Delta P}\,(\Delta\theta/360N) \tag{5-28}$$

where

C_D = discharge coefficient of injector
A_n = flow area of nozzle orifice(s)
ρ_f = density of fuel
ΔP = pressure differential across injector

$\Delta\theta$ = crank angle through which injection takes place (in degrees)
N = engine speed

Pressure differential ΔP is about equal to the injection pressure:

$$P_{inj} \approx \Delta P \tag{5-29}$$

It is desirable that the crank angle of rotation through which injection takes place be almost constant for all speeds. To accomplish this, as engine speed changes, requires that injection pressure be related to speed as

$$P_{inj} \propto N^2 \tag{5-30}$$

Satisfying this relationship can require very high injector pressure at higher engine speeds. On some modern injectors, the orifice flow area A_n can be varied somewhat to allow greater flow at higher speeds.

Big, slow engines with large open combustion chambers have low air motion and turbulence within the cylinder. The injector is mounted near the center of the chamber, often with five or six orifices to spray over the entire chamber. Because of the low turbulence, evaporation and mixing are slower and real time between start of injection and start of combustion is longer. However, engine speed is slower, so injection timing in cycle time is about the same. Large engines must have very high injection pressure and high spray velocity. With lower air motion and turbulence, high liquid spray velocity is needed to enhance evaporation and mixing. Also, high velocity is needed to assure that some spray reaches fully across the large combustion chamber. Injectors with multiple orifices require higher pressure to obtain the same injection velocity and penetration distance. Fuel velocity leaving the injector can be as high as 250 m/sec. However, viscous drag and evaporation reduce this very quickly (Fig. 5-27).

For optimum fuel viscosity and spray penetration, it is important to have fuel at the correct temperature. Often, engines are equipped with temperature sensors and means of heating or cooling the incoming fuel. Many large truck engines are equipped with heated fuel filters. This allows the use of cheaper fuel that has less viscosity control.

Small high-speed engines need much faster evaporation and mixing of the fuel due to the shorter real time available during the cycle. This occurs because of the high turbulence and motion within the cylinder caused by high engine speed. As speed is increased, the level of turbulence and air motion increases. This increases evaporation and mixing and shortens ignition delay, resulting in fairly constant injection timing for all speeds. Part of the fuel spray is often directed against hot cylinder walls to speed evaporation. More costly, lower viscosity fuel is often required in smaller engines. Injection pressure in high-speed engines can be lower, evaporation and mixing being enhanced by air motion and not spray velocity. Shorter spray travel distance is also acceptable in the smaller combustion chambers.

Example Problem 5-9

An automobile has a 3.2-liter, five-cylinder, four-stroke cycle diesel engine operating at 2400 RPM. Fuel injection occurs from 20° bTDC to 5° aTDC. The engine has a volumetric efficiency of 0.95 and operates with fuel equivalence ratio of 0.80. Light diesel fuel is used.

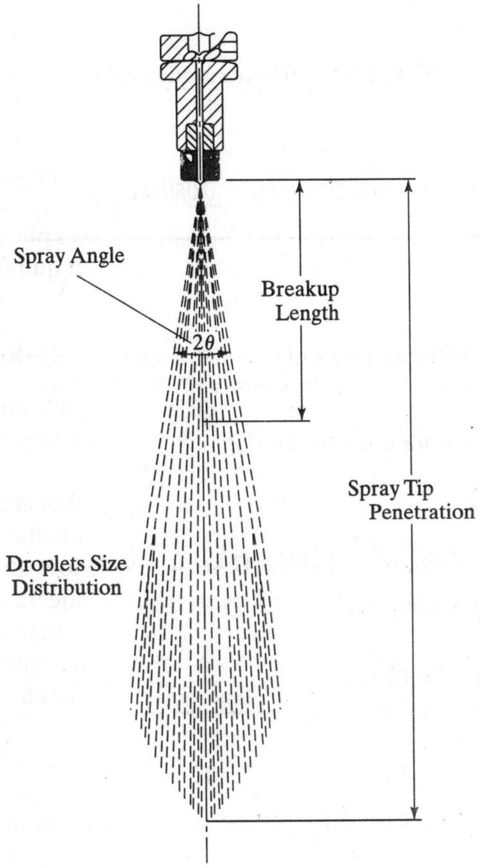

FIGURE 5-27

Fuel spray from injector in CI engine. The liquid spray breaks into individual droplets of varying sizes in the breakup length. The droplets then atomize, vaporize, and mix with air that has been pulled into the spray by the liquid flow. It is generally desired for the spray tip penetration length to reach the farthest distance in the combustion chamber. In most engines, spray will be distorted by swirl and tumble. Reprinted with permission from SAE Paper No. 840275 © 1984 SAE International, [144].

Calculate:

1. time for one injection
2. fuel flow rate through an injector

For one cylinder for one cycle,

$$V_d = (0.0032\ \text{m}^3)/5 = 0.00064\ \text{m}^3$$

Equation (2-70) gives the mass of air:

$$m_a = \eta_v \rho_a V_d = (0.95)(1.181\ \text{kg/m}^3)(0.00064\ \text{m}^3) = 0.000718\ \text{kg}$$

Equations (2-56) and (2-57) give the mass of fuel needed:

$$m_f = \phi m_a/(\text{AF})_{\text{stoich}} = (0.80)(0.000718\ \text{kg})/(14.5) = 0.0000396\ \text{kg}$$

$$\text{Engine time} = (60\ \text{sec/min})/(2400\ \text{rev/min}) = 0.025\ \text{sec/rev}$$

$$= (0.025\ \text{sec/rev})/(360°/\text{rev}) = 6.9 \times 10^{-5}\ \text{sec/degree}$$

1. Time for injection:

$$t = (25°/\text{injection})(6.9 \times 10^{-5}\ \text{sec/degree}) = \underline{0.00173\ \text{sec/injection}}$$

2. Injection rate:

$$\dot{m}_f = (0.0000396\ \text{kg})/(0.00173\ \text{sec}) = \underline{0.0229\ \text{kg/sec} = 0.050\ \text{lbm/sec}}$$

Example Problem 5-10

A fuel injector in a CI automobile engine has an orifice diameter of 0.31 mm, a discharge coefficient of 0.85, and an operating pressure differential of 110 MPa. Density of the diesel fuel is 750 kg/m^3.
Calculate the mass flow rate of fuel through the injector using Eq. (5-27):

$$\dot{m}_f = C_D A_n \sqrt{2\rho_f \Delta P}$$
$$= (0.85)[(\pi/4)(0.00031\ \text{m})^2)]\sqrt{(2)(750\ \text{kg/m}^3)(110\ \text{MPa})(1\ \text{kg-m/N-sec}^2)}$$
$$= \underline{0.0261\ \text{kg/sec} = 26.1\ \text{gm/sec} = 0.0574\ \text{lbm/sec}}$$

Compare this flow rate with that found in Example Problem 5-9.

5.11 NITROUS OXIDE

A unique way of ingesting a greater amount of oxygen into an engine, and thus producing more power, is to input the oxygen in the form of liquid nitrous oxide N_2O. By injecting oxygen as a liquid, a much greater amount can be input during each cycle without the normal restriction of limited volumetric efficiency. With more oxygen added, it is a fairly simple process to add more fuel, with a net result of a greater amount of combustible mixture in the cylinder to produce power during each cycle.

Development of the use of nitrous oxide in engines first occurred during World War II in the 1940s to produce added power in reciprocating engines of war planes. With the advent of the jet engine, this development for aircraft ceased, but was taken up by race car enthusiasts and automobile hot-rodders.

Nitrous oxide (laughing gas) was discovered in 1772 and was used as an anesthetic for medical surgery of short duration [185, 224]. Since then, it has been used for a number of other applications, some of which can be dangerous. In the 1970s it became popular with some builders of drag-racing automobiles, and is still available for such use today.

One of the major problems of using nitrous oxide in a reciprocating engine is that it can produce enough power to destroy the engine. Power increases of 100–300% are possible, and unless the mechanical structure of the engine is reinforced, most engines would not survive this kind of operation. When a burst of power is desired, such as during a quarter mile drag race, a charge of liquid N_2O, as well as a corresponding charge of liquid fuel, are injected into the engine cylinders. This produces a short-term surge of power output.

In addition to requiring much less gaseous air input, there is an additional gain caused by the evaporative cooling of the nitrous oxide. When liquid N_2O changes to vapor in the cylinders, the evaporative cooling makes the incoming air denser, which allows more air to be ingested. When used for power boost in an engine, nitrous oxide is normally stored as a liquid at about 6000 kPa pressure.

Example Problem 5-11

A woman adds a nitrous oxide system to her drag racer which normally had a brake power output of 200 kW using stoichiometric isooctane fuel. At the advise of the dealer, she decides to add N_2O at a rate of about 4 to 1 by mass (oxidizer/fuel). The temperature of the air–fuel into the cylinders can be considered 25°C, while the exhaust gas leaves the cylinders at a temperature of 1000 K. It can be assumed that the thermal efficiency remains the same, with the same percent of the heat input being converted to power output. Calculate the percent increase in heat that can be added to each cycle (and thus the increase in power output).

One kgmole of isooctane has a mass of (1 kgmole)(114 kg/kgmole) = 114 kg

Amount of nitrous oxide to be added = (114 kg)(4)

= (456 kg)/(44 kg/kgmole) ≈ 10 kgmoles

Without the nitrous oxide, the chemical reaction equation for one mole of fuel is

$$C_8H_{18} + 12.5\,O_2 + 12.5(3.76)\,N_2 \rightarrow 8\,CO_2 + 9\,H_2O + 12.5\,(3.76)\,N_2$$

Equation (4-5) is used to find the heat input for one kgmole of fuel (enthalpy values are from [90]):

$$\begin{aligned} Q_{in} &= \textstyle\sum_{PROD} N_i(h_f{}^\circ + \Delta h)_i - \sum_{REACT} N_i(h_f{}^\circ + \Delta h)_i \\ &= 8[(-393{,}522) + (33{,}397)] + 9[(-241{,}826) + (26{,}000)] + 12.5(3.76)[(0) + (21{,}463)] \\ &\quad - 1[(-259{,}280) + (0)] - 12.5[(0) + (0)] - 12.5(3.76)[(0) + (0)] \\ &= -3{,}555{,}000 \text{ kJ/kgmole} \end{aligned}$$

With nitrous oxide, the chemical reaction equation for one kgmole of fuel is

$$C_8H_{18} + 10\,N_2O + 7.5\,O_2 + 7.5(3.76)\,N_2 \rightarrow 8\,CO_2 + 9\,H_2O + 38.2\,N_2$$

$$\begin{aligned} Q_{in} &= 8[(-393{,}522) + (33{,}397)] + 9[(-241{,}826) + (26{,}000)] + 38.2[(0) + (21{,}463)] \\ &\quad - 1[(-259{,}280) + (0)] - 10[(-81{,}600) + (0)] - 7.5[(0) + (0)] - 7.5(3.76)[(0) + (0)] \\ &= -2{,}928{,}000 \text{ kJ/kgmole} \end{aligned}$$

This amount of heat is increased as more fuel can be put in, the fuel requiring less air because of the oxygen supplied by the N_2O:

$$Q_{in} = (-2{,}928{,}000 \text{ kJ/kgmole})[(12.5)/(7.5)] = -4{,}880{,}000 \text{ kJ/kgmole}$$

This amount of heat input will be further increased because of the cooling effect of the nitrous oxide evaporating. This gives a denser air–fuel input, increasing the volumetric efficiency. Heat of vaporization of nitrous oxide h_{fg} = 11,037 kJ/kgmole.

Change in the temperature of the gas input due to evaporative cooling (with all gases treated as air in air-standard analysis and using a low temperature value of c_p) is given by

$$mc_p\Delta T = h_{fg}(\text{number of moles})$$

$$(7.5)(4.76)\text{ kgmoles }(29\text{ kg/kgmole})(1.005\text{ kJ/kg-K})\Delta T = (11{,}037\text{ kJ/kgmole})(10\text{ kgmoles})$$

$$\Delta T = 106\text{ K} \qquad T_{final} = 298\text{ K} - 106 = 192\text{ K}$$

Density of air is proportional to absolute temperature, and heat (fuel) input is proportional to density of air input:

$$Q_{in} = (-4{,}880{,}000\text{ kJ/kgmole})[(298)/(192)] = -7{,}574{,}000\text{ kJ/kgmole}$$

Percent increase in power is

$$\Delta\% = [(7{,}574{,}000 - 3{,}555{,}000)/(3{,}555{,}000)](100) = \underline{113\%\text{ increase}}$$

$$\underline{200\text{ kW becomes }426\text{ kW}}$$

5.12 CONCLUSIONS

Correct, consistent induction of air and fuel into an engine is one of the more important and difficult processes to obtain in engine design. High volumetric efficiency of intake systems, giving a maximum flow of air, is important to supply the oxygen needed to react with the fuel. Ideally, the engine should receive a consistent amount of air cylinder-to-cylinder and cycle-to-cycle. This does not happen due to turbulence and other flow inconsistencies, and engine operation must be limited by statistical averages.

Equally important and equally difficult is the supplying of the correct amount of fuel to the engine. Again, the goal is to supply an equal amount to each cylinder with no variation cycle-to-cycle. This is limited by the quality and control of the fuel injectors or carburetor.

Air is supplied through an intake manifold, with flow rate controlled on most SI engines by a throttle butterfly valve and uncontrolled on CI engines. Inlet air pressure is ambient or is increased with a supercharger, a turbocharger, or crankcase compression. Fuel is added in SI engines with throttle body injectors located upstream in the intake manifold, with port injectors at the intake valve, or by injection directly into the cylinder. Carburetors are used for fuel input on many small, less costly engines and on older automobile engines. CI engines inject fuel directly into the combustion chamber and control engine speed by injection amount.

Lean-burning engines, stratified charge engines, dual-chamber engines, dual-fuel engines, and two-stroke cycle automobile engines all have unique and more complicated induction systems. These require special combinations and design of carburetors, fuel injectors, valves, and valve timing.

PROBLEMS

5.1 A five-cylinder, four-stroke cycle SI engine has a compression ratio $r_c = 11{:}1$, bore $B = 5.52$ cm, stroke $S = 5.72$ cm, and connecting rod length $r = 11.00$ cm. Cylinder inlet conditions are 63°C and 92 kPa. The intake valve closes at 41° aBDC and the spark plug is fired at 15° bTDC.

Calculate:

(a) Temperature and pressure in the cylinder at ignition, assuming Otto cycle analysis (i.e., assume the intake valve closes at BDC and ignition is at TDC). [K, kPa]

(b) Effective compression ratio (i.e., actual compression of the air–fuel mixture before ignition).

(c) Actual temperature and pressure in the cylinder at ignition. [K, kPa]

5.2 Two engine options are to be offered in a new automobile model. Engine A is naturally aspirated with a compression ratio of 10.5:1 and cylinder inlet conditions of 60°C and 96 kPa. Engine B is supercharged with aftercooling and has cylinder inlet conditions of 80°C and 130 kPa. To avoid knock problems, it is desirable to have the air–fuel temperature at the start of combustion in engine B be the same as in engine A.

Calculate:

(a) Temperature at start of combustion in engine A, using air-standard Otto cycle analysis. [°C]

(b) Compression ratio of engine B that would give the same temperature at the start of combustion.

(c) Temperature reduction in the aftercooler of engine B if the compressor has an isentropic efficiency of 82% and inlet conditions are the same as in engine A. [°C]

5.3 Air enters the intake manifold at 74°F and 14.7 psia on a V12 airplane engine with throttle body injection, using gasoline at an equivalence ratio of $\phi = 0.95$. It can be assumed that all the fuel evaporates in the adiabatic manifold.

Calculate:

(a) Temperature of the air–fuel mixture after fuel evaporation [°F]

(b) Percent loss or gain in engine volumetric efficiency due to fuel evaporation. [%]

(c) Temperature in the cylinder at the start of compression, after inlet air–fuel mixes with the 5% exhaust residual from the previous cycle, which is at 900° R. [°F]

5.4 Water injection is added to the engine in Problem 5-3 which delivers 1 lbm of water for every 30 lbm of gasoline used. Heat of vaporization of water $h_{fg} = 1052$ BTU/lbm.

Calculate:

(a) Temperature of the air–fuel mixture after evaporation of fuel and water. [°F]

(b) Percent loss or gain in engine volumetric efficiency due to evaporation of fuel and water. [%]

5.5 (a) Why is the compression ratio of an SI engine often reduced when the engine is redesigned to be used with a turbocharger? (b) Is brake power increased or decreased? (c) Is thermal efficiency increased? (d) Why isn't reducing the compression ratio as important when a turbocharger is added to a CI engine design?

5.6 A 2.4-liter, four-cylinder, four-stroke cycle engine is equipped with multipoint port fuel injection, having one injector per cylinder. The injectors are constant-flow devices, so the fuel flow rate into the engine is controlled by injection pulse duration. Maximum power is obtained at WOT when injection duration is continuous. At this condition, engine speed is 5800 RPM with stoichiometric gasoline and an inlet pressure of 101 kPa. At idle condition, the engine speed is 600 RPM with stoichiometric gasoline and an inlet pressure of 30 kPa. Volumetric efficiency can be considered 95% at WOT.

Calculate:

(a) Fuel flow rate through an injector. [kg/sec]
(b) Injection pulse duration in seconds at idle conditions.
(c) Injection pulse duration in degrees of engine rotation at idle conditions.

5.7 A six-cylinder, four-stroke cycle SI engine with multipoint fuel injection has a displacement of 2.4 liters and a volumetric efficiency of 87% at 3000 RPM, and operates on ethyl alcohol with an equivalence ratio of 1.06. Each cylinder has one port injector which delivers fuel at a rate of 0.02 kg/sec. The engine also has an auxiliary injector upstream in the intake manifold which delivers fuel at a rate of 0.003 kg/sec to change the air–fuel ratio and give a richer mixture when needed. When in use, the auxiliary injector operates continuously and supplies all cylinders.
Calculate:

(a) Time of one injection pulse for one cylinder for one cycle. [sec]
(b) AF if the auxiliary injector is not being used.
(c) AF if the auxiliary injector is being used.

5.8 As speed increases in an engine with throttle body fuel injection, does the temperature of the air–fuel mixture at the intake manifold exit increase or decrease? Explain what parameters affect your answer.

5.9 A 6.2-liter, V8, four-stroke cycle SI engine is designed to have a maximum speed of 6500 RPM. At this speed, volumetric efficiency is 88%. The engine is equipped with a four-barrel carburetor, each barrel having a discharge coefficient of $C_{Dt} = 0.95$. The fuel used is gasoline at AF = 15:1 (density of gasoline $\rho_g = 750$ kg/m^3).
Calculate:

(a) Minimum throat diameter needed in each carburetor venturi. [cm]
(b) Fuel capillary tube diameter needed for each venturi throat if tube discharge coefficient $C_{Dc} = 0.85$ and the capillary tube height differential is small. [mm]

5.10 (a) Explain how a carbureted automobile engine is started on a cold winter morning; tell what must be done, why, and how. (b) Why is there an accelerating pump on an automobile carburetor? (c) Explain what happens in the engine cylinders when the throttle on a carburetor is suddenly closed to decelerate an automobile traveling at high speed.

5.11 A V8 engine with 7.5-cm bores is redesigned from two valves per cylinder to four valves per cylinder. The old design had one inlet valve of 34 mm diameter and one exhaust valve of 29 mm diameter per cylinder. These are replaced with two inlet valves of 27 mm diameter and two exhaust valves of 23 mm diameter. Maximum valve lift equals 22% of the valve diameter for all valves.
Calculate:

(a) Increase of inlet flow area per cylinder when the valves are fully open. [cm^2]
(b) Give advantages and disadvantages of the new system.

5.12 A CI engine with bore $B = 8.2$ cm has the fuel injectors mounted in the center of the cylinder head. The injectors have a nozzle diameter of 0.073 mm, a discharge coefficient of 0.72, and an injection pressure of 50 MPa. Average cylinder pressure during injection can be considered 5000 kPa. Density of the diesel fuel is 860 kg/m^3.

Calculate:

(a) Average velocity of the fuel jet as it leaves the injector. [m/sec]

(b) Time for a fuel particle to reach the cylinder wall if it travels at average exit velocity. [sec]

5.13 A 3.6-liter, V6 SI engine is designed to have a maximum speed of 7000 RPM. There are two intake valves per cylinder, and valve lift equals one-fourth valve diameter. Bore and stroke are related as $S = 1.06\ B$. Design temperature of the air–fuel mixture entering the cylinders is 60°C.

Calculate:

(a) Ideal theoretical valve diameter. [cm]

(b) Maximum flow velocity through intake valve. [m/sec]

(c) Whether the valve diameters and bore size seem compatible

5.14 The volume of the average diesel fuel droplet in Example Problem 5-9 is $3 \times 10^{-14}\ m^3$. The compression ratio of the engine is 18:1. As a rough approximation, it can be assumed that all fuel droplets have the same volume and are equally spaced throughout the combustion chamber at TDC. Density of the diesel fuel is $\rho = 860\ kg/m^3$

Calculate:

(a) Number of fuel droplets in one injection.

(b) Approximate distance between droplets in the combustion chamber at TDC. [mm]

5.15 A small in-line, four-cylinder, 2.2 L, CI automobile engine equipped with cylinder cutout, operates on a four-stroke Dual cycle using light diesel fuel with AF = 21. When very little power is required the engine converts to a two-cylinder 1.1 L engine and operates at a higher speed. As a four-cylinder, the engine runs at 2100 RPM, with a volumetric efficiency of 61%, and a brake thermal efficiency of 45%. As a two-cylinder, the engine has a volumetric efficiency of 82%, and a brake thermal efficiency of 42%. Combustion efficiency is 98% at all times.

Calculate:

(a) Flow rate of air into four-cylinder engine at 2100 RPM. [kg/sec]

(b) Flow rate of air into two-cylinder engine to produce same brake power output. [kg/sec]

(c) Speed needed as two-cylinder engine to produce same brake power. [RPM]

5.16 A truck engine has variable valve timing on its intake valves. The exhaust valves open at 31° bBDC and close 20° aTDC at all engine speeds. It is desired to have 0.004 seconds of valve overlap at 3000 RPM and 0.002 seconds of valve overlap at 1200 RPM.

Calculate:

(a) Crank angle when intake valve should open at 3000 RPM. [° bTDC]

(b) Crank angle when intake valve should open at 1200 RPM. [° aTDC]

5.17 A large twelve-cylinder, 460 L, two-stroke cycle engine operates using dual fuel. 92% of the intake air is used for stoichiometric combustion of methanol, while 8% is used for stoichiometric combustion of light diesel fuel for ignition. The engine operatcs at 195 RPM with a volumetric efficiency of 93%. The methanol is input during the intake stroke, while diesel fuel is injected into each cylinder by a single injector from 15° bTDC to 6° aTDC.

Calculate:

(a) Mass flow rate of air into engine. [kg/sec]
(b) Mass flow rate of methanol into engine. [kg/sec]
(c) Mass flow rate of diesel fuel through an injector. [kg/sec]

5.18 As a three-cylinder, 1.5 liter, two-stroke cycle, spark ignition engine runs at 3400 RPM, there are 0.000440 kg of gases trapped in each cylinder during the cycle. This includes 4.60% exhaust residual from the preceding cycle. At this condition, the engine has a trapping efficiency $\lambda_{te} = 0.760$.

Calculate:

(a) Delivery ratio.
(b) Charging efficiency.
(c) Scavenging efficiency.
(d) Relative charge.

5.19 An experimental V6, two-stroke cycle, 3 liter, SI engine uses crankcase compression. Stoichiometric gasoline is used and oil is added to the fuel at a ratio of 25:1. When the engine is running at 3000 RPM, the delivery ratio $\lambda_{dr} = 0.95$ and the trapping efficiency $\lambda_{te} = 0.85$. All fuel and oil present at combustion is burned.

Calculate:

(a) Mass flow rate of air into engine. [kg/sec]
(b) Mass flow rate of oil into engine. [kg/sec]
(c) Mass flow rate of unburned oil in exhaust. [kg/sec]

5.20 Fearful that the increase in power gained by the engine in Example Problem 5-11 will structurally damage the engine, it is decided to use only half the amount of nitrous oxide, i.e., 5 kgmole of N_2O for each kgmole of C_8H_{18}. Thermal efficiency, and temperatures into and out off the cylinders are the same.

Using the method of Example Problem 5-11, approximate the percentage increase in power gained by using this amount of nitrous oxide.

DESIGN PROBLEMS

5.1D An in-line, straight 8, four-stroke cycle SI engine has throttle body fuel injection using two injectors. Each injector supplies fuel for four of the cylinders. The firing order of the engine is 1-3-7-5-8-6-2-4. Design an intake manifold for this engine with consideration to maintaining consistent AF to each cylinder and overall engine cycle smoothness.

5.2D A 2.5-liter, four-stroke cycle SI engine using multipoint port fuel injection has an idling speed of 300 RPM (AF = 13.5 and $\eta_v = 0.12$) and a maximum WOT speed of 4800 RPM (AF = 12 and $\eta_v = 0.95$). Injectors have a constant mass flow rate at all conditions. Design an injection system for this engine, giving the number of injectors per cylinder and the flow rate through each injector in kg/sec. What is the duration of injection for one cycle, and when should injection start relative to the intake valve opening? Give your answers in seconds, in degrees of engine rotation at idle, and at WOT. State all assumptions you make.

5.3D Design a fuel intake system for a large dual-fuel CI engine to be used in a poor underdeveloped country. The engine is to use only enough diesel oil to promote ignition, while using some less costly main fuel. Draw a schematic, and give engine values of displacement, speed, and volumetric efficiency. Choose an appropriate main fuel and give the flow rates of both fuels. What is the overall air–fuel ratio? State all assumptions you make.

CHAPTER 6

Fluid Motion within Combustion Chamber

"The ordinary accounts of this vortex had by no means prepared me for what I saw."

A Descent into the Maelstrom

By *Edgar Allan Poe* (1841)

This chapter discusses air, fuel, and exhaust gas motion that occurs within the cylinders during the compression stroke, combustion, and power stroke of the cycle. It is important to have this motion to speed evaporation of the fuel, to enhance air–fuel mixing, and to increase combustion speed and efficiency. In addition to the normal desired turbulence, a rotational motion called **swirl** is generated in the air–fuel mixture during intake. Near the end of the compression stroke, two additional mass motions are generated: **squish** and **tumble**. Squish is a radial motion towards the centerline of the cylinder, while tumble is a rotational motion around a circumferential axis. One additional flow motion will be discussed: that of crevice flow and blowby. This is the flow into the very small crevices of the combustion chamber due to the very high pressures generated during compression and combustion.

6.1 TURBULENCE

Due to the high velocities involved, all flows into, out of, and within engine cylinders are turbulent flows. The only exceptions to this are those flows in the corners and small crevices of the combustion chamber where the close proximity of the walls dampens out turbulence. As a result of turbulence, thermodynamic transfer rates within an engine are increased by an order of magnitude. Heat transfer, evaporation, mixing, and combustion rates all increase. As engine speed increases, flow rates increase, with a corresponding increase in swirl, squish, and turbulence. This increases the real-time rate of fuel evaporation, mixing of the fuel vapor and air, and combustion.

When flow is turbulent, particles experience random fluctuations in motion superimposed on their main bulk velocity. These fluctuations occur in all directions, perpendicular to the flow and in the flow direction. This makes it impossible to predict the exact flow conditions at any given time and position. Statistical averaging over many engine cycles gives accurate average flow conditions, but cannot predict the exact flow of any one cycle. The result is cyclic variations in operating parameters within an engine (e.g., cylinder pressure, temperature, burn angle, etc.).

A number of different models for turbulence can be found in fluid mechanics literature, which can be used to predict flow characteristics [59]. One simple model uses fluctuation velocities of u' in the X coordinate direction, v' in the Y direction, and w' in the Z direction. These are superimposed on the average bulk velocities of u, v, and w in the X, Y, and Z directions, respectively. The level of turbulence is then calculated by taking the root-mean-square average of u', v', and w'. The linear average of u', v', or w' will be zero.

There are many levels of turbulence within an engine. Large-scale turbulence occurs with eddies on the order of the size of the flow passage (e.g., valve opening, diameter of intake runner, height of clearance volume, etc.). These fluctuations are random but have a directionality controlled by the passage of the flow. At the other extreme, the smallest-scale turbulence is totally random and homogeneous, with no directionality, and is controlled by viscous dissipation. There are all levels of turbulence in between these extremes, with characteristics ranging from those of small-scale turbulence to those of large-scale turbulence. References [58, 163] examine the role of turbulence in internal combustion engines in great detail and are highly recommended for a more in-depth study of this subject.

Turbulence in a cylinder is high during intake, but then decreases as the flow rate slows near BDC. It increases again during compression as swirl, squish, and tumble increase near TDC. Swirl makes turbulence more homogeneous throughout the cylinder.

The high turbulence near TDC when ignition occurs is very desirable for combustion. It breaks up and spreads the flame front many times faster than that of a laminar flame. The air–fuel is consumed in a very short time, and self-ignition and knock are avoided. Local flame speed depends on the turbulence immediately in front of the flame. This turbulence is enhanced by the expansion of the cylinder gases during the combustion process. The shape of the combustion chamber is extremely important in generating maximum turbulence and increasing the desired rapid combustion.

Turbulence intensity is a strong function of engine speed (Fig. 6-1). As speed is increased, turbulence increases, and this increases the rate of evaporation, mixing, and combustion. One result of this is that all engine speeds have about the same burn angle (i.e., the crank angle through which the engine turns as combustion takes place). The one phase of this process that is not totally changed by the increase in turbulence is ignition delay. This is compensated for by advancing ignition spark timing (initiate the spark earlier) as the engine speed is increased.

To maximize volumetric efficiency, the inside surface of most intake manifolds is made as smooth as possible. An exception to this concept is applied to the intake manifolds of the engines on some economy vehicles where high power is not desired. The

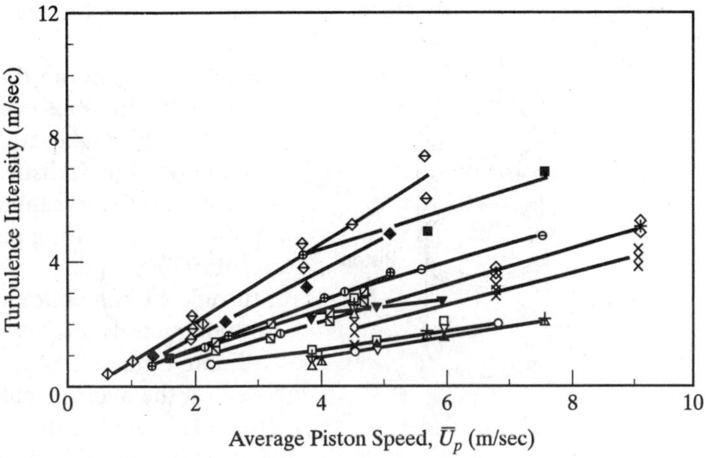

FIGURE 6-1

Turbulence intensity near TDC in combustion chamber as a function of average piston speed. Data represent experiments by several researchers using different engines, both with and without swirl. Turbulence was generally greater with swirl. Turbulence intensity increases with engine speed in most fluid flows within the engine. Reprinted with permission from SAE Paper No. 840375 © 1984 SAE International, [195].

inside surfaces of these manifolds are roughened to promote higher turbulence levels to enhance evaporation and air–fuel mixing.

One place turbulence is detrimental is in the scavenging process of a two-stroke cycle engine. Because of turbulence, the incoming air mixes more with the exhaust gases, and a greater exhaust residual will remain in the cylinder. Another negative result occurs during combustion when high turbulence enhances the convection heat transfer to the walls in the combustion chamber. This higher heat loss lowers the thermal efficiency of the engine.

6.2 SWIRL

The main bulk mass motion within the cylinder is a rotational motion called swirl. It is generated by constructing the intake system to give a tangential component to the intake flow as it enters the cylinder (see Fig. 6-2). This is done by shaping and contouring the intake manifold, valve ports, and sometimes even the piston face. Swirl greatly enhances the mixing of air and fuel to give a homogeneous mixture in the very short time available for this in modern high-speed engines. It is also a main mechanism for very rapid spreading of the flame front during the combustion process.

Swirl ratio is a dimensionless parameter used to quantify rotational motion within the cylinder. It is defined in two different ways in the technical literature:

$$(SR)_1 = (\text{angular speed})/(\text{engine speed}) = \omega/N \tag{6-1}$$

$$(SR)_2 = (\text{swirl tangential speed})/(\text{average piston speed}) = u_t/\overline{U}_p \tag{6-2}$$

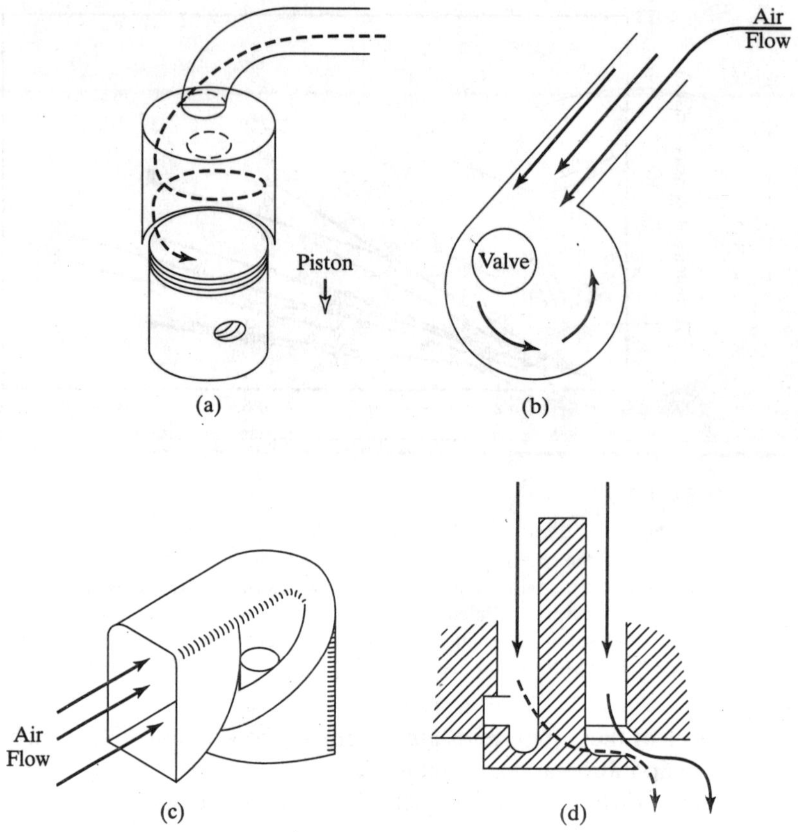

FIGURE 6-2

(a) Swirl motion within engine cylinder. Methods to generate swirl include (b) air entering cylinder from tangential direction, (c) contoured intake runner, (d) contoured valve.

Average values of either the angular speed or the tangential speed should be used in these equations. Angular motion is very nonuniform within the cylinder, being a maximum away from the walls and being much less near the walls due to viscous drag. The nonuniformity is both in the radial direction, due to drag with the cylinder walls, and in the axial direction, due to drag with the piston face and cylinder head.

Swirl ratio continuously changes during the cycle of a reciprocating engine. During intake it is high, decreasing after BDC in the compression stroke due to viscous drag with the cylinder walls. Combustion expands the gases and increases swirl to another maximum part way into the power stroke. Expansion of the gases and viscous drag quickly reduce this again before blowdown occurs. Maximum swirl ratio as defined by Eq. (6-1) can be on the order of 5 to 10 for a modern engine. One-fourth to one-third of angular momentum will be lost during the compression stroke.

One simple way of modeling cylinder swirl is the *paddle wheel* model [58]. The volume within the cylinder is idealized to contain an imaginary paddle wheel that has

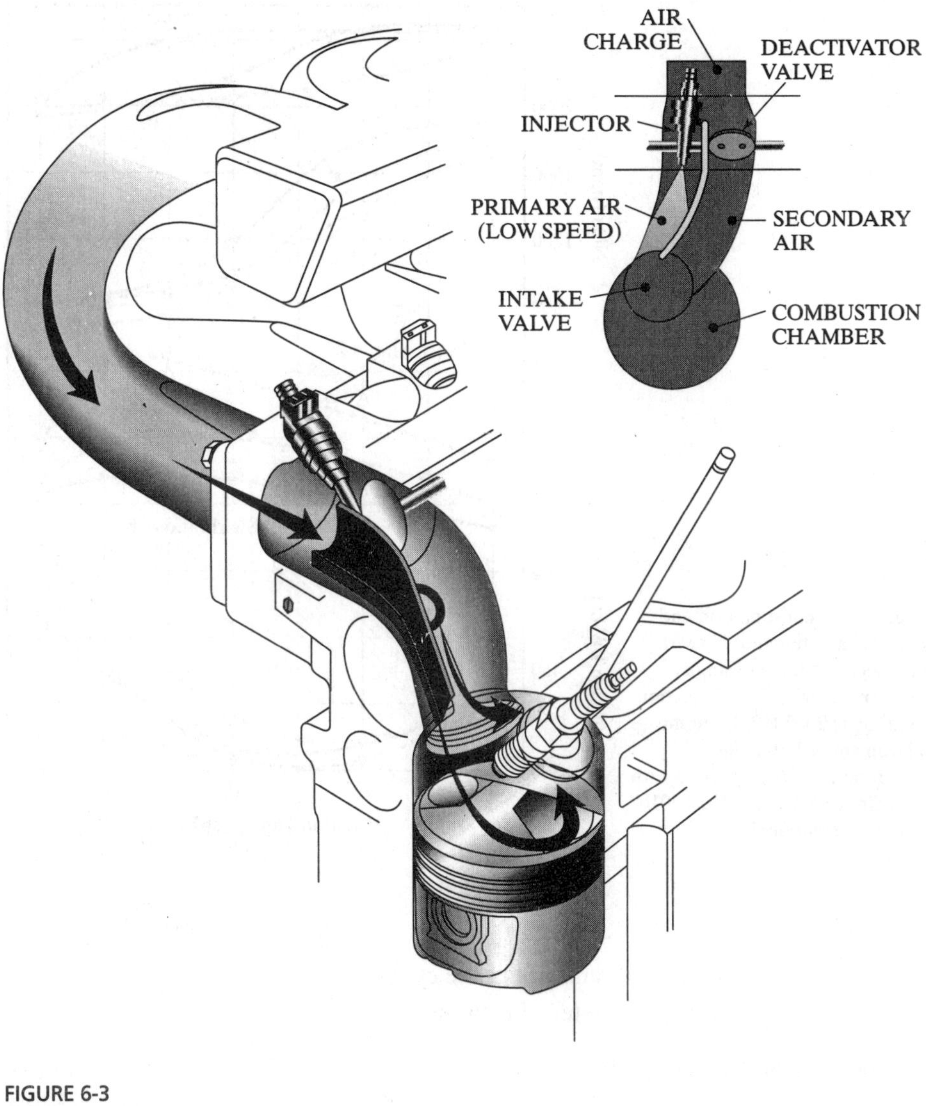

FIGURE 6-3

Intake system of spark-ignition engine equipped with multipoint port injection. Injector is positioned to spray fuel into intake runner directly behind intake valve. Runner is designed to promote swirl motion in cylinder. System is from four-cylinder, two-valve-per-cylinder, 1.9-liter 1995 Ford engine. Courtesy Ford Motor Company.

no mass. As the paddle wheel turns, the gas between the blades turns with it, resulting in a cylinder of gas all rotating at one angular velocity. The mass moment of inertia of this cylinder of gas is

$$I = mB^2/8 \tag{6-3}$$

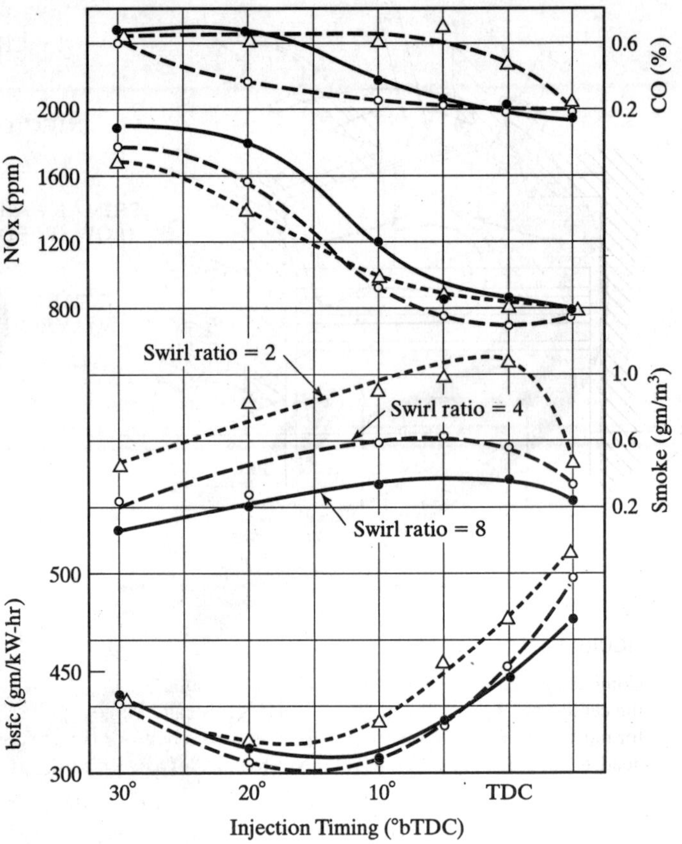

FIGURE 6-4

Brake-specific fuel consumption and emissions level as a function of swirl ratio and injection timing of single-cylinder CI engine with direct injection (DI) operating at 2000 RPM. Engine has displacement of 1.36 L and compression ratio of 16. Reprinted with permission from SAE Paper No. 730169 © 1973 SAE International, [193].

where

m = mass of gas mixture in the cylinder

B = bore = diameter of rotating mass

The angular momentum is

$$\Gamma = I\omega \tag{6-4}$$

where ω = solid-body angular velocity

Combustion chambers of most modern engines are shaped like those in Fig. 6-5, with most of the clearance volume close to the cylinder centerline. The reason for this is to reduce the flame travel distance for most of the air–fuel mixture as it combusts near TDC. The clearance volume can be in the cylinder head as in Fig. 6-5 (a), in the crown of the piston as in Fig. 6-5 (b), or in a combination of the two. With this kind of combustion chamber, as the piston nears TDC, the radius of the rotating cylinder of air–fuel is suddenly greatly reduced. This results in a large increase in angular velocity due to conservation of angular momentum. It is common to have angular velocity

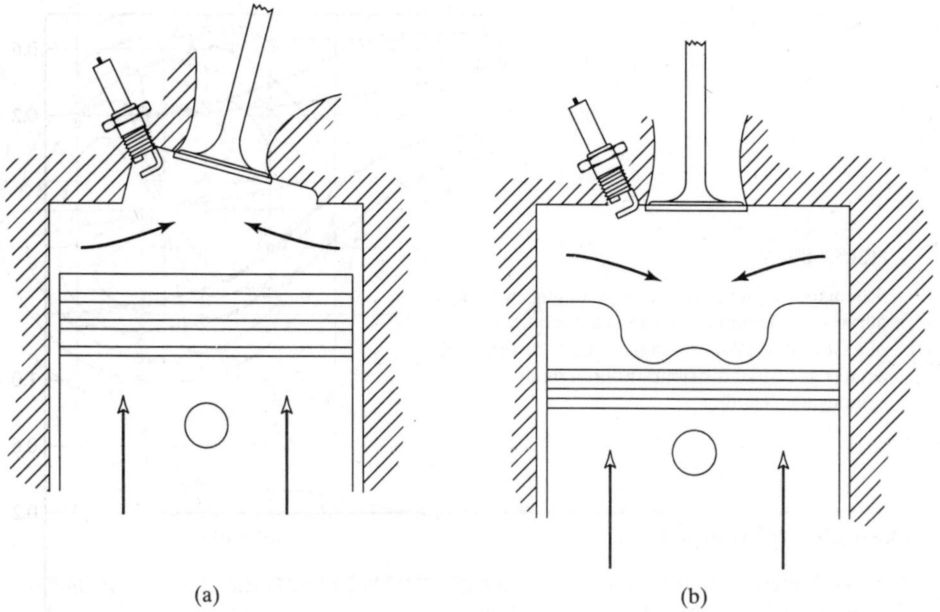

FIGURE 6-5

Combustion chamber geometry of modern automobile engines, with most of the clearance volume near the centerline of the cylinder. This increases squish and tumble, and decreases the flame travel distance for most of the combustion process. Engines can be built with (a) the clearance volume in the cylinder head, (b) as a bowl in the crown of the piston face, or as a combination of these.

increase by a factor of three to five at TDC, even though viscous drag with the walls is very great at this point. High angular velocity at TDC is very desirable because it spreads the flame front through the combustion chamber very quickly. In some engines, burn time is decreased by positioning the spark plug so that it is offset from center to take advantage of high swirl.

In two-stroke cycle engines with intake ports in the cylinder walls, swirl is generated by shaping the edges of the ports and direction of the intake runners. Swirl greatly reduces dead spots in the scavenging process but also increases mixing of the inlet charge with exhaust residual. The shaping of inlet ports and runners to promote swirl reduces the volumetric efficiency of all engines.

For CI engines and SI engines with direct injection, the period of one swirl rotation and the number of holes in the injector nozzle should be related to injection time as

$$\text{injection time} = (\text{period of swirl})/(\text{number of holes}) \tag{6-5}$$

This will assure fuel distribution throughout the entire combustion chamber as shown in Fig. 6-6

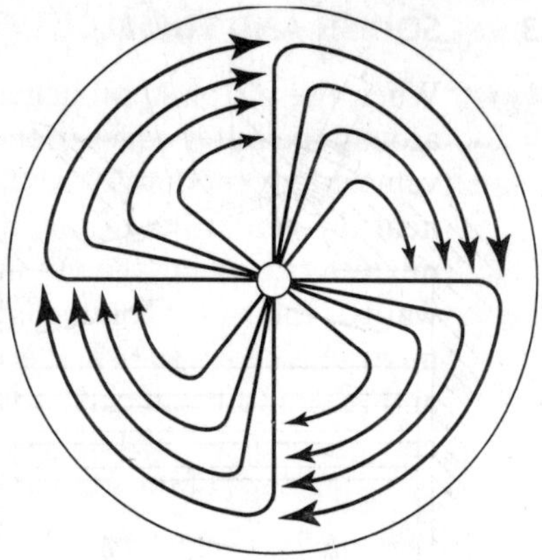

FIGURE 6-6

Schematic showing relationships among swirl, injection duration, and number of orifice holes in the fuel nozzle of a CI engine. Proper design, as given in Eq. (6-5), assures good fuel distribution throughout the entire combustion chamber.

Example Problem 6-1

A four-cylinder, 3.2-liter engine running at 4500 RPM has a swirl ratio of 6 as defined by Eq. (6-1). The stroke and bore are related as $S = 1.06\,B$.
Calculate:

1. angular velocity of gas mixture in the cylinder using the paddle wheel model
2. swirl ratio as defined by Eq. (6-2)

(1) Equation (6-1) is used to find angular velocity:

$$(\mathrm{SR})_1 = \omega/N = 6 = \omega/(4500/60\ \mathrm{rev/sec})$$
$$\underline{\omega = 450\ \mathrm{rev/sec}}$$

(2) For one cylinder

$$V_d = (3.2\ \mathrm{L})/4 = 0.8\ \mathrm{L} = 0.0008\ \mathrm{m}^3$$
$$V_d = (\pi/4)B^2S = (\pi/4)(1.06)B^3 = 0.0008\ \mathrm{m}^3$$
$$B = 0.0987\ \mathrm{m} = 9.87\ \mathrm{cm}$$
$$S = (1.06)(9.87\ \mathrm{cm}) = 10.46\ \mathrm{cm} = 0.1046\ \mathrm{m}$$

Equation (2-2) is used to find average piston speed:

$$\overline{U}_p = 2SN = (2\ \mathrm{strokes/rev})\ (0.1046\ \mathrm{m/stroke})\ (4500/60\ \mathrm{rev/sec})$$
$$= 15.7\ \mathrm{m/sec}$$

Tangential speed of rotating gas is

$$u_t = 2\pi\ \omega r = (2\pi \mathrm{radians/rev})(450\ \mathrm{rev/sec})(0.0987/2\ \mathrm{m}) = 139.5\ \mathrm{m/sec}$$

Use Eq. (6-2) to find swirl ratio:

$$(\mathrm{SR})_2 = u_t/\overline{U}_p = (139.5)/(15.7) = \underline{8.9}$$

6.3 SQUISH AND TUMBLE

When the piston approaches TDC at the end of the compression stroke, the volume around the outer edges of the combustion chamber is suddenly reduced to a very small value. Many modern combustion chamber designs have most of the clearance volume near the centerline of the cylinder (Fig. 6-5). As the piston approaches TDC, the gas mixture occupying the volume at the outer radius of the cylinder is forced radially inward as this outer volume is reduced to near zero. This radial inward motion of the gas mixture is called **squish**. It adds to other mass motions within the cylinder to mix the air and fuel, and to quickly spread the flame front. Maximum squish velocity usually occurs at about 10° bTDC.

During combustion, the expansion stroke begins and the volume of the combustion chamber increases. As the piston moves away from TDC, the burning gases are propelled radially outward to fill the now-increasing outer volume along the cylinder walls. This **reverse squish** helps to spread the flame front during the latter part of combustion.

As the piston nears TDC, squish motion generates a secondary rotational flow called **tumble**. This rotation occurs about a circumferential axis near the outer edge of the piston bowl as shown in Fig. 6-7.

The role of tumble in the combustion process of a modern engine has become much more important in recent years, and much R & D has been devoted to understanding and promoting it. It is one of the important parameters in establishing the stratification of the air–fuel mixture in those engines which operate with that type of combustion. Special contouring of the piston face and variable timing and lift of the intake valves are used to create different tumble characteristics for various engine speeds and load conditions. To maximize thermal efficiency while minimizing fuel consumption and emissions, these engines sometimes operate with a stratified charge, and

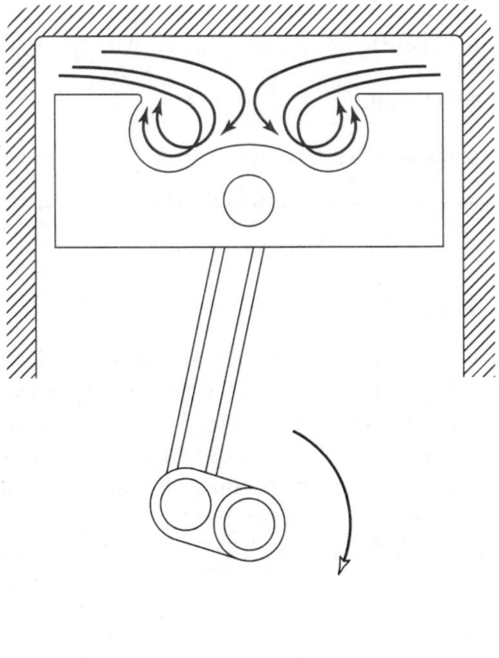

FIGURE 6-7

Tumble action caused by squish as piston approaches TDC. Tumble is a rotational motion about a circumferential axis near the edge of the clearance volume in the piston bowl or in the cylinder head.

at other times operate with a stoichiometric homogeneous charge. Tumble is one of the parameters that is varied to accomplish this.

Tumble ratio is the dimensionless parameter used to characterize the magnitude of tumble:

$$TR = (\text{angular speed of tumble})/(\text{engine speed}) = \omega_t/N \tag{6-6}$$

If the paddlewheel model is used as with swirl, tumble ratio will usually have values of about 1 to 2 as the piston approaches TDC. Near TDC, the TR will generally increase, and then, during combustion, tumble energy will be converted to additional turbulence.

Example Problem 6-2

An engine with pistons as shown in Fig. (6-7) operates at 3500 RPM, with each cylinder containing 0.0014 kg of air–fuel. When a piston approaches TDC, the gas inward squish velocity equals 7.66 m/sec. At TDC half of the cylinder gases then create a tumble rotation of 2.2 cm diameter. Calculate:

1. Angular momentum of gases in tumble rotation
2. Tumble ratio, assuming a paddlewheel model for the rotation

(1) Rotational velocity of gases in tumble rotation is

$$\omega_t = u_t/r = (7.66\ \text{m/sec})/(0.011\ \text{m}) = 696\ \text{radians/sec}$$

Use Eq. (6-3) for the mass moment of inertia of a rotating gas:

$$I = mB^2/8 = (0.0014/2\ \text{kg})(0.022\ \text{m})^2/8 = 4.235 \times 10^{-8}\ \text{kg-m}^2$$

Use Eq. (6-4) for the angular momentum:

$$\Gamma = I\omega_t = (4.235 \times 10^{-8}\ \text{kg-m}^2)(696\ \text{radians/sec}) = \underline{2.95 \times 10^{-5}\ \text{kg-m}^2/\text{sec}}$$

(2) Equation (6-6) gives the tumble ratio:

$$TR = \omega_t/N = (696\ \text{radians/sec})/[(3500/60\ \text{rev/sec})(2\pi\ \text{radians/rev})] = \underline{1.9}$$

6.4 DIVIDED COMBUSTION CHAMBERS

Some engines have divided combustion chambers, usually with about 80% of the clearance volume in the main chamber above the piston and about 20% of the volume as a secondary chamber connected through a small orifice (Fig. 6-8). Combustion is started in the small secondary chamber, and the flame then passes through the orifice, where it ignites the main chamber. Intake swirl is not as important in the main chamber of this type of engine, so the intake system can be designed for greater volumetric efficiency. It is desirable to have very high swirl in the secondary chamber, and the orifice between

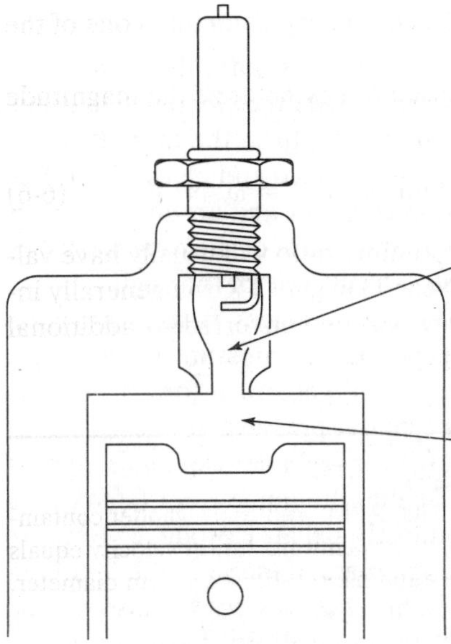

FIGURE 6-8

Divided combustion chamber of an SI engine. Secondary chamber will typically contain about 20% of the total clearance volume. Combustion is generally initiated in the secondary chamber by positioning of the spark plug. The main air–fuel mixture in the primary chamber is ignited by torch ignition as the flame expands through the orifice between the chambers. Often, divided combustion chamber engines are also stratified charge engines, with a rich charge in the secondary chamber for good ignition and a lean charge in the primary chamber for good fuel economy.

the chambers is shaped to supply this; often, the secondary chamber is called a **swirl chamber.** As the gases in the secondary chamber are consumed by combustion, the pressure rises and flaming gas expands back through the orifice and acts as a *torch ignition* for the main chamber. The expanding gas rushing back through the orifice creates a large secondary swirl in the main chamber, which enhances the combustion there. Some energy is lost in forcing the gases into and out of the secondary chamber, and there are additional heat losses due to the large surface area.

Often, a divided chamber engine will also be a stratified charge engine. The intake system is designed to supply a rich mixture in the secondary chamber and a lean mixture in the main chamber. The rich mixture with very high swirl in the secondary chamber will ignite readily and combust very quickly. The flaming gases expanding back through the orifice will then ignite the lean mixture in the main chamber, a mixture often so lean that it would be difficult to ignite with a spark plug alone. The net result is an engine that has good ignition and combustion, yet operates mostly lean to give good fuel economy. Placement and timing of intake valves and injectors to supply the proper air and fuel to all parts of this engine are extremely important.

A variation of this type of combustion chamber on some CI engines is one with a totally passive secondary chamber, with all valves and injectors located in the main chamber. When combustion occurs in the main chamber, high pressure forces gas through the very small orifice and raises the pressure in the secondary chamber also. When the pressure in the main chamber is reduced during the power stroke, the high-pressure gases in the secondary chamber flow back into the main chamber. This holds the pressure in the main chamber to a higher level for a short time and gives a smooth, slightly greater force on the piston during the power stroke. This kind of secondary chamber usually consists of about 5–10% of the clearance volume (Fig. 7-15).

6.5 CREVICE FLOW AND BLOWBY

In the combustion chamber of an engine there are tiny crevices that fill with air, fuel, and exhaust gas during the engine cycle. These crevices include the clearance between the piston and cylinder walls (about 80% of total), imperfect fit in the threads of the spark plug or fuel injector (5%), gaps in the gasket between head and block (10–15%), and unrounded corners at the edge of the combustion chamber and around the edges of valve faces. Although this volume is on the order of only 1–3% of the total clearance volume, the flow into and out of it greatly affects the overall cycle of the engine.

In an SI engine air–fuel mixture is forced into these crevices, first during compression and then more so during combustion when cylinder pressure is increased. During combustion, when cylinder pressure is very high, gases are forced into the crevices and the pressure will be about the same as cylinder pressure. In those regions away from the spark plug in front of the flame front (piston–cylinder clearance), a mixture of air and fuel will be forced into the crevice. Behind the flame front (spark plug threads), the gases forced into the crevice will be mostly exhaust products. Because crevice volume is so small and is surrounded by a large mass of metal at the temperature of the combustion chamber wall, the gases forced into the crevices will also remain at about wall temperature. The air–fuel within the crevices will not burn due to the close proximity of the walls. Flame cannot propagate through tiny metal passages. Heat released by the flame is conducted away by the metal walls faster than it can be generated by the tiny flame front. There is not enough energy to keep combustion occurring, and the flame dies.

Because pressure in the crevices is high and the temperature is about the same as that of the much cooler walls, density in the crevices is very high. Therefore, even with crevice volume equaling only a few percent of the total volume, as much as 20% of the total mass of air–fuel can be trapped in the crevices at peak pressure (see Review Problem 6-6). As the power stroke occurs and pressure is reduced in the cylinder, the high crevice pressure forces gases in the crevice back into the combustion chamber, where some of the trapped fuel is burned. Some of the reversed crevice flow occurs later in the power stroke, after combustion has stopped, and this fuel does not get burned. Some of the fuel, therefore, ends up in the engine exhaust, contributing to hydrocarbon emissions and lowering combustion efficiency and engine thermal efficiency. Because fuel is not added until immediately before combustion in a CI engine, less fuel gets into the crevice volume and these resulting problems are greatly reduced.

Most pistons have two or more compression rings and at least one oil ring. Compression rings seal the clearance gap between the piston and cylinder walls. They are made with highly polished surfaces and are spring loaded against the hard, polished cylinder walls. As the piston moves towards TDC in the compression stroke, the compression rings are forced to the bottom surface of the ring grooves, and some gases leak into the groove at the top (Fig. 6-9). Then, as the piston reverses direction and starts on the power stroke, the compression rings are forced to the top of the ring grooves, and the trapped gas can flow out of the groove and further along the piston. The second compression ring is to stop some of the gases that have leaked pass the first ring. Another path from which gases leak past the piston rings is the gap where the two ends meet. Figure 6-10 shows various configurations to minimize this flow. The oil ring is for

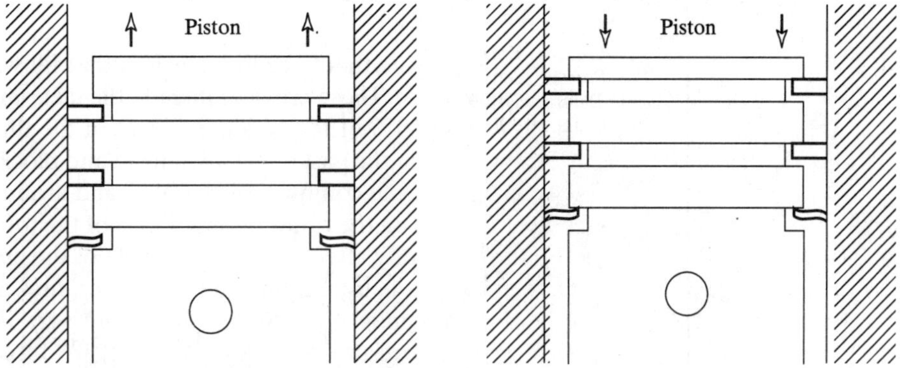

FIGURE 6-9

Schematic showing how blowby occurs when combustion chamber gases are forced past the compression rings of a piston. As the piston moves up in the compression stroke the compression rings are forced to the bottom of the ring grooves, and gas is forced into the crevice volume between the piston and cylinder walls and into the piston ring grooves. When the piston reverses direction for the power stroke the piston rings are forced to the top of the grooves and the gas in the grooves can flow past the piston. Gas also leaks past the piston rings through the gap where the ends meet.

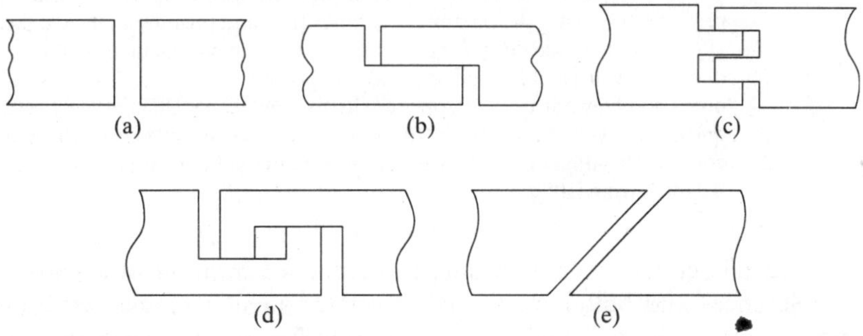

FIGURE 6-10

Various designs used on the ends of piston rings to reduce blowby flow. Adapted from [34].

lubrication and offers no resistance to gas leakage. However, in addition to lubricating, the film of oil between the piston and cylinder walls is a major gas sealant restricting gas flow past the piston. The gas that gets totally past the piston and ends up in the crankcase is called **blowby**.

Figure 6-11 shows how the pressure in the combustion chamber, between the compression rings, and in the crankcase varies with crank angle in an engine cycle. There is a time delay in the pressure change from one chamber to the next due to the restricted flow passage created by the compression rings. Late in the power stroke, when the exhaust valve opens, pressure between the compression rings will be greater than in the combustion chamber, and some gases will be forced back into the chamber. This is called **reverse blowby**.

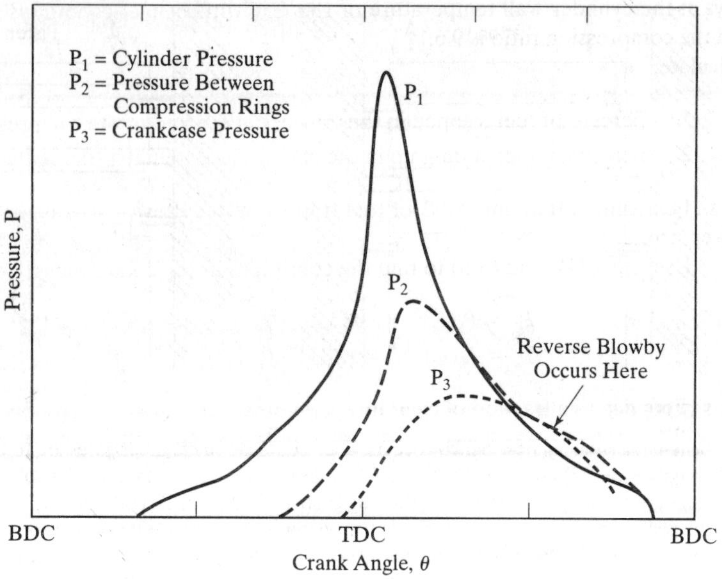

FIGURE 6-11

Engine pressures as a function of crank angle, showing cylinder pressure (P_1), pressure between piston compression rings (P_2), and pressure in the crankcase (P_3). There is a time delay for pressure change from one chamber to the next due to the restricted flow passage past the pistons. When the exhaust valve opens and blowdown occurs, pressure in the combustion chamber decreases quickly and $P_2 > P_1$ can occur. This is when reverse blowby occurs. The need for crankcase ventilation can be seen by the pressure buildup in the crankcase. Adapted from [105].

Ideally, crevice volume should be kept at a minimum. Modern engines with closer tolerances and better quality control have smaller crevice volumes. At the same time, however, clearance volumes are also smaller due to high compression ratios, and the percent crevice volume remains about the same. Iron pistons can have closer tolerances than aluminum pistons because of their lower thermal expansion. The top compression piston ring should be as close as structurally possible to the top of the piston.

Blowby raises the pressure in the crankcase and contaminates the oil with fuel and exhaust gases. As much as 1% of the fuel is forced into the crankcase in some engines. To keep crankcase pressure down, it must be ventilated. In older engines the crankcase was ventilated to the surroundings, wasting fuel and polluting the surroundings with fuel vapor. All modern automobile engines ventilate the crankcase back into the intake system, avoiding these problems. Some small engines still have crankcase ventilation to the surrounding air. Because of the oil contamination caused by blowby, oil filter systems and more frequent oil changes are necessary.

Example Problem 6-3

Crevice volume of an engine equals 2% of the total clearance volume. It can be assumed that pressure in the crevices is about the same as in the combustion chamber but the temperature

stays at the cylinder wall temperature of 180°C. Cylinder inlet conditions are 60°C and 98 kPa, and the compression ratio is 9.6:1.
Calculate:

1. percent of fuel trapped in the crevices at the end of the compression stroke
2. percent of fuel ending up in the exhaust due to being trapped in the crevice volume

It can be assumed that only 80% of fuel trapped in the crevice volume gets burned later in the power stroke.

Use Eqs. (3-4) and (3-5) to find the conditions at the end of the compression stroke:

$$P_2 = P_1(r_c)^k = (98\ \text{kPa})(9.6)^{1.35} = 2076\ \text{kPa}$$
$$T_2 = T_1(r_c)^{k-1} = (333\ \text{K})(9.6)^{0.35} = 735\ \text{K} = 462°\text{C}$$

Mass in crevice as a fraction of clearance volume is

$$m_{\text{crev}} = PV/RT = (2076\ \text{kPa})(0.02V_c\ \text{m}^3)/(0.287\ \text{kJ/kg-K})(453\ \text{K}) = 0.319V_c$$

Mass in the combustion chamber at TDC

$$m_{\text{chamb}} = (2076\ \text{kPa})(V_c)/(0.287\ \text{kJ/kg-K})(735\ \text{K}) = 9.841V_c$$

(1) Percent of fuel in crevice is

$$\% = [(m_{\text{crev}})/(m_{\text{total}})](100) = [(0.319\ V_c)(100)]/[(0.319\ V_c) + (9.841\ V_c)] = \underline{3.14\%}$$

(2) Twenty percent of this fuel does not get burned:

$$\text{percent of total not burned} = (0.20)(3.14) = \underline{0.63\%}$$

This is fuel lost in the exhaust due to crevice flow. Additional fuel is lost due to mixing and combustion inefficiencies.

6.6 MATHEMATICAL MODELS AND COMPUTER SIMULATION

Mathematical Models

Many mathematical models have been developed to help understand, correlate, and analyze the operation of engine cycles. These include combustion models, models of physical properties, and models of flow into, through, and out of the cylinders. Even though models often cannot represent processes and properties to the finest detail, they are a powerful tool in the understanding and development of engines and engine cycles. With the use of models and computers in the design of new engines and components, great savings are made in time and cost. Historically, new design was a costly, time-consuming practice of trial and error, requiring new part construction and testing for each change. Now engine changes and new designs are first developed on the computer, using the many models that exist. Often, only after a component is optimized on the computer is a part actually constructed and tested. Generally, only minor modifications must then be made to the actual component.

Models range from simple and easy to use, to very complex and requiring extensive computer usage. In general, the more useful and accurate models are quite complex. Models to be used in engine analysis are developed using empirical relationships and approximations, and often treat cycles as quasi-steady-state processes. Normal fluid flow equations are often used.

Some models will treat the entire flow through the engine as one unit, some will divide the engine into sections, and some will subdivide each section (e.g., divide the combustion chamber into several zones—burned and unburned, boundary layer near the wall, etc.). Most models deal only with one cylinder, which eliminates any interaction from multicylinders that can occur, mainly in the exhaust system.

Models for the combustion process address ignition, flame propagation, flame termination, burn rate, burned and unburned zones, heat transfer, emissions generation, knock, and chemical kinetics [51, 85, 114]. They are available for SI and CI engines with either direct injection or indirect injection. Values for properties are obtained from standard thermodynamic equations of state and relationships for thermophysical and transport properties.

Models are available for flow into, within, and out of the combustion chamber. These include turbulence models [16, 91, 118, 119, 127]; models of the flow of swirl, squish, and tumble in the cylinders [6, 18, 21, 54, 55, 60, 66, 72, 109, 128, 129, 134, 142]; and fuel injection spray models [7, 17, 53, 137].

Computer Simulation

At least three methods of computer use are utilized in the operation, testing, and development of engines. Automobile engines are equipped with their own controlling computer that optimizes smoothness of operation, fuel consumption, emission control, diagnostics, and a number of other facets of operation. This is done as a response to inputs from thermal, electrical, chemical, mechanical, and optical sensors located throughout the engine.

Testing for maintenance or experimentation is done by connecting the engine to an external computer, often larger and with more elaborate sensing equipment and instrumentation. This can be done with the engine mounted in the automobile or on an external test stand. The amount and usefulness of the information gathered depends on many factors. These include the number of channels in the data acquisition equipment, resolution of data, sampling rate, and amount of data (size of computer).

For development work, elaborate mathematical models are often used on computers to simulate actual engine operation. The complexity and accuracy of the model often depend on the capacity of the computer, with some models requiring very large capacity. Commercial software is available for many operations within an engine, some made specifically for IC engines and others developed for more general use (e.g., heat transfer, chemical kinetics, property values, combustion analysis [6, 49, 52, 54, 67, 75, 96, 117]). With a computer of adequate capacity and available software, detailed combustion analysis can be done, including burn rate, dissociation, changing composition, heat release analysis, heat transfer, and chemical equilibrium analysis, as well as accurate determinations of thermodynamic and transport properties for burned and unburned gases. Reference [40] describes computer programs for a number of engine processes.

Automobile companies use very elaborate programs in their engine development work. Usually these have been generated in-house and are highly confidential. They allow for much faster development of new engines and modifications and improvements in existing designs. As an example of what these programs are capable of doing, one such program is discussed in the next section. This is a reduced version of a program used by General Motors that has been released for use by educational institutions. A quotation from the user's guide for this engine simulation program puts modeling and computer use in its proper perspective [87]: "Some people do not believe computer models at all. I trust you are not one of them. Other people believe computer models completely. I trust you are not one of them. The engine simulation is only as good as its assumptions and the input data YOU provide. Always be suspicious of the input data, always ask the question 'Is this a good problem to address with the simulation?' And, especially be sure to ask these questions before drawing conclusions. The greatest danger of complex computer models is that they can give plausible but incorrect results."

6.7 INTERNAL COMBUSTION ENGINE SIMULATION PROGRAM

The General Motors Internal Combustion Engine Simulation Program analyzes what occurs within the combustion chamber and exhaust system during an engine cycle, starting at the close of the intake valve and proceeding through compression, combustion, and into the exhaust process. It is applicable to single-cylinder engines operating on a four-stroke, spark ignition, homogeneous charge cycle. It has limited capability of analyzing effects from various combustion chamber geometries. The fact that it is written for single-cylinder engines precludes its use in the analysis of tuning effects in the exhaust system from interactions among multiple cylinders.

Scope of the Program

The program uses the first law of thermodynamics by integrating various models for combustion, heat transfer, and valve flow rates, using integration methods from [111]. Integration proceeds by crank angle position, allowing for various fuels, air–fuel ratios, EGR or other inert gas for charge dilution, and/or valve-opening profiles. The program is divided into three principal sections: compression, combustion, and gas exchange.

Compression: Integration starts at the closing of the intake valve and proceeds until the crank angle of ignition is reached. Residual gases from the previous cycle are included in the cylinder gas mixture, and a number of iterations are performed until the percentage and chemical content of the residual gases remain at a steady-state value after each cycle.

Combustion: A spherical flame front is assumed, spreading outward from the ignition point. This divides the cylinder into burned and unburned zones. The burned zone is further divided into an adiabatic core and a boundary layer subzone where heat transfer occurs. Heat transfer to the walls also occurs from the unburned zone. When the exhaust valve opens, the two zones are no longer distinguished and all gases are considered mixed. As integration continues, the values of properties are found using the methods for combustion calculations from [96].

Gas Exchange: Three control masses are considered during this process: cylinder gases, exhaust gases downstream of the exhaust valve, and gases that backflow into the intake system through the open intake valve. When backflow gases are carried back into the combustion chamber, they again become part of the cylinder gases. Under some conditions there is no backflow. These calculations require valve discharge coefficients and valve lift versus crank angle data.

Input Quantities

The program reads a data input file of engine and operating parameters. This list shows the wide range of parameters that can be selected for the simulated run:

Input Section 1: Title Line Identifying Engine Geometry

bore
stroke
piston pin offset
connecting rod length
compression ratio
valve head diameters
valve seat angles
piston surface area/bore area
head surface area/bore area
TDC wall area/bore area
fuel identification
combustion table selection
heat transfer calculation multiplier
boundary zone weighted factor for three-zone model

Input Section 2: Tables

rocker ratio for intake valve train
intake valve opening position
intake valve closing position
crank angles versus intake cam lifts
rocker ratio for exhaust valve train
exhaust valve opening position
exhaust valve closing position
crank angles versus exhaust cam lifts
valve lift/diameter versus discharge coefficient for intake
valve lift/diameter versus discharge coefficient for exhaust
burned volume ratios versus wetted area ratios (see equations after list)
combustion crank angle fraction versus mass fraction burned

Input Section 3: Operating Conditions

engine speed
air–fuel ratio
integration tolerance
crank angle steps for output
number of cycles to run
intake pressure
exhaust pressure
shift intake valve lifts
shift exhaust valve lifts
intake temperature of air–fuel mixture
temperature of EGR
mass fraction of EGR
crank angle at ignition
combustion efficiency
combustion duration in crank angle degrees
firing or motoring of engine
piston surface temperature
head temperature
cylinder wall temperature
intake valve temperature
exhaust valve temperature

$$\text{burned volume ratio} = (\text{burned volume behind flame})/(\text{total cylinder volume})$$

$$\text{wetted area ratio} = (\text{cylinder surface area behind flame})/(\text{total cylinder area})$$

Output File

Output variables of interest, such as temperature, pressure, volume, mass burned, etc., are all listed versus crank angle. The user can specify the crank angle step size for output except during combustion, when unit crank angle steps (every degree) are used. At certain crank angles, the program lists the components in the combustion gas. CO and NO data are given for when the intake valve closes and when the exhaust valve opens. Indicated, brake, and frictional power are given. For reference, the output file also includes the input file.

The output file can be read into a spreadsheet program that can be used to generate tables or curve plotting. Figures 6-12 through 6-15 show results from a demonstration run. Figure 6-12 shows the effect of changing the intake pressure on the resulting pressure–volume indicator diagram. Figure 6-13 shows how the generation of NO and CO are affected by the input air–fuel ratio. Figure 6-14 plots adiabatic flame temperature versus input equivalence ratio, and Fig. 6-15 plots various power versus intake pressure.

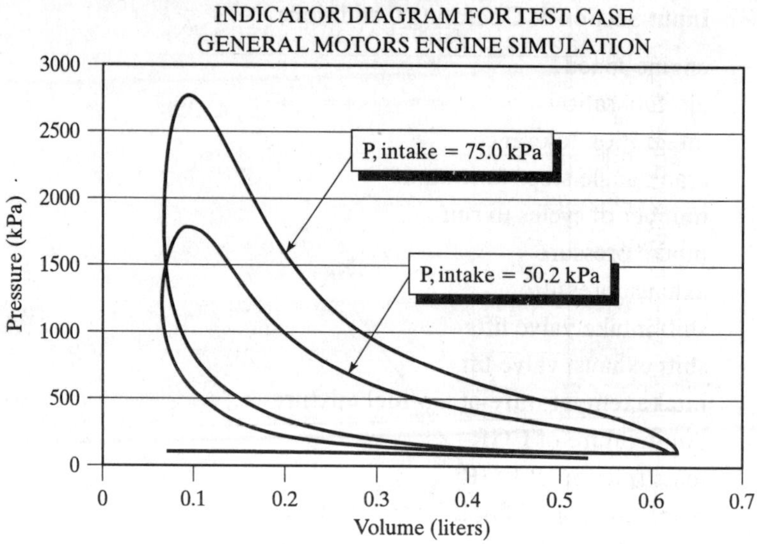

FIGURE 6-12

Output from General Motors simulation program showing how intake pressure affects pressure–volume indicator diagram of cycle. Reprinted with permission from [41].

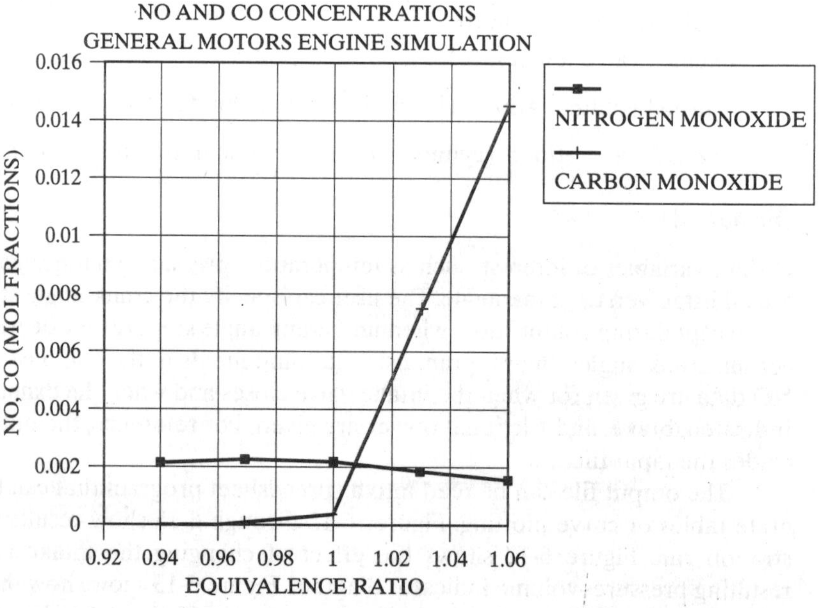

FIGURE 6-13

Output from General Motors simulation program showing how NO and CO exhaust concentrations are affected by intake equivalence ratio. Reprinted with permission from [41].

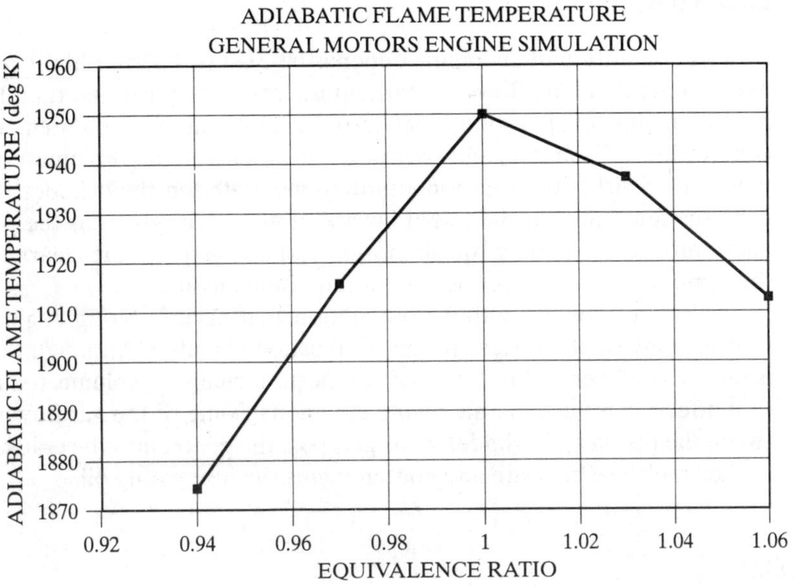

FIGURE 6-14

Output from General Motors simulation program showing adiabatic flame temperature as a function of intake equivalence ratio. Reprinted with permission from [41].

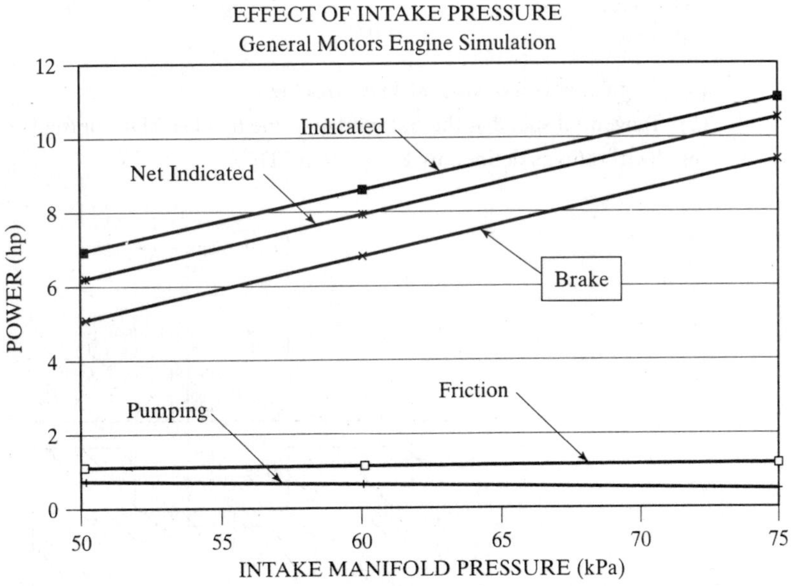

FIGURE 6-15

Output from General Motors simulation program showing output power as a function of intake pressure. Reprinted with permission from [41].

6.8 CONCLUSIONS

Efficient operation of an engine depends on high turbulence in the air–fuel flows, plus major generated bulk flows of swirl, squish, reverse squish, and tumble. Turbulence enhances mixing, evaporation, heat transfer, and combustion. High turbulence during combustion is desirable, and part of combustion chamber geometry design is to promote this. Swirl is the rotational motion generated in the cylinder during intake and compression, squish is the radial inward motion that occurs as the piston nears TDC, and tumble is created by squish motion and the shape of the clearance volume. All of these motions enhance proper operation of the engine.

Crevice flow is another flow motion that occurs during engine operation—the flow into the small crevices of the combustion chamber. Although the crevice volume is only a small percent of the total combustion chamber volume, the flow into and out of it affects combustion and engine emissions. Some of the gas flow in the crevice between the piston and cylinder walls gets past the piston into the crankcase. Here it raises the crankcase pressure and contaminates the lubricating oil.

PROBLEMS

6.1 A 2.4-liter, three-cylinder, four-stroke cycle SI engine with a 9.79-cm stroke is running at 2100 RPM. During the compression stroke, the air–fuel mixture has a swirl ratio, as defined by Eq. (6-1), of 4.8. At TDC the mixture, which consists of 0.001 kg in each cylinder, is compressed into a clearance volume that can be approximated as a cylindrical bowl in the face of the piston, as shown in Fig. 6-16. It can be assumed that angular momentum is conserved.

Calculate:

(a) Angular speed of swirl at TDC. [rev/sec]

(b) Tangential speed at the outer edge of the bowl at TDC. [m/sec]

(c) Swirl ratio as defined by Eq. (6-2) at TDC.

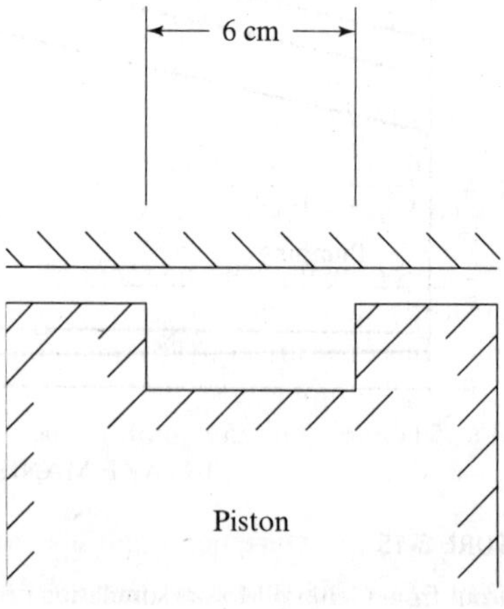

FIGURE 6-16

6.2 A 150-in.^{3}, four-cylinder, four-stroke cycle, high-swirl CI engine is running at 3600 RPM. Bore and stroke are related by $S = 0.95\ B$. During the compression stroke, the cylinder air has a swirl ratio, as defined by Eq. (6-2), of 8.

Calculate:

(a) Swirl tangential speed. [ft/sec]

(b) Angular velocity of cylinder air using the paddle wheel model. [rev/sec]

(c) Swirl ratio as defined by Eq. (6-1).

6.3 A five-cylinder SI engine with bore $B = 8.56$ cm and stroke $S = 0.92$ B, operates at 2800 RPM on an air-standard Otto cycle. During the compression stroke, the air in each cylinder is rotating at an angular velocity of 250 rev/sec, using the paddlewheel model. At TDC, the gas mixture is compressed into a clearance volume that can be approximated as a 5-cm-diameter cylindrical bowl in the face of the piston.

Calculate:

(a) Swirl ratio during compression using Eq. (6-1).

(b) Swirl ratio during compression using Eq. (6-2).

(c) Angular velocity of gases in bowl at TDC, assuming angular momentum is conserved. [rev/sec]

6.4 As an engine operates at 3200 RPM, 0.0012 kg of air--fuel-exhaust residual experience tumble in each cylinder. The radius of tumble rotation is 0.90 cm, and tumble ratio is 1.78.

Calculate:

(a) Tangential velocity at edge of tumble rotation. [m/sec]

(b) Rotational kinetic energy of tumble in one cylinder. [J]

6.5 An SI engine operates at 3400 RPM on an air-standard Otto cycle at WOT. The compression ratio is 10.2, and the temperature and pressure at the start of compression are 60°C and 100 kPa, respectively. The crevice volume is equal to 3% of the clearance volume and has pressure equal to cylinder pressure and temperature equal to 180°C. The cylinder air–fuel mixture is homogeneous with AF = 15. The mass flow rate of fuel into the engine equals 0.0042 kg/sec. Eighty-five percent of fuel trapped in the crevice volume at the end of compression eventually gets burned during combustion, with the rest being exhausted.

Calculate:

(a) Percent of fuel trapped in crevice volume at end of compression stroke. [%]

(b) Mass flow rate of unburned fuel in exhaust due to crevice volume. [kg/hr]

(c) Chemical energy lost in exhaust due to this unburned fuel. [kW]

6.6 A 3.3-liter, V6 SI engine with a compression ratio of 10.9:1 operates on an Otto cycle at 2600 RPM using stoichiometric gasoline. Crevice volume, which equals 2.5% of the clearance volume, has pressure equal to the pressure in the combustion chamber but remains at the cylinder wall temperature of 190°C. Conditions at the start of compression are 65°C and 98 kPa. Assume complete combustion.

Calculate:

(a) Percent of gas mixture mass that is in the crevice volume at the start of combustion.

(b) Percent of gas mixture mass that is in the crevice volume at the end of combustion.

6.7 A 6.8-liter, in-line, eight-cylinder CI engine has a compression ratio $r_c = 18.5$ and a crevice volume equal to 3% of the clearance volume. During the engine cycle, pressure in the crevice volume equals combustion chamber pressure while remaining at the cylinder wall temperature of 190°C. Cylinder conditions at the start of compression are 75°C and 120 kPa, and peak pressure is 11,000 kPa. Cutoff ratio is $\beta = 2.3$.

Calculate:

(a) Crevice volume of one cylinder. [cm^3]
(b) Percent of air–fuel mixture in the crevice volume at the end of compression. [%]
(c) Percent of air–fuel mixture in the crevice volume at the end of combustion. [%]

6.8 A 292-in.^{3}, V8, four-stroke cycle CI engine operates at 1800 RPM, using light diesel fuel at AF = 24 and a volumetric efficiency of 94%. Injection timing is from 22° bTDC to 4° aTDC. Swirl ratio, as defined by Eq. (6-1), equals 2.8 during fuel injection.

Calculate:

(a) Time of one injection. [sec]
(b) Period of swirl (one rotation). [sec]
(c) Number of orifice holes needed in each injector, with one injector per cylinder.

6.9 A 2.6-liter, four-cylinder, stratified charge SI engine with a compression ratio of 10.5:1 operates on an Otto cycle. The engine has divided combustion chambers, with a secondary chamber containing 18% of the clearance volume in each cylinder. A 1-cm^2 orifice connects the secondary chamber with the main combustion chamber. AF = 13.2 in the secondary chamber where the spark plug is located, and AF = 20.8 in the main chamber. The fuel is gasoline with a 98% combustion efficiency. When operating at 2600 RPM, the conditions in both chambers at the start of combustion are 700 K and 2100 kPa. Combustion can be modeled as an instantaneous heat addition in the secondary chamber, followed by a gas expansion into the main chamber which lasts for about 7° of engine rotation. Additional heat is then added from combustion in the main chamber.

Calculate:

(a) Overall AF.
(b) Peak temperature and pressure in the secondary chamber. [°C, kPa]
(c) Approximate velocity of gas flow into the main chamber immediately after combustion in the secondary chamber. [m/sec]

DESIGN PROBLEMS

6.1D An automobile has a 3-liter V6 SI engine. At slow engine speeds it is desired to have low cylinder swirl to reduce the flame speed during combustion. At fast engine speeds high swirl is desired. To accomplish this, the cylinders will each have two (or three) intake valves, using only one at low speeds and all valves at high speeds. Design the intake manifold, valve system, combustion chambers, and camshaft(s) for this engine. Describe the operation at low and high speeds.

6.2D It has been suggested that to reduce crevice volume in a cylinder, the top piston compression ring should be located at the top of the piston (i.e., the top of the compression ring is flush with the piston face). Design a piston–ring–groove system in which this is possible. Give careful attention to reducing crevice volume and blowby.

6.3D Design a method to measure swirl in the cylinder of an operating engine.

CHAPTER 7

Combustion

"The normal rate of burning (as distinct from the detonation rate) of any stagnant fuel/air mixture is so low as to be practically useless so far as any internal-combustion engine is concerned. We must look, therefore, entirely to turbulence or the mechanical distribution of the flame to spread combustion throughout the whole mass of the working fluid, and, since this is the case, it follows that the normal rate of burning of any fuel is practically without influence on the speed at which an engine will run."

The High-Speed Internal-Combustion Engine

by *Harry R. Ricardo* (1923)

This chapter examines the combustion process that occurs in the combustion chamber of an IC engine. Combustion in an engine is a very complex process that is not completely understood. Simplified models are used to describe this not-so-simple phenomenon. Although these models do not always explain the fine details of the combustion process, they do a fairly accurate job of correlating the important broad operating parameters such as pressure, temperature, fuel, knock, engine speed, etc.

For over 100 years the combustion process in most SI engines occurred in a homogeneous air–fuel mixture with an equivalence ratio close to one (stoichiometric). Many modern SI engines still use this mode of combustion, but some employ one or both of two new combustion strategies: **stratified charge** and **lean-burn**. In stratified charge engines, the combustible mixture has a different air–fuel ratio at various locations in the combustion chamber. This creates changing combustion characteristics as the flame front passes through the chamber. The combustion process in a lean-burn engine takes place in an air–fuel mixture with an equivalence ratio much less than one, either as a homogeneous mixture or as a stratified charge mixture.

Combustion in an SI engine is quite different from combustion in a CI engine, and the two types are studied separately.

7.1 COMBUSTION IN SI ENGINES WITH HOMOGENEOUS AIR–FUEL MIXTURES

The combustion process of SI engines can be divided into three broad regions: (1) ignition and flame development, (2) flame propagation, and (3) flame termination. Flame development is generally considered the consumption of the first 5% of the air–fuel mixture (some sources use the first 10%). During the flame development period, ignition occurs and the combustion process starts, but very little pressure rise is noticeable and little or no useful work is produced (Fig. 7-1). Just about all useful work produced in an engine cycle is the result of the flame propagation period of the combustion process. This is the period when the bulk of the fuel and air mass is burned (i.e., 80–90%, depending on how defined). During this time, pressure in the cylinder is greatly increased, providing the force to produce work in the expansion stroke. The final 5% (some sources use 10%) of the air–fuel mass that burns is classified as flame termination. During this time, pressure quickly decreases and combustion stops.

In an SI engine, combustion ideally consists of an exothermic subsonic flame progressing through a premixed air–fuel mixture, which is locally homogeneous. The spread of the flame front is greatly increased by induced turbulence, swirl, and squish within the cylinder. The right combination of fuel and operating characteristics is such that knock is avoided or almost avoided.

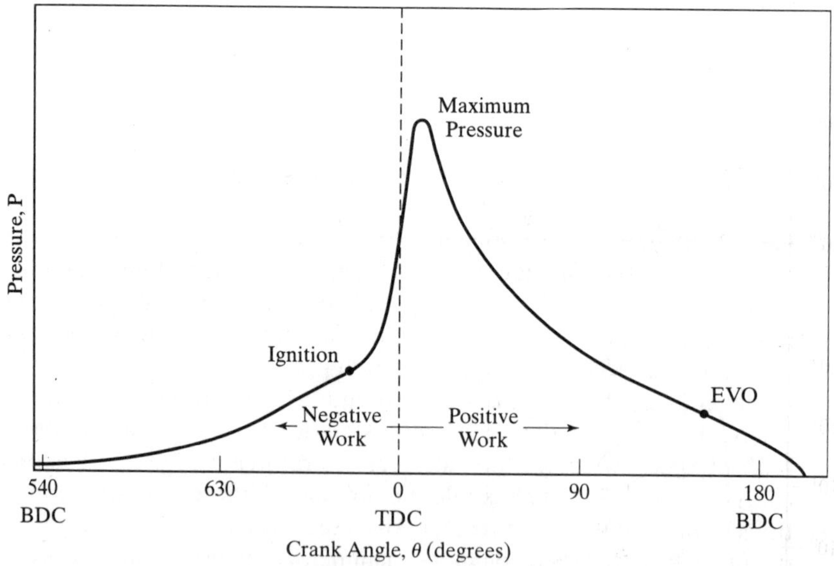

FIGURE 7-1

Cylinder pressure in the combustion chamber of an SI engine as a function of crank angle. The increase in pressure rise is very slow after ignition during the flame development period. This results in a slow pressure force increase on the piston and a smooth engine cycle. Maximum pressure occurs 5° to 10° aTDC. Adapted from [93].

Ignition and Flame Development

Combustion is initiated by an electrical discharge across the electrodes of a spark plug. This occurs anywhere from 10° to 30° before TDC, depending on the geometry of the combustion chamber and the immediate operating conditions of the engine. This high-temperature *plasma* discharge between the electrodes ignites the air–fuel mixture in the immediate vicinity, and the combustion reaction spreads outward from there. Combustion starts very slowly because of the high heat losses to the relatively cold spark plug and gas mixture. Flame can generally be detected at about 6° of crank rotation after spark plug firing.

Energy dissipation versus time across the electrodes of a typical spark plug is shown in Fig. 7-2. Applied potential is generally 25,000–40,000 volts, with a maximum current on the order of 200 amps lasting about 10 nsec (1 nsec $= 10^{-9}$ sec). This gives a peak temperature on the order of 60,000 K. Overall spark discharge lasts about 0.001 second, with an average temperature of about 6000 K. A stoichiometric mixture of

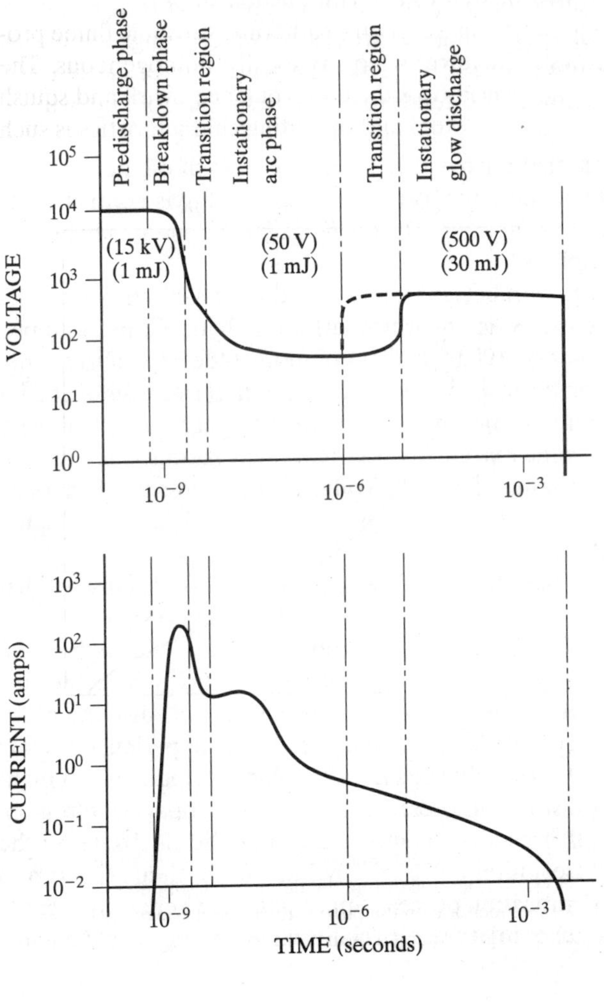

FIGURE 7-2

Spark plug voltage and current as a function of time for ignition in a typical SI engine. Maximum voltage can be greater than 40,000 volts, with a maximum current on the order of 200 amps lasting for about 10^{-8} seconds. The total energy delivered during one discharge is generally about 30 to 50 mJ. Reprinted by permission of Elsevier Science Inc. from "Initiation and Propagation of Flame Fronts in Lean CH_4–Air Mixtures by the Three Modes of Ignition Spark," by Maly and Vogel, copyright 1976 by The Combustion Institute, Ref. [83].

hydrocarbon fuel requires about 0.2 mJ (0.2×10^{-3} J) of energy to ignite self-sustaining combustion. This varies to as much as 3 mJ for nonstoichiometric mixtures. The discharge of a spark plug delivers 30 to 50 mJ of energy, most of which, however, is lost by heat transfer.

Several different methods are used to produce the high voltage potential needed to cause electrical discharge across spark plug electrodes. One common system is a battery–coil combination. Most automobiles use a 12-volt electrical system, including a 12-volt battery. This low voltage is multiplied many times by the coil that supplies the very high potential delivered to the spark plug. Some systems use a capacitor to discharge across the spark plug electrodes at the proper time. Most small engines and some larger ones use a magneto driven off the engine crankshaft to generate the needed spark plug voltage. Some engines have a separate high-voltage generation system for each spark plug, while others have a single system with a distributor that shifts from one cylinder to the next.

The gap distance between electrodes on a modern spark plug is about 0.7 to 1.7 mm. Smaller gaps are acceptable if there is a rich air–fuel mixture or if the pressure is high (i.e., high inlet pressure by turbocharging or a high compression ratio). Normal quasi-steady-state temperature of spark plug electrodes between firings should be about 650° to 700°C. A temperature above 950°C risks the possibility of causing surface ignition, and a temperature below 350°C tends to promote surface fouling over extended time.

For older engines with worn piston rings that burn an excess of oil, hotter plugs are recommended to avoid fouling. The temperature of a spark plug is controlled by the heat-loss path manufactured into the plug. Hotter plugs have a greater heat conduction resistance than do colder plugs.

Modern spark plugs are made with better materials and have a much greater life span than those of a few decades ago. Some quality spark plugs with platinum-tipped electrodes are made to last 160,000 km (100,000 miles) or more. One reason this is desirable is the difficulty of replacing plugs in some modern engines. Because of the increased amount of engine equipment and smaller engine compartments, it is very difficult to change spark plugs. In some extreme cases on modern automobiles, the engine must be partially removed to change the plugs. Voltage, current, electrode material, and gap size must be compatible if long-life plugs are to be used (e.g., too high a current will wear spark plug electrodes).

When a spark plug fires, the plasma discharge ignites the air–fuel mixture between and near the electrodes. This creates a spherical flame front that propagates outward into the combustion chamber. At first, the flame front moves very slowly because of its small original size. It does not generate enough energy to quickly heat the surrounding gases and thus propagates very slowly. This, in turn, does not raise the cylinder pressure very quickly, and very little compression heating is experienced. Only after the first 5–10% of the air–fuel mass is burned does the flame velocity reach higher values with the corresponding fast rise in pressure—the flame propagation region.

It is desirable to have a slightly rich air–fuel mixture around the electrodes of the spark plug at ignition. A rich mixture ignites more readily, has a faster flame speed, and gives a better start to the overall combustion process. Spark plugs are generally located near the intake valves to assure a richer mixture, especially when starting a cold engine.

Work to develop better ignition systems continues. Spark plugs with several electrodes and two or more simultaneous sparks are now available. These give a more consistent ignition and quicker flame development. One modern experimental system gives a continuing arc after the initial discharge. It is reasoned that this additional spark will speed combustion and give more complete combustion as the air–fuel mixture is swirled through the combustion chamber. This system is similar to methods tried over a hundred years ago. Development work has been done to create a spark plug with a variable electrode gap size. This would allow flexibility in ignition for different operating conditions. At least one automobile manufacturer is experimenting with engines that use a point on the top of the piston as one of the spark electrodes [70]. With this system, spark ignition can be initiated across gaps of 1.5 to 8 mm, with a reported lowering of fuel consumption and emissions.

HISTORIC—IGNITION SYSTEMS

During early engine development in the 19th century, many types of ignition systems were tried. One method was to use torch holes. At the proper time in the cycle when ignition was desired, a small hole in the side of the combustion chamber would be opened, exposing the air–fuel mixture to a continuous external flame. The flame would pass through the hole and ignite the gas mixture in the cylinder. This is somewhat like the method used to ignite the main chamber in a modern divided-chamber engine described later in this chapter. Another early method employed a small ignition rod protruding through the wall of the combustion chamber. The rod was heated by a continuous flame outside the chamber. Heat conduction kept the end of the rod inside the chamber hot, and this hot surface ignited the air–fuel mixture. No ignition timing control existed with this type of system. Glow plug ignition used with model airplane engines is a modern variation of this method.

Electric ignition was a major breakthrough that gave early engine development the spark (pun intended) it needed to create manageable cycle operation. Some early systems gave a semicontinuous spark, but the single spark per cycle evolved as the standard method. Soon, 6 volts became the common voltage, with batteries supplying the needed energy. Transformer-type coils were used to convert low battery voltage to the high potential needed to create the discharge across the spark plug electrodes. DC battery charging appeared in 1912 and voltage regulators in 1930. In the mid-1950s, the standard voltage of automobile electrical systems was changed to 12 volts because of the larger energy needs for starter motors, the ignition, lights, etc. Early in the 1960s, DC generators were replaced with AC alternators that allowed better charging at low speeds. During the electronic revolution, at the end of the 20th century and early 21st century, many forms of electronic ignition systems have been appearing on the market [45].

Flame Propagation in SI Engines

By the time the first 5–10% of the air–fuel mass has been burned, the combustion process is well established and the flame front moves very quickly through the combustion chamber. Due to induced turbulence, swirl, and squish, flame propagation speed is about 10 times faster than if there were a laminar flame front moving through a stationary gas mixture. In addition, the flame front, which would expand spherically from the spark plug in stationary air, is greatly distorted and spread by these motions.

As the gas mixture burns, the temperature, and consequently the pressure, rises to high values. Burned gases behind the flame front are hotter than the unburned gases before the front, with all gases at about the same pressure. This decreases the density of the burned gases and expands them to occupy a greater percent of the total combustion chamber volume. Figure 7-3 shows that, when only 30% of the gas mass is burned, the burned gases already occupy almost 60% of the total volume, compressing 70% of the mixture that is not yet burned into 40% of the total volume. Compression of the unburned gases raises their temperature by compressive heating. In addition, radiation heating emitted from the flame reaction zone, which is at a temperature on the order of 3000 K, further heats the gases, unburned and burned, in the combustion chamber. A temperature rise from radiation then further raises the pressure. Heat transfer by conduction and convection is minor compared with that from radiation, due to the very short real time involved in each cycle. As the flame moves through the combustion chamber, it travels through an environment that is progressively increasing in temperature and pressure. This causes the chemical reaction time to decrease and the flame front speed to increase, a desirable result. Because of radiation, the temperature of the

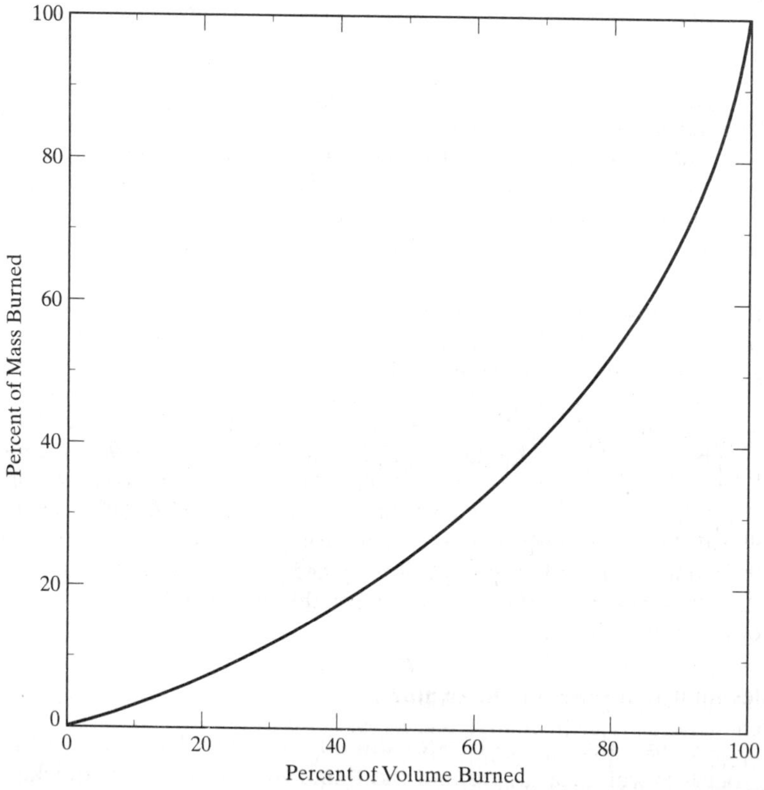

FIGURE 7-3

Mass percent burned vs. volume percent burned in the combustion chamber of a typical SI engine with homogeneous air–fuel mixture. Adapted from [93].

unburned gases behind the flame front continues to increase, reaching a maximum at the end of the combustion process. Temperature of the burned gases is not uniform throughout the combustion chamber, but is higher near the spark plug where combustion started. This is because the gas there has experienced a greater amount of radiation energy input from later flame reaction.

Ideally the air–fuel mixture should be about two-thirds burned at TDC and almost completely burned at about 15° aTDC. This causes maximum temperature and maximum pressure of the cycle to occur somewhere between 5° and 10° aTDC, about optimum for a four-stroke cycle SI engine. Using Eq. (2-14) at 15° aTDC and an R value of 4 gives $V/V_c = 1.17$. Thus, combustion in a real four-stroke cycle SI engine is almost, but not exactly, a constant-volume process, as approximated by the ideal air-standard Otto cycle. The closer the combustion process is to constant volume, the higher will be the thermal efficiency. This can be seen in the comparison of the thermal efficiencies of the Otto, Dual, and Diesel cycles in Chapter 3. However, in a real engine cycle, constant-volume combustion is not the best way to operate. Figure 7-1 shows how pressure rises with engine rotation for a well-designed four-stroke cycle engine. During combustion, a maximum pressure rise of about 240 kPa per degree of engine rotation is desirable for a smooth transfer of force to the face of the piston [58]. True constant-volume combustion would give the pressure curve an infinite upward slope at TDC, with a corresponding rough engine operation.

A lesser pressure rise rate gives lower thermal efficiency and danger of knock (i.e., a slower rise in pressure means slower combustion and the likelihood of knock). The combustion process is thus a compromise between the highest thermal efficiency possible (constant volume) and a smooth engine cycle with some loss of efficiency.

In addition to the effects of turbulence, swirl, and squish, the flame speed depends on the type of fuel and the air–fuel ratio. Lean mixtures have slower flame speeds, as shown in Fig. 7-4. Slightly rich mixtures have the fastest flame speeds, with the maximum for most fuels occurring at an equivalence ratio near 1.2. Exhaust residual and recycled exhaust gas slow the flame speed. Flame speed increases with engine speed due to the higher turbulence, swirl, and squish (Fig. 7-5).

The typical **burn angle**, the angle through which the crankshaft turns during the flame propagation mode of combustion, is about 25° for most engines (Fig. 7-6). If combustion is to be completed at 15° aTDC, then ignition should occur at about 10° bTDC. If ignition is too early, the cylinder pressure will increase to undesirable levels before TDC, and work will be wasted in the compression stroke. If ignition is late, peak pressure will not occur early enough, and work will be lost at the start of the power stroke due to lower pressure. Actual ignition timing is typically anywhere from 10° to 30° bTDC, depending on the fuel used, engine geometry, and engine speed. For any given engine, combustion occurs faster at higher engine speed. Real time for the combustion process is therefore less, but real time for the engine cycle is also less, and the burn angle is only slightly changed. This slight change is corrected by advancing the spark as the engine speed is increased. This initiates combustion slightly earlier in the cycle, with peak temperature and pressure remaining at about 5° to 10° aTDC. At part throttle, ignition timing is advanced to compensate for the resulting slower flame speed. Modern engines automatically adjust ignition timing with electronic controls. These not only use engine speed to set timing, but also sense and make fine adjustments for knock and incorrect exhaust emissions.

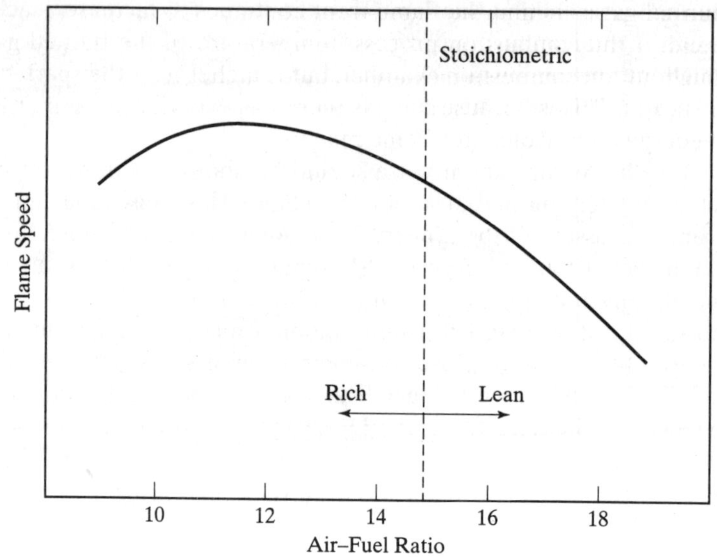

FIGURE 7-4

Average flame speed in the combustion chamber of an SI engine as a function of the air–fuel ratio for gasoline-type fuels. Lean air–fuel mixtures have slower flame speeds, with maximum speed occurring when slightly rich at an equivalence ratio near $\phi = 1.2$. Adapted from [93].

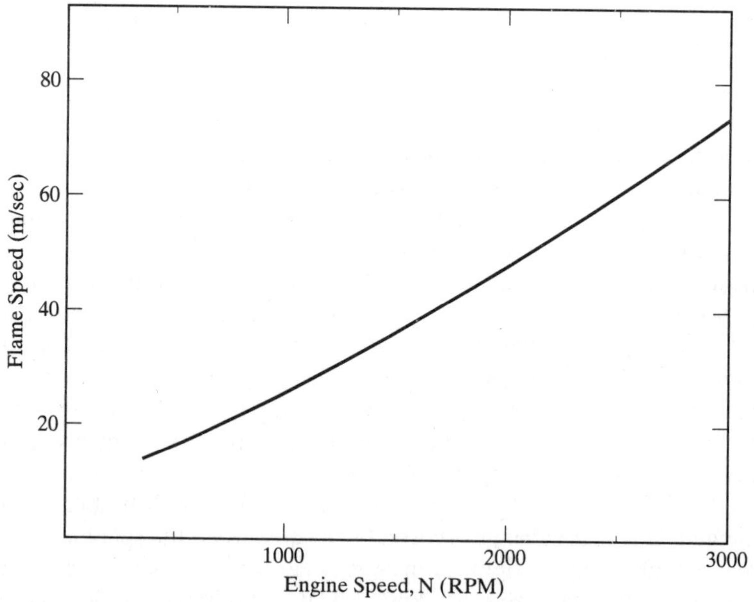

FIGURE 7-5

Average combustion chamber flame speed as a function of engine speed for a typical SI engine. As engine speed increases, the intensity of turbulence, swirl, squish, and tumble all increase, resulting in a faster flame speed.

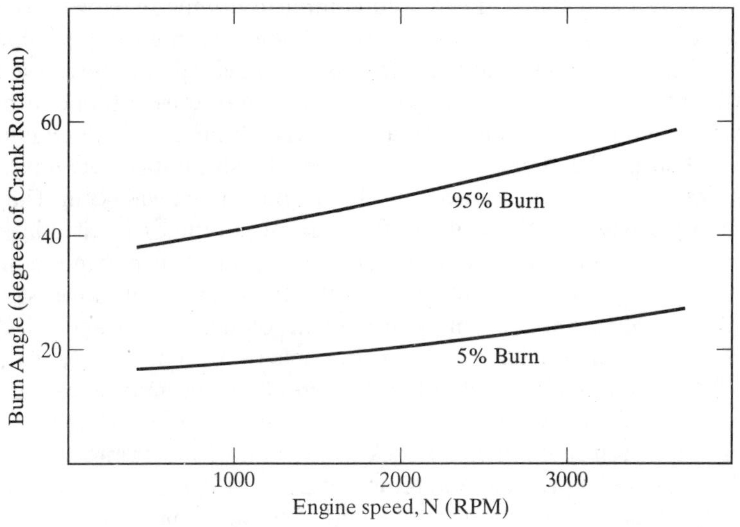

FIGURE 7-6

Burn angle as a function of engine speed for a typical modern SI engine with fast-burn combustion chambers and homogeneous air–fuel mixture. Burn angle is the angle through which the crankshaft turns during combustion. The increase in the angle of the ignition and flame development period (5% burn) is due mainly to the almost constant real time of the spark ignition process. During flame propagation (5% burn to 95% burn), both combustion speed and engine speed increase, resulting in a fairly constant burn angle of about 25° for the main part of combustion. Adapted from [61].

Earlier engines used a mechanical timing adjustment system that consisted of a spring-loaded ignition distributor that changed with engine speed due to centrifugal forces. Ignition timing on many small engines is set at an average position with no adjustment possible.

Flame Termination

At about 15° to 20° aTDC, 90–95% of the air–fuel mass has been combusted and the flame front has reached the extreme corners of the combustion chamber. Figure 7-3 shows that the last 5% or 10% of the mass has been compressed into a few percent of the combustion chamber volume by the expanding burned gases behind the flame front. Although, at this point, the piston has already moved away from TDC, the combustion chamber volume has increased only on the order of 10–20% from the very small clearance volume. This means that the last mass of air and fuel will react in a very small volume in the corners of the combustion chamber and along the chamber walls.

Due to the closeness of the combustion chamber walls, the last *end gas* that reacts does so at a very reduced rate. Near the walls, turbulence and mass motion of the gas mixture have been dampened out, and there is a stagnant boundary layer. The large mass of the metal walls also acts as a heat sink and conducts away much of the energy being released in the reaction flame. Both of these mechanisms reduce the rate of reaction and flame speed, and combustion ends by slowly dying away. Although very little additional work is delivered by the piston during this flame termination period due to the slow reaction rate, it is still a desirable occurrence. Because the rise in cylinder pressure tapers off slowly towards zero during flame termination, the forces transmitted to the piston also taper off slowly, and smooth engine operation results.

During the flame termination period, self-ignition will sometimes occur in the end gas in front of the flame front, and engine knock will occur. The temperature of the unburned gases in front of the flame front continues to rise during the combustion process, reaching a maximum in the last end gas. This maximum temperature is often above self-ignition temperature. Because the flame front moves slowly at this time, the gases are often not consumed during ignition delay time, and self-ignition occurs. The resulting knock is usually not objectionable or even noticeable. This is because there is so little unburned air–fuel left at this time that self-ignition can only cause very slight pressure pulses. Maximum power is obtained from an engine when it operates with very slight self-ignition and knock at the end of the combustion process. This occurs when maximum pressure and temperature exist in the combustion chamber and knock gives a small pressure boost at the end of combustion.

Example Problem 7-1

The spark plug is fired at 18° bTDC in an engine running at 1800 RPM. It takes 8° of engine rotation to start combustion and get into flame propagation mode. Flame termination occurs at 12° aTDC. Bore diameter is 8.4 cm, and the spark plug is offset 8 mm from the centerline of the cylinder. The flame front can be approximated as a sphere moving out from the spark plug. Calculate the effective flame front speed during flame propagation.

Rotational angle during flame propagation is from 10° bTDC to 12° aTDC, which equals 22°. Time of flame propagation is

$$t = (22°)/[(360°/\text{rev})(1800/60\ \text{rev/sec})] = 0.00204\ \text{sec}$$

Maximum flame travel distance is

$$D_{max} = \text{bore}/2 + \text{offset} = (0.084/2) + (0.008) = 0.050\ \text{m}$$

Effective flame speed is

$$V_f = D_{max}/t = (0.050\ \text{m})/(0.00204\ \text{sec}) = \underline{24.5\ \text{m/sec}}$$

Example Problem 7-2

The engine in Example Problem 7-1 is now run at 3000 RPM. As speed is increased in this engine, greater turbulence and swirl increase the flame front speed at a rate such that $V_f \propto 0.85\ N$. Flame development after spark plug firing still takes 8° of engine rotation. Calculate how much ignition timing must be advanced such that flame termination again occurs at 12° aTDC.

Flame speed is

$$V_f = (0.85)(3000/1800)(24.5\ \text{m/sec}) = 34.7\ \text{m/sec}$$

With flame travel distance the same, the time of flame propagation is

$$t = D_{max}/V_f = (0.050\ \text{m})/(34.7\ \text{m/sec}) = 0.00144\ \text{sec}$$

Rotational angle during flame propagation is

$$\text{angle} = (3000/60\ \text{rev/sec})(360°/\text{rev})(0.00144\ \text{sec}) = 25.92°$$

Flame propagation starts at 13.92° bTDC, and spark plug firing is at 21.92° bTDC. Therefore, ignition timing must be advanced 3.92°.

Variations in Combustion

Ideally, combustion in every cylinder of an engine would be exactly the same, and there would be no cycle-to-cycle variation in any one cylinder. This does not happen due to variations that occur in the intake system and within the cylinder. Even if no variations occurred before combustion, the turbulence within the cylinder would cause statistical variations to occur during combustion.

Differences in length and geometry of the intake manifold runners leading to the different cylinders causes cylinder-to-cylinder variations in the volumetric efficiency and air–fuel delivered. Temperature differences in the runners cause variations in the evaporation rates, and this causes variations in the air–fuel ratio. More fuel vapor in a hotter runner will displace more air and give a richer mixture and lower volumetric efficiency. Evaporative cooling causes temperature differences and, consequently, density differences. Because gasoline is a mixture with components that evaporate at different temperatures, the component mixture in each cylinder will not be exactly the same. The vapor of components that evaporate early in the intake manifold will not follow exactly the same paths and distribution as those of the still-liquid particles of the components that evaporate later at higher temperatures. This is less of a problem in engines with port injectors than in those with throttle body injectors or carburetors. Fuel additives evaporate at different temperatures and so end up in different cylinder-to-cylinder concentrations and even in different cycle-to-cycle concentrations for any single cylinder. Time and spacial variations will occur when EGR is added to the intake system (Fig. 7-7). Passage of air around the throttle plate breaks into two flows, causing vortices and other variations that will then affect all downstream flow. Because of imperfect quality control in the manufacturing of fuel injectors, each injector does not deliver exactly the same quantity of fuel, and there will be cycle-to-cycle variations from any one injector. The standard deviation of AF within a cylinder is typically on the order of 2–6% of the average (Fig. 7-8).

Within the cylinder, variations that already exist in the air–fuel ratio, amount of air, fuel components, and temperature, along with normal turbulence, will cause slight variations in swirl and squish, cylinder-to-cylinder and cycle-to-cycle. The variations in turbulence and mass motion within the cylinder affect the flame that occurs, and this results in substantial combustion variations, as shown in Fig. 7-9.

Local variations and incomplete mixing, especially near the spark plug, cause the initial discharge across the electrodes to vary from the average, which then initiates cycle-to-cycle combustion differently. Once there is a difference in the start of

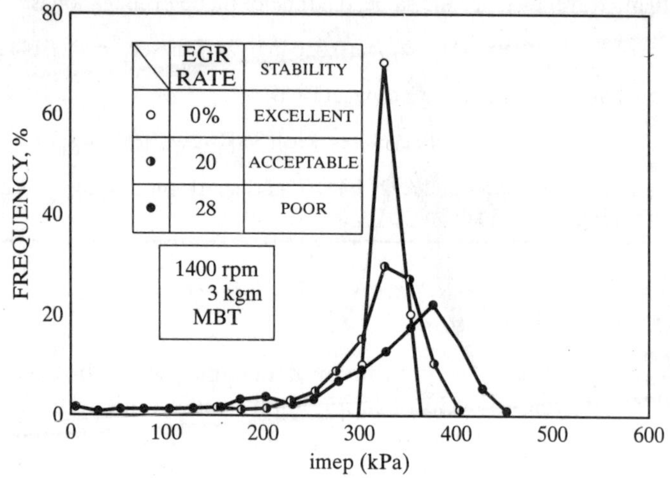

FIGURE 7-7

Effect of EGR on the consistency of combustion in the cylinder of an SI engine. Ideally, the value of the indicated mean effective pressure (X axis) would be the same for all cycles (100%). With no EGR, the frequency of average imep is very high, with some variation due to inconsistency in turbulence, AF, etc. As EGR is added, more variation in combustion occurs. This results in a larger spread of imep experienced and average imep occurring less often. Reprinted with permission from SAE Paper 780006 © 1978, SAE International [77].

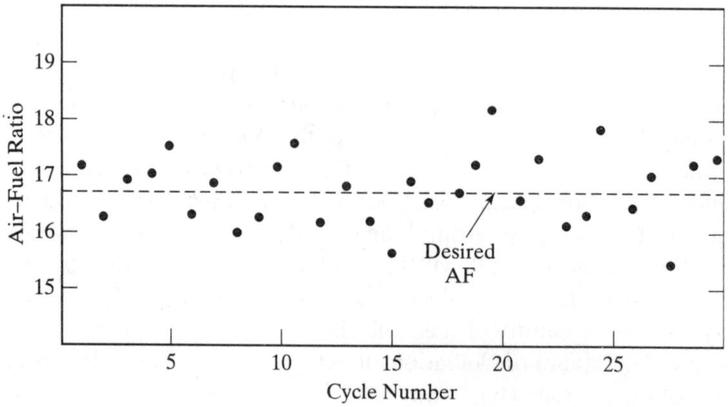

FIGURE 7-8

Typical variation of air–fuel ratio that may be delivered to a single cylinder in 30 consecutive engine cycles. Adapted from [84].

combustion, the entire following combustion process will be changed. Figure 7-10 shows how turbulence can change the way the same spark plug initiates combustion in two different cycles. The *kernel* of combusting gases can even become detached from the spark plug by high swirl or tumble action at the start of combustion. When this

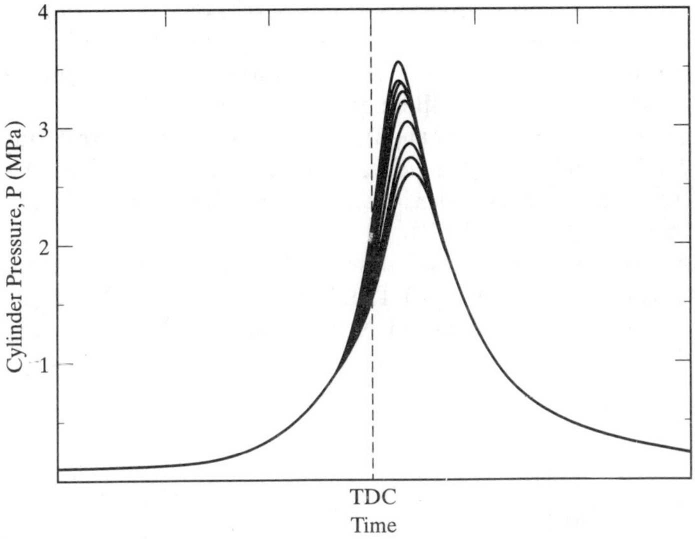

FIGURE 7-9

Pressure as a function of time for 10 consecutive cycles in a single cylinder of an SI engine, showing variation that occurs due to inconsistency of combustion. Similar variation would be obtained if the pressure of the Y coordinate were replaced with temperature. Adapted from [42].

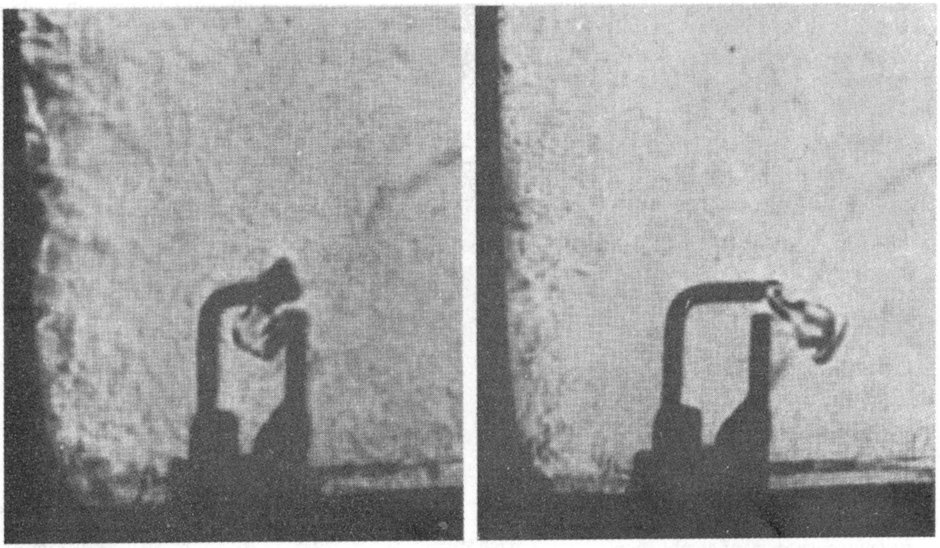

FIGURE 7-10

Schlieren photographs of the start of combustion for two different cycles in the same test engine cylinder, using propane fuel at 1400 RPM and a spark plug gap of 0.8 mm. Variations result from randomness of cylinder turbulence and cycle-to-cycle inconsistencies in swirl, squish, and tumble. Once there is variation at the start of combustion, the entire combustion process will be different for the two cycles. Reprinted by permission of Elsevier Science Inc. from "Flame Photographs in a Spark-Ignition Engine," by Gatowski, Heyward, and Deleplace, *Combustion and Flame*, vol. 56, pp. 71–81, copyright 1984 by The Combustion Institute, Ref. [43].

happens, the entire combustion process is changed. If this kernel of startup combustion is pushed against the combustion chamber wall, the added heat loss will cause very slow reaction during that cycle. When two cycles have a difference in ignition, the ensuing combustion process for those cycles will be quite different.

The fastest burn time within a cylinder is about twice as fast as the slowest burn time within the same cylinder, the difference being due to random variations that occur. The greatest percentage differences occur at light engine loads and low speeds, with idle being the worst condition (Fig. 7-11).

As a compromise, the average burn time is used to set the engine operating conditions (i.e., spark timing, AF, compression ratio, etc.). This lowers the output of the engine from that which could be obtained if all cylinders and all cycles had exactly the same com-

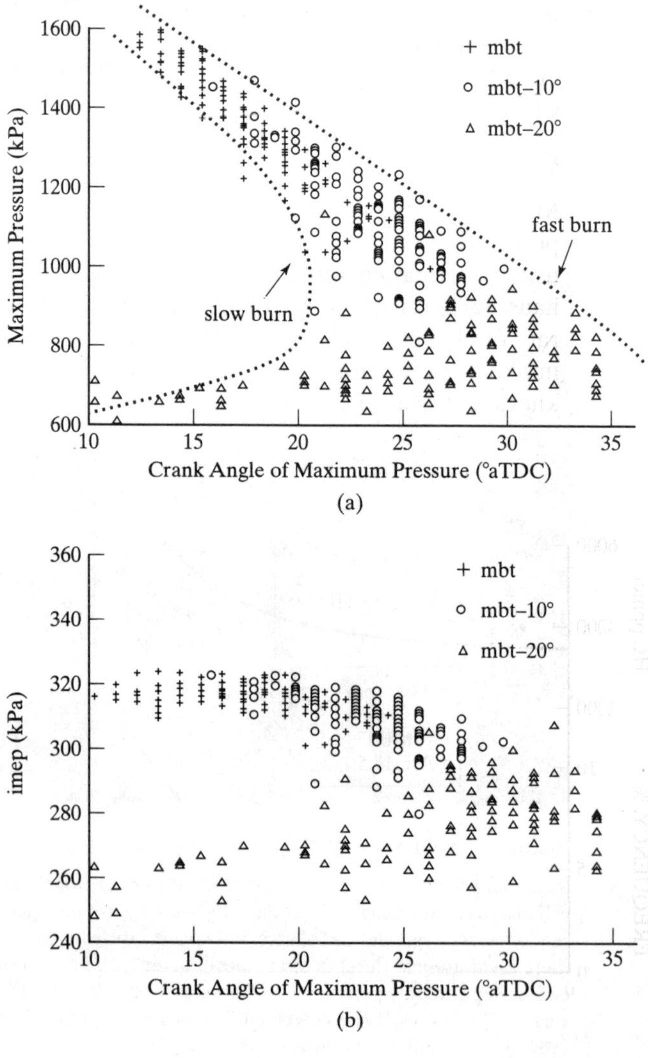

FIGURE 7-11

Cyclic pressure variation in same cylinder for three different spark timings, showing cycle-to-cycle inconsistency of combustion in running engine. Spark plug was mounted in center of combustion chamber and AF = 20. (a) Maximum cycle pressure vs. crank angle of maximum pressure, (b) Indicated mean effective pressure vs. crank angle of maximum pressure. Reprinted with permission from SAE Paper No. 830337 © 1983 SAE International [201].

bustion process. A cycle in which fast burn occurs is like a cycle with an over-advanced spark. This happens when there is a rich AF ratio, higher than average turbulence, and good initial combustion start-up. The result of this condition is a temperature and pressure rise too early in the cycle, with a good chance of knock occurring. This limits the compression ratio and fuel octane number that can be tolerated for a given engine. A cycle with a slower-than-average burn time is like a cycle with a retarded spark. This occurs when there is a lean mixture and higher-than-average EGR. The result of this condition is a flame lasting well into the power stroke and thus hot exhaust and hot exhaust valves. This is when partial burns and misfires occur (Fig. 7-12). There is also a power loss due to higher-than-average heat loss during these cycles. Higher heat loss occurs because of the longer combustion time and because the flame front is wider with the slower burning lean mixture. Slow burn limits the EGR setting for an engine and the acceptable lean setting for good fuel economy at cruise conditions. For smooth operation, engine conditions must be set for the worst cyclic variations in the worst cylinder. If all cylinders had exactly the same combustion process cycle after cycle, a higher engine compression ratio could be tolerated, and the air–fuel ratio could be set for higher power and greater fuel economy. Cheaper, lower octane fuel could be used.

Controls and Sensors

Modern smart engines continuously adjust combustion to give optimum output of power, fuel economy, and emissions. This is done with programed electronic controls using information input from sensors located in appropriate engine, intake, and exhaust locations. Among other things, these sensors measure throttle position, throttle rate of change, intake manifold pressure, atmospheric pressure, coolant temperature, intake temperature, EGR valve position, crank angle, O_2 and CO in the exhaust, knock, etc. Methods used by these sensors are mechanical, thermal, electronic, optical,

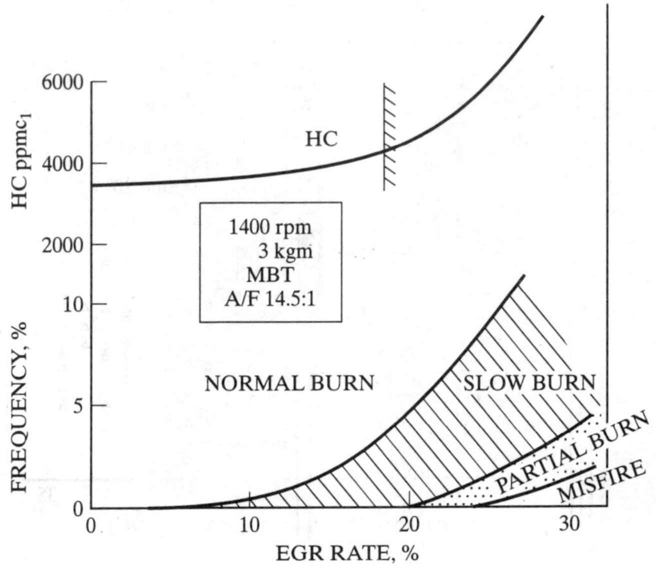

FIGURE 7-12

Effect of EGR on quality of combustion in an SI engine and hydrocarbon emissions in the exhaust. With no EGR, most cycles will have normal burn time. As the percent of EGR is increased, there will be an increase in cycles with slow burn or partial burn. With too much EGR, combustion in some cycles will die out, resulting in a misfire. Reprinted with permission from SAE Paper 780006 © 1978 SAE International [77].

chemical, and combinations of these. The controlled variables include ignition timing, valve timing, fuel injection duration, exhaust air pump actuation, air–fuel ratio, transmission shifting, turning on of warning lights, repair diagnostic recording, reprogramming of computer, etc.

On some engines, control of things such as ignition timing and injection duration are adjusted for the entire engine. On other engines, these adjustments are made separately for a bank of cylinders or even for a single cylinder. The fewer cylinders controlled by a separate control unit, the more optimum engine operation can be made. However, this requires more sensors, a larger control computer, potentially greater maintenance, and higher cost.

7.2 COMBUSTION IN DIVIDED–CHAMBER ENGINES AND STRATIFIED CHARGE LEAN-BURN ENGINES

Some engines have the combustion chamber divided into a main chamber and a smaller secondary chamber as shown in Figs. 6-8 and 7-13. This is done to create different intake and combustion characteristics in the two chambers, with a consequent gain in power and/or fuel economy. Often, these will be stratified charge engines with different air–fuel ratios at different locations in the combustion chamber.

The volume of the small secondary chamber is typically up to about 20% of the total clearance volume. It is designed with different goals in different engines. On some engines, the secondary chamber is designed primarily to provide very high swirl. The orifice between the chambers is contoured to create high swirl during the compression stroke as the intake air–fuel passes from the main chamber into the *swirl chamber* where the spark plug is located. This promotes good combustion ignition in the swirl chamber. As the air–fuel burns in the swirl chamber, it expands back through the orifice, creating a secondary swirl in the main chamber and acting as a *jet ignition* or *torch*

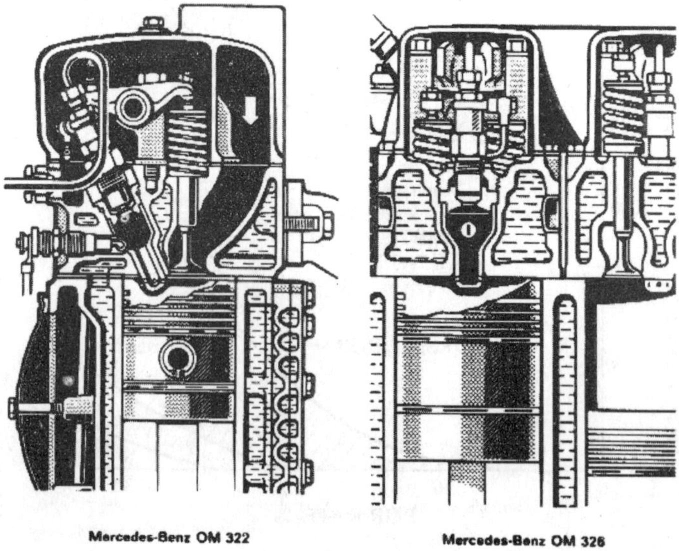

FIGURE 7-13
Divided combustion chambers of CI engines with fuel injector mounted in secondary chamber. Reprinted with permission from SAE Paper 710558 © 1971 SAE International [62].

ignition for the gas mixture there. This type of swirl chamber eliminates the need to create a primary swirl in the main chamber. The intake manifold and valves can be designed with a smoother straight-in flow, and higher volumetric efficiency is achieved.

Stratified charge engines often have divided chambers. These engines do not have a homogeneous air–fuel mixture throughout the combustion chamber, but have a rich mixture around the spark plug and a leaner mixture away from the plug. The rich mixture around the spark plug assures a good start and early spread of combustion, while the lean mixture in the rest of the combustion chamber gives good fuel economy. Often, the air–fuel mixture occupying the greater part of the combustion chamber is too lean to consistently ignite from a spark plug, but burns adequately when ignited by the small rich mixture near the plug. Dual-chamber stratified charge engines have a rich mixture in the small secondary chamber where the spark plug is located and a lean mixture in the main chamber (Fig. 6-8). Most of the power of the engine is generated in the large primary chamber using an economical lean mixture. Some *super economy lean-burn engines* on the market use an overall air–fuel ratio of 25:1. This mixture is leaner than that which could be combusted in a homogeneous mixture engine. High swirl and squish, a rich mixture around the spark plug, and a very high voltage spark plug with a larger than normal electrode gap promote good starting of combustion. Experimental SI engines that can operate on overall air–fuel ratios up to 50:1 have been developed.

Lean mixtures have low flame speeds, which, under normal conditions, would create a potential knock problem. However, lean mixtures have lower combustion temperatures, caused by the excess nonreacting gases, and this eases the knock problem. Some modern combustion technologies use very large amounts of recycled exhaust gases (EGR) in lean-burn engines. The actual air and fuel combination used is then near stoichiometric, which helps keep harmful emissions to a minimum.

Some engines have one intake valve in the main chamber and one in the secondary chamber of each cylinder. These engines supply air and fuel at different air–fuel ratios, with a rich mixture in the secondary chamber (Fig. 7-14). The extreme of this approach

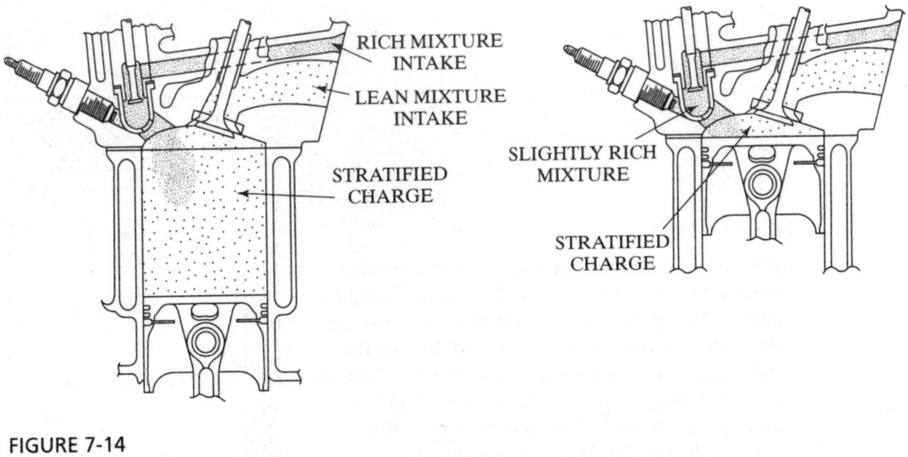

FIGURE 7-14

Divided–chamber stratified charge engine at end of intake stroke (left) and end of compression stroke (right). Reprinted with permission from SAE Paper No. 740605 © 1974 SAE International [168].

is when one intake valve supplies only air with no fuel added. Some engines have only one intake valve per cylinder operating at low engine speeds and two valves operating at higher engine speeds, supplying different air–fuel ratios. Some engines use a combination of intake valves and an in-cylinder fuel injector to create a stratified charge in the combustion chamber. Some stratified charge engines do not use a divided chamber, but have only normal, single open-chamber geometry.

A variation of stratified charge engines is dual-fuel engines. These engines use two types of fuel simultaneously, one usually a less expensive fuel, with a lesser amount of a better fuel used to assure ignition. These engines can be of a divided–chamber or normal open-chamber design. Fuel is supplied, and various air–fuel ratios are obtained, by a combination of multiple intake valves, fuel injectors, and/or proper contouring of the intake flow. Natural gas is often the main fuel used in dual-fuel engines. This is especially true in underdeveloped third-world countries, where natural gas is more available than other fuels.

Another type of divided–chamber engine is shown schematically in Fig. 7-15. This engine essentially has a small passive secondary chamber off the side of the main combustion chamber containing no intake, ignition, or special swirl. When combustion occurs in the main chamber, high-pressure gases are forced through the very small orifice into the secondary chamber. When the pressure in the main chamber falls during the power stroke, these high-pressure gases flow slowly back into the main chamber, slightly increasing the pressure pushing on the piston face and producing more work. Depending on the design, these backflowing gases may contain a combustible mixture and extend combustion time (and, consequently, work output).

Many combinations and variations of divided chambers and stratified charge engines have been tried, and a number of these exist in modern automobiles.

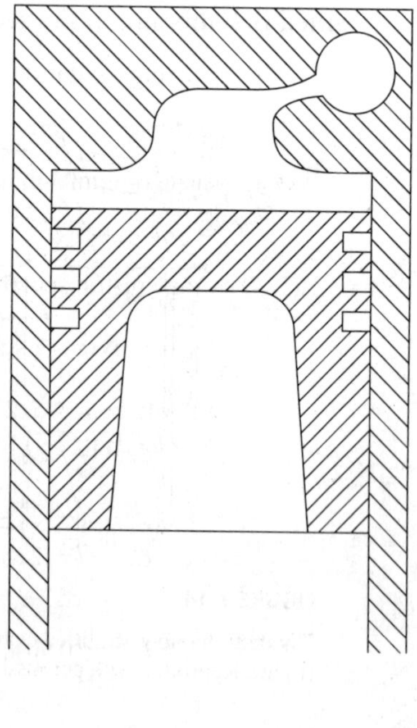

FIGURE 7-15

Schematic of a combustion chamber with a passive secondary *air chamber* as is found on some SI and CI engines. During the high pressure of combustion, gas (air, air–fuel, and/or exhaust) is forced through the small orifice and fills the secondary chamber. Then, as cylinder pressure is reduced during the expansion stroke, gas in the secondary chamber slowly flows back into the main chamber, slightly extending combustion and the power stroke process.

Example Problem 7-3

A six-cylinder SI engine with a total displacement of 1.86 liters operates at 2400 RPM using gasoline direct injection (GDI) with two injections per cycle in each cylinder. The spark plug is fired at 19° bTDC, and there is an ignition delay of 0.0015 seconds before combustion is established. During combustion, there is a rich air–fuel mixture around the spark plug of AF = 11:1 and a lean mixture in the rest of the combustion chamber of AF = 20:1. The rich zone can be modeled as a 2-cm–diameter hemisphere around the spark plug with a flame speed of 32 m/sec. The lean zone then fills the rest of the combustion chamber to the outer edge of the bore and a flame speed of 19 m/sec. The engine has a compression ratio of 9.8, a stroke of 7.20 cm, and a connecting rod length of 13.3 cm, with the spark plug at the center of the combustion chamber. Calculate:

1. crank angle position at the end of combustion
2. piston speed at the end of combustion
3. volume in combustion chamber of one cylinder at end of combustion

(1) Use Eq. (2-9) to find bore:

$$V_d = N_c(\pi/4)B^2S = 0.00186\ \text{m}^3 = (6)(\pi/4)B^2(0.0720\ \text{m})$$
$$B = 0.0740\ \text{m} = 7.40\ \text{cm}$$

Time for flame to reach outer edge of fuel-rich zone is

$$t_1 = (\text{distance})/(\text{velocity}) = (0.020/2\ \text{m})/(32\ \text{m/sec}) = 0.000313\ \text{sec}$$

Time for flame to travel through fuel-lean zone to outer edge of bore is

$$t_2 = \{[(0.0740/2) - (0.020/2)]\text{m}\}/(19\ \text{m/sec}) = 0.00142\ \text{sec}$$

Total time between ignition and the end of combustion is

$$t_{\text{total}} = t_{\text{id}} + t_1 + t_2 = (0.0015) + (0.000313) + (0.00142) = 00323\ \text{sec}$$

Crank angle rotation during combustion is

$$\text{CA} = (2400/60\ \text{rev/sec})(360°/\text{rev})(0.00323\ \text{sec}) = 46.5°$$

Position of crank angle at end of combustion is

$$19°\ \text{bTDC} + 46.5° = \underline{27.5°\ \text{aTDC}}$$

(2) Average piston speed using Eq. (2-2):

$$\overline{U}_p = 2SN = (2\ \text{strokes/rev})(0.0720\ \text{m/stroke})(2400/60\ \text{rev/sec}) = 5.76\ \text{m/sec}$$

Crank offset $a = S/2 = (7.20\ \text{cm})/2 = 3.60\ \text{cm}$

Use Eq. (2-5) to find piston speed at end of combustion:

$$R = r/a = (13.3)/(3.60) = 3.69$$
$$U_p/\overline{U}_p = (\pi/2)\sin\theta[1 + (\cos\theta/\sqrt{R^2 - \sin^2\theta})]$$
$$= (\pi/2)\sin(27.5°)\{1 + [\cos(27.5°)/\sqrt{(3.69)^2 - \sin^2(27.5)}]\} = 0.90$$
$$U_p = (0.90)\overline{U}_p = (0.90)(5.76\ \text{m/sec}) = \underline{5.18\ \text{m/sec} = 17.0\ \text{ft/sec}}$$

(3) Displacement volume of one cylinder is

$$V_d = (1.86\ \text{L})/(6\ \text{cylinders}) = 0.31\ \text{L} = 310\ \text{cm}^3$$

Use Eq. (2-12) to find the clearance volume of one cylinder:

$$r_c = (V_c + V_d)/V_c = 9.8 = (V_c + 310)/V_c$$
$$V_c = 35.2\ \text{cm}^3$$

Use Eq. (2-14) to find the combustion chamber volume at end of combustion:

$$V/V_c = 1 + \tfrac{1}{2}(r_c - 1)[R + 1 - \cos\theta - \sqrt{R^2 - \sin^2\theta}]$$
$$= 1 + \tfrac{1}{2}(9.8 - 1)[(3.69) + (1) - \cos(27.5°) - \sqrt{(3.69)^2 - \sin^2(27.5°)}] = 1.62$$
$$V = (1.62)V_c = (1.62)(35.2\ \text{cm}^3) = \underline{57.0\ \text{cm}^3 = 0.057\ \text{L} = 0.000057\ \text{m}^3 = 3.48\ \text{in.}^3}$$

7.3 ENGINE OPERATING CHARACTERISTICS

Power Operation

For maximum power at WOT (e.g., fast start-up, accelerating up a hill, an airplane taking off), fuel injectors and carburetors are adjusted to give a rich mixture, and ignition systems are set with retarded spark (spark later in cycle). This gives maximum power at a sacrifice of fuel economy. The rich mixture burns faster and allows the pressure peak to be more concentrated near TDC, with the probable compromise of rougher operation. At high engine speeds, there is less time for heat transfer to occur from the cylinders, and exhaust gases and exhaust valves will be hotter. To maximize flame speed at WOT, no exhaust gas is recycled, resulting in higher levels of NOx.

Interestingly, another way of obtaining added power from an engine is to operate with a lean mixture. Race cars are sometimes operated this way. In a lean mixture, flame speed is slow and combustion lasts well past TDC. This keeps the pressure high well into the power stroke, which produces a greater power output. This way of operation produces very hot exhaust gases due to the late combustion. This hot exhaust, combined with the unused oxygen of the lean mixture, oxidizes the exhaust valves and seats very quickly. This possibly requires changing of the exhaust valves quite often, something unacceptable except for race cars. Ignition timing must be set specially for this kind of operation.

Cruising Operation

For cruising operation such as steady freeway driving or long-distance airplane travel, less power is needed and brake-specific fuel consumption becomes important. For this type of operation, a lean mixture is supplied to the engine, high EGR is used, and ignition timing is advanced to compensate for the resulting slower flame speed. Fuel usage efficiency (miles/liter) will be high, but thermal efficiency of the engine will be lower. This is because the engine will be operating at a lower speed, which gives more time per cycle for heat losses from the combustion chamber.

Idle and Low Engine Speed

At very low engine speeds, the throttle will be almost closed, resulting in a high vacuum in the intake manifold. This high vacuum and low engine speed generate a large exhaust residual during valve overlap. This creates poor combustion, which must be compensated for by supplying a rich mixture to the engine. The rich mixture and poor combustion contribute to high exhaust emissions of HC and CO. Misfires and cycles during which only partial combustion occurs in some cylinders are more common at idle speeds. A 2% misfire rate would cause exhaust emissions to exceed acceptable standards by 100–200%.

Closing Throttle at High Engine Speed

When quick deceleration is desired and the throttle is closed at high engine speed, a very large vacuum is created in the intake system. High engine speed wants a large inflow of air, but the closed throttle allows very little air flow. The result is a high–intake vacuum, high exhaust residual, a rich mixture, and poor combustion. Misfires and high exhaust emissions are very common with this kind of operation.

Engines with carburetors give especially poor combustion under these conditions. Due to the high vacuum, the carburetor produces a large fuel flow through both the normal orifice and the idle valve. This, combined with the restricted air flow rate, creates an overrich mixture with poor combustion and high exhaust pollution of HC and CO. The controls on engines with fuel injectors shut the fuel flow down under these conditions, and this results in much smoother operation.

Starting a Cold Engine

When a cold engine is started, an overrich supply of fuel must be provided to assure enough fuel vapor to create a combustible gas mixture. When the walls of the intake system and cylinders are cold, a much smaller percentage of the fuel will vaporize than in normal steady-state operation. The fuel is also cold and does not flow as readily. The engine turns very slowly, being driven only by the starting motor, and a greater amount of the compressive heating during compression is lost by heat transfer to the cold walls. This is made worse by the cold viscous lubricating oil that resists motion and slows the starting speed even more. All of these factors contribute to the need for a very rich air–fuel ratio when starting a cold engine. Air–fuel ratios as rich as 1:1 are sometimes used.

Even when everything is very cold, a small percentage of fuel vaporizes and a combustible air and vapor mixture can be obtained. This mixture is ignited, and after only a few cycles of combustion, the engine begins to heat up. Within a few seconds, it starts to operate in a more normal mode, but it can take many minutes before fully warmed steady-state operation is reached. Once the engine starts to warm, all of the excess fuel that was originally input vaporizes and a short period of overrich operation is experienced. During this period, there is a large excess of HC and CO emissions in the exhaust. To compound this problem, the catalytic converter is also cold at start-up and does not remove these excess emissions. This problem of excess air pollution at cold start-up is addressed in Chapter 9.

Special starting fluids can be purchased for aiding engine start-up in extremely cold temperatures. Substances such as diethyl ether with very high vapor pressures evaporate more readily than gasoline and give a richer air–fuel vapor mixture for initiating combustion. These fluids generally are obtained in pressurized containers and are sprayed into the engine air intake before starting.

7.4 MODERN FAST-BURN COMBUSTION CHAMBERS

The combustion chamber for a modern high-speed SI engine must be able to burn the contained air–fuel mixture very rapidly without creating excess exhaust emissions. It must provide a smooth power stroke, low specific fuel consumption, and maximum thermal efficiency (a high compression ratio). Two general designs for such a combustion chamber are shown in Fig. 6-5. Many modern engines have combustion chambers that are a variation of one or both of these designs. As a comparison, Fig. 7-16 shows the general design of a combustion chamber found in historic, L head, valve-in-block engines.

It is desirable to have the minimum combustion time possible without actually having an instantaneous constant-volume reaction (detonation). If the combustion time is less than the ignition delay time of the air–fuel mixture after the temperature has been raised above self-ignition temperature, knock is avoided (see Chapter 4). The faster the burn time, the higher the allowable compression ratio and the lower the octane number required in the fuel.

For the fastest burn time, a minimum flame travel distance is desired with maximum turbulence, swirl, and squish. The two chambers in Fig. 6-5 satisfy these requirements, while the older chamber in Fig. 7-16 does not. As the piston approaches TDC in the chambers in Fig. 6-5, the air–fuel mixture is compressed towards the centerline of the cylinder. Conservation of angular momentum will cause a large increase in swirl

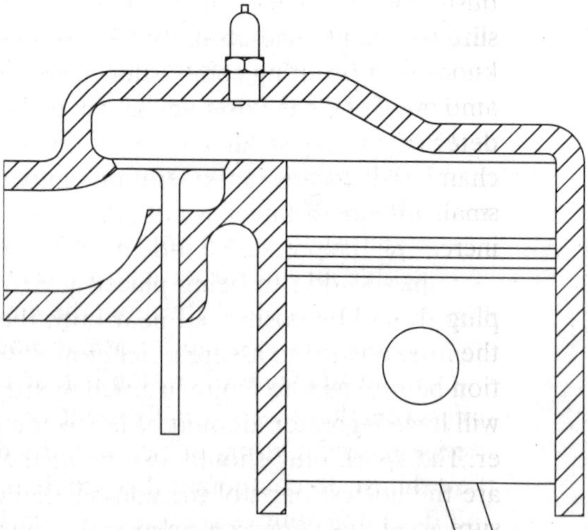

FIGURE 7-16

Combustion chamber of L head, valve-in-block engine. For several decades from around 1910 to the 1950s, this was the standard geometry of many engines. With a few exceptions to the general design, this type of combustion chamber generally did not promote high levels of swirl, squish, or tumble, considered very desirable in modern combustion philosophy. Flame travel distance was also long compared with that of modern combustion chambers. All of these factors restricted these early engines to much lower compression ratios than those common today.

rotation as the average mass radius is decreased. Some momentum will be lost via viscous friction with the walls. This inward compression also causes a large squish velocity in the radial direction towards the cylinder centerline. Both these motions greatly increase flame front speed and decrease combustion time. There is also a reverse outward squish that further increases the spread of the flame front. This occurs early in the power stroke when the piston starts to move away from TDC. In a modern combustion chamber, these motions, along with turbulence, increase flame velocity by a factor of about 10 over a flame passing through a stagnant air–fuel mixture. It is assumed that the intake systems for the cylinders shown in Fig. 6-5 are made to provide high turbulence and high inlet swirl.

The spark plug is placed near the centerline of the cylinder, so the flame must travel only about one-fourth of the bore diameter before most of the air–fuel mixture is consumed. Some engines have two spark plugs per cylinder. This allows the air–fuel mixture to be consumed by two flame fronts and, with proper placement, can almost decrease combustion time by a factor of two. Dual plugs can be designed to be fired either simultaneously or sequentially. At least one automobile manufacturer has experimented with a prototype four-cylinder engine with four spark plugs per cylinder, one in the center and three at the outer periphery. Most aircraft engines have two spark plugs per cylinder. However; this is more a safety feature than a design to improve combustion. Many aircraft systems have redundancy in case one spark plug fails.

In addition to fast combustion time, the combustion chambers shown in Fig. 6-5 would provide smooth engine operation during the power stroke. With the spark plug located near the center of the clearance volume, pressure buildup at the start of combustion will be slow due to the large volume of surrounding gas that must be compressed. Placing the spark plug near the edge of the combustion chamber would give a quicker early pressure rise, because there would be less gas to compress in the immediate vicinity. This would create a rougher engine cycle. Near the end of combustion, the flame front exists in the small volume of gas at the edges of the combustion chamber. This allows the pressure rise to die away slowly and contributes again to a smooth power stroke. If combustion were to end with the flame front in a large part of the combustion chamber, pressure rise would end abruptly and the end of the power stroke would be less smooth. If knock occurs during combustion, it will occur in the last end gas to burn. If only a small amount of end gas exists, any knock that occurs can be tolerated and probably will not be detected. A very small amount of knock is even desirable with this kind of combustion chamber. It means that operating temperature and pressure are at a maximum, and the small amount of knock is not detectable. It can even raise power output very slightly by increasing the pressure a little near the end of the combustion process.

In addition to being placed near the center of the clearance volume, the spark plug should be positioned near both the intake and exhaust valves. It should be near the intake valve to assure a richer mixture between the spark plug electrodes, combustion being easier to ignite in a rich mixture. The gas mixture away from the intake valve will have a greater amount of leftover exhaust residual and will consequently be leaner. The spark plug should also be near the exhaust valve. The exhaust valve and port are the hottest parts of the combustion chamber, and this higher temperature will assure good fuel vaporization near the spark plug. This also keeps the exhaust valve away from the hot end gas, where the higher surface temperature could cause surface ignition and knock.

To keep the size of the combustion chamber at a minimum, most modern SI engines have overhead valves. This requires overhead camshafts or a hydromechanical linkage between the valves and a camshaft mounted in the engine block. Another way of decreasing combustion chamber size is to have more cylinders for a given displacement volume.

These combustion chambers offer less heat loss, less force on the head bolts, less wall deposit buildup, and less exhaust emissions. There is less heat loss due to the smaller wall surface area-to-volume ratio than that which existed in the earlier valve-in-block engines. This, in turn, provides better thermal efficiency. There is less force on the head bolts holding the head to the engine block because of the smaller head surface area of the combustion chamber. For a given cylinder pressure, total force will be proportional to the surface area on which that pressure is applied. There will be less wall deposit buildup with time due to the hotter temperatures and high swirl motion, which cleans the walls in this type of chamber. There will be less exhaust emissions because of the smaller flame quench volume and fewer wall deposits. These will be discussed in greater detail in Chapter 9.

Probably the greatest disadvantage of this type of combustion chamber is the limited design flexibility it offers. Because of the limited wall surface area, it is extremely difficult to fit the needed valves, spark plugs, and fuel injectors. Often, valve sizing and gas flow control contouring must be compromised because of space considerations. Cylinders with multiple intake and exhaust valves decrease flow resistance, but increase design complexity. Often, surface areas must be cut away to allow for clearance between valves and piston face. This compromises the desire for minimum corner spaces in the chamber. Mechanical strength cannot be compromised, and enough surface material must be allowed between valves to assure structural stability.

Some modern engines have divided combustion chambers, as described earlier. These offer high volumetric efficiency, good fuel economy, and cycle operation flexibility. Two of their main disadvantages are greater heat loss, due to high surface area, and higher cost and difficulty in manufacturing.

Combustion chambers in older automobile engines, especially the flat-head valve-in-block type shown in Fig. 7-16, had a much longer flame travel distance and combustion time. Inlet systems were not designed to create swirl motion, and any inlet swirl that might exist would be greatly dampened out near TDC, when the air–fuel mixture is forced away from the cylinder centerline. Little squish motion is promoted. Some mass motion and some turbulence are present, but at lower levels, because of the slower engine speeds. Because of the much greater resulting combustion times, compression ratios had to be much lower. In the early years of this type of engine (the 1920s), compression ratios were in the range of four or five, increasing to about seven in later years (the 1950s).

Very large engines are almost always CI engines. Because of their large combustion chambers and corresponding long flame travel distance, combined with slow engine speed, they would require very high octane fuel and/or very low compression ratios if operated as an SI engine. With the very long real time of combustion in a cylinder, it would be impossible to avoid serious knock problems.

7.5 COMBUSTION IN CI ENGINES

Combustion in a compression ignition engine is quite different from that in an SI engine. Whereas combustion in an SI engine is essentially a flame front moving through a locally homogeneous mixture, combustion in a CI engine is an unsteady process occurring simultaneously at many spots in a very nonhomogeneous mixture at a rate controlled by fuel injection. Air intake into the engine is unthrottled, with engine torque and power output controlled by the amount of fuel injected per cycle. Because the incoming air is not throttled, pressure in the intake manifold is consistently at a value close to one atmosphere. This makes the pump work loop of the engine cycle shown in Fig. 3-10 very small, with a corresponding better thermal efficiency compared with that of an SI engine. This is especially true at low speeds and low loads, when an SI engine would be at part throttle with a large pump work. For CI engines

$$w_{net} = w_{gross} - w_{pump} \approx w_{gross} \tag{7-1}$$

Only air is contained in the cylinder during the compression stroke, and much higher compression ratios are used in CI engines. Compression ratios of modern CI engines range from 12 to 24. Unlike thermal efficiencies in normal SI engines, in CI engines high thermal efficiencies (fuel conversion efficiencies) are obtained when these compression ratios are used in Eqs. (3-73) and (3-89). However, because the overall air–fuel ratio on which CI engines operate is quite lean (equivalence ratio $\phi \approx 0.8$), less brake power output is often obtained for a given engine displacement.

Fuel is injected into the cylinders late in the compression stroke by one or more injectors located in each cylinder combustion chamber. Injection time is usually about 20° of crankshaft rotation, starting at about 15° bTDC and ending about 5° aTDC. Ignition delay is fairly constant in real time, so at higher engine speeds fuel injection must be started slightly earlier in the cycle.

In addition to the swirl and turbulence of the air, a high injection velocity is needed to spread the fuel throughout the cylinder and cause it to mix with the air. After injection, the fuel must go through the following series of events to assure the proper combustion process:

1. *Atomization.* Fuel drops break into very small droplets. The smaller the original drop size emitted by the injector, the quicker and more efficient will be this atomization process (Fig. 5-27).
2. *Vaporization.* The small droplets of liquid fuel evaporate to vapor. This occurs very quickly due to the hot air temperatures created by the high compression of CI engines. High air temperature needed for this vaporization process requires a minimum compression ratio in CI engines of about 12:1. About 90% of the fuel injected into the cylinder has been vaporized within 0.001 seconds after injection. As the first fuel evaporates, the immediate surroundings are cooled by evaporative cooling. This greatly affects subsequent evaporation. Near the core of the fuel jet, the combination of high fuel concentration and evaporative cooling will cause adiabatic saturation of fuel to occur. Evaporation will stop in this region, and only after additional mixing and heating will this fuel be evaporated.

3. *Mixing.* After vaporization, the fuel vapor must mix with air to form a mixture within the AF range that is combustible. This mixing comes about because of the high fuel injection velocity added to the swirl and turbulence in the cylinder air. Figure 7-17 shows the nonhomogeneous distribution of air–fuel ratio that develops around the injected fuel jet. Combustion can occur within the equivalence ratio limits of $\phi = 1.8$ (rich) and $\phi = 0.8$ (lean).
4. *Self-Ignition.* At about 8° bTDC, 6–8° after the start of injection, the air–fuel mixture starts to self-ignite. Actual combustion is preceded by secondary reactions, including breakdown of large hydrocarbon molecules into smaller species and some oxidation. These reactions, caused by the high-temperature air, are exothermic and further raise the air temperature in the immediate local vicinity. This finally leads to an actual sustained combustion process.
5. *Combustion.* Combustion starts from self-ignition simultaneously at many locations in the slightly rich zone of the fuel jet, where the equivalence ratio is $\phi = 1$ to 1.5 (zone B in Fig. 7-17). At this time, somewhere between 70% and 95% of the fuel in the combustion chamber is in the vapor state. When combustion starts, multiple flame fronts spreading from the many self-ignition sites quickly consume all the gas mixture that is in a correct combustible air–fuel ratio, even where self-ignition wouldn't occur. This produces a very quick rise in temperature and pressure within the cylinder, shown in Fig. 7-18. The higher temperature and pressure reduce the vaporization time and ignition delay time for additional fuel particles and cause more self-ignition points to further increase the combustion process. Liquid fuel is still being injected into the cylinder after the first fuel is already burning. After the initial start of combustion, when all the air–fuel mixture that is in a combustible state is quickly consumed, the rest of the combustion

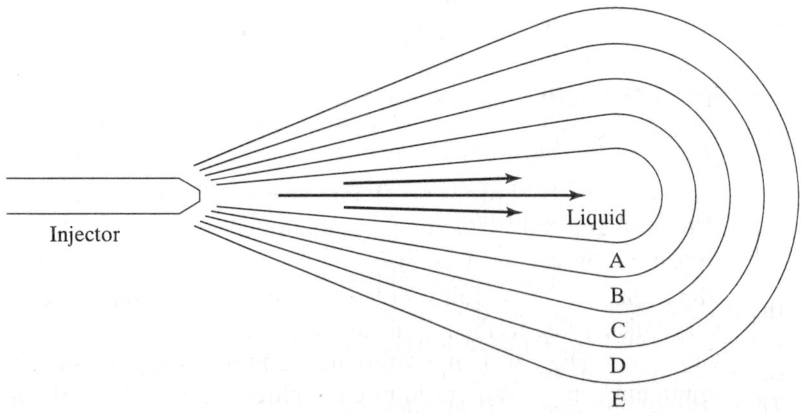

FIGURE 7-17

Fuel jet of a CI engine showing air–fuel vapor zones around the inner liquid core. The liquid core is surrounded by successive zones of vapor that are (A) too rich to burn, (B) rich combustible, (C) stoichiometric, (D) lean combustible, and (E) too lean to burn. Self-ignition starts mainly in zone B. Solid carbon soot is generated mostly in zones A and B.

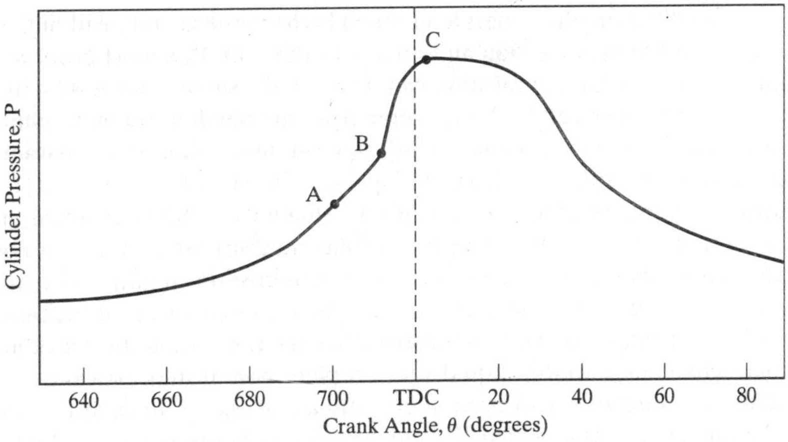

FIGURE 7-18
Cylinder pressure as a function of crank angle for a CI engine. Point A is where fuel injection starts, A to B is ignition delay, and point C is the end of fuel injection. If the cetane number of the fuel is too low, a greater amount of fuel will be injected during ignition delay time. When combustion then starts, the additional fuel will cause the pressure at point B to increase too fast, resulting in a rough engine cycle. Adapted from [10].

process is controlled by the rate at which fuel can be injected, atomized, vaporized, and mixed into the proper AF. This rate of combustion, now controlled by injection rate, can be seen in Fig. 7-18 in the slower pressure rise that occurs after the initial fast rise. Combustion lasts for about 40° to 50° of engine rotation, much longer than the 20° of fuel injection. This is because some fuel particles take a long time to mix into a combustible mixture with the air, and combustion therefore lasts well into the power stroke. This can be seen in Fig. 7-18, where the pressure remains high until the piston is 30°–40° aTDC. About 60% of the fuel is burned in the first third of combustion time. Burning rate increases with engine speed, so the burn angle remains about constant. During the main part of the combustion process, anywhere from 10% to 35% of the fuel vapor in the cylinder will be in a combustible AF.

In Chapter 2, it was reasoned that average engine speed strongly correlates with the inverse of stroke length. This puts the average piston speed for all engines in the range of about 5 to 20 m/sec. Large, slow engines have adequate real time to inject, atomize, vaporize, and mix the fuel for combustion to occur in 40°–50° of engine rotation. These *direct injection* (DI) engines have large open chambers without the need for high swirl. They generally have very high injection pressures that give the fuel jets high velocity. This assures that the penetration of the jet reaches across the large combustion chamber and greatly assists in the mixing of the fuel and air. Large DI engines generally have higher brake thermal efficiency because they operate slower, which reduces friction losses, and they have lower combustion chamber surface area-to-volume ratios, which reduces heat losses.

Small CI engines operate at much higher speeds and need high swirl to enhance and speed the vaporization and mixing of the fuel. This must occur at speeds up to 10 times faster so that combustion can occur in the same desired 40°–50° of engine rotation. Special intake and cylinder geometries are needed to generate this necessary high swirl. These geometries can include special swirl chambers separate from the main combustion chamber, as shown in Fig. 7-13. These *indirect injection* (IDI) engines with divided chambers inject the fuel into the smaller secondary chamber and can use much lower injection pressures. This type of engine gives lower fuel jet velocities, which are adequate to penetrate across the smaller combustion chamber. The high swirl generated in the secondary chamber provides the needed mixing of fuel and air. As the gas mixture in the secondary chamber combusts, it expands through the orifice into the main chamber, carrying liquid fuel droplets with it and providing swirl in the main chamber. Here, the main portion of combustion occurs much as it does in an SI engine. The higher speeds at which IDI engines generally operate make them better automobile engines. Because of the large surface area-to-volume ratio in the combustion chambers, there is a greater heat loss, and this typically requires a higher compression ratio. Starting a cold engine is also more difficult because of this factor.

Example Problem 7-4

The diesel engine of Example Problem 5-9 has a compression ratio of 18:1 and operates on an air-standard Dual cycle. At 2400 RPM, combustion starts at 7° bTDC and lasts for 42° of engine rotation. The ratio of connecting rod length to crank offset is $R = 3.8$. Calculate:

1. ignition delay
2. cycle cutoff ratio

(1) Combustion starts at 7° bTDC and fuel injection starts at 20° bTDC (from Example Problem 5-9). Ignition delay in degrees of engine rotation is

$$\underline{\text{ID} = 13° \text{ of engine rotation}}$$

Ignition delay in seconds is

$$\text{ID} = (13°)/[(2400/60 \text{ rev/sec})(360°/\text{rev})] = \underline{0.0009 \text{ sec}}$$

(2) Combustion stops at 35° aTDC. Equation (2.14) is used to find cutoff ratio:

$$\beta = V/V_{TDC} = V/V_c = 1 + \tfrac{1}{2}(r_c - 1)[R + 1 - \cos\theta - \sqrt{R^2 - \sin^2\theta}]$$
$$= 1 + \tfrac{1}{2}(18 - 1)[3.8 + 1 - \cos(35°) - \sqrt{(3.8)^2 - \sin^2(35)°}]$$
$$\underline{= 2.91}$$

Example Problem 7-5

It is desired to have combustion in a medium-size CI engine operating on a Diesel cycle start at 1° aTDC, using fuel with a cetane number of 41. The engine is a straight six operating at 980 RPM, with a total displacement of 15.6 liters and bore and stroke related as $S = 2.02\ B$. The compression ratio is 16.5:1, and the temperature and pressure of air entering the cylinders are 41° C and 0.98 bar.

Calculate:

1. crank angle at which fuel injection should start
2. ignition delay in milliseconds

(1) Use Eq. (2-9) to find bore and stroke:

$$V_d = N_c(\pi/4)B^2S = 15.6\text{ L} = 0.0156\text{ m}^3 = (6\text{ cyl})(\pi/4)B^2(2.02\,B)$$
$$B = 0.1179\text{ m} \qquad S = (2.02)B = (2.02)(0.1179\text{ m}) = 0.2382\text{ m}$$

Equation (2-2) gives average piston speed:

$$\overline{U}_p = 2SN = (2\text{ strokes/rev})(0.2382\text{ m/stroke})(980/60\text{ rev/sec}) = 7.78\text{ m/sec}$$

Equation (4-14) gives ignition delay in crank angle degrees.

Fuel activation energy is calculated as follows:

$$E_A = (618{,}840)/(\text{CN} + 25) = (618{,}840)/(41 + 25) = 9376$$
$$\text{ID(ca)} = (0.36 + 0.22\,\overline{U}_p)\exp\{E_A[1/(R_uT_ir_c^{\,k-1}) - (1/17{,}190)][(21.2)/(P_ir_c^{\,k} - 12.4)]^{0.63}\}$$
$$[1/(R_uT_ir_c^{\,k-1}) - (1/17{,}190)] = [1/\{(8.314)(314)(16.5)^{1.35-1}\} - (1/17{,}190)] = 0.0000854$$
$$[(21.2)/(P_ir_c^{\,k} - 12.4)]^{0.63} = \{(21.2)/[(0.98)(16.5)^{1.35} - (12.4)]\}^{0.63} = 0.7914$$
$$\text{ID(ca)} = [(0.36) + (0.22)(7.78)]\exp[(9376)(0.0000854)(0.7914)] = 3.91°$$

Crank angle at which fuel injection should start is

$$1°\text{ aTDC} - 3.91° = \underline{2.91°\text{ bTDC}}$$

(2) Equation (4-15) gives ignition delay in milliseconds:

$$\text{ID(ms)} = \text{ID(ca)}/(0.006\,N) = (3.91)/[(0.006)(980)] = \underline{0.665\text{ msec}}$$

Fuel Injection

The nozzle diameter of a typical fuel injector is 0.2–1.0 mm. An injector may have one nozzle or several (Fig. 6-6).

The velocity of liquid fuel leaving a nozzle is usually about 100 to 200 m/sec. This is quickly reduced by viscous drag, evaporation, and combustion chamber swirl. The vapor jet extends past the liquid jet and, ideally, just reaches the far walls of the combustion chambers. Evaporation occurs on the outside of the fuel jet while the center remains liquid. Figure 7-17 shows how the inner liquid core is surrounded by the following successive vapor zones of air–fuel:

(A) too rich to burn
(B) rich combustible
(C) stoichiometric
(D) lean combustible
(E) too lean to burn

Liquid drop diameter size leaving the injector is on the order of 10^{-5} m (10^{-2} mm) and smaller, generally with some normal distribution of sizes. Factors that affect droplet size include pressure differential across the nozzle, nozzle size and geometry, fuel properties, and air temperature and turbulence. Higher nozzle pressure differentials produce smaller droplets.

Injectors on some small engines with high swirl are designed to spray the fuel jet against the cylinder wall. This speeds the evaporation process, but can be done only in those engines that operate with very hot walls. This is necessary because of the limited real time of each cycle in small engines that operate at high speeds. This practice is not needed and should not be done with large engines operating at slower speeds. These have low swirl and cooler walls, which would not evaporate the fuel efficiently. This would lead to high specific fuel consumption and high HC emissions in the exhaust.

Ignition Delay and Cetane Number

Once the air–fuel mixture is in a combustible air–fuel ratio and the temperature is hot enough for self-ignition, ignition delay will still be in the range of 0.4 to 3 msec (0.0004 to 0.003 sec). An increase in temperature, pressure, engine speed, or compression ratio will decrease ignition delay time. Fuel droplet size, injection velocity, injection rate, and physical characteristics of the fuel seem to have little or no effect on ignition delay time. At higher engine speeds, turbulence is increased, wall temperature is higher, and ignition delay is decreased in real time. However, ID is almost constant in cycle time, which results in a fairly constant crankshaft angle position for the combustion process at all speeds.

If injection occurs too early, ignition delay time will increase because temperature and pressure will be lower. If injection occurs late, the piston will move past TDC, pressure and temperature will decrease, and again ignition delay time will increase. It is important to use fuel with the correct cetane number for a given engine. Cetane number is a measure of ignition delay and must be matched to a given engine cycle and injection process. If the cetane number is low, ignition delay will be too long, and a more-than-desirable amount of fuel will be injected into the cylinder before combustion starts. Then, when combustion does start, a greater amount of fuel will be quickly consumed, and the initial cylinder pressure rise will be greater. This causes a very large initial force to be applied to the piston face and a rough engine cycle. If the cetane number is high, combustion will start too early before TDC, with a resulting loss in engine power.

Normal cetane numbers of commonly used fuels are in the range of 40–60. In this range, ignition delay time is inversely proportional to cetane number.

$$\text{ID} \propto 1/\text{CN} \tag{7-2}$$

Cetane number and octane number also have a strong inverse correlation for most fuels, as shown in Fig. 4-7:

$$\text{CN} \propto 1/\text{ON} \tag{7-3}$$

Cetane number can be changed by blending small amounts of certain additives to the fuel. Additives that accelerate ignition include nitrites, nitrates, organic peroxides, and some sulfur compounds.

Alcohol, with its high octane number, is a poor fuel for CI engines.

Soot

The flame in a CI engine is highly nonuniform. When self-ignition occurs, the flame will quickly engulf all parts of the combustion chamber that have an air–fuel mixture in a combustible ratio. Mixtures with an equivalence ratio in the range of 0.8 to 1.5 will support combustion, so some of the reaction will be in a lean mixture, some will be at or close to stoichiometric, and some will be in a rich mixture. In the combustion zone, where the mixture is rich, there is not enough oxygen to form stoichiometric CO_2. Instead, some carbon monoxide and some solid carbon will be formed in the reaction products, yielding

$$C_xH_y + a\,O_2 + a(3.76)N_2 \rightarrow b\,CO_2 + c\,CO + d\,C(s) + e\,H_2O + a(3.76)N_2$$

where

$$x = b + c + d$$
$$a = b + c/2 + e/2$$
$$e = y/2$$

These solid carbon particles are the black *smoke* seen in the exhaust of large trucks and railroad locomotives.

HISTORIC—HIGH COMPRESSION RATIOS

Experimental military vehicles with CI engines having compression ratios as high as 50:1 have been tested. At these very high compression ratios, just about any liquid that is combustible, theoretically, could be self-ignited and burned in these engines. This would be very advantageous for military vehicles in the field during combat, when fuel supplies may be interrupted and it would be necessary to scrounge fuel from local supplies. Just about anything that burns could be used in an emergency. Very rugged high-pressure fuel injectors would be necessary. Starting a cold engine with such a high compression ratio would be very difficult.

In the very fuel-rich zones where AF is just marginally combustible, very large amounts of solid carbon particles are generated. As the combustion process proceeds and the air–fuel mixture in the combustion chamber is further mixed by swirl and turbulence, most of the carbon particles further react, and only a very small percentage of them eventually reach the surrounding environment. Solid carbon particles are a fuel and react with oxygen when the proper mixture is obtained:

$$C(s) + O_2 \rightarrow CO_2 + \text{heat}$$

In that the overall air–fuel ratio is lean in a CI engine, most of the carbon will find and react with the excess oxygen. Even after the mixture leaves the combustion chamber, additional reactions take place in the exhaust system, further reducing the amount of solid carbon. In addition, most CI engine exhaust systems have a particulate trap that filters out a large percentage of the remaining solid carbon. Only a small percentage of the original solid carbon particles that were generated in the combustion chamber is exhausted to the environment.

To keep exhaust smoke (soot) within tolerable limits. CI engines are operated with an overall lean AF ($\phi < 0.8$). If these engines were to operate with an overall stoichiometric AF, the amount of exhaust smoke would be unacceptable. Even with lean operation, many metropolitan areas are very concerned with *diesel exhaust* from trucks and buses. In many locations, very stringent laws are being imposed on bus and truck operation, and major improvements must be made to reduce exhaust emissions from these vehicles.

A CI engine operates with unthrottled intake air, controlling engine power by the amount of fuel injected. When a truck or railroad locomotive is under heavy load, such as when it accelerates from a stop or going up a hill, more than the normal amount of fuel is injected into the cylinders. This results in a richer mixture that generates a higher amount of solid carbon soot. A large amount of exhaust smoke is very noticeable under these conditions.

Because CI engines operate lean overall, they have a high combustion efficiency—generally around 98%. Of the 2% combustion inefficiency, about half appears as HC emissions in the exhaust. This is in the form of solid carbon and other HC components. Some HC components are absorbed on the carbon particles and carried out in the exhaust. If 0.5% of the carbon in the fuel were exhausted as solid particles, the resulting smoke would be unacceptable. This means that the amount of solid carbon being exhausted must be kept well below this 0.5%.

Because of the 98% fuel conversion efficiency and the high compression ratios, the thermal efficiency of CI engines is high compared with that of Otto cycle SI engines. However, because of their fuel-lean operation, their power output per unit displacement volume is not as good.

Cold-Weather Problems

Starting a cold CI engine can be very difficult. The air and fuel are cold, so fuel evaporation is very slow and ignition delay time is lengthened. Lubrication oil is cold, its viscosity is high, and distribution is limited. The starter motor has to turn the cold engine that is poorly lubricated with very high viscosity oil. This results in a slower-than-normal turnover speed to start the engine. Because of very low engine speed, a greater amount of blowby past the piston occurs, reducing the effective compression. As the starter motor turns the engine to start it, air within the cylinders must compress enough to raise the temperature well above self-ignition temperature. This does not readily happen. The slower-than-normal rotation of the engine, combined with the cold metal cylinder walls, promotes a large heat loss to the walls and keeps the air temperature below that needed to self-ignite the fuel. To overcome this problem, a *glow plug* is used when starting some CI engines. A glow plug is a simple resistance heater connected to a battery with the heated surface located within the combustion chamber of the engine. For a short time, 10–15 seconds, before starting the engine, the glow plug is turned on and the resistor becomes red hot. Now, when the engine is started, combustion in the first few cycles is not ignited by compressive heating, but by surface ignition off the glow plug. After just a few cycles, the cylinder walls and lubricant are warmed enough and more normal operation of the engine is possible. Then, the glow plug is turned off and self-ignition caused by compressive heating occurs. Another cold-weather method used to aid starting on some engines is an electrically heated intake manifold that heats the air entering the cylinders. Due to their larger wall surface area, engines with divided combustion chambers have a greater heat loss problem than those with single open chambers and are generally more difficult to start.

Because of the large amount of energy needed to rotate and start very large CI engines that are cold, using an electric motor powered from a battery is sometimes not practical. Instead, a small internal combustion engine can be used as the starting *motor* for the larger engine. This *pony engine*, usually having two or four cylinders, is first started and then used to turn over the large engine by engaging it to the flywheel of the large engine. The pony engine is then disengaged when the large engine is started.

To aid in cold starting, many medium-size CI engines are built with a higher compression ratio than would otherwise be needed. Some are also given a larger flywheel for this purpose. Preheating the lubricating oil electrically is done on some engines to help the starting process; some systems even distribute the oil throughout the engine before starting by means of an electric oil pump. This oil not only lubricates the engine parts and makes starting easier, but also reduces the high engine wear that occurs at this time. Late injection and a richer air–fuel mixture are also used to aid starting.

It is not an uncommon practice to leave large CI engines running continuously during cold weather to avoid the problem of restarting them. Truck engines are often left running at highway truck stops in northern climates during winter. This is undesirable in that it wastes fuel and adds air pollution to the environment.

Another cold-weather problem encountered with cold CI engines in trucks and automobiles is pumping the fuel from the fuel tank to the engine. Often the fuel tank is located some distance from the engine, and the fuel supply lines run outside the warm engine compartment. The high viscosity of cold fuel oil makes it very difficult to pump it through the long, often small-diameter, fuel lines. Some diesel fuels will even gel in the fuel tank in cold weather. Many vehicles overcome this problem with an electric fuel tank heater and/or by recirculating the fuel through the warm engine compartment. As much as twice the needed fuel is pumped to the engine compartment. The excess fuel, after being warmed in the engine compartment, is recirculated back to the fuel tank, where it mixes with and warms the rest of the fuel. It is often necessary to change to a higher grade fuel oil for winter operation of an automobile. The more costly high-grade fuel has lower viscosity and is more easily pumped. It also works better through the fuel injectors.

7.6 HOMOGENEOUS CHARGE COMPRESSION IGNITION—HCCI

Recent development work has been done on engines, using a new combustion philosophy for CI engines. These engines add fuel with the input air much as an SI engine does. This results in an almost homogeneous air–fuel mixture filling the combustion chamber before combustion. Compression ignition is still used, but the resulting combustion is a combination of diffusion flame combustion and homogeneous mixture combustion. Some fuel is added with intake port injectors, while ignition occurs from normal CI injection. The engine can operate on dual fuel, with the homogeneous charge of fuel filling the combustion chamber being one fuel (e.g., natural gas, methanol) and diesel fuel supplying ignition.

7.7 VARIABLE COMPRESSION RATIO

Combustion characteristics in an engine depend on many parameters, air–fuel ratio and compression ratio being two important ones. During the 1990s, the Saab Motor Company introduced an experimental engine with a unique, very promising design, which allows to change the compression ratio as the engine is running [148, 149, 151, 219]. This five-cylinder, 1.6-liter SI engine has an engine block horizontally split at the center. The upper half of the block, called the *monohead*, contains the cylinders cast as a single unit with the head. The lower half contains the crankcase, crankshaft, and connected pistons. The two halves of the engine are connected on one side with a hinge connection, which allows the upper half to rotate up to 4° relative to the lower half. As the monohead rotates, it changes the size of the clearance volume of the cylinders at TDC and thus changes the compression ratio. With the engine running, the compression ratio can be varied anywhere from 8:1 for heavy loads to 14:1 for cruising at light load. The upper half of the engine is rotated with a large cam and a rubber bellows seals the two halves together.

A small displacement engine is generally overall more efficient than a large displacement engine. At heavy load, a large engine can run efficiently, but at light load the throttle is closed, and this results in a large negative work pumping loop. A small displacement engine will run efficiently at open throttle when less power is needed, but then must be able to supply more power when that is needed. To do this, Saab's small engine is equipped with a high-pressure supercharger having a pressure boost of up to 280 kPa (40 psia). This is a much higher pressure boost than normal, which generally is about 25 to 50 kPa. High pressure boost is essential so that, when needed, high power output matching that of larger engines is possible. A supercharger must be used, as a turbocharger would not be able to supply a pressure boost of this magnitude.

The engine is constructed with a computer-controlled hydraulic actuator and camshaft that can rotate the upper half of the engine up to 4° relative to the stationary lower half. At the highest load requirements, the EMS puts a maximum rotation on the monohead, resulting in a slightly larger clearance volume in each cylinder. This produces a compression ratio of 8:1, which is correct for a heavy load using gasoline that is close to stoichiometric. At light load, such as when cruising on level roadway, much less power is needed and the EMS rotates the monohead back to zero. At this position, the compression ratio is 14:1 and, with a fuel-lean mixture, can supply the needed power efficiently without any knock problems.

The Saab engine produces up to 105 kW/L of brake power and maximum torque of 190 N-m/L. It is claimed that it uses 30% less fuel, which also means 30% less CO_2 produced. It can pass all emission standards of NOx, HC, and CO. The EMS, with input from various sensors, can adjust the compression ratio anywhere between 8:1 and 14:1 to obtain the most efficient and cleanest operation for any condition. Many different fuels can be used, with automatic adjustment being made for air–fuel ratio, octane number, etc.

With the engine head and upper block constructed as one piece, better coolant flow can be obtained, as there is no need for head bolts or gaskets. The small engine also has less mass, and small engines in general have less frictional losses.

Another engine system with the capability of changing compression ratio as the engine is running was introduced early in the 21st century [205]. With this system, there is a pivoted lever arm between the crankshaft and connecting rod of each piston. The lever arm is also connected to the engine block through an actuator. By pivoting the lever arm as the engine is running, the circular rotation of the large end of the connecting rod can be modified, changing the stroke length and thus changing the compression ratio. In addition, this system can modify piston motion and give an elliptical path to the large end of the connecting rod. This causes the piston to slow down immediately after ignition, resulting in combustion being closer to constant volume, the most efficient form of combustion. It is claimed that this system can be used with either four-stroke or two-stroke cycles, SI or CI engines, with or without supercharging, and with variable valve timing.

A third method of changing the compression ratio of a running engine is to use an **Alvar cycle engine** [238]. This is an engine that uses small secondary pistons that reciprocate in secondary chambers which are housed in the cylinder head and connected to the primary combustion chambers as shown in Fig. 7-19. By phasing the motion of the secondary pistons relative to the motion of the primary pistons, the displacement and the compression ratio of the cylinders can be changed. This can be done with a movable idler pulley in the belt drive of the secondary pistons.

Example Problem 7-6

A small airplane has a modified supercharged 2.4–liter SI engine equipped with variable compression ratio. The Otto cycle engine is operated in two different modes, high-load takeoff and low-load cruising, with the following conditions:

	Takeoff	Cruising
Engine speed	3600 RPM	2200 RPM
Compression ratio	8:1	14:1
Volumetric efficiency	120%	88%
Fuel	stoichiometric gasoline	gasoline with AF = 22:1
Combustion efficiency	97%	99%

Calculate:

1. indicated thermal efficiency at takeoff and at cruising
2. rate of fuel flow into engine at takeoff and at cruising
3. indicated power at takeoff and at cruising
4. indicated specific fuel consumption at takeoff and at cruising

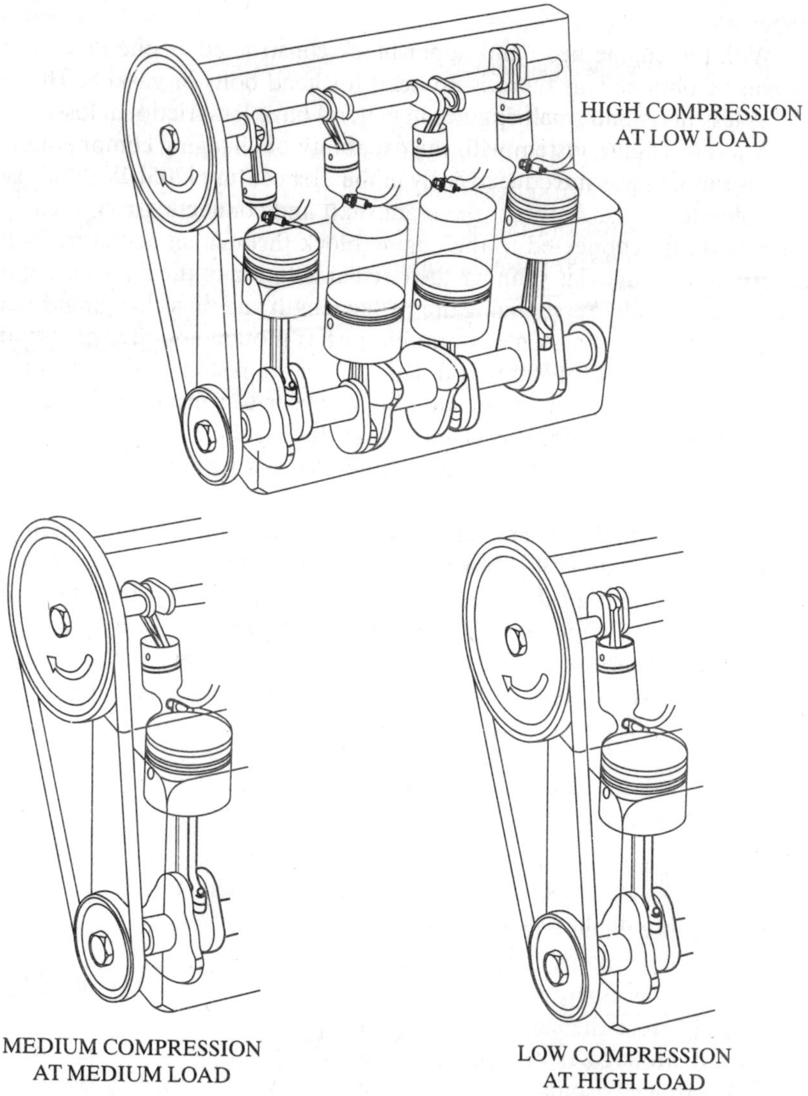

FIGURE 7-19

Alvar cycle engine system with small pistons in secondary combustion chamber of each cylinder. Secondary pistons can be run in phase or out of phase with large primary pistons by changing phase angle of secondary crankshaft relative to primary crankshaft. This changes the clearance volume at TDC of primary pistons, which changes the compression ratio. Reprinted with permission from SAE Paper No. 981027 © 1998 SAE International [238].

(1) Equation (3-31) gives indicated thermal efficiencies:

$$\eta_t = 1 - (1/r_c)^{k-1} = 1 - (1/8)^{1.35-1} = 0.517 = 51.7\% \text{ at takeoff}$$
$$= 1 - (1/14)^{1.35-1} = 0.603 = 60.3\% \text{ at cruising}$$

(2) Mass flow rate of air into the engine is calculated using Eq. (2-71):

$$\dot{m}_a = \eta_v \rho_a V_d N/n$$
$$= (1.20)(1.181\ \text{kg/m}^3)(0.0024\ \text{m}^3/\text{cycle})(3600/60\ \text{rev/sec})/(2\ \text{rev/cycle})$$
$$= 0.1020\ \text{kg/sec at takeoff}$$
$$= (0.88)(1.181\ \text{kg/m}^3)(0.0024\ \text{m}^3/\text{cycle})(2200/60\ \text{rev/sec})/(2\ \text{rev/cycle})$$
$$= 0.0457\ \text{kg/sec at cruising}$$

Equation (2-55) gives mass flow rate of fuel into engine:

$$\dot{m}_f = \dot{m}_a/(\text{AF}) = (0.1020\ \text{kg/sec})/(14.6) = 0.00699\ \text{kg/sec at takeoff}$$
$$= (0.0457\ \text{kg/sec})/(22) = 0.00208\ \text{kg/sec at cruising}$$

(3) Equation (2-65) gives indicated power:

$$\dot{W}_i = \eta_t \dot{m}_f Q_{HV} \eta_c = (0.517)(0.00699\ \text{kg/sec})(43{,}000\ \text{kJ/kg})(0.97)$$
$$= 151\ \text{kW at takeoff}$$
$$= (0.603)(0.00208\ \text{kg/sec})(43{,}000\ \text{kJ/kg})(0.99) = 53\ \text{kW at cruising}$$

(4) Equation (2-61) gives indicated specific fuel consumption:

$$\text{isfc} = \dot{m}_f/\dot{W}_i$$
$$= [(0.00699\ \text{kg/sec})(3600\ \text{sec/hr})(1000\ \text{gm/kg})]/(151\ \text{kW}) = 167\ \text{gm/kW-hr at takeoff}$$
$$= [(0.00208\ \text{kg/sec})(3600\ \text{sec/hr})(1000\ \text{gm/kg})]/(53\ \text{kW}) = 141\ \text{gm/kW-hr at cruising}$$

7.8 SUMMARY

Combustion in an SI engine consists of ignition and flame development, flame propagation, and flame termination. Combustion is initiated with a spark plug that ignites the air–fuel mixture in the immediate vicinity of the spark plug electrodes. There is essentially no pressure rise or work done at first, and 5–10% of the air–fuel mixture is consumed before the combustion process is fully developed. When the flame is fully developed it, propagates very rapidly across the combustion chamber, accelerated and spread by turbulence and mass motions within the cylinder. This raises the temperature and pressure in the cylinder, and the piston is forced down in the power stroke. By the time the flame front reaches the corners of the combustion chamber, only a small percentage of the air–fuel mixture remains, and the flame is terminated by heat transfer and viscous drag with the wall.

Combustion in engines with divided combustion chambers is a two-step process: fairly normal ignition and flame development in the secondary chamber, and main flame propagation in the primary chamber ignited by a flame jet through the separating orifice. Often the air–fuel ratio is fuel rich in the secondary chamber and lean in the primary main chamber.

Combustion in a CI engine starts with injection of fuel late in the compression stroke. The liquid fuel droplets atomize, evaporate, mix with air, and then, after an ignition delay time, self-ignite simultaneously at many sites. The flame then consumes all fuel that is in a combustible state and continues to do this as injection continues. Combustion terminates when the last fuel droplets have reacted after evaporating and mixing with air to form a combustible mixture.

PROBLEMS

7.1 An SI engine operating at 1200 RPM has a 10.2-cm bore with the spark plug offset by 6 mm from center. The spark plug is fired at 20° bTDC. It takes 6.5° of engine rotation for combustion to develop and get into flame propagation mode, where the average flame speed is 15.8 m/sec.

Calculate:

(a) Time of one combustion process after flame has developed (i.e., time for flame front to reach the furthest cylinder wall). [sec]

(b) Crank angle position at the end of combustion.

7.2 It is desired that flame termination be at the same crank angle position when the speed of the engine in Problem 7-1 is increased to 2000 RPM. In this range, flame development takes the same amount of real time and flame speed is related to engine speed as $V_f \propto 0.92\, N$.

Calculate:

(a) Flame speed at 2000 RPM. [m/sec]

(b) Crank angle position when the spark plug should be fired.

(c) Crank angle position when flame propagation starts.

7.3 A CI engine with a 3.2-inch bore and 3.9-inch stroke operates at 1850 RPM. In each cycle fuel injection starts at 16° bTDC and lasts for 0.0019 second. Combustion starts at 8° bTDC. Due to the higher temperature, the ignition delay of any fuel injected after combustion starts is reduced by a factor of two from the original ID.

Calculate:

(a) ID of first fuel injected. [sec]

(b) ID of first fuel injected in degrees of engine rotation.

(c) Crank angle position when combustion starts on last fuel droplets injected.

7.4 A 3.2-liter SI engine is to be designed with bowl-in-piston combustion chambers as shown in Fig. 6-5(b). With a central spark plug and combustion at TDC, this gives a flame travel distance of $B/4$. The engine is to operate with an average piston speed of 8 m/sec and a burn angle of 25° of engine crank rotation. Stroke and bore will be related by $S = 0.95\, B$.

Calculate:

(a) Average flame speed if the design is for an in-line four-cylinder engine. [m/sec]

(b) Average flame speed if the design is for a V8 engine. [m/sec]

7.5 A large CI engine operating at 310 RPM has open combustion chambers and direct injection, with 26-cm bores, a 73-cm stroke, and a compression ratio of 16.5:1. Fuel injection in each cylinder starts at 21° bTDC and lasts for 0.019 second. ID is 0.0065 second.

Calculate:

(a) ID in degrees of engine rotation.

(b) Crank angle position when combustion starts.

(c) Crank angle position when injection stops.

7.6 Some Ford Thunderbird V8 engines have two spark plugs per cylinder. If everything else is kept the same, list three advantages and three disadvantages this gives for modern engine operation and design.

7.7 The divided combustion chambers of a two-liter, four-cylinder, lean-burn, stratified charge SI engine have 22% of clearance volume in the secondary chamber of each cylinder. The intake system delivers an air–gasoline mixture into the secondary chamber at an equivalence ratio of 1.2 and into the main chamber at an equivalence ratio of 0.75.

Calculate:

(a) Overall air–fuel ratio.

(b) Overall equivalence ratio.

7.8 The engine in Problem 7-7 has a volumetric efficiency of 92%, an overall combustion efficiency of 99%, an indicated thermal efficiency of 52%, and a mechanical efficiency of 86% when operating at 3500 RPM.

Calculate:

(a) Brake power at this condition. [kW]

(b) bmep. [kPa]

(c) Amount of unburned fuel exhausted from the engine [kg/hr]

(d) bsfc. [gm/kW-hr]

7.9 A two-liter, four-cylinder, open-chamber SI engine operates at 3500 RPM using stoichiometric gasoline. At this speed, volumetric efficiency is 93%, combustion efficiency is 98%, indicated thermal efficiency is 47%, and mechanical efficiency is 86%.

Calculate:

(a) Brake power. [kW]

(b) bmep. [kPa]

(c) Amount of unburned fuel exhausted from the engine. [kg/hr]

(d) bsfc. [gm/kW-hr]

Compare these results with those of Problem 7-8.

7.10 The spark plug in the cylinder of an SI engine is fired at 20° bTDC when the engine is running at 2400 RPM. Flame propagation mode starts at 10° bTDC and last for 0.001667 seconds.

Calculate:

(a) Time of ignition and flame development in degrees of engine rotation. [°]

(b) Time of ignition and flame development in real time. [sec]

(c) Crank angle when flame propagation mode ends. [° aTDC]

7.11 A large low-compression SI engine has the spark plug mounted at the center of a 24-cm bore (i.e., flame travel distance during flame propagation = 12 cm). At 1200 RPM, the spark plug is fired at 19° bTDC. Flame speed is proportional to engine speed as $V_f \propto 0.80\ N$,

while ignition and flame development takes 0.00125 seconds at all speeds. It can be assumed that all combustion takes place during flame propagation. The engine has a 35-cm stroke, a compression ratio of 8.2, and a connecting–rod length of 74 cm.
Calculate:

(a) Crank angle when flame propagation starts. [° bTDC]
(b) Crank angle when flame propagation ends, with flame speed V_f = 48 m/sec at 1200 RPM. [° aTDC]
(c) Crank angle when spark plug should be fired at 2400 RPM so that flame propagation will end at same crank angle as at 1200 RPM. [° bTDC]
(d) Piston speed when combustion ends at 1200 RPM. [m/sec]
(e) Volume in combustion chamber when combustion ends. [m^3]

7.12 A large truck has a supercharged V10 SI engine with a displacement of 380 in.^{3}, which is equipped with variable compression ratio. The Otto cycle engine is operated in two different modes, high load when starting up or going up a hill and low load when cruising on level roadway. The operating conditions are as follows:

	High Load	Low Load
Engine speed	3200 RPM	2100 RPM
Compression ratio	8.4:1	13.7:1
Volumetric efficiency	120%	78%
Gasoline fuel	AF = 13.5:1	AF = 22:1
Combustion efficiency	94%	99%

Calculate:

(a) Indicated thermal efficiency at high load and at low load. [%]
(b) Rate of fuel flow into engine at high load and at low load. [%]
(c) Indicated power at high load and at low load. [hp]
(d) The isfc at high load and at low load. [lbm/hp-hr]

DESIGN PROBLEMS

7.1D It is desired to build an engine with a low height for a high-speed sports car with a very low profile. Design the intake manifold and combustion chamber for a modern fast-burn valve-in-block engine. The engine must have high turbulence, swirl, squish, and tumble, with a short flame travel distance.

7.2D Design a fuel delivery system for a flexible-fuel automobile engine. The engine should be able to use any mixture combination of gasoline, ethanol, and methanol. Tell how engine variables will change for various fuel combinations (e.g., ignition timing, fuel injection, etc.). State all assumptions you make.

CHAPTER 8

Exhaust Flow

"The man who really and thoroughly understands the working of the petroleum gas engine must be a well-equipped mechanical engineer and something of an eletrician as well. The old dictum that 'knowledge is power' was never better exemplified than in the handling of a hydrocarbon motor."

Touring in Automobiles

by *Henry R. Sutphen* (1901)

After combustion is completed and the resulting high-pressure gases have been used to transfer work to the crankshaft during the expansion stroke, these gases must be removed from the cylinder to make room for the air–fuel charge of the next cycle. The exhaust process that does this occurs in two steps, exhaust blowdown followed by the exhaust stroke. The resulting flow out the exhaust pipe is a non-steady-state pulsing flow that is often modeled as pseudo-steady-state.

8.1 BLOWDOWN

Exhaust blowdown occurs when the exhaust valve starts to open towards the end of the power stroke, somewhere around 60° to 40° bBDC. At this time, pressure in the cylinder is still at about 4–5 atmospheres and the temperature is upwards of 1000 K. Pressure in the exhaust system is about one atmosphere, and when the valve is opened the resulting pressure differential causes a rapid flow of exhaust gases from the cylinder through the valve into the exhaust system (i.e., **exhaust blowdown**).

Flow at first will be choked, and the outflow velocity will be sonic. This occurs when the ratio of pressures across an orifice is greater than or equal to

$$(P_1/P_2) = [(k + 1)/2]^{k/(k-1)} \tag{8-1}$$

where

P_1 = upstream pressure
P_2 = downstream pressure
k = ratio of specific heats

This ratio is equal to about 2 for most gases. $P_1/P_2 = 1.86$ for air with $k = 1.35$. Sonic velocity is equal to

$$c = \sqrt{kRT} \tag{8-2}$$

where

R = gas constant
T = temperature

It is quite high at this point because of the high temperature of the gases in the cylinder.

As the gas flows from the cylinder into the exhaust system, it experiences a pressure drop and a corresponding temperature drop due to expansion cooling. A model often used to calculate the temperature in the exhaust system is the ideal–gas isentropic expansion relationship between temperature and pressure, namely,

$$T_{ex} = T_{EVO}(P_{ex}/P_{EVO})^{(k-1)/k} \tag{8-3}$$

where

T_{ex}, P_{ex} = exhaust temperature and pressure
T_{EVO}, P_{EVO} = cylinder temperature and pressure when exhaust valve opens

Although the gases are not truly ideal and the blowdown process is not isentropic due to heat losses, irreversibility, and choked flow, Eq. (8-3) gives a fairly good approximation to gas temperature entering the exhaust system.

In addition, the first gas leaving the cylinder will have a high velocity and a correspondingly high kinetic energy. This high kinetic energy will quickly be dissipated in the exhaust system, and the kinetic energy will be changed to additional enthalpy, raising the temperature above T_{ex} of Eq. (8-3). As the pressure in the cylinder decreases during the blowdown process, the gas leaving will have progressively lower velocity and kinetic energy. The first gas leaving the cylinder will thus have the highest temperature in the exhaust system, with any following gas having a lower temperature. The last elements of exhaust leaving the cylinder during blowdown will have very little velocity and kinetic energy and will be at a temperature about equal to T_{ex} in Eq. (8-3). If a turbocharger is located close to the engine near the exhaust valves, the kinetic energy gained in blowdown can be utilized in the turbine of the turbocharger. Heat transfer also contributes to the final pseudo-steady-state temperature found in the exhaust system.

In an ideal air-standard Otto cycle or Diesel cycle, the exhaust valve opens at BDC and blowdown occurs instantaneously at constant volume (process 4–5 in Fig. 8-1). This does not happen in a real engine, where blowdown takes a finite length of time. So that

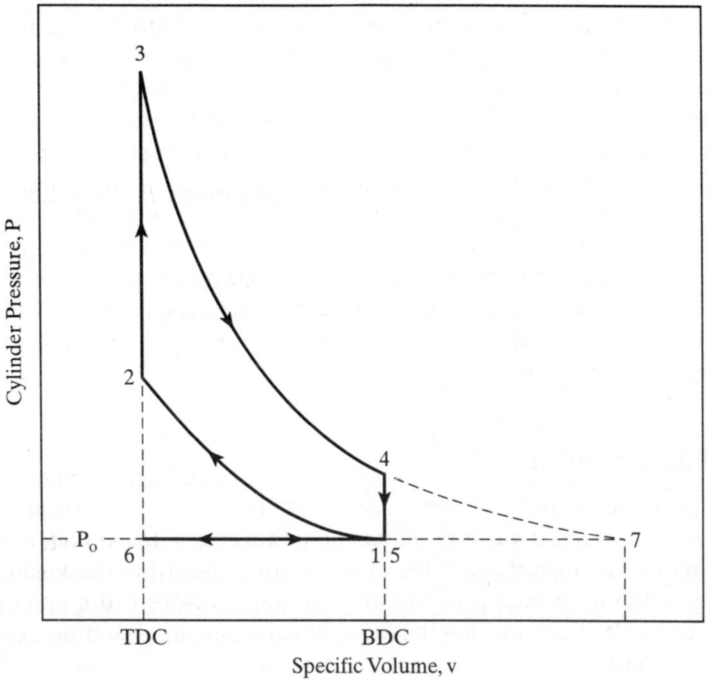

FIGURE 8-1

Air-standard Otto cycle on *P–v* coordinates showing exhaust gas after blowdown at hypothetical state 7.

the pressure in the cylinder has been fully reduced by BDC when the exhaust stroke starts, the exhaust valve starts to open somewhere around 60° to 40° bBDC. When this happens, the pressure is quickly reduced, and what would have been additional useful work is lost during the last part of the expansion stroke. Because of the finite time required, the exhaust valve is not fully open until BDC or slightly before. The timing when the exhaust valve is opened (in most engines using a camshaft) is critical. If the valve opens too early, more than the necessary amount of work is lost in the latter stages of the power stroke. If it opens late, there is still excess pressure in the cylinder at BDC. This pressure resists the piston movement early in the exhaust stroke and increases the negative pumping work of the engine cycle.

The ideal time to open the exhaust valve depends on engine speed. The finite real time of blowdown is fairly constant, mainly because of the choked flow condition that occurs at the start (i.e., sonic velocity is the same regardless of engine speed). The lobe on a camshaft can be designed to open the valve at one given crankshaft angle, which can be picked to be optimum at one engine speed. Once this compromise speed has been decided in the design and manufacture of the camshaft, all other engine speeds will have less-than-optimum timing for the opening of the exhaust valve. At higher speeds the valve will be opening late, and at lower speeds the valve will be opening early. Many automobile engines with variable valve control can lessen this problem somewhat, depending on the sophistication of the control system.

The exhaust valve should be as large as possible, considering all other demands in the design of the combustion chamber. A larger valve gives a greater flow area and reduces the time of blowdown. This allows for a later exhaust valve opening and a longer expansion stroke with less lost work. Many modern engines have two exhaust valves per cylinder, with the flow area of the two smaller valves being greater than the flow area of one larger valve. This gives added design flexibility (and complexity) for fitting the exhaust valves into the existing combustion chamber space.

Some industrial and other constant-speed engines can be designed with valve timing optimized for that speed. Vehicle engines that do not have variable control can be designed for the most-used condition (e.g., cruising speed for trucks and airplanes, red line maximum speed for a drag racer). Slow-speed engines can have very late exhaust valve opening.

8.2 EXHAUST STROKE

After exhaust blowdown, the piston passes BDC and starts towards TDC in the exhaust stroke. The exhaust valve remains open. Pressure in the cylinder resisting the piston in this motion is slightly above the atmospheric pressure of the exhaust system. The difference between cylinder pressure and exhaust pressure is the small pressure differential caused by the flow through the exhaust valves as the piston pushes the gases out of the cylinder. The exhaust valve is the greatest source of flow restriction in the entire exhaust system and is the location of the only appreciable pressure drop during the exhaust stroke.

The exhaust stroke can best be approximated by a constant-pressure process, with gas properties remaining constant at the conditions of point 7 in Fig. 8-1. Pressure remains about constant, slightly above atmospheric, with temperature and density constant at values consistent with Eq. (8-3).

Ideally, at the end of the exhaust stroke when the piston reaches TDC, all the exhaust gases have been removed from the cylinder and the exhaust valve closes. One reason this does not actually happen is the finite time it takes to close the exhaust valve. The lobe on the camshaft is designed to give a smooth closing of the valve and to keep wear at a minimum. One cost of doing this is a slightly longer time required to close the valve. To have the valve totally closed at TDC requires the closing process to start at least 20° bTDC. This is unacceptable in that the valve would be partially closed during the last segment of the exhaust stroke. Closing can start only at or very close to TDC, which means that total closing doesn't occur until 8°–50° aTDC.

When the exhaust valve is finally closed, there is still a residual of exhaust gases trapped in the clearance volume of the cylinder. The higher the compression ratio of the engine, the less clearance volume exists to trap this exhaust residual.

The valve problem is compounded by the fact that the intake valve should be totally open at TDC when the intake stroke starts. Because of the finite time required to open this valve, it must start to open 10°–25° bTDC. There is, therefore, a period of 15°–50° of engine rotation when both intake and exhaust valves are open. This is called valve overlap.

During valve overlap, there can be some reverse flow of exhaust gas back into the intake system. When the intake process starts, this exhaust is drawn back into the cylinder along with the air–fuel charge. This results in a larger exhaust residual during the rest of the cycle. This backflow of exhaust gases is a greater problem at low engine speeds, being worst at idle conditions. At most low engine speeds, the intake throttle is at least partially closed, creating low pressure in the intake manifold. This creates a greater pressure differential, forcing exhaust gas back into the intake manifold. Cylinder pressure is about one atmosphere, while intake pressure can be quite low. In addition, real time of valve overlap is greater at low engine speed, allowing more backflow. Some engines are designed to use this small backflow of hot exhaust gas to help vaporize the fuel that has been injected directly behind the intake valve face.

Some engines have a one-way reed valve at the exhaust port to keep exhaust gas from flowing back from the exhaust manifold into the cylinder and intake system during valve overlap.

Engines equipped with turbochargers or superchargers often will have intake pressures above one atmosphere and are not subject to exhaust backflow.

Another negative result of valve overlap is that some intake air–fuel mixture can short-circuit through the cylinder when both valves are open, with some fuel ending up as pollution in the exhaust system.

Variable valve timing, which is becoming common in automobile engines, decreases the problem of valve overlap. At low engine speeds, the exhaust valve can be closed earlier and the intake valve can be opened later, resulting in less overlap.

If the exhaust valve is closed too early, an excess of exhaust gases is trapped in the cylinder. Also, cylinder pressure will go up near the end of the exhaust stroke, causing loss of net work from the engine cycle. If the exhaust valve is closed late, there is an excess of overlap, with more backflow of exhaust gas into the intake.

Figure 8-2 shows the flow of gases through the exhaust valve out of the cylinder. When the valve is first opened, blowdown occurs with a very high flow rate due to the large pressure differential. Choked flow will occur (sonic velocity) at first, limiting the maximum flow rate. By the time the piston reaches BDC, blowdown is complete, and flow out of the exhaust valve is now controlled by the piston during the exhaust stroke. The piston reaches maximum speed about halfway through the exhaust stroke, and this is reflected in the rate of exhaust flow. Towards the end of the exhaust stroke, near TDC, the intake valve opens and valve overlap is experienced. Depending on engine operating conditions, a momentary reverse flow of exhaust gas back into the cylinder can occur at this point.

Example Problem 8-1

A 6.4-liter V8 engine with a compression ratio of 9:1 operates on an air-standard cycle and has the exhaust process shown in Fig. 8-1. Maximum cycle temperature and pressure are 2550 K and 11,000 kPa when operating at 3600 RPM. The exhaust valve effectively opens at 52° bBDC. Calculate:

1. time of exhaust blowdown
2. percent of exhaust gas that exits the cylinder during blowdown
3. exit velocity at the start of blowdown, assuming choked flow occurs

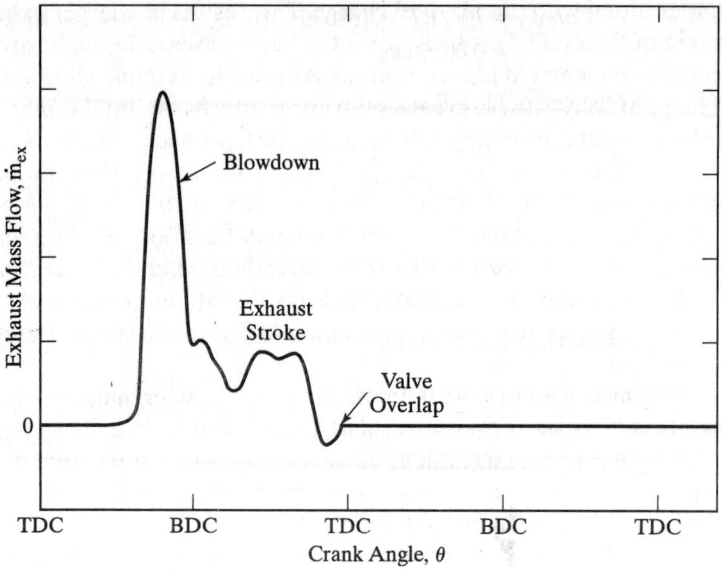

FIGURE 8-2

Exhaust gas flow out of cylinder through the exhaust valve(s), showing blowdown and exhaust stroke. Possible reverse flow back into the cylinder may occur during valve overlap. Adapted from [28].

(1) Blowdown will occur between 52° bBDC and BDC. This is 52/360 = 0.1444 of a revolution.

$$\text{Time of Blowdown} = (0.1444\ \text{rev})/(3600/60\ \text{rev/sec}) = \underline{0.0024\ \text{sec}}$$

(2) For one cylinder,

$$V_d = (6.4\ \text{L})/8 = 0.80\ \text{L} = 0.0008\ \text{m}^3$$

Use Eq. (2-12) to find clearance volume:

$$r_c = V_{BDC}/V_{TDC} = (V_d + V_c)/V_c = 9 = (0.0008 + V_c)/V_c$$
$$V_c = 0.0001\ \text{m}^3$$

Use Eq. (2-14) to find volume when the exhaust valve opens (let $R = 4$):

$$V_{EVO}/V_c = 1 + \tfrac{1}{2}(r_c - 1)[R + 1 - \cos\theta - \sqrt{R^2 - \sin^2\theta}]$$
$$= 1 + \tfrac{1}{2}(9 - 1)[(4) + (1) - \cos(128°) - \sqrt{(4)^2 - \sin^2(128°)}]$$
$$= 7.78$$
$$V_{EVO} = 7.78\ V_c = (7.78)(0.0001\ \text{m}^3) = 0.000778\ \text{m}^3$$

Equations (3-16) and (3-17) give the temperature and pressure when the exhaust valve opens:

$$T_{EVO} = T_3(V_3/V_{EVO})^{k-1} = (2550\ \text{K})(0.0001/0.000778)^{0.35} = 1244\ \text{K}$$
$$P_{EVO} = P_3(V_3/V_{EVO})^{k} = (11{,}000\ \text{kPa})(0.0001/0.000778)^{1.35} = 690\ \text{kPa}$$

$$m_{\text{EVO}} = PV/RT = (690 \text{ kPa})(0.000778 \text{ m}^3)/(0.287 \text{ kJ/kg-K})(1244 \text{ K}) = 0.00150 \text{ kg}$$

At the end of blowdown, gases are at hypothetical state 7, but volume is V_4 or V_1:

$$P_7 = P_o = 101 \text{ kPa}$$
$$T_7 = T_3(P_7/P_3)^{(k-1)/k} = (2550 \text{ K})(101/11{,}000)^{(1.35-1)/1.35} = 756 \text{ K}$$
$$V_{\text{BDC}} = V_4 = V_1 = (V_c + V_d) = (0.0001 + 0.0008) = 0.0009 \text{ m}^3$$
$$m_7 = PV/RT = (101 \text{ kPa})(0.0009 \text{ m}^3)/(0.287 \text{ kJ/kg-K})(756 \text{ K}) = 0.00042 \text{ kg}$$
$$(\Delta m)_{\text{blowdown}} = m_{\text{EVO}} - m_7 = 0.00150 - 0.00042 = 0.00108 \text{ kg}$$

This is [(0.00108)/(0.00150)](100) $=$ 72.0% of total mass.

Note: If Otto cycle analysis is used and the exhaust valve is assumed to open at point 4, then using Eqs. (3-16) and (3-17) gives

$$T_4 = T_3(1/r_c)^{k-1} = (2550 \text{ K})(1/9)^{0.35} = 1182 \text{ K}$$
$$P_4 = P_3(1/r_c)^k = (11{,}000 \text{ kPa})(1/9)^{1.35} = 566 \text{ kPa}$$
$$m_4 = PV/RT = (566 \text{ kPa})(0.0009 \text{ m}^3)/(0.287 \text{ kJ/kg-K})(1182 \text{ K}) = 0.00150 \text{ kg}$$

This is the same result as the one obtained previously. Calculation of the amount of mass in the cylinder would be the same using any point along process line 3–4. This means that the same percentage of the exhaust flow must occur during blowdown regardless of when the exhaust valve is opened. Size of the exhaust valve(s) and the corresponding mass flow rate through the valve(s) then dictate when the valve should be opened.

(3) If flow is choked at the start of blowdown, velocity will be sonic. Using Eq. (8-2) yields,

$$V_{\text{EVO}} = c = \sqrt{kRT_{\text{EVO}}} = \sqrt{(1.35)(287 \text{ J/kg-K})(1244 \text{ K})} = 694 \text{ m/sec}$$

8.3 EXHAUST VALVES

Exhaust valves are made smaller than intake valves, although the same amount of mass must flow through each. The pressure differential across the intake valves of a naturally aspirated engine is less than one atmosphere, while the pressure differential across the exhaust valves during blowdown can be as high as three or four atmospheres. In addition, if and when choked flow is occuring, sonic velocity through the exhaust valve is higher than sonic velocity through the intake valve. This can be seen in Eq. (8-2), with the exhaust gas being much hotter than the intake air–fuel mixture. Using Eq. (5-4) to size valves, we have for intake

$$A_i = (\text{constant}) B^2 (\overline{U}_p)_{\max}/c_i \tag{8-4}$$

where

A_i = area of inlet valve(s)

$(\overline{U}_p)_{\max}$ = average piston speed at maximum engine speed

c_i = sonic velocity at inlet temperature
B = bore

Using this same equation to size exhaust valves yields

$$A_{ex} = (\text{constant})B^2(\overline{U}_p)_{max}/c_{ex} \tag{8-5}$$

where c_{ex} is sonic velocity at exhaust temperature.

In multivalve engines, A_i and A_{ex} are the total valve areas of the intake valves of one cylinder and the exhaust valves of one cylinder, respectively. If Eq. (8-5) is divided by Eq. (8-4), the ratio of the valve areas is obtained. Everything cancels except the speed of sound, and by using Eq. (8-2), the ratio of the exhaust valve area to the intake valve area can be approximated as

$$\alpha = A_{ex}/A_i = c_i/c_{ex} = \sqrt{kRT_i}/\sqrt{kRT_{ex}} = \sqrt{T_i/T_{ex}} \tag{8-6}$$

In actual engines, α usually has a value of about 0.8 to 0.9. To find the valve diameters, we use the relationship

$$A/x = (\pi/4)d_v^2 \tag{8-7}$$

where

d_v = valve diameter
x = number of intake valves or number of exhaust valves

Valves can be sized by making them as large as possible on the basis of Eq. (8-6). There will probably not be enough room in a modern combustion chamber to make them large enough to totally satisfy Eqs. (8-4) and (8-5). To reduce noise, some engines with multiple exhaust valves have only one valve activated at low speeds.

8.4 EXHAUST TEMPERATURE

Both temperature and mass flow rate vary greatly with time in the exhaust system of an engine. Cyclic variations will occur at engine cycle times (e.g., 0.04 second at 3000 RPM). A temperature sensor, such as a thermocouple, with a time constant much greater than this will give a pseudo-steady-state temperature of the flow. This thermocouple temperature will be approximately an enthalpy average temperature and not necessarily a true time average

$$T_{\text{thermocouple}} = [\int \dot{m}c_pT\,dt]/[\int \dot{m}c_p\,dt] \tag{8-8}$$

where

$\dot{m}$ = mass flow rate of exhaust
t = time

c_p = specific heat
T = temperature

Temperature of the gases in the exhaust system of a typical SI engine will average 400°C to 600°C. This drops to about 300°C to 400°C at idle conditions and goes up to about 900°C at maximum power. This is about 200°C to 300°C cooler than the exhaust gases in the cylinder when the exhaust valve opens. The difference is because of expansion cooling. All temperatures will be affected by the equivalence ratio of the original combustion mixture.

The average temperature in the exhaust system of a typical CI engine will be 200°–500°C. This is lower than SI engine exhaust because of the larger expansion cooling that occurs due to the higher compression ratios of CI engines. If the maximum temperature in a CI engine is about the same as in an SI engine, the temperature when the exhaust valve opens can be several hundred degrees less. The overall lean equivalence ratio of a CI engine also lowers all cycle temperatures from combustion on.

Exhaust temperature of an engine will go up with higher engine speed or load, with spark retardation, or with an increase in equivalence ratio. Things that are affected by exhaust temperature include turbochargers, catalytic converters, and particulate traps.

Example Problem 8-2

When the exhaust valve opens and blowdown occurs, the first elements of flow have high sonic velocity and high kinetic energy. The high velocity is quickly dissipated in the exhaust manifold, where flow velocity is relatively low. The kinetic energy of the gas is converted to additional enthalpy with an increase in temperature. Calculate the theoretical maximum temperature in the exhaust flow of Example Problem 8-1.

Applying conservation of energy, we have

$$\Delta KE = V^2/2g_c = \Delta h = c_p \Delta T$$

Using values from Example Problem 8-1, we obtain

$$\Delta T = V^2/2g_c c_p = (694\ \text{m/sec})^2/[2(1\ \text{kg-m/N-sec}^2)(1.108\ \text{kJ/kg-K})] = 217°$$

$$T_{max} = T_{ex} + \Delta T = 756 + 217 = \underline{973\ \text{K} = 700°\text{C}}$$

Actual maximum temperature would be less than this, due to heat losses and other irreversibilities. Only a very small percentage of the exhaust flow would have maximum kinetic energy and would reach maximum temperature. The time-averaged temperature of the exhaust would be more consistent with T_7 in Fig. 8-1 and Example Problem 8-1.

8.5 EXHAUST MANIFOLD

After leaving the cylinders by passing out of the exhaust valves, exhaust gases pass through the exhaust manifold, a piping system that directs the flow into one or more exhaust pipes. Exhaust manifolds are usually made of cast iron and are sometimes designed to have close thermal contact with the intake manifold. This is to provide heating and vaporization in the intake manifold.

Chemical reactions are still occurring in the exhaust flow as it enters the exhaust manifold, with carbon monoxide and fuel components reacting with unreacted oxygen. These reactions are greatly reduced because of heat losses and the lower temperature after blowdown. Some modern engines have insulated exhaust manifolds that are designed to operate at much hotter temperatures and act as a **thermal converter** to reduce unwanted emissions in the exhaust gas. Some of these are equipped with electronically controlled air intake to provide additional oxygen for reaction. This is discussed in the next chapter.

Modern smart engines have a number of sensors in the exhaust manifold to provide input to engine controls. These can be some combination of thermal, chemical, electrical, and/or mechanical in nature and can supply information about levels of O_2, HC, NOx, CO, CO_2, particulates, temperature, and knock. This information is then used by the engine management system (EMS) to adjust engine parameters such as AF, injection timing, ignition timing, and EGR rate.

From the exhaust manifold, the gases flow through an exhaust pipe to the emission control system of the engine, which generally consists of thermal and/or catalytic converters. One argument says these converters should be as close to the engine as space allows to minimize heat losses. On the other hand, this setup creates high–temperature problems in the engine compartment. These converters promote reduction of emissions in the exhaust gases by additional chemical reaction. They are discussed in the next chapter.

Tuning of Exhaust Manifold

As with intake manifolds, the runner lengths of the exhaust manifold can be tuned to give an assist to the gas flow. Because the flow is a pulsed flow, pressure waves are set up in the manifold runners. When a wave reaches the end of a passage or a restriction, a reflected wave is generated that travels back in the opposite direction. When the reflected wave is in phase with the primary wave, the pulses reinforce and there is a slight increase in total pressure. At those points where the waves are out of phase, they cancel each other and there is a slight decrease in total pressure. An exhaust manifold runner is tuned when the reflected wave is out of phase with the primary wave at the exit of the exhaust valve. This causes a slight decrease in pressure at that point, increasing the pressure differential across the valve and giving a small increase in flow. The pressure pulse wavelength is determined by the frequency, so a runner length can be designed to give a tuned exhaust at only one engine speed. Exhaust manifolds can therefore be effectively tuned on engines that run at one engine speed. Race cars that often run at a constant WOT speed and need all possible power can very successfully use exhaust tuning. Trucks and airplanes can have tuned exhaust systems for cruising conditions. It is very difficult to have a tuned exhaust system on a standard automobile engine that operates over a large speed range. Often, space limitations in the engine compartment are the dominant factor that dictates the exhaust manifold design, and effective tuning is not possible.

On the other hand, with a number of state-of-the-art engines, exhaust tuning is considered in the engine design. On some high-performance engines, variable tuning is used, with runner length dynamically adjusted as engine speed changes. Other engines use dual runners of different lengths, automatically switching the exhaust flow to the runner that is best tuned for the current speed.

HISTORIC—EXHAUST REDUCTION

To reduce exhaust and emissions, and to save fuel, some automobiles have been designed to automatically turn the engine off when the vehicle stops, such as at a stoplight. A light touch on the accelerator pedal restarts the engine when the driver desires to move on. This method of operation will become more efficient when 42-volt electrical systems become standard and the starter is integrated with the engine flywheel. Some manufacturers have developed engine drive systems that shift into neutral gear when the engine is at idle. By thus reducing the speed and load of the engine, fuel use and exhaust emissions are reduced. When engine speed is increased, the system automatically shifts back into drive gear.

8.6 TURBOCHARGERS

In turbocharged engines, exhaust gases leaving the exhaust manifold enter the turbine of the turbocharger, which drives the compressor that compresses the incoming air. Pressure of the exhaust gas entering the turbine is only slightly higher than atmospheric, and only a very small pressure drop is possible through the turbine. In addition, this non-steady-state pulsed flow varies widely in kinetic energy and enthalpy, due to the velocity and temperature differences that occur during blow down and the following exhaust stroke. A pseudo-steady-state flow is assumed, with

$$\dot{W}_t = \dot{m}(h_{in} - h_{out}) = \dot{m}c_p(T_{in} - T_{out}) \tag{8-9}$$

where

$\dot{W}_t$ = time-averaged turbine power
$\dot{m}$ = time-averaged exhaust mass flow rate
h = specific enthalpy
c_p = specific heat
T = temperature

Because of the limited pressure drop through the turbine, it must operate at speeds upward of 100,000 RPM to generate enough power to drive the compressor. These high speeds, along with the high-temperature corrosive gases within which the turbine operates, create major mechanical and lubrication design challenges.

Turbochargers should be mounted as close as possible to the cylinder exhaust ports so that turbine inlet pressure, temperature, and kinetic energy can be as high as possible.

One problem associated with turbocharging is the slow response time experienced when the throttle is opened quickly. It takes several engine cycles before the increased exhaust flow can accelerate the turbine rotor and give the desired pressure boost to the inlet air–fuel mixture. To minimize this **turbo lag**, lightweight ceramic rotors with small rotational moments of inertia that can be accelerated more quickly are used. Ceramic is also an ideal material because of the high temperatures.

Many turbochargers have a bypass that allows exhaust gas to be routed around the turbine. This is to keep intake flow from being overcompressed when engine operating conditions are less demanding. The amount of gas passing through the turbine is controlled according to engine needs.

Some experimental exhaust turbines have been used to drive small high-speed generators instead of intake compressors. The electrical energy output from these systems can be utilized in various ways, such as driving the engine cooling fan.

8.7 EXHAUST GAS RECYCLE—EGR

Modern automobile engines use exhaust gas recycle (EGR) to reduce nitrogen oxide emissions. Some gas is routed from the exhaust system back into the intake system. This dilutes the intake gas mixture with noncombustibles, which then lowers the maximum combustion temperature and consequently reduces the generation of nitrogen oxides. The amount of EGR can be as high as 15–20% of the total mass and is regulated according to engine operating conditions. Under some conditions, such as starting or WOT, no EGR is used. In addition to reducing the maximum combustion temperature, EGR increases the intake mixture temperature and affects fuel evaporation.

8.8 TAILPIPE AND MUFFLER

After exiting the catalytic converter, exhaust gases flow through a **tailpipe** that ducts the flow away from the passenger compartment of the vehicle and vents it to the surroundings. This is usually under and out the back (or side) of an automobile and often upward behind the cab of large tracks.

Somewhere in the tailpipe section there is usually a larger flow chamber called the **muffler**. This is a sound chamber designed to reduce the operating noise of the engine, most of which is carried out with the exhaust flow. Mufflers use two general methods of sound reduction. One method absorbs the energy of sound pulses by flow through a porous medium. Other mufflers reduce sound by the cancellation of waves. Instead of fully dampening all engine noise, some mufflers are designed to give a louder, sporty sound. Mufflers and tailpipes are generally made of steel, with stainless steel or titanium used on some more expensive vehicles.

Some automobiles with air-cooled engines, such as the early models of VW Beetles, use hot exhaust gas for heating the passenger compartment in cold weather. The exhaust flow is ducted through one side of a heat exchanger, while passenger compartment air is circulated through the other side. This works fine when all equipment is in good condition. However, as automobiles age, many components suffer from oxidation, rust, and leakage. Any leakage in the heat exchanger allowing exhaust gas into the circulating passenger air would be very dangerous.

8.9 TWO-STROKE CYCLE ENGINES

The exhaust process of a two-stroke cycle engine differs from that of a four-stroke cycle engine in that there is no exhaust stroke. Blowdown is the same, occurring when the exhaust valve opens or when the exhaust slot is uncovered near the end of the power stroke. This is immediately followed with an intake process of compressed air or air–fuel mixture. As the air enters the cylinder at a pressure usually between 1.2 and 1.8 atmospheres, it pushes the remaining lower pressure exhaust gas out the still-open exhaust port in a scavenging process. There is some mixing of intake and exhaust, with some exhaust residual staying in the cylinder and some intake gas passing into the exhaust system. For those engines which use direct fuel injectors (CI engines and larger modern SI engines) and have only air in the intake system, this intake gas that gets into the exhaust system during valve overlap does not contribute to emission problems. However, it does reduce engine volumetric efficiency and/or trapping efficiency. For those engines which intake an air–fuel mixture, any intake gas that gets into the exhaust system adds to hydrocarbon pollution and reduces fuel economy. Some two-stroke cycle engines have a one-way reed valve at the exhaust port to stop gases from flowing back into the cylinder from the exhaust system.

Example Problem 8-3

The experimental two-stroke cycle automobile engine of Example Problem 5-8 burns stoichiometric gasoline mixed with oil at a ratio of 15 to 1. The engine has a combustion efficiency of 98% for gasoline, but only 82% for the oil. Peak temperature and pressure in the cycle are 2550°C and 9610 kPa.

Calculate:

1. total mass flow rate of exhaust
2. mass flow rate of unburned gasoline in the exhaust
3. mass flow rate of unburned oil in the exhaust
4. approximate exhaust temperature

From Example Problem 5-8, we have the following:

$$\text{engine speed } N = 3700 \text{ RPM}$$
$$\text{displacement volume } V_d = 0.0004095 \text{ m}^3 \text{ (one cylinder)}$$
$$\text{delivery ratio } \lambda_{dr} = 0.880$$
$$\text{charging efficiency } \lambda_{ce} = 0.641$$
$$\text{trapping efficiency } \lambda_{te} = 0.728$$

(1) Use a time-rate form of Eq. (5-22) to find mass flow rate of air into engine:

$$\dot{m}_a = \lambda_{dr} V_d \rho_a N/n$$
$$= [(0.880)(0.0004095 \text{ m}^3\text{/cycle})(1.181 \text{ kg/m}^3)(3700/60 \text{ rev/sec})/(1 \text{ rev/cycle})] \times (6 \text{ cylinders})$$
$$= 0.1575 \text{ kg/sec}$$

Equation (2-55) gives mass flow rate of gasoline into engine:

$$\dot{m}_f = \dot{m}_a/\text{AF} = (0.1575 \text{ kg/sec})/(14.6) = 0.0108 \text{ kg/sec}$$

Mass flow rate of oil into engine is

$$\dot{m}_o = \dot{m}_f/15 = (0.0108)/(15) = 0.00072 \text{ kg/sec}$$

Total mass flow rate of exhaust equals total mass flow rate in:

$$\dot{m}_{ex} = \dot{m}_{in} = \dot{m}_a + \dot{m}_f + \dot{m}_o = (0.1575) + (0.0108) + (0.00072)$$
$$= \underline{0.1690 \text{ kg/sec} = 608.5 \text{ kg/hr} = 0.373 \text{ lbm/sec}}$$

(2) Of the mass flow rates in, 72.8% gets trapped in the cylinder and 27.2% passes through the cylinder during valve overlap.
Mass flow rate of fuel not trapped is

$$(\dot{m}_f)_{nt} = (0.0108 \text{ kg/sec})(0.272) = 0.0029 \text{ kg/sec}$$

Mass flow rate of trapped fuel that does not get burned is

$$(\dot{m}_f)_{nb} = (\dot{m}_f)_{in}\lambda_{te}(1 - \eta_c) = (0.0108 \text{ kg/sec})(0.728)(1 - 0.98) = 0.00016 \text{ kg/sec}$$

Total mass flow rate of unburned fuel in exhaust is

$$(\dot{m}_f)_{ex} = (\dot{m}_f)_{nt} + (\dot{m}_f)_{nb} = (0.0029) + (0.00016)$$
$$= \underline{0.0031 \text{ kg/sec} = 11.0 \text{ kg/hr} = 0.0068 \text{ lbm/sec}}$$

(3) Mass flow rate of oil not trapped

$$(\dot{m}_o)_{nt} = (0.00072 \text{ kg/sec})(0.272) = 0.00020 \text{ kg/sec}$$

Mass flow rate of trapped oil that does not get burned is

$$(\dot{m}_o)_{nb} = (\dot{m}_o)_{in}\lambda_{te}(1 - \eta_c)$$
$$= (0.00072 \text{ kg/sec})(0.728)(1 - 0.82) = 0.000094 \text{ kg/sec}$$

Total mass flow rate of unburned oil in exhaust is

$$(\dot{m}_o)_{ex} = (\dot{m}_o)_{nt} + (\dot{m}_o)_{nb} = (0.00020) + (0.000094)$$
$$= \underline{0.000294 \text{ kg/sec} = 1.06 \text{ kg/hr} = 0.00065 \text{ lbm/sec}}$$

(4) After combustion, the cylinder gas pressure is reduced during the power stroke and then is further reduced during blowdown to a final pressure of one atmosphere. Using Eq. (3-37) and Fig. 3-17, we have

$$T_{ex} = T_{max}(P_o/P_{max})^{(k-1)/k} = (2823 \text{ K})(101/9610)^{(1.35-1)/1.35}$$
$$= \underline{867 \text{ K} = 594°\text{C} = 1101°\text{F} = 1561°\text{R}}$$

8.10 SUMMARY AND CONCLUSIONS

The exhaust process of a four-stroke cycle IC engine is a two-step process: blowdown and exhaust stroke. Blowdown occurs when the exhaust valve opens late in the expansion stroke and the remaining high pressure in the cylinder forces the exhaust gases through the open valve into the exhaust manifold. Because of the large pressure differential across the valve, sonic velocity occurs and the flow is choked. As the exhaust gas experiences blowdown, the temperature decreases, due to expansion cooling. The high kinetic energy of the gas during blowdown is dissipated quickly in the exhaust manifold, and there is a momentary rise in the temperature again from the resulting increase in specific enthalpy. The exhaust valve must open soon enough for blowdown to be complete when the piston reaches BDC. At this point, the cylinder is still filled with exhaust gas at about atmospheric pressure, and most of this is now expelled during the exhaust stroke.

Two-stroke cycle engines experience exhaust blowdown, but have no exhaust stroke. Most of the gas that fills the cylinder after blowdown is expelled by a scavenging process when inlet air enters at elevated pressure.

To reduce the generation of nitrogen oxides, many engines have exhaust gas recycling, with some of the exhaust flow ducted back into the intake system. Those engines equipped with turbochargers use the exhaust flow to drive the turbine, which in turn drives the inlet compressor.

PROBLEMS

8.1 A six-cylinder SI engine, with a compression ratio of $r_c = 8.5$, operates on an air-standard Otto cycle at WOT. Cylinder temperature and pressure when the exhaust valve opens are 1000 K and 520 kPa. Exhaust pressure is 100 kPa, and air temperature in the intake manifold is 35°C.

Calculate:

(a) Exhaust temperature during exhaust stroke. [°C]
(b) Exhaust residual. [%]
(c) Cylinder temperature at the start of compression. [°C]
(d) Peak temperature of cycle. [°C]
(e) Cylinder temperature when the intake valve opens. [°C]

8.2 A four-cylinder SI engine, with a compression ratio of $r_c = 9$, operates on an air-standard Otto cycle at part throttle. Conditions in the cylinders when the exhaust valve opens are 70 psia and 2760°F. Exhaust pressure is 14.6 psia, and conditions in the intake manifold are 8.8 psia and 135°F.

Calculate:

(a) Exhaust temperature during exhaust stroke. [°F]
(b) Exhaust residual. [%]
(c) Cylinder temperature and pressure at the start of compression stroke. [°F, psia]

8.3 A three-cylinder, two-stroke cycle SI automobile engine, operating at 3600 RPM, has peak cycle operating conditions of 2900°C and 9000 kPa. Cylinder temperature when the exhaust port opens is 1275°C.

Calculate:

(a) Cylinder pressure when the exhaust port opens. [kPa]
(b) Maximum flow velocity through the exhaust port. [m/sec]

8.4 An SI Otto cycle engine has a compression ratio of $r_c = 8.5$, and a CI diesel cycle engine has a compression ratio of $r_c = 20.5$. Both engine cycles have a maximum temperature of 2400 K and maximum pressure of 9800 kPa. The diesel engine has a cutoff ratio $\beta = 1.95$. Calculate the cylinder temperature when the exhaust valve opens on each engine. [°C]

8.5 Give two reasons that exhaust valves are smaller than intake valves.

8.6 A 1.8-liter, three-cylinder SI engine produces brake power of 42 kW at 4500 RPM, with a compression ratio r_c = 10.1:1 and bore and stroke related by $S = 0.85\ B$. Connecting–rod length is 16.4 cm. Maximum temperature in the cycle is 2700 K, and maximum pressure is 8200 kPa. Exhaust pressure is 98 kPa. The exhaust valve effectively opens at 56° bBDC.

Calculate:

(a) Time of exhaust blowdown. [sec]
(b) Percent of exhaust gas that exits cylinder during blowdown. [%]
(c) Exit velocity at the start of blowdown, assuming choked flow occurs. [m/sec]

8.7 Exhaust manifold pressure of the engine in Problem 8-6 is 98 kPa. In the manifold, the high kinetic energy of the exhaust flow during blowdown is quickly dissipated and converted to an increase in specific enthalpy.

Calculate:

(a) Pseudo-steady-state exhaust temperature in the exhaust stroke. [°C]
(b) Theoretical peak temperature experienced in the exhaust flow. [°C]

8.8 A four-cylinder, 2.5-liter, four-stroke cycle SI engine with a compression ratio r_c = 9.6 operates at 3200 RPM. Peak cycle temperature is 2227°C, peak cycle pressure is 6800 kPa, and exhaust pressure is 101 kPa. The engine operates at part throttle with inlet air at 60°C and 75 kPa. An exhaust residual remains in the cylinders at the end of the exhaust stroke. In addition, 12% EGR at exhaust temperature and pressure is diverted back into the intake manifold, where it mixes with the inlet air before the intake valve.

Calculate:

(a) Exhaust temperature during exhaust stroke. [°C]
(b) Exhaust residual before EGR is added. [%]
(c) Cylinder temperature at the start of compression stroke. [°C]
(d) Theoretical design ratio of exhaust valve diameter to intake valve diameter.

8.9 A CI engine with a compression ratio of 18:1 operates on an air-standard diesel cycle at 3200 RPM. Maximum temperature and pressure in the cycle are 2527°C and 6500 kPa, and exhaust pressure is 100 kPa. Cylinder pressure is 530 kPa when the exhaust valve opens at 48° bBDC.

Calculate:

(a) Time of one blowdown process. [sec]
(b) Average temperature of exhaust during exhaust stroke. [°C]
(c) Velocity of exhaust gas through the exhaust valve at start of blowdown. [m/sec]

8.10 An SI engine operates on an Otto cycle at 3800 RPM, with maximum cycle temperature and pressure of 3100 K and 7846 kPa. The engine has a compression ratio of 9.8:1, and immediately after combustion there is 0.000622 kg of exhaust gas in each cylinder. During exhaust blowdown, the average mass flow rate through the single exhaust valve is 0.218 kg/sec. At the end of blowdown, cylinder pressure is reduced to exhaust pressure of 101 kPa.

Calculate:

(a) Temperature in cylinder at end of exhaust blowdown. [K]
(b) Mass of exhaust gas in cylinder at end of blowdown. [kg]
(c) Crank angle when exhaust valve opens. [° bBDC]

8.11 A 5.6-liter V8 engine, with a compression ratio of $r_c = 9.4:1$, operates on an air-standard Otto cycle at 2800 RPM, with a volumetric efficiency $\eta_v = 90\%$ and a stoichiometric air–fuel ratio using gasoline. The exhaust gas flow undergoes a temperature drop of 44°C as it passes through the turbine of the turbocharger.
Calculate:

(a) Mass flow rate of exhaust gas. [kg/sec]
(b) Power available to drive the turbocharger compressor. [kW]

8.12 A turbocharged, three-cylinder, four-stroke cycle, 1.5-liter, multipoint port-injected SI engine using stoichiometric gasoline operates at 2400 RPM with a volumetric efficiency of 108%. The turbocharger has a turbine isentropic efficiency of 80% and a compressor isentropic efficiency of 78%. Exhaust flow enters the turbine at 770 K and 119 kPa and exits at 98 kPa. Air enters the compressor at 27°C and 96 kPa and exits at 120 kPa.
Calculate:

(a) Mass flow rate through the turbocharger compressor. [kg/sec]
(b) Mass flow rate through the turbocharger turbine. [kg/sec]
(c) Inlet air temperature at turbocharger compressor exit. [°C]
(d) Exhaust temperature at turbocharger turbine exit. [°C]

DESIGN PROBLEMS

8.1D Design a variable valve-timing system to be used on an in-line four-cylinder SI engine.

8.2D Design a turbine–generator system that can be driven by the exhaust flow of a four-cylinder SI farm tractor engine. The output of the generator can be used to power the engine cooling fan and other accessories.

8.3D Design exhaust valves for the engine in Problem 5-13. Decide on the number of valves, valve diameter, valve lift, and valve timing. Flow though the valves must be such that blowdown occurs before BDC at high engine speed. Address the problem of fitting valves into the combustion chamber.

CHAPTER 9

Emissions and Air Pollution

"I began work on exhaust analysis in about 1898, and I have been following it through pretty thoroughly ever since."

paper on carburetor testing

by *E. R. Hewitt* (1913)

This chapter explores the undesirable emissions generated in the combustion process of automobile and other IC engines. These emissions pollute the environment and contribute to global warming, acid rain, smog, odors, and respiratory and other health problems. The major causes of these emissions are nonstoichiometric combustion, dissociation of nitrogen, and impurities in the fuel and air. The emissions of concern are hydrocarbons (HC), carbon monoxide (CO), oxides of nitrogen (NOx), sulfur, and solid carbon particulates (part). Ideally, engines and fuels could be developed such that very few harmful emissions are generated, and these could be exhausted to the surroundings without a major impact on the environment. With present technology, this is not possible, and aftertreatment of the exhaust gases to reduce emissions is very important. This consists mainly of the use of thermal or catalytic converters and particulate traps.

9.1 AIR POLLUTION

Until the middle of the 20th century, the number of IC engines in the world was small enough that the pollution they emitted was tolerable, and the environment, with the help of sunlight, stayed relatively clean. As world population grew, power plants, factories, and an ever-increasing number of automobiles began to pollute the air to the extent that it was no longer acceptable. During the 1940s, air pollution was first recognized as a problem in the Los Angeles basin in California. Two causes of this were the large population density and the natural weather conditions of the area. The large population created many factories and power plants, as well as one of the largest automobile densities in the world. Smoke and other pollutants from the many factories and automobiles combined with the fog that was common in this ocean area, and *smog* resulted. During the 1950s, the smog problem increased along with the population density and automobile density. It was recognized that the automobile was one of the major contributors to the problem, and by the 1960s emission standards were beginning to be enforced in California. During the next decades, emission standards were adopted in the rest of the United States and in Europe and Japan. By making engines more fuel efficient, and with the use of exhaust aftertreatment, emissions per vehicle of HC, CO, and NOx were reduced by about 95% during the 1970s and 1980s. Lead, one of the major air pollutants, was phased out as a fuel additive during the 1980s. More fuel-efficient engines were developed, and by the 1990s the average automobile consumed less than half the fuel used in 1970. However, during this time, the number of automobiles greatly increased, resulting in no overall decrease in fuel usage. In 1999, petroleum consumption in the United States amounted to 16,500 L/sec (4350 gal/sec), a large percentage of which was fuel for internal combustion engines [214].

Additional reduction will be difficult and costly. As world population grows, emission standards become more stringent out of necessity. The strictest laws are generally initiated in California, with the rest of the United States and the world following. Although air pollution is a global problem, some regions of the world still have no emission standards or laws.

Table 9-1 gives some of the emission standards established by the Environmental Protection Agency (EPA). Tier 1 consists of Standards that were phased in during the 1990s and that apply in the early years of the 21st century. Tier 2 consists of the Standards

TABLE 9-1 EPA Standards - Tier 1

	50,000 miles/5 years					100,000 miles/10 years				
	NMHC	CO	NOx (gasoline)	NOx (diesel)	PM	NMHC	CO	NOx (gasoline)	NOx (diesel)	PM
Passenger Cars	0.25	3.4	0.4	1.0	0.08	0.31	4.2	0.6	1.25	0.10
Light-Light-Duty Trucks	0.25	3.4	0.4	1.0	0.08	0.31	4.2	0.6	1.25	0.10
Heavy-Light-Duty Trucks	0.32	4.4	0.7		0.08	0.40	5.5	0.97	0.97	0.10

EPA Standards - Tier 2

	50,000 miles					120,000 miles				
	NMOG	CO	NOx (all fuels)	PM	HCHO	NMOG	CO	NOx (all fuels)	PM	HCHO
All Light Vehicles	0.075–0.125	3.4	0.05–0.4	—	0.015	0.01–0.156	4.2	0.02–0.9	0.01–0.12	0.004–0.032

CO = carbon monoxide
HCHO = formaldehyde
NOx = nitrogen oxides
NMHC = nonmethane hydrocarbons
NMOG = nonmethane organic gases
PM = particulate matter

All numbers have units of gm/mile.

from Refs. [171, 175]

phased in during 2004 to 2009. During the Tier 1 period, light vehicles are divided into (1) passenger cars, (2) light-light-duty trucks, and (3) heavy-light-duty trucks, depending on weight. There are also separate standards for NOx levels for gasoline-fueled vehicles and diesel oil–fueled vehicles. In Tier 2 Standards, there is only one light–vehicle classification, and all fuels (gasoline, diesel oil, or others) have the same NOx requirements. Tier 2 Standards have several levels of compliance under which certain vehicles can be certified. During the phase-in period the entire fleet average of any manufacturer must meet a NOx standard of 0.30 gm/mile. After the phase-in period, the fleet average must then meet the NOx standard of 0.07 gm/mile. In that the standards are given in units of gm/mile, they become more difficult to comply with as the size of the vehicle and engine increases. In the United States, all states accept these Standards except California. California has its own Standards, which are generally more stringent then those of the EPA or any other country's Standards. One-tenth of new vehicle sales in the United States occurs in California, constituting about 100 billion dollars per year early in the 21st century. Because of this, automobile manufacturers must give major consideration to California emission Standards in designing their engine and vehicle products [171, 175].

9.2 HYDROCARBONS (HC)

Exhaust gases leaving the combustion chamber of an SI engine contain up to 6000 ppm of hydrocarbon components, the equivalent of 1–1.5% of the fuel. About 40% of this is unburned gasoline fuel components. The other 60% consists of partially reacted components that were not present in the original fuel. These consist of small nonequilibrium molecules that are formed when large fuel molecules break up (thermal cracking) during the combustion reaction. It is often convenient to treat these molecules as if they contained one carbon atom, as CH_1.

The makeup of HC emissions will be different for each gasoline blend, depending on the original fuel components. Combustion chamber geometry and engine operating parameters also influence the HC component spectrum.

When hydrocarbon emissions get into the atmosphere, they act as irritants and odorants; some are carcinogenic. All components except CH_4 react with atmospheric gases to form photochemical smog.

Causes of HC Emissions

Nonstoichiometric Air–Fuel Ratio. Figure 9-1 shows that HC emission levels are a strong function of AF. With a fuel-rich mixture, there is not enough oxygen to react with all the carbon, resulting in high levels of HC and CO in the exhaust products. This is particularly true in engine start-up, when the air–fuel mixture is purposely made very rich. It is also true to a lesser extent during rapid acceleration under load. If AF is too lean, poorer combustion occurs, again resulting in HC emissions. The extreme of poor combustion for a cycle is total misfire. This occurs more often as AF is made more lean. One misfire out of 1000 cycles gives exhaust emissions of 1 gm/kg of fuel used.

Incomplete Combustion. Even when the fuel and air entering an engine are at the ideal stoichiometric mixture, perfect combustion does not occur and some HC ends

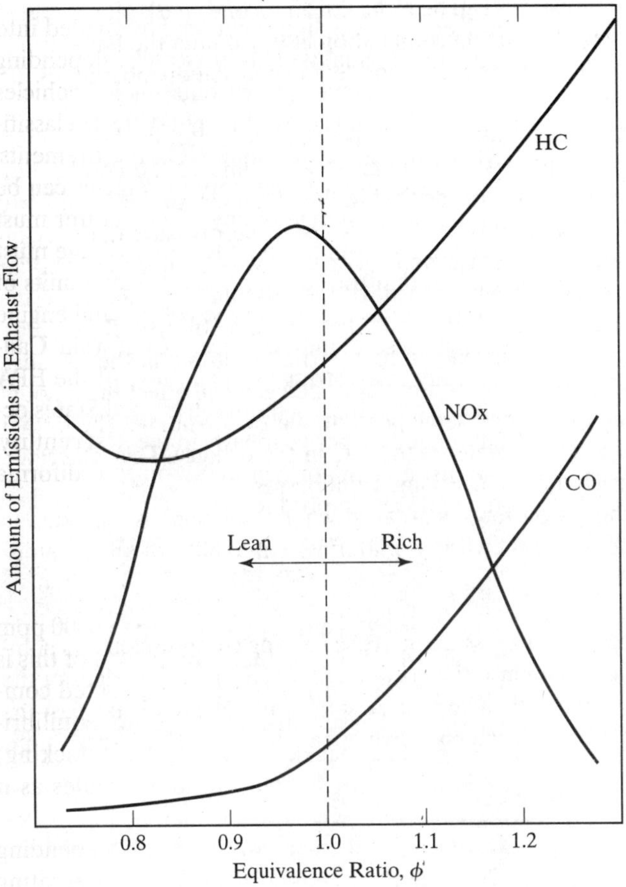

FIGURE 9-1

Emissions from an SI engine as a function of equivalence ratio. A fuel–rich air–fuel ratio does not have enough oxygen to react with all the carbon and hydrogen, and both HC and CO emissions increase. HC emissions also increase at very lean mixtures due to poor combustion and misfires. The generation of nitrogen oxide emissions is a function of the combustion temperature, being greatest near stoichiometric conditions when temperatures are the highest. Peak NOx emissions occur at slightly lean conditions, where the combustion temperature is high and there is an excess of oxygen to react with the nitrogen. Adapted from [58].

up in the exhaust. There are several causes of this. Incomplete mixing of the air and fuel results in some fuel particles not finding oxygen to react with. Flame quenching at the walls leaves a small volume of unreacted air-and-fuel mixture. The thickness of this unburned layer is on the order of tenths of a mm. Some of this mixture, near the wall that does not originally get burned as the flame front passes, will burn later in the combustion process as additional mixing occurs due to swirl and turbulence.

Another cause of flame quenching is the expansion that occurs during combustion and power stroke. As the piston moves away from TDC, expansion of the gases lowers both temperature and pressure within the cylinder. This slows combustion and finally quenches the flame somewhere late in the power stroke. This leaves some fuel particles unreacted.

High exhaust residual causes poor combustion and a greater likelihood of expansion quenching. This is experienced at low load and idle conditions. High levels of EGR will also cause this.

It has been found that HC emissions can be reduced if a second spark plug is added to an engine combustion chamber. By starting combustion at two points, the flame travel distance and total reaction time are both reduced, and less expansion quenching results.

Crevice Volumes. During the compression stroke and early part of the combustion process, air and fuel are compressed into the crevice volume of the combustion chamber at high pressure. As much as 3% of the fuel in the chamber can be forced into this crevice volume. Later in the cycle during the expansion stroke, pressure in the cylinder is reduced below crevice volume pressure, and reverse blowby occurs. Fuel and air flow back into the combustion chamber, where most of the mixture is consumed in the flame reaction. However, by the time the last elements of reverse blowby flow occur, flame reaction has been quenched and unreacted fuel particles remain in the exhaust. Location of the spark plug relative to the top compression ring gap will affect the amount of HC in engine exhaust, the ring gap being a large percent of crevice volume. The farther the spark plug is from the ring gap, the greater is the HC in the exhaust. This is because more fuel will be forced into the gap before the flame front passes.

Crevice volume around the piston rings is greatest when the engine is cold, due to the differences in thermal expansion of the various materials. Up to 80% of all HC emissions can come from this source.

Leak Past the Exhaust Valve. As pressure increases during compression and combustion, some air–fuel mixture is forced into the crevice volume around the edges of the exhaust valve and between the valve and valve seat. A small amount even leaks past the valve into the exhaust manifold. When the exhaust valve opens, the air–fuel mixture that is

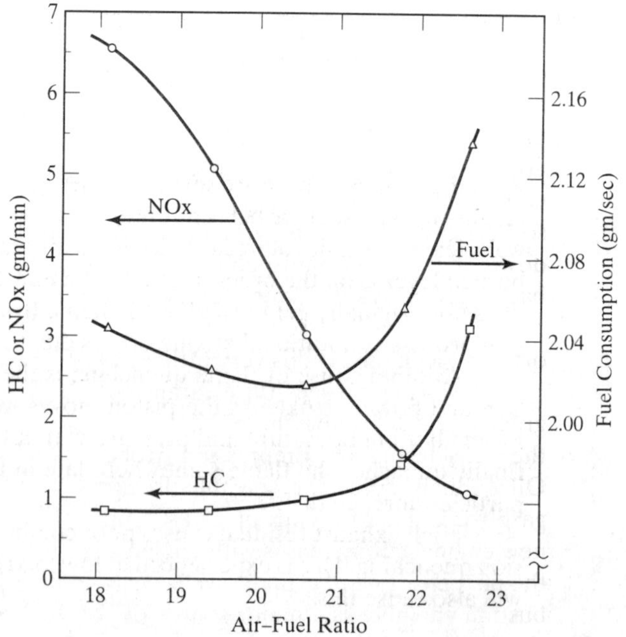

FIGURE 9-2

Fuel consumption, and HC and NOx emissions, as a function of air–fuel ratio for a 5.7-liter, eight-cylinder SI engine running with maximum torque (MBT) at 1400 RPM, using a homogeneous air–fuel mixture. Reprinted with permission from SAE Paper No. 760288 © 1976 SAE International [180].

still in this crevice volume gets carried into the exhaust manifold, and there is a momentary peak in HC concentration at the start of blowdown.

Valve Overlap. During valve overlap, both the exhaust and intake valves are open, creating a path where air–fuel intake can flow directly into the exhaust. A well-designed engine minimizes this flow, but a small amount can occur. The worst condition for this is at idle and low speed, when real time of overlap is greatest.

Deposits on Combustion Chamber Walls. Gas particles, including those of fuel vapor, are absorbed by the deposits on the walls of the combustion chamber. The amount of absorption is a function of gas pressure, so the maximum occurs during compression and combustion. Later in the cycle, when the exhaust valve opens and cylinder pressure is reduced, absorption capacity of the deposit material is lowered and gas particles are desorbed back into the cylinder. These particles, including some HC, are then expelled from the cylinder during the exhaust stroke. This problem is greater in engines with higher compression ratios, due to the higher pressure these engines generate. More gas absorption occurs as pressure goes up. Clean combustion chamber walls with minimum deposits will reduce HC emissions in the exhaust. Most gasoline blends include additives to reduce deposit buildup in engines.

Older engines will typically have a greater amount of wall deposit buildup and a corresponding increase of HC emissions. This is due both to age and to the fact that less swirl was generally found in earlier engine design. High swirl helps to keep wall deposits to a minimum. When lead was eliminated as a gasoline additive, HC emissions from wall deposits became more severe. When leaded gasoline is burned, the lead treats the metal wall surfaces, making them harder and less porous to gas absorption.

Oil on Combustion Chamber Walls. A very thin layer of oil is deposited on the cylinder walls of an engine to provide lubrication between them and the moving piston. During the intake and compression strokes, the incoming air and fuel come in contact with this oil film. In much the same way as wall deposits build up, this oil film absorbs and desorbs gas particles, depending on gas pressure. During compression and combustion, when cylinder pressure is high, gas particles, including fuel vapor, are absorbed into the oil film. When pressure is later reduced during expansion and blowdown, the absorption capability of the oil is reduced and fuel particles are desorbed back into the cylinder. Some of this fuel ends up in the exhaust.

Propane is not soluble in oil, so in propane-fueled engines the absorption–desorption mechanism adds very little to HC emissions.

As an engine ages, the clearance between piston rings and cylinder walls becomes greater, and a thicker film of oil is left on the walls. Some of this oil film is scraped off the walls during the compression stroke and ends up being burned during combustion. Oil is a high-molecular-weight hydrocarbon compound that does not burn as readily as gasoline. Some of it ends up as HC emissions. This happens at a very slow rate with a new engine, but increases with engine age and wear. Oil consumption also increases as the piston rings and cylinder walls wear. In older engines, oil being burned in the combustion chamber is a major source of HC emissions. Figure 9-3 shows how HC emissions go up as oil consumption increases.

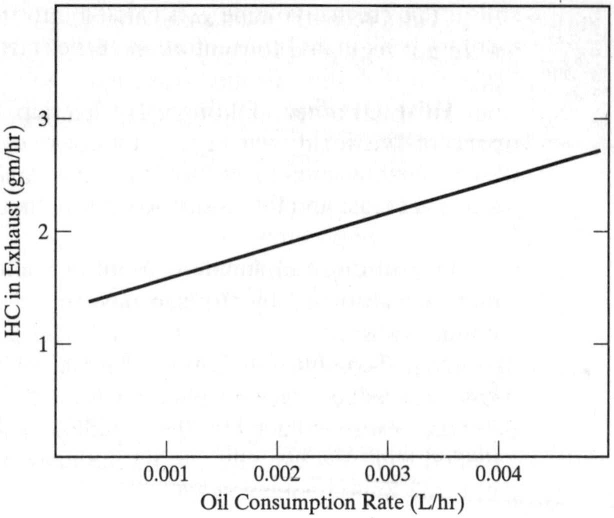

FIGURE 9-3

HC exhaust emissions as a function of engine oil consumption. Often, as an engine ages, clearance between the pistons and cylinder walls increases due to wear. This increases oil consumption and contributes to an increase in HC emissions in three ways: There is added crevice volume, there is added absorption–desorption of fuel in the thicker oil film on cylinder walls, and there is more oil burned in the combustion process. Adapted from [138].

In addition to oil consumption going up as piston rings wear, blowby and reverse blowby also increase. The increase in HC emissions is, therefore, due to both combustion of oil and the added crevice volume flow.

Two-Stroke Cycle Engines. Older two-stroke cycle SI engines and many modern small two-stroke cycle SI engines add HC emissions to the exhaust during the scavenging process. The air–fuel intake mixture is used to push exhaust residual out the open exhaust port. When this is done, some of the air and fuel mixes with the exhaust gases and is expelled out of the cylinder before the exhaust port closes. This can be a major source of HC in the exhaust and is one of the major reasons that there have been no modern two-stroke cycle automobile engines. They could not meet antipollution requirements. Some experimental automobile two-stroke cycle engines and just about all small engines use crankcase compression, and this is a second source of hydrocarbon emissions. The crankcase area and pistons of these engines are lubricated by adding oil to the inlet air–fuel mixture. The oil is vaporized with the fuel and lubricates the surfaces that come in contact with the air–fuel–oil mixture. Some of the oil vapor is carried into the combustion chamber and burned with the air–fuel mixture. Lubricating oil is composed mostly of hydrocarbon components and acts like additional fuel. However, due to the high molecular weight of its components, oil does not fully combust as readily as fuel, and this adds to HC emissions in the exhaust.

Modern experimental two-stroke cycle automobile engines do not add fuel to the intake air, but scavenge the cylinders with pure air, avoiding the placement of HC into the exhaust. After the exhaust port closes, fuel is added by fuel injection directly into the cylinder. This creates a need for very fast and efficient vaporization and mixing of the air and fuel, but it eliminates a major source of HC emissions. Some automobile engines use superchargers instead of crankcase compression, and this eliminates HC pollution from that source.

Until recently, most small engines, such as those used in lawn mowers and boats, were not regulated for pollution control. Many of these engines are still being manufactured with uncontrolled scavenging and oil vapor lubrication, contributing to serious HC (and other) pollution. This problem is starting to be addressed, and in some parts of the world (starting in California) emission laws and standards are starting to be applied to lawn mowers, boats, and other small engines. This will probably phase out, or at least greatly reduce, the number of small two-stroke cycle engines. Low cost is a major requirement for small engines, and fuel injection systems are much more costly than the very simple carburetors found on older engines. Many small engines now operate on a cleaner four-stroke cycle, but still use a less costly carburetor for fuel input.

In the 1990s, there were an estimated 83 million lawn mowers in the United States, producing as much air pollution as 3.5 million automobiles. Government studies of equipment using small engines give the following pollution comparison between that equipment and automobiles (the numbers represent one hour of operation in terms of miles traveled in an average automobile):

Riding mower—20 miles
Garden tiller—30 miles
Lawn mower—50 miles
String trimmer—70 miles
Chain saw—200 miles
Forklift—250 miles
Agricultural tractor—500 miles
Outboard motor—800 miles

CI Engines. Because they operate with an overall fuel-lean equivalence ratio, CI engines have only about one-fifth the HC emissions of SI engines.

The components in diesel fuel have higher molecular weights on average than those in a gasoline blend, and this results in higher boiling and condensing temperatures. This allows some HC particles to condense onto the surface of the solid carbon soot that is generated during combustion. Most of this is burned as mixing continues and the combustion process proceeds. Only a small percentage of the original carbon soot that is formed is exhausted out of the cylinder. The HC components condensed on the surface of the carbon particles, in addition to the solid carbon particles themselves, contribute to the HC emissions of the engine.

In general, a CI engine has about a 98% combustion efficiency, with only about 2% of the HC fuel being emissions (Fig. 4-1). Some local spots in the combustion chamber will be too lean to combust properly, and other spots will be too rich, with not enough oxygen to consume all the fuel. Less than total combustion can be caused by undermixing or overmixing. Unlike the homogeneous air–fuel mixture of an SI engine that essentially has one flame front, the air–fuel mixture in a CI engine is very much nonhomogeneous, with fuel still being added during combustion. Local spots range from very rich to very lean, and many flame fronts exist at the same time. With undermixing, some fuel particles in fuel-rich zones never find oxygen to react with. In fuel-lean zones, combustion is limited and some fuel does not get burned. With over-mixing, some fuel particles will be mixed with already burned gas and will therefore not combust totally.

It is important that injectors be constructed such that when injection stops, there is a minimum of dribble from the nozzle. A small amount of liquid fuel will be trapped on the tip of the nozzle, however. This very small volume of fuel is called **sac volume**, its size depending on the nozzle design. This sac volume of liquid fuel evaporates very slowly because it is surrounded by a fuel-rich environment, and, once the injector nozzle closes, there is no pressure pushing it into the cylinder. Some of this fuel does not evaporate until combustion has stopped, and this results in added HC particles in the exhaust.

CI engines also have HC emissions for some of the same reasons as SI engines do (i.e., wall deposit absorption, oil film absorption, crevice volume, etc.).

Example Problem 9-1

As the flame front reaches the wall of a combustion chamber, reaction stops due to the closeness of the wall, which dampens out all fluid motion and conducts heat away. This unburned boundary layer can be considered a volume 0.1 mm thick along the entire combustion chamber surface. The combustion chamber consists mainly of a bowl in the face of the piston, which can be approximated as a 3-cm-diameter hemisphere. Fuel is originally distributed equally throughout the chamber. Calculate the percentage of fuel that does not get burned due to being trapped in the surface boundary layer.

Volume of the combustion chamber is

$$V_{CC} = (\pi/12)d^3 = (\pi/12)(3.0\ \text{cm})^3 = 7.0686\ \text{cm}^3$$

Volume of the boundary layer in the bowl is

$$V = (\pi/2)d^2(\text{thickness}) = (\pi/2)(3.0\ \text{cm})^2(0.01\ \text{cm}) = 0.1414\ \text{cm}^3$$

Volume of the boundary layer on the top of combustion chamber

$$V = (\pi/4)d^2(\text{thickness}) = (\pi/4)(3.0\ \text{cm})^2(0.01\ \text{cm}) = 0.0707\ \text{cm}^3$$

Total volume of the boundary layer is

$$V_{BL} = 0.1414 + 0.0707 = 0.2121\ \text{cm}^3$$

Percentage of total volume that does not get burned is

$$\%\ \text{not burned} = (V_{BL}/V_{CC})(100) = (0.2121/7.0686)(10) = \underline{3.0\%}$$

9.3 CARBON MONOXIDE (CO)

Carbon monoxide, a colorless, odorless, poisonous gas, is generated in an engine when it is operated with a fuel-rich equivalence ratio, as shown in Fig. 9-1. When there is not enough oxygen to convert all carbon to CO_2, some fuel does not get burned and some carbon ends up as CO. Typically, the exhaust of an SI engine will be about 0.2% to 5%

carbon monoxide. Not only is CO considered an undesirable emission, but it also represents lost chemical energy that was not fully utilized in the engine. CO is a fuel that can be combusted to supply additional thermal energy:

$$CO + \tfrac{1}{2}O_2 \rightarrow CO_2 + \text{heat} \tag{9-1}$$

Maximum CO is generated when an engine runs rich, such as when starting or when accelerating. Even when the intake air–fuel mixture is stoichiometric or lean, some CO will be generated in the engine. Poor mixing, local rich regions, and incomplete combustion will create some CO.

A well-designed SI engine operating under ideal conditions can have an exhaust mole fraction of CO as low as 10^{-3}. CI engines that operate in a lean manner overall generally have very low CO emissions (see Fig. 9-1).

9.4 OXIDES OF NITROGEN (NOx)

Exhaust gases of an engine can have up to 2000 ppm of oxides of nitrogen. Most of this will be nitrogen oxide (NO), with a small amount of nitrogen dioxide (NO_2), and traces of other nitrogen–oxygen combinations. These are all grouped together as NOx (or NO_x, with x representing some suitable number). NOx is a very undesirable emission, and regulations that restrict the allowable amount continue to become more stringent. Released NOx reacts in the atmosphere to form ozone and is one of the major causes of photochemical smog.

NOx is created mostly from nitrogen in the air. Nitrogen can also be found in fuel blends, which may contain trace amounts of NH_3, NC, and HCN, but this would contribute only to a minor degree. There are a number of possible reactions that form NO, all of which are probably occurring during the combustion process and immediately after. These include but are not limited to,

$$O + N_2 \rightarrow NO + N \tag{9-2}$$

$$N + O_2 \rightarrow NO + O \tag{9-3}$$

$$N + OH \rightarrow NO + H \tag{9-4}$$

NO, in turn, can then further react to form NO_2 by various means, including the following:

$$NO + H_2O \rightarrow NO_2 + H_2 \tag{9-5}$$

$$NO + O_2 \rightarrow NO_2 + O \tag{9-6}$$

Atmospheric nitrogen exists as a stable diatomic molecule at low temperatures, and only very small trace amounts of oxides of nitrogen are found. However, at the very high temperatures that occur in the combustion chamber of an engine, some diatomic nitrogen (N_2) breaks down to monatomic nitrogen (N), which is reactive:

$$N_2 \rightarrow 2\,N \tag{9-7}$$

Table A-3 in the Appendix shows that the chemical equilibrium constant for Eq. (9-7) is highly dependent on temperature, with a much more significant amount of N generated in the 2500–3000 K temperature range that can exist in an engine. Other gases that are stable at low temperatures, but become reactive and contribute to the formation of NOx at high temperatures, include oxygen and water vapor, which break down as follows:

$$O_2 \rightarrow 2\,O \tag{9-8}$$

$$H_2O \rightarrow OH + \tfrac{1}{2}H_2 \tag{9-9}$$

Examination of Table A-3 and more elaborate chemical equilibrium constant tables found in chemistry handbooks show that chemical Eqs. (9-7)–(9-9) all react much further to the right as high combustion chamber temperatures are reached. The higher the combustion reaction temperature, the more diatomic nitrogen, N_2, will dissociate to monatomic nitrogen, N, and the more NOx will be formed. At low temperatures, very little NOx is created.

Although maximum flame temperature will occur at a stoichiometric air–fuel ratio ($\phi = 1$), Fig. 9-1 shows that maximum NOx is formed at a slightly lean equivalence ratio of about $\phi = 0.95$. At this condition, the flame temperature is still very high, and in addition, there is an excess of oxygen that can combine with the nitrogen to form various oxides.

In addition to its dependence on temperature, the formation of NOx depends on pressure, air–fuel ratio, and combustion time within the cylinder, chemical reactions not being instantaneous. Figure 9-4 shows the NOx-versus-time relationship and supports the finding that NOx is reduced in modern engines with fast-burn combustion chambers. The amount of NOx generated also depends on the location within the combustion chamber. The highest concentration is formed around the spark plug, where the highest temperatures occur. Because they generally have higher compression ratios and higher temperatures and pressure, CI engines with divided combustion chambers and indirect injection (IDI) tend to generate higher levels of NOx.

Figure 9-5 shows how NOx can be correlated with ignition timing. If ignition spark is advanced, the cylinder temperature will be increased and more NOx will be created.

Photochemical Smog. NOx is one of the primary causes of photochemical smog, which has become a major problem in many large cities of the world. Smog is formed by the photochemical reaction of automobile exhaust and atmospheric air in the presence of sunlight. NO_2 decomposes into NO and monatomic oxygen:

$$NO_2 + \text{energy from sunlight} \rightarrow NO + O + \text{smog} \tag{9-10}$$

Monatomic oxygen is highly reactive and initiates a number of different reactions, one of which is the formation of ozone:

$$O + O_2 \rightarrow O_3 \tag{9-11}$$

Ground-level ozone is harmful to lungs and other biological tissue. It is harmful to trees and causes billions of dollars of crop loss each year. Damage is also caused through reaction with rubber, plastics, and other materials. Ozone also results from

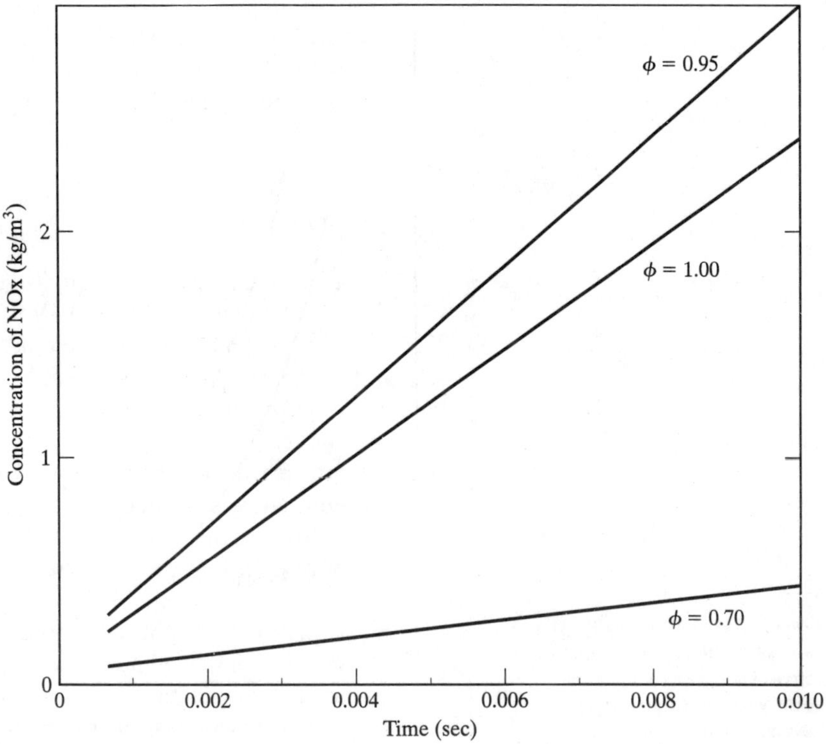

FIGURE 9-4

Generation of NOx in an engine as a function of combustion time. Many modern engines produce lower NOx emissions due to fast-burn combustion chamber design. Adapted from [92].

atmospheric reactions with other engine emissions such as HC, aldehydes, and other oxides of nitrogen.

Example Problem 9-2

To reduce the amount of reactive monatomic nitrogen in the engine of Example Problem 4-2, EGR is added to the engine operation, which reduces cylinder temperature at the end of combustion from 3500 K to 2500 K. Calculate the percent reduction of the approximate amount of N produced by using Equation (4-4), assuming pressure remains the same.

Chemical equilibrium constant K_e is obtained from Table A-3 in the Appendix at $T = 2500$ K:

$$\log_{10} K_e = -13.06 \qquad K_e = 8.710 \times 10^{-14}$$

Equation (4-4) is then

$$K_e = 8.710 \times 10^{-14} = [(2x)^2(1 - x)]/[(104)/(2x + (1 - x))]^{2-1}$$

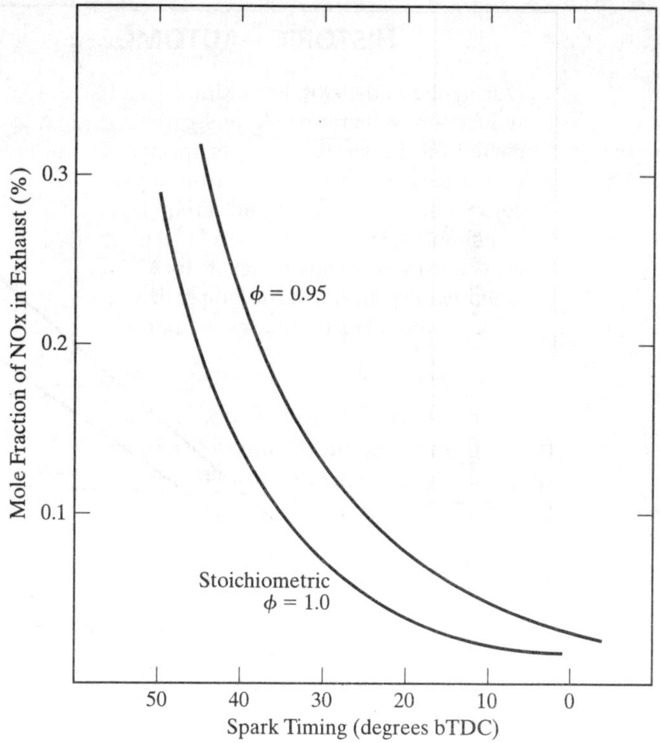

FIGURE 9-5

Generation of NOx in an SI engine as a function of spark timing. Earlier spark ignition creates a higher combustion temperature, which generates higher levels of NOx. Adapted from [108].

or

$$\text{extent of reaction } x = 1.450 \times 10^{-8} = 0.00000145\%$$

Percent of original amount, from Example Problem 4-2, is

$$(x_{2500}/x_{3500})(100) = [(1.450 \times 10^{-8})/(1.041 \times 10^{-5})](100) = 0.14\%$$

Percent reduction of monatomic N produced $= \underline{99.86\%}$.

9.5 PARTICULATES

The exhaust of CI engines contains solid carbon *soot* particles that are generated in the fuel-rich zones within the cylinder during combustion. These are seen as exhaust smoke and are an undesirable odorous pollution. Maximum density of particulate emissions occurs when the engine is under load at WOT. At this condition, maximum fuel is injected to supply maximum power, resulting in a rich mixture and poor fuel economy. This can be seen in the heavy exhaust smoke emitted when a truck or railroad locomotive accelerates up a hill or from a stop.

HISTORIC—AUTOMOBILES THAT CLEAN THE AIR

During the mid-1990s, Ford Motor Company began testing a catalytic system that could reduce atmospheric ozone and carbon monoxide. A platinum-based catalyst would be coated on the air–flow side of specially built automobile radiators and air conditioner condensers. As air passes over these surfaces, a reaction would be promoted that would convert ozone to oxygen and carbon monoxide to carbon dioxide. It is estimated that the 9 million vehicles in Los Angeles that travel a total of 266 million miles per day could reduce carbon monoxide levels by 12%. Even when a vehicle is not being operated, it could help clean the surrounding air. Air would be circulated through the radiator by a solar-powered fan, programmed to turn on when ozone levels are high [30, 132].

Soot particles are clusters of solid carbon spheres. These spheres have diameters from 10 nm to 80 nm ($1 \text{ nm} = 10^{-9}$ m), with most within the range of 15–30 nm. The spheres are solid carbon with HC and traces of other components absorbed on the surface. A single soot particle will contain up to 4000 carbon spheres [58].

Carbon spheres are generated in the combustion chamber in the fuel-rich zones where there is not enough oxygen to convert all carbon to CO_2:

$$C_xH_y + z\,O_2 \rightarrow a\,CO_2 + b\,H_2O + c\,CO + d\,C(s) \qquad (9\text{-}12)$$

Then, as turbulence and mass motion continue to mix the components in the combustion chamber, most of these carbon particles find sufficient oxygen to further react and are consumed to CO_2:

$$C(s) + O_2 \rightarrow CO_2 \qquad (9\text{-}13)$$

Over 90% of carbon particles originally generated within an engine are thus consumed and never get exhausted (Fig. 9-6). If CI engines were to operate with an overall stoichiometric air–fuel mixture, instead of the overall lean mixture they do operate with, particulate emissions in the exhaust would far exceed acceptable levels.

Up to about 25% of the carbon in soot comes from lubricating–oil components, which vaporize and then react during combustion. The rest comes from the fuel and amounts to 0.2–0.5% of the fuel. Because of the high compression ratios of CI engines, a large expansion occurs during the power stroke, and the gases within the cylinder are cooled by expansion cooling to a relatively low temperature. This causes the remaining high-boiling-point components found in the fuel and lubricating oil to condense on the surface of the carbon soot particles. This absorbed portion of the soot particles is called the **soluble organic fraction** (SOF), and the amount is highly dependent on cylinder temperature. At light loads, cylinder temperatures are reduced and can drop to as low as 200°C during final expansion and exhaust blowdown. Under these conditions, SOF can be as high as 50% of the total mass of soot. Under other operating conditions when temperatures are not so low, very little condensing occurs and SOF can be as low as 3% of total soot mass. SOF consists mostly of hydrocarbon components with some hydrogen, SO_2, NO, NO_2, and trace amounts of sulfur, zinc, phosphorus, calcium, iron, silicon, and chromium. Diesel fuel contains sulfur, calcium, iron, silicon, and chromium, while lubricating–oil additives contain zinc, phosphorus, and calcium.

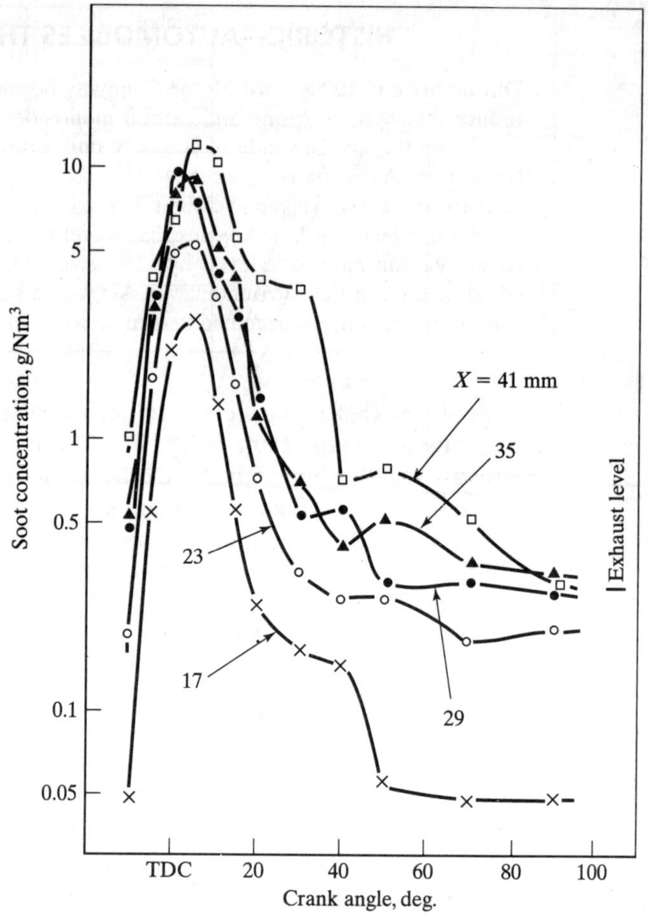

FIGURE 9-6
Particulate concentrations at various locations in combustion chamber during combustion and power stroke of large direct injection diesel engine. A large percentage of the particulates generated early in the combustion process is consumed later during the latter part of the combustion and power stroke. Only a small percentage of the original particulates is exhausted from the combustion chamber; the level is shown at the right of the figure. Engine running at 500 RPM with compression ratio of 12.9. Reprinted with permission from SAE Paper No. 820464 © 1982 SAE International [202].

Particulate generation can be reduced by engine design and control of operating conditions, but quite often this will create other adverse results. If the combustion time is extended by combustion chamber design and timing control, particulate amounts in the exhaust can be reduced. Soot particles originally generated will have a greater time to be mixed with oxygen and combusted to CO_2. However, a longer combustion time means a high cylinder temperature and more NOx generated (Fig. 9-7). Dilution with EGR lowers NOx emissions, but increases particulate and HC emissions. Higher injection pressure gives a finer droplet size, which reduces HC and particulate emissions, but increases cylinder temperature and NOx emissions. Engine management systems are programmed to minimize NOx, HC, CO, and particulate emissions by controlling ignition timing, injection pressure, injection timing, and/or valve timing. Obviously, compromise is necessary. In most engines, exhaust particulate amounts cannot be reduced to acceptable levels solely by engine design and control.

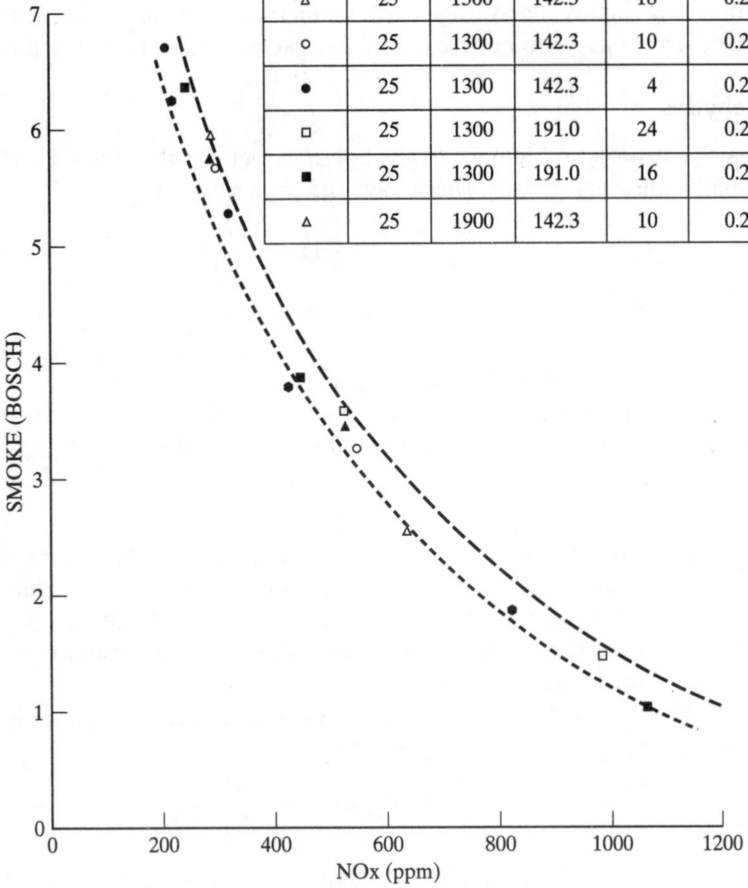

Symbol	A/F	Speed rev/min	Fuel Rate mm^3	Timing °aTDC	EGR Rate Vol. Pct.
●	25	1300	109.5	9	0.20
△	25	1300	142.3	18	0.20
○	25	1300	142.3	10	0.20
●	25	1300	142.3	4	0.20
□	25	1300	191.0	24	0.20
■	25	1300	191.0	16	0.20
△	25	1900	142.3	10	0.20

FIGURE 9-7

Nitrogen oxide (NOx)–smoke (particulates) trade-off at various engine operating conditions. Reprinted with permission from SAE Paper No. 811234 © 1981 SAE International [240].

9.6 OTHER EMISSIONS

Carbon Dioxide (CO_2)

At moderate levels of concentration, carbon dioxide is not considered an air pollutant. However, it is considered a major **greenhouse gas** and, at higher concentrations, is a major contributor to global warming. CO_2 is a major component of the exhaust in the combustion of any hydrocarbon fuel. Because of the growing number of motor vehicles, along with more factories and other sources, the amount of carbon dioxide in the atmosphere continues to grow. At upper elevations in the atmosphere, this higher concentration of carbon dioxide, along with other greenhouse gases, creates a thermal radiation shield. This shield reduces the amount of thermal radiation energy allowed to escape from the earth, raising slightly the average earth temperature. The most efficient way of reducing the amount of CO_2 is to burn less fuel (i.e., use engines with higher thermal efficiency).

Aldehydes

A major emission problem when alcohol fuel is used is the generation of aldehydes, an eye and respiratory irritant. These have the chemical formula

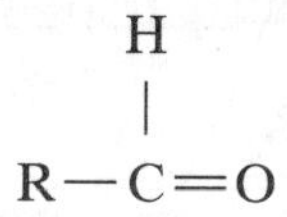

where R denotes various chemical radicals.

This is a product of incomplete combustion and would be a major problem if alcohol fuel were used to the extent that gasoline presently is.

Sulfur

Many fuels used in CI engines contain small amounts of sulfur, which, when exhausted, contributes to the acid rain problem of the world. Unleaded gasoline generally contains 150–600 ppm sulfur by weight. Some diesel fuels contain up to 5000 ppm by weight, but in the United States and some other countries sulfur content is restricted by law to a tenth of this value or less.

At high temperatures, sulfur combines with hydrogen to form H_2S and with oxygen to form SO_2:

$$H_2 + S \rightarrow H_2S \tag{9-14}$$
$$O_2 + S \rightarrow SO_2 \tag{9-15}$$

Engine exhaust can contain up to 20 ppm of SO_2. SO_2 then combines with oxygen in the air to form SO_3:

$$2\,SO_2 + O_2 \rightarrow 2\,SO_3 \tag{9-16}$$

These molecules combine with water vapor in the atmosphere to form sulfuric acid (H_2SO_4) and sulfurous acid (H_2SO_3), which are ingredients in acid rain:

$$SO_3 + H_2O \rightarrow H_2SO_4 \tag{9-17}$$
$$SO_2 + H_2O \rightarrow H_2SO_3 \tag{9-18}$$

Many countries have laws restricting the amount of sulfur allowed in fuel, and these laws are continuously being made more stringent. During the 1990s, the United States reduced acceptable sulfur levels in diesel fuel from 0.05% by weight to 0.01%.

The amount of sulfur in natural gas can range from small (sweet) to large (sour) amounts. This can be a major emissions problem when this fuel is used in an IC engine or any other combustion system.

When the allowable sulfur level in diesel fuel was lowered, a new problem surfaced in CI engines. It was found that fuel with very low levels of sulfur lost its lubricating ability, resulting in sticking fuel pumps and injectors [184]. In addition, there was abnormal wear on cylinder surfaces and rapid pressure buildup in some particulate traps. To overcome these problems, additives are put into low–sulfur fuels. These additives include aliphatic ester derivatives and carboxylic acids.

A more serious effect of sulfur, in addition to being a harmful emission, is that it poisons most emissions aftertreatment systems. Catalyst materials in catalytic converters and regenerating particulate traps deteriorate in the presence of sulfur, lead, or phosphorus.

Lead

Lead was a major gasoline additive from its introduction in 1923 to when it was phased out in the 1980s. The additive TEL (tetraethyl lead) was effectively used to increase gasoline octane number, which allowed higher compression ratios and more efficient engines. However, the resulting lead in the engine exhaust was a highly poisonous pollutant. During the first half of the 1900s, due to the lower number of automobiles and other engines, the atmosphere was able to absorb these emissions of lead without noticeable problems. As population and automobile density increased, the awareness of air pollution and its danger also increased. The danger of lead emissions was recognized, and a phaseout occurred during the 1970s and 1980s.

The use of lead could not be stopped immediately, but had to be phased out over a number of years. First, low-lead gasoline was introduced, followed, years later, by no-lead gasoline. Lead was still the major additive to raise the octane number of gasoline, and alternate octane raisers had to be developed as lead was phased out. Millions of modern high-compression engines could not use low-octane fuel. Metals used in engines also had to be changed as lead in gasoline was phased out. When leaded fuel is burned, it hardens the surfaces in the combustion chamber and those of the valves and valve seats. Engines designed to use leaded fuel had softer metal surfaces initially and relied on surface hardening effects that occurred during use. If these engines are used with unleaded fuel, surface hardening is not realized and serious wear is quickly experienced. Catastrophic failures of valve seats or piston faces are common in a short period of time (i.e., 10,000–20,000 miles in an automobile). Harder metals and added surface treatments are used for engines designed to use unleaded fuel. It was necessary to phase out leaded gasoline over a period of time as older automobiles wore out and were taken out of operation.

Leaded gasoline contains about 0.15 gm/liter of lead in the fuel. Between 10% and 50% of this gets exhausted out with the other combustion products. The remaining lead gets deposited on the walls of the engine and exhaust system. The hardened combustion chamber surfaces that resulted from the burning of leaded gasoline were quite impervious to the absorption of gases such as fuel vapor. HC emissions were also, therefore, slightly reduced in these engines.

HISTORIC—LEAD AT THE SOUTH POLE

An indication that air pollution from automobile emissions and other sources is a global problem is that lead was found in the snow strata deposited each year in the Antarctic. Lead from automobile exhaust was carried by wind currents and deposited throughout the world, including in the snow that fell near the South Pole. As lead was phased out of automobile fuel, a corresponding decrease in lead was detected each year in the Antarctic snowfall.

Phosphorus

Small amounts of phosphorus are emitted in engine exhaust. These come from impurities in the air and small amounts found in some fuel blends and lubricating oil. In addition to being an air pollutant, phosphorus poisons the catalytic materials in catalytic converters.

9.7 AFTERTREATMENT

After the combustion process stops, those components in the cylinder gas mixture that have not fully burned continue to react during the expansion stroke, during exhaust blowdown, and into the exhaust process. Up to 90% of the HC remaining after combustion reacts during this time either in the cylinder, near the exhaust port, or in the upstream part of the exhaust manifold. CO and small component hydrocarbons react with oxygen to form CO_2 and H_2O and reduce undesirable emissions. The higher the exhaust temperature, the more these secondary reactions occur and the lower are the engine emissions. Higher exhaust temperature can be caused by stoichiometric air–fuel combustion, high engine speed, retarded spark, and/or a low expansion ratio.

Thermal Converters

Secondary reactions occur much more readily and completely if the temperature is high, so some engines are equipped with thermal converters as a means of lowering emissions. Thermal converters are high-temperature chambers through which the exhaust gas flows. They promote oxidation of the CO and HC which remain in the exhaust. For the CO, we have

$$CO + \tfrac{1}{2}O_2 \rightarrow CO_2 \qquad (9\text{-}19)$$

For this reaction to occur at a useful rate, the temperature must be held above 700°C [58]. For HC, we have

$$C_xH_y + z\,O_2 \rightarrow x\,CO_2 + \tfrac{1}{2}y\,H_2O \qquad (9\text{-}20)$$

where $z = x + \frac{1}{4}y$. This reaction needs a temperature above 600°C for at least 50 msec to substantially reduce HC. It is therefore necessary for a thermal converter not only to operate at a high temperature, but to be large enough to provide adequate dwell time to promote the occurrence of these secondary reactions. Most thermal converters are essentially enlarged exhaust manifolds connected to the engine immediately outside the exhaust ports. This location is necessary to minimize heat losses and keep the exhaust gases from cooling to nonreacting temperatures. However, in automobiles, this creates two very serious problems for the engine compartment. In modern, low-profile, aerodynamic automobiles, space in the engine compartment is very limited, and fitting in a large, usually insulated, thermal converter chamber is almost impossible. Secondly, because the converter must operate above 700°C to be efficient, even if it is insulated the heat losses create a serious temperature problem in the engine compartment.

Some thermal converter systems include an air intake that provides additional oxygen to react with the CO and HC. This increases the complexity, cost, and size of the system. Flow rate of air is controlled by the EMS as needed. Air addition is especially necessary during rich operating conditions such as start-up. Because exhaust from engines is often at a lower temperature than is needed for efficient operation of a thermal converter, it is necessary to sustain the high temperatures by the reactions within the system. Adding outside air, which is at a lower temperature, compounds this problem of maintaining the necessary operating temperature.

NOx emissions cannot be reduced with a thermal converter alone.

9.8 CATALYTIC CONVERTERS

The most effective aftertreatment system for reducing engine emissions is the catalytic converter found on most automobiles and other modern engines of medium or large size. HC and CO can be oxidized to H_2O and CO_2 in exhaust systems and thermal converters if the temperature is held at 600°–700°C. If certain catalysts are present, the temperature needed to sustain these oxidation processes is reduced to 250°–300°C, making for a much more attractive system. A catalyst is a substance that accelerates a chemical reaction by lowering the energy needed for it to proceed. The catalyst is not consumed in the reaction and so functions indefinitely unless degraded by heat, age, contaminants, or other factors. Catalytic converters are chambers mounted in the flow system through which the exhaust gases flow. These chambers contain catalytic material, which promotes the oxidation of the emissions contained in the exhaust flow.

Generally, catalytic converters are called three-way converters because they promote the reduction of CO, HC, and NOx. There are basically two types of construction. Some consist of a stainless steel container mounted somewhere along the exhaust pipe of the engine. Inside the container is a porous ceramic structure through which the gas flows. In most converters, the ceramic is a single honeycomb structure with many flow passages (see Fig. 9-8). Some converters use loose granular ceramic with the gas passing between the packed spheres. The volume of the ceramic structure of a converter is generally about half the displacement volume of the engine. This results in a volumetric flow rate of exhaust gas such that there are 5 to 30 changeovers of gas each second through the converter. Catalytic converters for CI engines need larger flow passages because of the solid soot in the exhaust gases.

The surface of the ceramic passages contains small embedded particles of catalytic material that promote the oxidation reactions in the exhaust gas as it passes. Aluminum oxide (alumina) is the base ceramic material used for most catalytic converters. Alumina can withstand the high temperatures, it remains chemically neutral, it has very low thermal expansion, and it does not thermally degrade with age. The catalyst

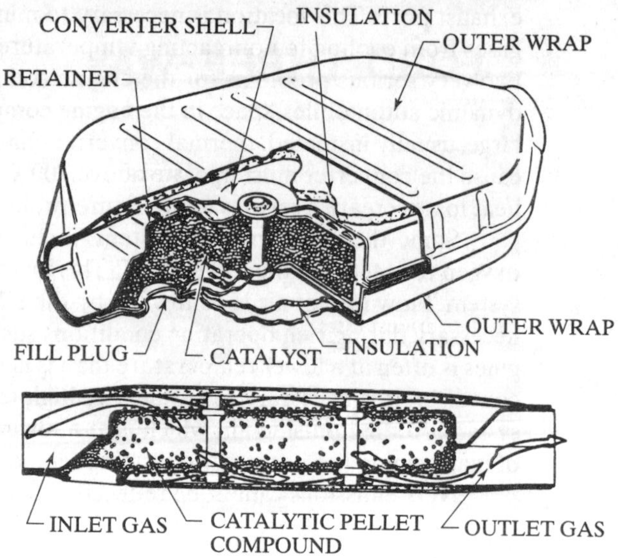

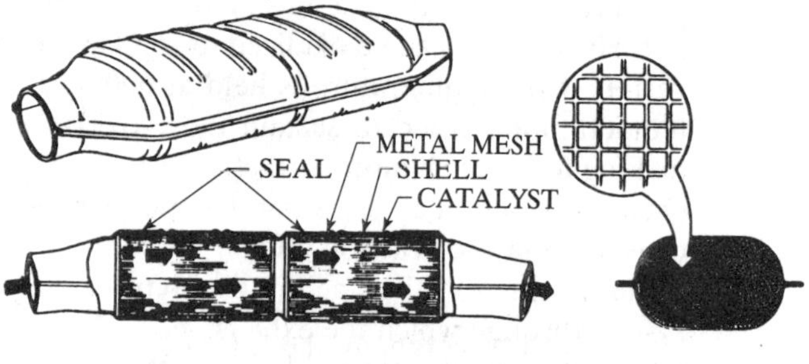

FIGURE 9-8
Catalytic converters for SI engines: (a) packed spheres; (b) honeycomb structure. Reprinted with permission from SAE Paper 801440 © 1980 SAE International [4].

materials most commonly used are platinum and rhodium, with other noble metals (e.g., palladium and iridium) sometimes used as well.

Palladium and platinum promote the oxidation of CO and HC as in Eqs. (9-19) and (9-20), with platinum especially active in the hydrocarbon reaction. Rhodium promotes the reaction of NOx in one or more of the following reactions:

$$NO + CO \rightarrow \tfrac{1}{2}N_2 + CO_2 \quad (9\text{-}21)$$

$$2\,NO + 5\,CO + 3\,H_2O \rightarrow 2\,NH_3 + 5\,CO_2 \quad (9\text{-}22)$$

$$2\,NO + CO \rightarrow N_2O + CO_2 \quad (9\text{-}23)$$

$$NO + H_2 \rightarrow \tfrac{1}{2}N_2 + H_2O \quad (9\text{-}24)$$

$$2\,NO + 5\,H_2 \rightarrow 2\,NH_3 + 2\,H_2O \quad (9\text{-}25)$$

$$2\,NO + H_2 \rightarrow N_2O + H_2O \quad (9\text{-}26)$$

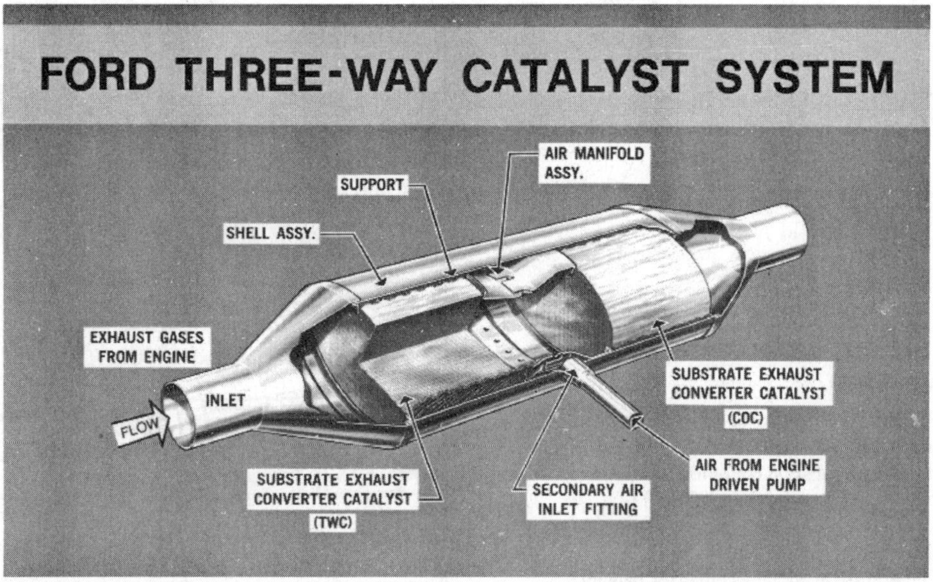

FIGURE 9-9

Cutaway drawing of three-way catalytic converter system, showing main components. This system was used on Ford automobiles during the 1990s. Courtesy Ford Motor Company.

Also often used is cerium oxide, which promotes the so-called water–gas shift:

$$CO + H_2O \rightarrow CO_2 + H_2 \tag{9-27}$$

This reduces CO by using water vapor as an oxidant instead of O_2, which is very important when the engine is running in a rich manner.

Instead of ceramics, the interior chamber of some catalytic converters is filled with a very thin corrugated metal foil wrapped into a monolith structure. The exhaust gases pass between the rolls of foil, the surface of which is embedded with catalytic material. The active surface area of a larger converter of this type can be as great as 70,000 m^2.

Figure 9-10 shows that the efficiency of a catalytic converter is very dependent on temperature. When a converter in good working order is operating at a fully warmed temperature of 400°C or above, it will remove 98–99% of CO, 95% of NOx, and more than 95% of HC from exhaust flow emissions. Figure 9-11 shows that it is also necessary to be operating at the proper equivalence ratio to get high converter efficiency. Effective control of HC and CO occurs with stoichiometric or lean mixtures, while control of NOx requires near–stoichiometric conditions. Very poor NOx control occurs with lean mixtures.

Because an engine has a number of cyclic variations occurring—including AF—the exhaust flow will also show variation. It has been found that this cyclic variation lowers the peak efficiency of a catalytic converter, but spreads the width of the equivalence–ratio envelope of operation in which there is acceptable emissions reduction.

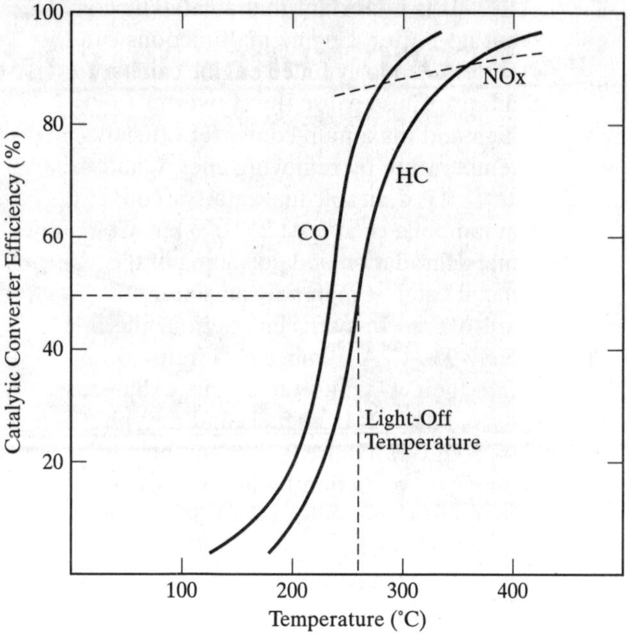

FIGURE 9-10

Conversion efficiency of catalytic converters as a function of converter temperature. A converter in good condition will generally reduce emissions by over 90% if it is at normal operating temperature. When cold, a converter is very inefficient. The temperature at which a converter becomes 50% efficient is often called light-off temperature. Adapted from [4].

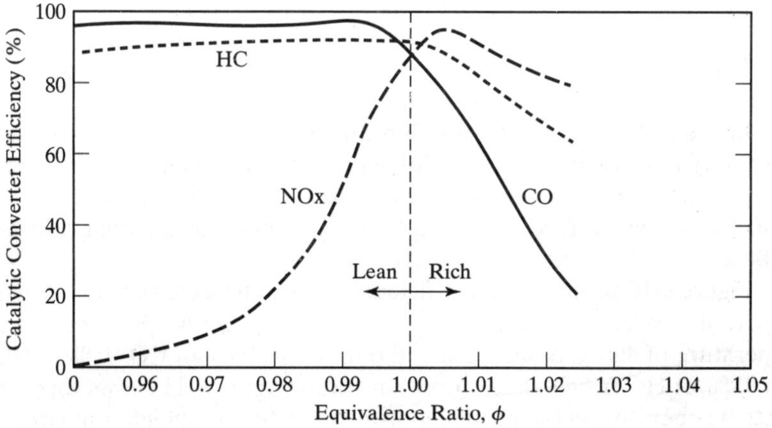

FIGURE 9-11

Conversion efficiency of catalytic converters as a function of fuel equivalence ratio. Greatest efficiency occurs when engines operate near stoichiometric conditions. Converters are very inefficient for NOx conversion when an engine operates in a lean manner. This creates a greater problem for modern CI engines and stratified charge SI engines, which generally operate in a very lean manner overall. Adapted from [76].

It is important that a catalytic converter be operated hot enough to be efficient, but no hotter. Engine malfunctions can cause poor efficiency and overheating of converters. A poorly tuned engine can have misfires and periods of too lean or too rich conditions. These cause the converter to be inefficient at the very time emissions are very high and maximum converter efficiency is needed. A turbocharger lowers the exhaust temperature by removing energy, and this can make a catalytic converter less efficient.

It is desirable that catalytic converters have an effective lifetime equal to that of the automobile or at least 200,000 km. Converters lose their efficiency with age due to thermal degradation and poisoning of the active catalyst material. At high temperatures, the metal catalyst material can sinter and migrate together, forming larger active sites that are, overall, less efficient. Serious thermal degrading occurs in the temperature range of 500°–900°C. A number of different impurities contained in fuel, lubricating oil, and air find their way into the engine exhaust and poison the catalyst material. These include lead and sulfur from fuels, and zinc, phosphorus, antimony, calcium, and magnesium from oil additives. Figure 9-12 shows how just a small amount of lead on a catalyst site can reduce HC reduction by a factor of two or three. Small amounts of lead impurities are found in some fuels, and 10–30% of this ends up in the catalytic converter. Up until the early 1990s, leaded gasoline was quite common, and it was imperative that it not be used in engines equipped with catalytic converters; such use was unlawful. Two fuel tanks of leaded gasoline would completely poison a converter and make it totally useless.

Sulfur

Sulfur offers unique problems for catalytic converters. Some catalysts promote the conversion of SO_2 to SO_3, which eventually gets converted to sulfuric acid. This degrades

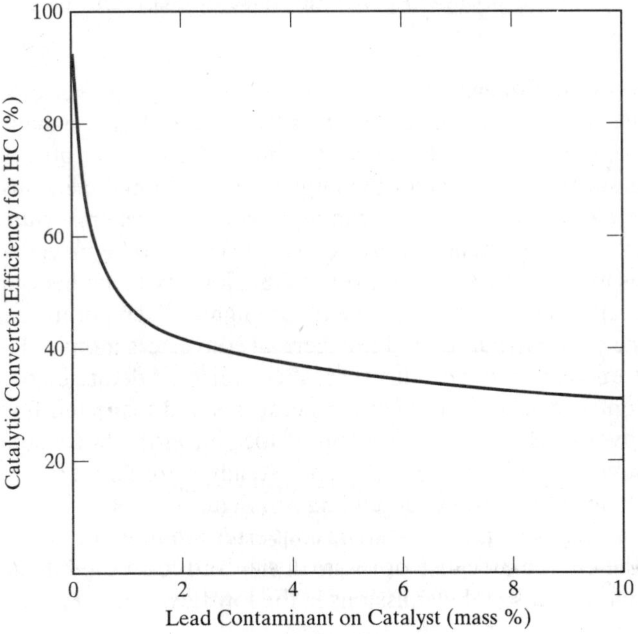

FIGURE 9-12

Reduction of catalytic converter efficiency due to contamination by lead. Some of the lead contained in fuels gets deposited on the catalyst material in a converter, greatly reducing converter efficiency. It is imperative (and legally required) that leaded gasoline not be used in automobiles equipped with catalytic converters. To reduce the chances of accidently using leaded gasoline with a catalytic converter, the fuel pump nozzle size and the diameter of the fuel tank inlet are made smaller for nonleaded gasoline. Adapted from [76].

the catalytic converter and contributes to acid rain. New catalysts are being developed that promote the oxidation of HC and CO, but do not change SO_2 to SO_3. Some of these create almost no SO_3 if the temperature of the converter is kept at 400°C or lower.

Cold Start-Ups

Figure 9-10 shows that catalytic converters are very inefficient when they are cold. When an engine is started after not being operated for several hours, it takes several minutes for the converter to reach an efficient operating temperature. The temperature at which a converter becomes 50% efficient is defined as the **light-off temperature**, and this is in the range of about 250°–300°C. A large percentage of automobile travel is for short distances where the catalytic converter never reaches efficient operating temperature, and therefore emissions are high. Some studies suggest that half of the fuel used by automobiles in the United States is on trips of less than 10 miles distance. Unfortunately, most short trips occur in cities where high emissions are more harmful. Add to this the fact that most engines use a rich mixture when starting, and it can be seen that cold start-ups pose a major problem. It is estimated that cold start-ups are the source of 70–90% of all HC emissions. A major reduction in emissions would therefore be possible if catalytic converters could be preheated, at least to light-off temperature, before engine start-up. Preheating to full steady-state operating temperature would be even better. Several methods of preheating have been tried with varying success. Because of the time involved and amount of energy needed, most of these methods preheat only a small portion of the total converter volume. This small section is large enough to treat the low exhaust flow rate that usually occurs at start-up and immediately following. By the time higher engine speeds are used, more of the catalytic converter has been heated by the hot exhaust gas, and the higher flow rates are fully treated. Methods of catalytic converter preheating include those discussed in the subsections that follow.

Locate Converter Close to the Engine. One method used to heat a converter as quickly as possible is to locate it in the engine compartment very close to the exhaust ports. This method does not actually preheat the converter, but does heat it as quickly as possible after the engine is started. It eliminates the large heat loss from the exhaust pipe that occurs between the engine and a converter in more common systems where the converter is located away from the engine. These converters can also be insulated to reduce early heat loss. This method does reduce overall emissions by quick heat-up of the converter, but there is still a short period of time before light-off temperature is reached. In addition, the same problems described for thermal converters mounted in the engine compartment are encountered with this type of converter. Adequate cooling of the engine compartment because of the high temperatures and restricted flow rate of air caused by the converter is a serious problem. If located in the hot engine compartment, a catalytic converter will also have a higher steady-state temperature, and this will cause a greater long-term thermal–degrading problem.

Some automobiles use a small secondary catalytic converter mounted in the engine compartment close to the engine. Because of its small size and location, it heats up very quickly and is sufficient to oxidize the emissions in the low flow rates at engine start-up. There is also a normal full-size catalytic converter mounted away from the engine compartment that supplies the catalytic action for the larger flow rates of normal operation. This converter is heated by the first exhaust flow and ideally reaches efficient operating temperature before the engine is speeded up and higher flow rates are experienced. These small preconverters restrict flow in the exhaust manifold and add some back pressure to the engine. This results in a slight reduction in engine power output.

Superinsulation. Some experimental systems with superinsulated catalytic converters have been developed. These do not actually preheat the converter on first engine start-up, but they do accelerate the temperature rise to steady-state conditions. They also keep the converter at an elevated temperature for up to a day after the engine has been turned off. The converter is therefore preheated for subsequent engine starts [131].

The converter is double walled with a vacuum between the walls. This provides superinsulation characteristics much like those of a vacuum bottle. When the engine is cold or not running, the vacuum is sustained. When the engine is running and the converter is at operating temperature, the vacuum is eliminated and the space between the walls is filled with a gas. This allows for normal heat losses during operation and keeps the catalytic converter from overheating.

Electric Heating. Some systems use electric preheating, usually by resistance heating. Heating resistors are embedded in the preheat zone of the converter, and an electrical discharge is initiated before the engine is started. The preheat zone can be a separate, small preconverter, or it can be the front end of the normal catalytic converter. This method works best with converters having an interior structure of metal foil. Some systems replace the ceramic honeycomb solid in the preheat zone with a metal structure of multiple flow passages. This allows a much quicker heating of the flow passage walls by means of thermal conduction, metal having much higher thermal conductivity than ceramic. Electrical energy for this type of system usually comes from a battery that is recharged when the engine is running. Typical values for preheating are 24 volts and 500–700 amps.

There is some time delay between heating the electrical elements and reaching light-off temperature, due to the conduction needed. The most serious problem, however, is the inability of a normal-sized battery to deliver the amount of energy needed for such a system.

Flame Heating. A catalytic converter can be heated with a flame from a burner nozzle mounted within the structure of the converter [57]. Before the engine is started (for instance, when the ignition key is inserted), a flame is initiated in the burner, using fuel and air pumped from external sources. Concern must be given to what emissions this flame would contribute to the overall air pollution problem. A fuel like propane burned with the correct amount of air would create very little pollution. However, this would require an axillary propane fuel tank on the automobile, something that would be undesirable unless the automobile engine were also fueled with propane. In a gasoline-fueled engine, it would be logical to use gasoline in the converter preheater. However, it would be more difficult to get clean burning with gasoline. Cost, complexity, and some time delay are disadvantages of this type of system.

A variation of this system used by at least one major automobile manufacturer is an *afterburner* mounted directly before the catalytic converter. A very rich air–fuel mixture is used at start-up, which leaves excess fuel in the first exhaust flow. Air is added to this exhaust by an electric pump, and the resulting mixture is combusted in the afterburner, preheating the catalytic converter.

Thermal Battery. Energy from a thermal storage system can be used to preheat a catalytic converter if the engine is started within about three days of last being used (see Chapter 10). With present technology, only partial preheating to a temperature around 60°C is possible, which is still below light-off temperature and well below normal operating temperatures. In addition, the limited amount of energy available in a thermal battery is often distributed between preheating the engine, warming the passenger compartment, and preheating the catalytic converter.

Chemical Reaction Preheating. A possible method for preheating a catalytic converter using the heat liberated from an exothermic chemical reaction has been suggested. When the ignition key is inserted, a small amount of water is introduced into the converter from an injector mounted through the side of the converter housing. The water spray reacts with a salt imbedded in the surface of the ceramic honeycomb. This exothermic reaction releases enough energy to heat the surrounding ceramic structure to a temperature above light-off temperature, and the converter is ready for efficient use in a matter of seconds. When the engine is then started, the hot exhaust gases dry the imbedded salt by evaporating away the water. The water vapor is carried away with the exhaust gas, and the system is ready for the next cold start. One major problem with this method is degradation of the salt with age. Also, there is a freezing problem with the water reservoir in cold climates. No practical system using this method has so far appeared on the market [98].

Absorption of HC When Cold. Some converter systems use surface materials that, when cold, absorb the HC in the exhaust flow. Then, when a higher steady-state temperature is reached, the HC desorbs back into the flow and is reduced in the normal way.

Dual-Fuel Engines

Some engines are made to run on a combination of gasoline and methanol, with the percent volume of methanol ranging from 0% to 85%. The engine control systems on these engines are capable of adjusting the air and fuel flow to give optimum combustion and minimum emissions with any combination of these fuels. However, this creates a unique problem for a catalytic converter. Each of these fuels requires separate catalysts. Incomplete combustion of methanol produces formaldehyde, which must be removed from the exhaust. To effectively reduce the formaldehyde and any remaining methanol, a catalytic converter must be operated above 300°C. Preheating of the converter on these systems is very important.

Lean-Burn Engines

A large number of automobiles on the market obtain high fuel efficiency by use of **lean-burn engines**. Lean-burn combustion is becoming a common philosophy for reducing fuel consumption, but it creates special problems in reducing NOx in catalytic converters. By using a stratified charge, engines obtain efficient combustion with overall air–fuel ratios of 20 or 21 ($\phi \approx 0.7$), with some operating as high as AF = 40 ($\phi \approx 0.4$). Figure 9-11 shows that normal catalytic converters will work in reducing HC and CO at these lean conditions, but are very inefficient at reducing NOx.

Some automobiles using lean-burn engines have special converters with interior surfaces that absorb the NOx which does not get treated when operating lean. Then, when the vehicle accelerates, works under load, or operates in any mode requiring stoichiometric combustion, the absorbed NOx desorbs off the surface and is treated with the high converter efficiency that occurs under these conditions. When the automobile does not accelerate for several minutes, the EMS is programed to periodically inject a few seconds of rich combustion to desorb the catalytic converter. For instance, when the Toyota Opa is traveling at a steady 60 km/hr (37 mph), one to two seconds of rich injection occurs every two minutes [239]. Another method used to eliminate this lean-burn problem is to use high levels of EGR, which helps in two ways. It dilutes the air–fuel mixture and lowers the combustion temperature, which is what normal lean-burn does. The lower combustion temperature then generates less NOx. In addition, it allows the actual air and fuel to be added in a near-stoichiometric ratio, which allows the catalytic converter to operate efficiently. Catalysts using platinum, rhodium, palladium, iridium, and other noble metals, combined with alkaline rare earths, have been developed for lean-burn engines.

Two-Stroke Cycle Engines

Modern two-stroke cycle engines that use fuel injectors have cooler exhaust because of their high efficiency and lean operation. Both the lower exhaust temperature and lean operation make the typical catalytic converter less efficient and create a more difficult emissions problem with these engines.

HISTORIC—AIRPLANE ENGINE EXHAUST & WEATHER

It has long been known that the large amounts of water vapor put into the upper atmosphere by aircraft engine exhaust affect the weather. Time-lapse photographs from satellites have shown how large concentrations of vapor trails sometimes merge together and created rain clouds. The days following the terrorist tragedy of September 11, 2001, supplied some unique additional data on this subject. Airline flights and most flying were stopped in the United States for four days. When weather data for these days were checked, it was found that there was an unnaturally small amount of water vapor in the upper atmosphere. Scientists gave credit to this lessening of the atmospheric thermal radiation shield for an abnormal surface temperature shift in the United States of about three degrees Fahrenheit, warmer during the day and cooler at night.

9.9 CI ENGINES

Catalytic converters are used with CI engines, but are not efficient at reducing NOx, due to their overall lean operation. HC and CO can be adequately reduced, although there is greater difficulty because of the cooler exhaust gases of a CI engine (because of the larger expansion ratio). This is counterbalanced by the fact that less HC and CO are generated in the lean burn of the CI engine. NOx is reduced in a CI engine by the use of EGR, which keeps the maximum temperature down. EGR and lower combustion temperatures, however, contribute to an increase in solid soot.

Platinum and palladium are two main catalyst materials used for converters on CI engines. They promote the removal of 30–80% of the gaseous HC and 40–90% of the CO in the exhaust. The catalysts have little effect on solid carbon soot, but do remove 30–60% of the total particulate mass by oxidizing a large percent of the HC absorbed on the carbon particles. Diesel fuel contains sulfur impurities, and this leads to poisoning of the catalyst materials. However, this problem is being reduced as legal levels of sulfur in diesel fuels continue to be lowered.

Particulate Traps

Compression ignition engine systems are equipped with particulate traps in their exhaust flow to reduce the amount of particulates released to the atmosphere. Traps are filterlike systems often made of ceramic in the form of a monolith or mat, or sometimes made of metal wire mesh. Traps typically remove 60–90% of particulates in the exhaust flow. As traps catch the soot particles, they slowly fill up with the particulates. This restricts exhaust gas flow and raises the back pressure of the engine. Higher back pressure causes the engine to run hotter, the exhaust temperature to rise, and fuel consumption to increase. To reduce this flow restriction, particulate traps are regenerated when they begin to become saturated. Regeneration consists of combusting the particulates in the excess oxygen contained in the exhaust of the lean-operating CI engine.

Carbon soot ignites at about 550°–650°C, while CI engine exhaust temperature is 150°–350°C at normal operating conditions. As the particulate trap fills with soot and restricts flow, the exhaust temperature rises, but is still not high enough to ignite the soot and regenerate the trap. In some systems, automatic flame igniters are used. These igniters start combustion in the carbon when the pressure drop across the trap reaches a predetermined value. The igniters can be electric heaters or flame nozzles that use diesel fuel. If catalyst material is installed in the traps, the temperature needed to ignite the carbon soot is reduced to the 350°–450°C range. Some such traps can automatically regenerate by self-igniting when the exhaust temperature rises from increased back pressure. Other catalyst systems use flame igniters.

Another way of lowering the ignition temperature of the carbon soot and promoting self-regeneration in traps is to use catalyst additives in the diesel fuel. These additives generally consist of copper compounds or iron compounds, with about seven grams of additive in 1000 liters of fuel being normal.

To keep the temperatures high enough for self-regeneration in a catalytic system, traps can be mounted as close to the engine as possible, even before the turbocharger.

On some larger stationary engines and on some construction equipment and large trucks, the particulate trap is replaced when it becomes nearly filled. The removed trap is then regenerated externally, with the carbon being burned off in a furnace. The regenerated trap can then be used again.

Various methods are used to determine when soot buildup becomes excessive and regeneration is necessary. The most common method is to measure pressure drop in the exhaust flow as it passes through the trap. When a predetermined ΔP is reached, regeneration is initiated. Pressure drop is also a function of exhaust flow rate, and this must be programmed into the regeneration controls. Another method used to sense soot buildup is to transmit radio frequency waves through the trap and determine the percentage of the signal that is absorbed. Carbon soot absorbs radio waves, while the ceramic structure does not. The amount of soot buildup can therefore be determined by the percent decrease in radio signal. This method does not readily detect soluble organic fraction (SOF).

Modern particulate traps are not totally satisfactory, especially for automobiles. They are costly and complex when equipped for regeneration, and long-term durability does not exist. An ideal catalytic trap would be simple, economical, and reliable; it would be self-regenerating; and it would impose a minimum increase in fuel consumption.

Modern Diesel Engines

Carbon soot particulate generation has been greatly reduced in modern CI engines by advanced design technology in fuel injectors and combustion chamber geometry. With greatly increased mixing efficiency and speeds, large regions of fuel-rich mixtures can be avoided when combustion starts. These are the regions where carbon soot is generated, and by reducing their volume, far less soot is generated. Increased mixing speeds are obtained by a combination of indirect injection, better combustion chamber geometry, better injector design and higher pressures, heated spray targets, and air-assisted injectors. Indirect injection into a secondary chamber that promotes high turbulence and swirl greatly speeds the air–fuel mixing process. Better nozzle design and higher injection pressures create finer fuel droplets, which evaporate and mix more quickly. Injection against a hot surface speeds evaporation, as do air-assisted injectors.

Some modern, top-of-the-line CI automobile engines (e.g., Mercedes) have reduced particulate generation enough that they meet stringent standards without the need for particulate traps.

9.10 CHEMICAL METHODS TO REDUCE EMISSIONS

Development work has been done on large stationary engines using cyanuric acid to reduce NOx emissions. Cyanuric acid is a low-cost solid material that sublimes in the exhaust flow. The gas dissociates, producing isocyanide that reacts with NOx to form N_2, H_2O, and CO_2. Operating temperature is about 500°C. Up to 95% NOx reduction has been achieved with no loss of engine performance. At present, this system is not practical for vehicle engines because of its size, weight, and complexity.

Research is being done using zeolite molecular sieves to reduce NOx emissions. These are materials that absorb selected molecular compounds and catalyze chemical reactions. For both SI and CI engines, the efficiency of NOx reduction is being determined over a range of operating variables, including AF, temperature, flow velocity, and zeolite structure. At present, durability is a serious limitation with this method.

Various chemical absorbers, molecular sieves, and traps are being tested to reduce HC emissions. HC is collected during engine start-up time, when the catalytic converter is cold, and then later released back into the exhaust flow when the converter is hot. The converter then efficiently burns the HC to H_2O and CO_2. A 35% reduction of cold-start HC has been achieved.

H_2S emissions occur under rich operating conditions. Chemical systems are being developed that trap and store H_2S when an engine operates under rich conditions and then convert this to SO_2 when operation is lean and excess oxygen exists. The reaction equation is

$$H_2S + O_2 \rightarrow SO_2 + H_2 \tag{9-28}$$

Ammonia Injection Systems

Some large ship engines and some stationary engines reduce NOx emissions with an injection system that sprays NH_3 into the exhaust flow. In the presence of a catalyst, the following reactions occur:

$$4\,NH_3 + 4\,NO + O_2 \rightarrow 4\,N_2 + 6\,H_2O \tag{9-29}$$

$$6\,NO_2 + 8\,NH_3 \rightarrow 7\,N_2 + 12\,H_2O \tag{9-30}$$

Careful control must be exercised, as NH_3 itself is an undesirable emission.

Emissions from large ships were not restricted for many years, even after strict laws were enforced on other engines. It was reasoned that ships operated away from land masses most of the time, and the exhaust gases could be absorbed by the atmosphere without affecting human habitat. However, most seaports are in large cities, where emission problems are most critical, and pollution from all engines is now restricted, including ship engines. In addition to concern for NOx emissions from ships, there is also concern for emissions of sulfur oxides (SOx). This is because fuels used on large ships often have higher amounts of sulfur. New regulations will force ship owners to use higher priced low-sulfur fuels [236].

Ammonia systems are not practical in automobiles because of the required NH_3 and fairly complex injection and control system. However, there has been recent development work to put ammonia injection systems in large trucks [233]. The ammonia is obtained from urea $[CO(NH_2)_2]$ which is stored on the truck. The urea is mixed with water, and this generates the needed ammonia by the reaction $CO(NH_2)_2 + H_2O \rightarrow CO_2 + 2\,NH_3$. Long road tests have shown a 63–68% reduction of NOx, depending on driving conditions. HC emissions have been reduced to zero [215].

9.11 EXHAUST GAS RECYCLE—EGR

The most effective way of reducing NOx emissions is to hold combustion chamber temperatures down. Although practical, this is a very unfortunate method in that it also reduces the thermal efficiency of the engine. We have been taught since our first thermodynamics course that for maximum engine thermal efficiency, Q_{in} should be at the highest temperature possible.

Probably the simplest practical method of reducing maximum flame temperature is to dilute the air–fuel mixture with a nonreacting parasite gas. This gas absorbs energy during combustion without contributing any energy input. The net result is a lower flame temperature. Any nonreacting gas would work as a diluent, as shown in Fig. 9-13. Those gases with larger specific heats would absorb the most energy per unit mass, and less of those gases would therefore be required; thus, less CO_2 would be required than argon for the same maximum temperature. However, neither CO_2 nor argon is readily available for use in an engine. Air is available as a diluent, but is not totally nonreacting. Adding air changes the AF and combustion characteristics. The one nonreacting gas that is available to use in an engine is exhaust gas, and that is used in all modern automobile and other medium-size and large engines.

Exhaust gas recycle (EGR) is done by ducting some of the exhaust flow back into the intake system, usually immediately after the throttle. The amount of flow can be as high as 30% of the total intake. EGR gas combines with the exhaust residual left in the cylinder from the previous cycle to effectively reduce the maximum combustion temperature. The flow rate of EGR is controlled by the EMS. EGR is defined as a mass percentage of the total intake flow, that is,

$$\text{EGR} = [\dot{m}_{\text{EGR}}/\dot{m}_i](100) \tag{9-31}$$

where $\dot{m}_i$ is the total mass flow into the cylinders.

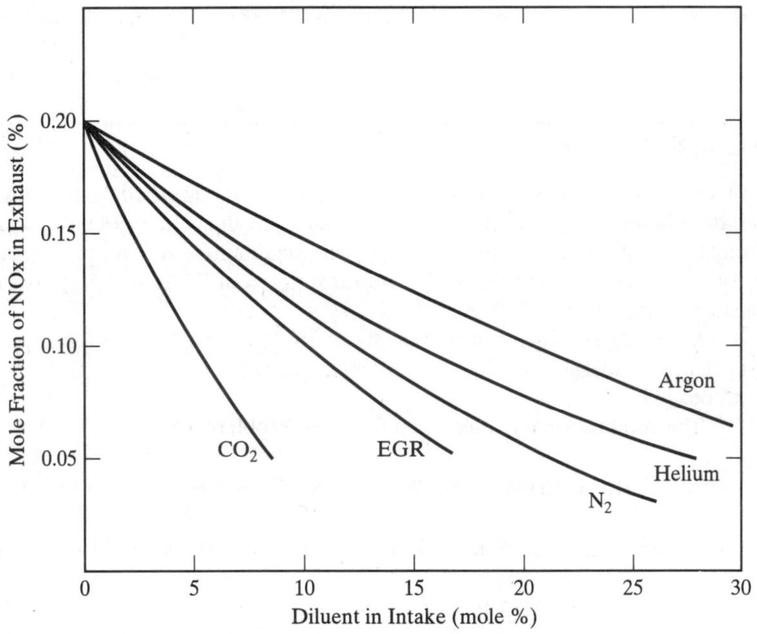

FIGURE 9-13
Reduction of NOx generation by adding non–combustible diluent gas to intake mixture. Adding any nonreacting neutral gas to the inlet air–fuel mixture reduces flame temperature and NOx generation. Exhaust gas is the one gas that is readily available for engine use. Adapted from [99].

After EGR combines with the exhaust residual left from the previous cycle, the total fraction of exhaust in the cylinder during the compression stroke is

$$x_{ex} = (\mathrm{EGR}/100)(1 - x_r) + x_r \tag{9-32}$$

where x_r is the exhaust residual from the previous cycle.

Not only does EGR reduce the maximum temperature in the combustion chamber, but it also slows combustion and lowers the overall combustion efficiency. Figure 7-12 shows that, as EGR is increased, the percentage of inefficient *slow-burn* cycles increases. Further increase in EGR results in some cycle *partial burns* and, in the extreme, total misfires. Thus, by using EGR to reduce NOx emissions, a costly price of increased HC emissions and lower thermal efficiency must be paid.

The amount of EGR is controlled by the EMS. By sensing inlet and exhaust conditions, the flow is controlled, ranging from 0 up to 15–30%. Lowest NOx emissions with relatively good fuel economy occur at about stoichiometric combustion, with as much EGR as possible without adversely affecting combustion. No EGR is used during WOT, when maximum power is desired. No EGR is used at idle, and very little is used at low speeds. Under these conditions, there is already a maximum exhaust residual and greater combustion inefficiency. Engines with fast-burn combustion chambers can tolerate a greater amount of EGR.

A problem unique to CI engines when using EGR is the solid carbon soot in the exhaust. The soot acts as an abrasive and breaks down the lubricant. Greater wear on the piston rings and valve train results.

Example Problem 9-3

In Example Problems 4-1 and 4-5, it was found that the theoretical maximum combustion temperature in an engine burning isooctane at an equivalence ratio of 0.833 was 2419 K. To reduce formation of NOx, it is desired to reduce this maximum temperature to 2200 K. This is done by exhaust gas recycling (EGR). Calculate the amount of EGR needed to reduce maximum combustion temperature to 2200 K.

Exhaust gas, which consists mostly of N_2, CO_2, and H_2O, will be approximated as all nitrogen at a temperature of 1000 K. Enthalpy values can be obtained from most thermodynamics textbooks [90].

The combustion equation from Example Problem 4-5 is

$$C_8H_{18} + 15\,O_2 + 15(3.76)\,N_2 \rightarrow 8\,CO_2 + 9\,H_2O + 2.5\,O_2 + 15(3.76)\,N_2$$

An unknown number of moles of EGR (x moles of N_2 at 1000 K) is added to the reactants:

$$C_8H_{18} + 15\,O_2 + 15(3.76)\,N_2 + x\,N_2 \rightarrow 8\,CO_2 + 9\,H_2O + 2.5\,O_2 + [15(3.76) + x]\,N_2$$

Using Eqs. (4-5) and (4-8) yields

$$\sum_{\substack{\text{PROD at}\\ 2200\ \text{K}}} N_i(h_f^\circ + \Delta h)_i = \sum_{\text{REACT}} N_i(h_f^\circ + \Delta h)_i$$

$$8[(-393{,}522) + (103{,}562)] + 9[(-241{,}826) + (83{,}153)]$$
$$+ 2.5[(0) + (66{,}770)] + [15(3.76) + x][(0) + (63{,}362)]$$
$$= [(-259{,}280) + (73{,}473)] + 15[(0) + (12{,}499)]$$
$$+ 15(3.76)[(0) + (11{,}937)] + x[(0) + (21{,}463)]$$

Solving for x gives

$$x = 16.28 \text{ moles} = (16.28 \text{ kgmoles})(28 \text{ kg/kgmole}) = 455.8 \text{ kg}$$

Mass of air in is

$$m_a = [15(4.76)\text{kgmoles}][29 \text{ kg/kgmole}] = 2070.6 \text{ kg}$$

Mass of fuel vapor in is

$$m_f = [1 \text{ kgmole}][114 \text{ kg/kgmole}] = 114 \text{ kg}$$

Total mass in is

$$m_i = m_a + m_f + m_{\text{EGR}} = (2070.6) + (114) + (455.8) = 2640.4 \text{ kg}$$

Equation (9-31) gives percent EGR

$$\text{EGR} = [m_{\text{EGR}}/m_i](100) = [455.8/2640.4](100) = \underline{17.3\%}$$

9.12 NONEXHAUST EMISSIONS

Engines and fuel supply systems also have sources of emissions other than exhaust flow. Historically, these were considered minor and were just released to the surrounding air.

A major source of HC emissions was the crankcase breather tube that was vented to the air in older automobiles. Blowby flow past the pistons ended up in the crankcase, and due to the higher pressure it created, it was then pushed out the breather vent tube. Blowby gas is very high in HCs, especially in SI engines. Also, in older engines with greater clearance between the piston and cylinder wall, blowby flow was much higher. As much as 1% of the fuel was vented to the atmosphere through the crankcase breather in some automobiles. This accounted for up to 20% of total emissions. A simple solution to this problem, which is used on all modern engines, is to vent the crankcase breather back into the intake system. This not only reduces emissions, but also increases fuel economy.

To keep the pressure at one atmosphere in the fuel tank and in the fuel reservoir of a carburetor, these systems are vented to the surroundings. Historically, these vents were an additional source of HC emissions when fuel evaporated from these fuel reservoirs. To eliminate these emissions, fuel vents now include some form of filter or absorption system that stops the HC vapor from escaping. One such system absorbs the HCs onto the surface of a carbon filter element. Then, when the engine is operating, the element is back flushed and the HC is desorbed off the surface. The recovered HC is ducted into the engine intake with no resulting emissions.

Many modern gasoline pumps and other fuel-dispensing systems are equipped with vapor-collecting nozzles that reduce HC vapor lost to the atmosphere during refueling.

9.13 NOISE POLLUTION

Since the 1990s, noise generated by engines and other systems has been considered a pollution. Noise is sometimes defined as "undesirable sound," and at high levels is recognized as a possible health hazard. Large engines produce high levels of sound, and many countries now have laws governing acceptable levels of noise allowed in closed engine rooms of ships and in stationary applications. Engine noise from automobiles is less of a problem, and existing technologies are adequate for controlling it.

Sound is caused by pressure waves in an elastic medium (i.e., air) generated by vibrations of the engine components. These vibrations cause pressure pulses in the air, and these pulses transfer energy to the human ear; the greater the energy, the louder is the sound. Fortunately, the ear is not very sensitive, and so the quantifying scale is a logarithmic scale with units of decibels (dB). Normal conversation has a sound level of about 55 dB, while the ear begins to feel pain at about 120 dB. Many engine room codes allow noise up to 110 dB. The sensitivity of the human ear is closely related to sound frequency, being less sensitive at low frequency. For this reason, international standards are divided into three categories—A, B, and C—each related to a range of frequencies. In the United States the EPA has an acceptable level of drive-by noise for vehicles of 74 dB (A). This standard often must be considered in automobile design in areas such as tailpipe and muffler placement.

Noise is generated in many ways in an operating IC engine: pressure pulsations in the gaseous flow of intake and exhaust, fuel injectors, superchargers, chain and belt drives, and vibration by many engine components. If exhaust systems did not include resonators and mufflers, the noise generated would be a serious pollution.

Reduction in engine and exhaust noise can be done in one of three ways: *passive, semiactive,* or *active.* Noise reduction is accomplished passively by correct design and use of proper materials. The use of ribs and stiffeners, composite materials, and sandwich construction is now routine. This type of construction reduces noise vibrations in the various engine components. Mufflers and resonators in the exhaust system dampen out the majority of exhaust noise. Engine mounts, which connect the engine to the vehicle body, are designed to dampen out vibrations that would transmit sound to the passenger compartment.

Hydraulics are often used in semiactive noise abatement systems (e.g., some engines are equipped with flywheels that have hydraulic passages through which fluid can flow). The flywheels are designed such that, at different speeds, the fluid flows to the specific locations that help to provide proper stiffness for absorbing engine vibrations at the frequencies present at that speed. Some automobiles have hydraulic engine mounts connecting the engine to the automobile body. Fluids in these mounts act to absorb and dampen engine vibrations and isolate them from the passenger compartment. Engine mounts using electrorheological fluids, which will allow better vibration dampening at all frequencies, are being developed. The viscosity of these fluids can be changed by as much as a factor of 50:1 with the application of an external voltage. Engine noise (vibration) is sensed by accelerometers that feed this information into the engine management system (EMS). Here, the vibration frequency is analyzed, and the proper voltage is applied to the engine mounts to best dampen that frequency [38]. Response time is on the order of 0.005 seconds.

Active noise abatement is accomplished by generating *antinoise* to cancel out engine exhaust noise. This is done by sensing the noise with a receiver, analyzing the frequency of the noise, and then generating noise of equal frequency, but out of phase with the original noise. If two noises are at the same frequency, but 180° out of phase, the wave fronts cancel each other and the noise is eliminated. This method works well with constant-speed engines and other rotating equipment, but is only partially successful with variable-speed automobile engines. It requires additional electronic equipment (receiver, frequency analyzer, transmitter) than that used with normal EMS computers. Some automobiles have receivers and transmitters mounted under the seats in the passenger compartment as an active engine noise abatement system. Similar systems are used near the end of the tailpipe, a major source of engine-related noise.

Engine Compression Brakes

A secondary use of variable valve timing in large trucks with CI engines is to use the engine as a vehicle brake by changing the exhaust valve timing and lift (sometimes called *Jake Brakes*). SI engines have a throttle valve that is closed when the vehicle decelerates, stopping most air flow through the engine and creating a vacuum in the intake stroke. This creates a large negative work in the pumping loop of the cycle, which absorbs some of the kinetic energy of the vehicle. CI engines do not have a throttle valve and under normal operating conditions would get very little energy absorption and braking effect from the engine when the vehicle decelerates. However, by shutting off the fuel supply and changing the exhaust valve timing, the engine can be used as a vehicle brake. The exhaust valve is kept closed during what would normally be the exhaust stroke. This changes the engine into a compressor which absorbs some vehicle kinetic energy as input work. The exhaust valve is opened when the piston nears TDC and the cylinder air is exhausted. Some engines are capable of increasing their exhaust valve lift to assist this blowdown process, limited only to the extent necessary to avoid piston–valve contact.

Unfortunately, engine compression brakes are very noisy, and their use is outlawed in many towns. A quieter method of using the engine as a vehicle brake is an *exhaust brake* which is sometimes used on medium-size trucks, [208]. This consists of a large butterfly valve installed in the exhaust system between the turbocharger and exhaust pipe. This takes the place of the butterfly throttle valve that SI engines have and CI engines do not have. To slow the vehicle, the butterfly valve is closed (throttle closed on SI automobiles), which restricts air flow through the engine and allows the engine to act as a brake.

PROBLEMS

9.1 A diesel truck uses 100 grams of light diesel fuel (assume $C_{12}H_{22}$) per mile of travel. One-half percent of the carbon in the fuel ends up as exhaust smoke. If the truck travels 15,000 miles per year, how much carbon is put into the atmosphere each year as smoke? [kg/year]

9.2 **(a)** Why isn't a normal three-way catalytic converter, as used with SI engines, as useful when used with a CI engine? **(b)** What main method is used to limit NOx emissions on a modern diesel truck or automobile? **(c)** Give at least three disadvantages to using this method.

9.3 **(a)** List five reasons that there are HC emissions in the exhaust of an automobile. **(b)** To reduce emissions from an SI engine, should AF be set at rich, lean, or stoichiometric? Explain the advantages and disadvantages of each. **(c)** Why is it good to place a catalytic converter as close to the engine as possible? Why is this bad?

9.4 A four-cylinder, 2.8-liter, four-stroke cycle SI engine operates at 2300 RPM with a volumetric efficiency of 88.5%. The fuel used is methyl alcohol at an equivalence ratio of $\phi = 1.25$. During combustion, all hydrogen is converted to water, and all carbon is converted to CO_2 and CO.
Calculate:

(a) Mole fraction of CO in the exhaust. [%]

(b) Energy lost in the exhaust due to CO. [kW]

9.5 The combustion chambers of a V8 Otto cycle engine with a 7.8:1 compression ratio, bore of 3.98 inches, and 410-cubic-inch displacement can be approximated as right circular cylinders. The engine operates at 3000 RPM using gasoline at an AF = 15.2 and a volumetric efficiency of 90%. When combustion occurs, the flame is dampened out near the walls and a boundary layer of air–fuel mixture does not get burned. Combustion is at constant volume at TDC, and the unburned boundary layer can be considered to be 0.004 inch thick over the entire combustion chamber surface. Fuel is originally distributed equally throughout the chamber.
Calculate:

(a) Percent of fuel that does not get burned due to being trapped in the surface boundary layer. [%]

(b) Amount of fuel lost in the exhaust due to this boundary layer. [lbm/hr]

(c) Chemical power of the fuel lost in the exhaust. [hp]

9.6 An older automobile using leaded gasoline gets 16 mpg fuel economy at 55 mph. The lead in the gasoline amounts to 0.15 gm/L. Forty-five percent of the lead in the fuel gets exhausted to the environment. Calculate the amount of lead exhausted to the environment in lbm/mile and lbm/day if the automobile is driven continuously.

9.7 A small truck has a four-cylinder, 2.2-liter CI engine that operates on an air-standard Dual cycle using light diesel fuel at an average AF = 21. At a speed of 2500 RPM, the engine has a volumetric efficiency $\eta_v = 92\%$. At this operating condition, 0.4% of the carbon in the fuel ends up as soot in the exhaust. In addition, there is 20% additional carbon soot from the lubricating oil. The amount of soot is then increased by 25% due to other components condensing on the carbon. Carbon density $\rho_c = 1400\ kg/m^3$.
Calculate:

(a) Rate of soot put into the environment. [kg/hr]

(b) Chemical power lost in the soot. (Consider the entire mass of soot as carbon.) [kW]

(c) Number of soot clusters exhausted per hour. Assume that an average cluster contains 2000 spherical carbon particles, and each particle has a diameter of 20 nm.

9.8 The engine in Problem 9-7 operating at 2500 RPM has an indicated thermal efficiency of 61%, combustion efficiency of 98%, and mechanical efficiency of 71%.
Calculate:

(a) Brake specific fuel consumption. [gm/kW-hr]

(b) Specific emissions of soot particulates. [gm/kW-hr]

(c) Emissions index of soot particulates. [gm/kg]

9.9 The clearance volume in the cylinder of a large stationary SI engine can be approximated as a right circular cylinder as shown in Fig. 6-16, with a diameter of 6 cm and a depth of 2 cm. The gas mixture in the combustion chamber is a homogeneous mixture of gasoline vapor and air at an AF = 16. It can be assumed that complete combustion occurs at TDC except for a thin boundary layer 0.10 mm thick on all clearance volume surfaces where the closeness of the walls damps out combustion and the air–fuel mixture does not get burned. Fuel flow into the entire engine equals 0.040 kg/sec.
Calculate:

(a) Percent of fuel that is in the boundary layer and so does not get burned. [%]
(b) Chemical energy lost in exhaust due to unburned fuel from the boundary layer. [kW]
(c) Emissions index of HC due to this unburned fuel. [gm_{HC}/kg_f]

9.10 A turbocharged, 6.4-liter, V8 SI engine operates on an air-standard Otto cycle at WOT with an engine speed of 5500 RPM. The compression ratio is r_c = 10.4:1, and conditions in the cylinders at the start of compression are 65°C and 120 kPa. Crevice volume is equal to 2.8% of clearance volume, pressure is equal to cylinder pressure, and temperature equal to 185°C.
Calculate:

(a) Total engine crevice volume. [cm^3]
(b) Percent of fuel that is trapped in the crevice volume at the start of combustion at TDC. [%]

9.11 The engine in Problem 9-10 has a volumetric efficiency of 89% and uses isooctane as fuel at an air–fuel ratio AF = 14.2. Sixty percent of the fuel that is trapped in the crevice volume at the start of combustion is later burned due to additional cylinder motion.
Calculate:

(a) HC emissions in the exhaust due to the 40% of crevice volume fuel that does not get burned. [kg/hr]
(b) Chemical power lost in these HC emissions of the exhaust. [kW]

9.12 A large supercharged, two-stroke cycle, diesel ship engine with a displacement of 196 liters operates at 220 RPM. The engine has a delivery ratio of λ_{dr} = 0.95 and uses fuel oil that can be approximated as $C_{12}H_{22}$, at an air–fuel ratio of AF = 22. The ship is equipped with an ammonia injection system to remove NOx from the exhaust.
Calculate:

(a) Amount of NO entering the exhaust system if 0.1% of the nitrogen in the air is converted to NO. (Assume no other forms of NOx are produced.) [kg/hr]
(b) Amount of ammonia to be injected to remove all NO in the exhaust by the reaction given in Eq. (9-29). [kg/hr]

9.13 It is desired to reduce NOx generation in an engine that burns stoichiometric ethanol by using exhaust gas recycling (EGR) to lower the peak combustion temperature. The temperature of the air and fuel at the start of combustion is 700 K, and the exhaust gas can be approximated as N_2 at 1000 K. The enthalpy of ethanol at 700 K is −199,000 kJ/kgmole.
Calculate:

(a) Theoretical maximum temperature with stoichiometric ethanol and no EGR. [K]
(b) Percent EGR needed to reduce maximum temperature to 2400 K. [%]

9.14 It is desired to use electricity to preheat the catalytic converter on a four-cylinder SI engine of 2.8-liter displacement. The preheat zone of the converter consists of 20% of the total alumina volume. The specific heat of the ceramic is 765 J/kg-K, and the density ρ = 3970 kg/m^3. Energy is obtained from a 24-volt battery supplying 600 amps.
Calculate:

(a) Electrical energy needed to heat the preheat zone from 25°C to a light-off temperature of 150°C. [kJ]

(b) Time needed to supply this amount of energy. [sec]

9.15 A 0.02-liter, two-stroke cycle SI lawn mower engine runs at 900 RPM, using gasoline at an equivalence ratio $\phi = 1.08$ and a fuel-to-oil ratio of 60:1 by mass. The engine is crankcase compressed and has a delivery ratio $\lambda_{dr} = 0.88$ and a charging efficiency $\lambda_{ce} = 0.72$. Combustion efficiency $(\eta_c)_{gasoline} = 0.94$ for the gasoline trapped in the cylinder, but $(\eta_c)_{oil}$ is only 0.72 for the oil trapped in the cylinder. There is no catalytic converter.
Calculate:

(a) HC from the fuel and oil exhausted to the environment due to valve overlap during scavenging. [kg/hr]

(b) HC in the exhaust from unburned fuel and oil due to combustion inefficiency. [kg/hr]

(c) Total HC in exhaust. [kg/hr]

9.16 A 3.2–liter, V6, four-stroke cycle, CI truck engine with a volumetric efficiency of 93% produces 92 kW of brake power at 3600 RPM, using light diesel fuel at AF = 22. The fuel contains 450 ppm of sulfur by mass, which is exhausted to the environment. In the surroundings the sulfur is converted to sulfurous acid by reacting with atmospheric oxygen and water vapor as given in Eqs. (9-15) and (9-18).
Calculate:

(a) Rate of fuel being used in engine. [kg/hr]

(b) Rate of sulfur in exhaust flow. [gm/hr]

(c) Specific emissions of sulfur in exhaust flow. [gm/kW-hr]

(d) Amount of sulfurous acid added to the environment. [kg/day]

9.17 A 5.2-liter, V8, four-stroke cycle CI truck engine with a volumetric efficiency of $\eta_v = 96\%$ operates at 2800 RPM, using light diesel fuel at AF = 20. The fuel contains 500 ppm of sulfur by mass, which is exhausted to the environment. In the surroundings, this sulfur is converted to sulfurous acid by reacting with atmospheric oxygen and water vapor as given in Eqs. (9-15) and (9-18).
Calculate:

(a) Amount of sulfur in the engine exhaust. [gm/hr]

(b) Amount of sulfurous acid added to the environment. [kg/hr]

9.18 An SI automobile engine produces 40 kW of brake power while using on average 8 kg of stoichiometric isooctane per 100 km traveled at 100 km/hr. Average emissions from the engine upstream of the catalytic converter include 12 grams of CO per kilogram of fuel used. The catalytic converter removes 95% of the exhaust emissions when it is at steady–state temperature. However, 10% of the time, the converter is cold at start-up and removes no emissions.

Calculate:

(a) Specific emissions of CO upstream of catalytic converter. [gm/kW-hr]

(b) Overall average of specific emissions of CO (cold and warmed) downstream of catalytic converter. [gm/kW-hr]

(c) Percentage of total CO emissions to the environment that occurs when converter is cold. [%]

9.19 A modern six-cylinder automobile CI engine is adjusted to operate properly using diesel fuel with a cetane number of 52. The vehicle is accidently fueled with a diesel fuel having a cetane number of 42. Would more or less exhaust smoke be expected? Explain.

9.20 In 1972, about 2.33×10^9 barrels of gasoline were consumed in the United States as fuel for internal combustion engines. The average automobile traveled 16,000 km, using gasoline at a rate of 15 L/100 km. The gasoline, on average, contained 0.15 gm/liter of lead, 35% of which was exhausted to the environment. One barrel = 160 liters.

Calculate:

(a) The amount of lead put into the atmosphere yearly by the average automobile. [kg]

(b) Total amount of lead put into the atmosphere in 1972. [kg]

9.21 A man wants to work on his automobile in his garage on a winter day. Having no heating system in the garage, he runs the automobile in the closed building to heat it. At idle speed, the engine burns 5 lbm of stoichiometric gasoline per hour, with 0.6% of the exhaust being carbon monoxide. The inside dimensions of the garage are 20 ft by 20 ft by 8 ft, and the temperature is 40°F. It can be assumed that 10 parts per million (ppm) of CO in the air is dangerous to health. Calculate the time to when the CO concentration in the garage is dangerous. [min]

9.22 An SI automobile engine produces 32 kW of brake power while using, on average, 6 kg of stoichiometric gasoline per 100 km traveled at 100 km/hr. Average emissions from the engine upstream of the catalytic converter are 1.1 gm/km of NO_2, 12.0 gm/km of CO, and 1.4 gm/km of HC. A catalytic converter removes 95% of the exhaust emissions when it is at steady-state temperature. However, 10% of the time, the catalytic converter is cold at start-up and removes no emissions.

Calculate:

(a) Specific emissions of HC upstream of the catalytic converter. [gm/kW-hr]

(b) Specific emissions of CO downstream of the catalytic converter, with the converter warmed. [gm/kW-hr]

(c) Concentration of NOx in the exhaust upstream of the catalytic converter. [ppm]

(d) Overall average (cold and warmed) of HC emissions to the atmosphere. [gm/km]

(e) Percentage of total HC emissions occurring when the converter is cold. [%]

DESIGN PROBLEMS

9.1D Design a catalytic converter preheater using solar energy. Decide if solar collectors should be on the vehicle or on a battery-recharging station. Calculate the sizes needed for the main components (e.g., battery, collector). Draw a simple schematic of the system.

9.2D Design a system to absorb the fuel vapors escaping from the vent on an automobile fuel tank. The system should have a method of regeneration, with all fuel eventually being input into the engine.

CHAPTER 10

Heat Transfer in Engines

"... it must be remembered that the reliability of an engine depends, not so much, it is true, on the proportion of the total heat converted into useful work, but rather upon the proportion of the total heat which is not so converted, and which is left over to make trouble."

The High-Speed Internal-Combustion Engine

by *Harry R. Ricardo* (1923)

This chapter examines the heat transfer that occurs within an IC engine, this being extremely important for proper operation. About 35% percent of the total chemical energy that enters an engine in the fuel is converted to useful crankshaft work, and about 30% of the fuel energy is carried away from the engine in the exhaust flow in the form of enthalpy and chemical energy. This leaves about one-third of the total energy that must be dissipated to the surroundings by some mode of heat transfer. Temperatures within the combustion chamber of an engine reach values on the order of 2700 K and up. Materials in the engine cannot tolerate this kind of temperature and would quickly fail if proper heat transfer did not occur. Removing heat is highly critical in keeping an engine and engine lubricant from thermal failure. On the other hand, it is desirable to operate an engine as hot as possible to maximize thermal efficiency.

Two general methods are used to cool combustion chambers of engines. The engine block of a **water-cooled** engine is surrounded with a **water jacket** that contains a coolant fluid which is circulated through the engine. An **air-cooled** engine has a finned outer surface on the block over which a flow of air is directed.

10.1 ENERGY DISTRIBUTION

The amount of energy (power) available for use in an engine is

$$\dot{W} = \dot{m}_f Q_{HV} \tag{10-1}$$

where

$\dot{m}_f$ = fuel flow rate into the engine
Q_{HV} = heating value of the fuel

The mass flow of fuel is limited by the mass flow of air that is needed to react with the fuel. Brake thermal efficiency gives the percentage of this total energy that is converted to useful output at the crankshaft:

$$(\eta_t)_{brake} = \dot{W}_b / \dot{m}_f Q_{HV} \eta_c \tag{10-2}$$

where

η_t = thermal efficiency
η_c = combustion efficiency
$\dot{W}_b$ = brake power

The rest of the energy can be divided into heat losses, parasitic loads, and energy that is lost in the exhaust flow. Figure 10-1 shows a typical distribution of energy use in an IC engine given as a percentage of total fuel energy. The total sums to over 100% because friction losses are counted twice, first as the original loss and then again in the resulting heat losses. For any engine,

$$\text{Power generated} = \dot{W}_{shaft} + \dot{Q}_{exhaust} + \dot{Q}_{loss} + \dot{W}_{acc} \tag{10-3}$$

where

$\dot{W}_{shaft}$ = brake output power off of the crankshaft
$\dot{Q}_{exhaust}$ = energy lost in the exhaust flow
$\dot{Q}_{loss}$ = all other energy lost to the surroundings by heat transfer
$\dot{W}_{acc}$ = power to run engine accessories

Depending on the size and geometry of an engine, as well as on how it is being operated, the shaft power output is

$$\dot{W}_{shaft} \approx 25\text{–}40\%$$

CI engines are generally on the high end of this range, and SI engines are on the lower end. Energy lost in exhaust flow is

$$\dot{Q}_{exhaust} \approx 20\text{–}45\%$$

A greater percentage of energy is lost in the exhaust of SI engines because of their higher exhaust temperatures. Lost exhaust energy is made up of two parts: enthalpy (heat) and chemical energy. When the engine is running rich at full load, chemical energy makes up about half of the exhaust loss. Under many operating conditions, lost exhaust energy exceeds the brake power output of the engine. Other heat losses are

$$\dot{Q}_{loss} \approx 10\text{–}35\%$$

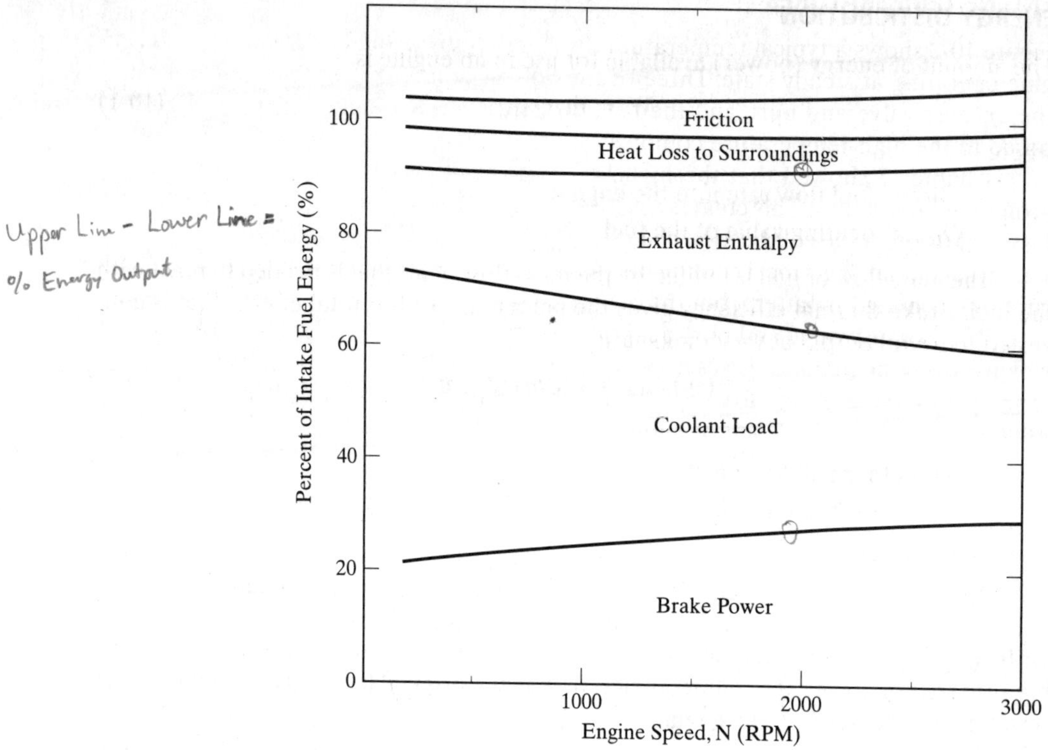

FIGURE 10-1

Distribution of energy in a typical SI engine as a function of engine speed. Friction losses, which are generally on the order of 10%, add to other heat losses and make the total energy distribution greater than 100%.

For many engines, the heat losses can be subdivided:

$$\dot{Q}_{loss} = \dot{Q}_{coolant} + \dot{Q}_{oil} + \dot{Q}_{ambient}$$

With CI engines on the high end, heat flow to the coolant is about

$$\dot{Q}_{coolant} \approx 10\text{–}30\%$$

At high load, energy lost to the coolant can amount to about half of the brake power output, increasing to about twice the brake power output at low load. Depending on the type of oil and engine speed,

$$\dot{Q}_{oil} \approx 5\text{–}15\%$$

Losses directly to the surroundings are

$$\dot{Q}_{ambient} \approx 2\text{–}10\%$$

Friction losses are on the order of

$$\dot{W}_{friction} \approx 10\%$$

10.2 ENGINE TEMPERATURES

Figure 10-2 shows a typical temperature distribution that would be found in an IC engine operating at steady state. Three of the hottest points are around the spark plug, the exhaust valve and port, and the face of the piston. Not only are these places exposed to the high-temperature combustion gases, but they are difficult places to cool.

Chapter 7 showed that the highest gas temperatures during combustion occur around the spark plug. This creates a critical heat transfer problem area. The spark plug fastened through the combustion chamber wall creates a disruption in the surrounding water jacket, causing a local cooling problem. On air-cooled engines, the spark plug disrupts the cooling fin pattern, but the problem may not be as severe.

The exhaust valve and port operate hot because they are located in the pseudo-steady flow of hot exhaust gases and create a difficulty in cooling similar to the one the spark plug creates. The valve mechanism and connecting exhaust manifold make it very difficult to route coolant or allow a finned surface to give effective cooling.

The piston face is difficult to cool because it is separated from the water jacket or outer finned cooling surfaces.

Engine Warm-up

As a cold engine heats up to steady-state temperature, thermal expansion occurs in all components. The magnitude of this expansion will be different for each component, depending on its temperature and the material from which it is made. Engine bore limits the thermal expansion of the pistons, and at operating temperatures of a newer engine,

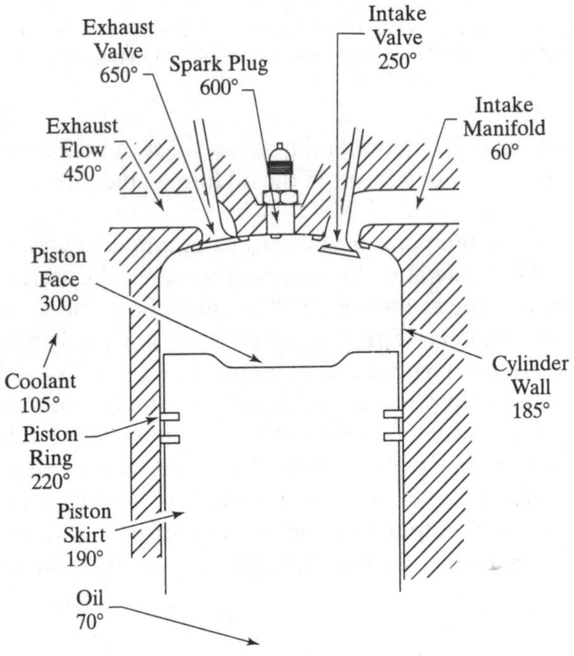

FIGURE 10-2

Typical temperature values found in an SI engine operating at normal steady state conditions. Temperatures are in degrees C.

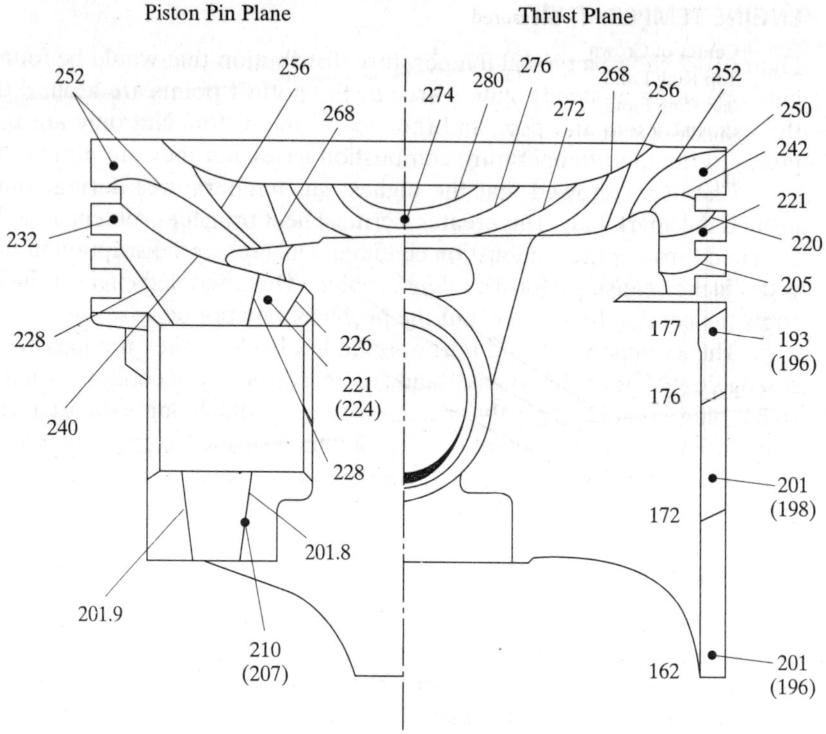

FIGURE 10-3

Steady-state temperatures of piston in plane of piston pin and in thrust plane. Engine is a four-cylinder, 2.5 liter, SI engine running at 4600 RPM with WOT. Temperatures are in °C. Reprinted with permission from SAE Paper No. 820086 © 1982 SAE International, [194].

there can be very high resulting forces between the piston rings and skirt and the walls of the cylinder. This causes high viscous heating in the oil film on the cylinder walls during engine operation.

Figure 10-5 shows how the temperature of various automobile components increases with time after a cold engine is started. In cold weather, the start-up time to reach steady-state conditions can be as high as 20–30 minutes. Some parts of the automobile reach steady state much sooner than this, but some do not. Fairly normal operating conditions may be experienced within a few minutes, but it can take as long as an hour to reach optimum fuel consumption rates. Engines are built to operate best at steady-state conditions, and full power and optimum fuel economy may not be realized until this state is reached. It would be poor practice to take off with an airplane, when full power is needed, before the engine is fully warmed up. This is not as critical with an automobile. A large percentage of automobile use is for short trips with engines that are not fully warmed up. In Chapter 9, it was found that this was also a major cause of air pollution.

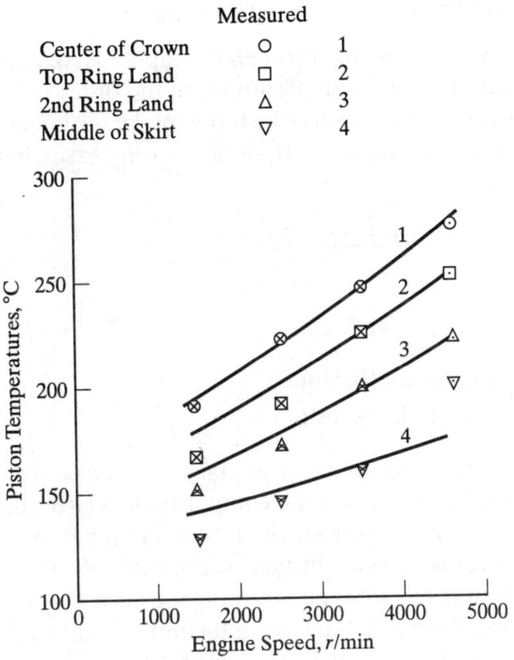

FIGURE 10-4

Piston temperatures as a function of engine speed at full load. Engine is a four-cylinder, 2.5 liter, SI engine running at 4600 RPM with WOT. Reprinted with permission from SAE Paper No. 820086 © 1982 SAE International, [194].

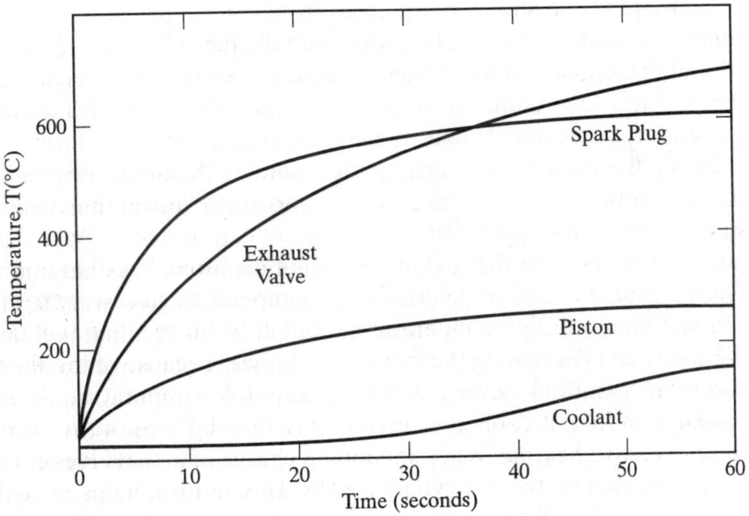

FIGURE 10-5

Temperatures of engine components of a typical SI engine as a function of time after cold start-up.

10.3 HEAT TRANSFER IN INTAKE SYSTEM

As the air or air–fuel mixture enters an engine through the intake system, its temperature increases from ambient conditions to a temperature on the order of 60°C. This is due to a number of thermodynamic processes to which the intake air is subjected.

The walls of the intake manifold are hotter than the flowing gases, heating them by convection:

$$\dot{Q} = hA(T_{wall} - T_{gas}) \tag{10-4}$$

where

T = temperature
h = convection heat transfer coefficient
A = inside surface area of intake manifold

The manifold is hot, either by design on some engines or just as a result of its location close to other hot components in the engine compartment. Carbureted engines and those with throttle body injection that introduce fuel early in the flow process purposely have heated intake manifolds to assist in the evaporation of the fuel. Various methods are used to heat these manifolds. Some are designed such that the flow passages of the runners come in close thermal contact with the hot exhaust manifold. Others use hot coolant flow through a surrounding water jacket. Electricity is used to heat some intake manifolds. Some systems have special localized hot surfaces, called *hot spots*, in optimum locations, such as immediately after fuel addition or at a tee where maximum convection occurs (Fig. 10-6).

There are several consequences from convective heating in the intake manifold, some good and some bad. The earlier that the fuel gets vaporized, the longer it is mixed with air, resulting in a more homogeneous mixture. However, increasing the temperature reduces the volumetric efficiency of the engine by two mechanisms. Higher temperature reduces the air density and added fuel vapor displaces some of the air, both reducing the mass of air reaching the cylinders. A compromise is to vaporize some of the fuel in the intake system and to vaporize the rest in the cylinder during compression, or even during combustion. With older carbureted engines, it was desirable to vaporize about 60% of the fuel in the intake manifold. Vaporization curves like those in Fig. 4-2 could be used to determine the temperature necessary to give this 60% evaporation. Often, a design temperature about 25°C higher than that determined from Fig. 4-2 was used in designing the manifold. This was because of the short time that the flow was in the manifold, never reaching steady-state temperature. As fuel vaporizes in the intake manifold, it cools the surrounding flow by evaporative cooling, counteracting the convective heating. When the intake charge of air and fuel enters the cylinder, it is further heated by the hot cylinder walls. This, in turn, helps to cool the cylinder walls and to keep them from overheating.

Another reason to limit the heating of inlet air is to keep temperature to a minimum at the start of the compression stroke. The higher the temperature at the start of compression, the higher will be all temperatures throughout the rest of the cycle, and the greater is the potential problem of engine knock.

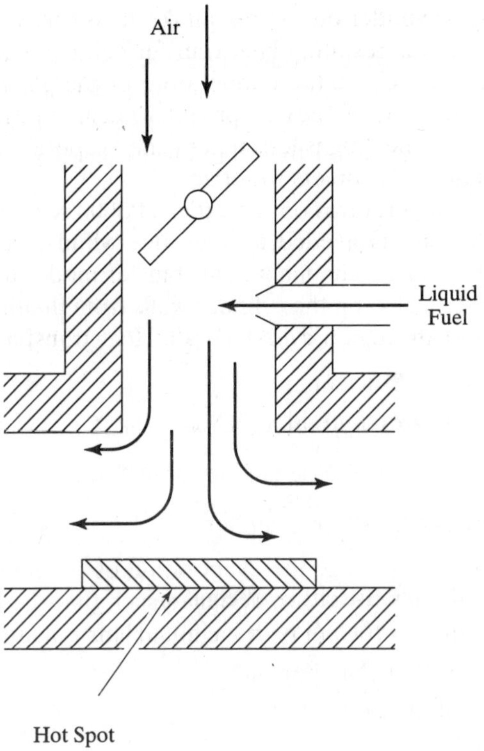

FIGURE 10-6
Hot spot in intake manifold to accelerate fuel evaporation. Localized sections of wall surface, called hot spots, can be heated by engine coolant, by conduction from the exhaust manifold, or by electrical heating. Heated sections are generally placed close after fuel addition, or at a tee where high convection occurs.

Engine systems using multipoint port injectors have less need for heating the intake manifold, relying on finer fuel droplets and higher temperature around the intake valve to assure necessary fuel evaporation. This results in higher volumetric efficiency for these engines. Often, the fuel is sprayed directly onto the back of the intake valve face. This not only speeds evaporation, but cools the intake valve, which can reach cyclic temperatures up to 400°C. Steady-state temperature of intake valves generally is in the 200°−300°C range.

If an engine is supercharged or turbocharged, the temperature of the inlet air is also affected by the resulting compressive heating. To avoid this, many of these systems are equipped with **aftercooling**, which again lowers the temperature. Aftercoolers are heat exchangers through which the compressed inlet air flows, using either engine coolant or external air flow as the cooling fluid.

10.4 HEAT TRANSFER IN COMBUSTION CHAMBERS

Once the air–fuel mixture is in the cylinders of an engine, the three primary modes of heat transfer (conduction, convection, and radiation) all play an important part for smooth steady-state operation. In addition, the temperature within the cylinders is affected by a phase change–evaporation of the remaining liquid fuel.

The air–fuel mixture entering a cylinder during the intake stroke may be hotter or cooler than the cylinder walls, with the resulting heat transfer being possible in either direction. During the compression stroke, the temperature of the gas increases, and by the time combustion starts, there is already a convective heat transfer to the cylinder walls. Some of this compressive heating is lessened by the evaporative cooling that occurs when the remaining liquid fuel droplets vaporize.

During combustion, peak gas temperatures on the order of 3000 K occur within the cylinders, and effective heat transfer is needed to keep the cylinder walls from overheating. Convection and conduction are the main heat transfer modes to remove energy from the combustion chamber and keep the cylinder walls from melting.

Figure 10-7 shows heat transfer through a cylinder wall. Heat transfer per unit surface area will be

$$\dot{q} = \dot{Q}/A = (T_g - T_c)/[(1/h_g) + (\Delta x/k) + (1/h_c)] \tag{10-5}$$

where

T_g = gas temperature in the combustion chamber
T_c = coolant temperature
h_g = convection heat transfer coefficient on the gas side
h_c = convection heat transfer coefficient on the coolant side
Δx = thickness of the combustion chamber wall
k = thermal conductivity of the cylinder wall

Heat transfer in Eq. (10-5) is cyclic. Gas temperature T_g in the combustion chamber varies greatly over an engine cycle, ranging from maximum values during combustion to

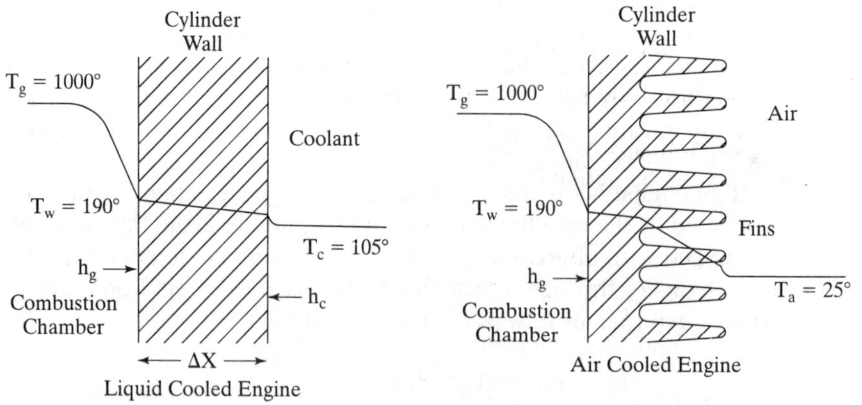

FIGURE 10-7
Heat transfer through the combustion chamber cylinder wall of an IC engine. The cylinder gas temperature T_g and convection heat transfer coefficient h_g vary over large ranges for each engine cycle, while the coolant temperature T_c (or air temperature T_a) and heat transfer coefficient h_c are fairly constant. As a result of this, heat conduction is cyclic for a small depth into the cylinder wall on the combustion chamber side. Temperatures are in degrees C.

minimum during intake. It can even be less than wall temperature early in the intake stroke, momentarily reversing heat transfer direction. Coolant temperature T_c is fairly constant, with any changes occurring over much longer cycle times. The coolant is air for air-cooled engines and antifreeze solution for water-cooled engines. The convection heat transfer coefficient h_g on the cylinder gas side of the wall varies greatly during an engine cycle due to changes in gas motion, turbulence, swirl, velocity, etc. This coefficient will also have large spatial variation within the cylinder for the same reasons. The convection heat transfer coefficient on the coolant side of the wall will be fairly constant, being dependant on coolant velocity. Thermal conductivity k of the cylinder wall is a function of wall temperature and will be fairly constant.

Convection heat transfer on the inside surface of the cylinder is

$$\dot{q} = \dot{Q}/A = h_g(T_g - T_w) \tag{10-6}$$

Wall temperature T_w should not exceed 180°–200°C to assure thermal stability of the lubricating oil and structural strength of the wall.

There are a number of ways of identifying a Reynolds number to use for comparing flow characteristics and heat transfer in engines of different sizes, speeds, and geometries. Choosing the best characteristic length and velocity is sometimes difficult [40, 120]. One way of defining a Reynolds number for engines [120] which correlates data fairly well is

$$\text{Re} = [(\dot{m}_a + \dot{m}_f)B]/(A_p\mu_g) \tag{10-7}$$

where

$\dot{m}_a$ = mass flow rate of air into the cylinder
$\dot{m}_f$ = mass flow rate of fuel into the cylinder
B = bore
A_p = area of piston face
μ_g = dynamic viscosity of gas in the cylinder

From this Reynolds number, a Nusselt number for the inside of the combustion chamber can be defined as

$$\text{Nu} = h_gB/k_g = C_1(\text{Re})^{C_2} \tag{10-8}$$

where

C_1 and C_2 = constants
k_g = thermal conductivity of cylinder gas
h_g = average value of the convection heat transfer coefficient to be used in Eqs. (10-5) and (10-6)

The Nusselt number and convection heat transfer coefficient on the coolant side of the cylinder walls can be approximated by conventional methods of forced convection heat transfer.

Radiation heat transfer between cylinder gas and the combustion chamber walls is

$$\dot{q} = \dot{Q}/A = [\sigma(T_g^4 - T_w^4)]/\{[(1 - \epsilon_g)/\epsilon_g] + [1/F_{1-2}] + [(1 - \epsilon_w)/\epsilon_w]\} \quad (10\text{-}9)$$

where

T_g = gas temperature
T_w = wall temperature
σ = Stefan–Boltzmann constant
ϵ_g = emissivity of gas
ϵ_w = emissivity of wall
F_{1-2} = view factor between gas and wall

Even though gas temperatures are very high, radiation to the walls amounts to only about 10% of the total heat transfer in SI engines. This is due to the poor emitting properties of gases, which emit only at specific wavelengths. N_2 and O_2, which make up the majority of the gases before combustion, radiate very little, while the CO_2 and H_2O of the combustion products do contribute more to radiation heat transfer.

The solid carbon particles that are generated in the combustion products of a CI engine are good radiators at all wavelengths, and radiation heat transfer to the walls in these engines is in the range of 20–35% of the total. A large percentage of radiation heat transfer to the walls occurs early in the power stroke. At this point, the combustion temperature is maximum, and with thermal radiation potential equal to T^4, a very large heat flux is generated. This is also the time when there is a maximum amount of carbon soot in CI engines, which further increases radiative heat flow. Instantaneous heat fluxes as high as 10 MW/m^2 can be experienced in a CI engine at this point of the cycle.

Because an engine operates on a cycle, the gas temperature T_g within the cylinder in Fig. 10-7 and Eq. (10-5) is pseudo-steady state. This cyclic temperature causes a cyclic heat transfer to occur in the cylinder walls. However, due to the very short cycle times, this cyclic heat transfer occurs only to a very small surface depth. At normal speeds, 90% of these heat transfer oscillations are dampened out within a depth of about 1 mm of the surface in engines with cast-iron cylinder walls. In engines with aluminum cylinders, this depth of 90% dampening is slightly over 2 mm, and in ceramic walls it is on the order of 0.7 mm. At surface depths greater than these, the oscillations in heat transfer are almost undetectable and conduction can be treated as steady state [40].

Heat transfer to cylinder walls continues during the expansion stroke, but the rate quickly decreases. Expansion cooling and heat losses reduce the gas temperature within the cylinder during this stroke from a maximum temperature on the order of 2700 K to an exhaust temperature of about 800 K. During the exhaust stroke, heat transfer to the cylinder walls continues but at a greatly reduced rate. At this time, cylinder gas temperature is much lower, as is the convection heat transfer coefficient. There is no swirl or squish motion at this time, and turbulence is greatly reduced, resulting in a much lower convection heat transfer coefficient.

Cycle-to-cycle variations in combustion that occur within a cylinder result in cycle-to-cycle variations in cylinder wall temperature (Fig. 10-8) and cylinder wall heat

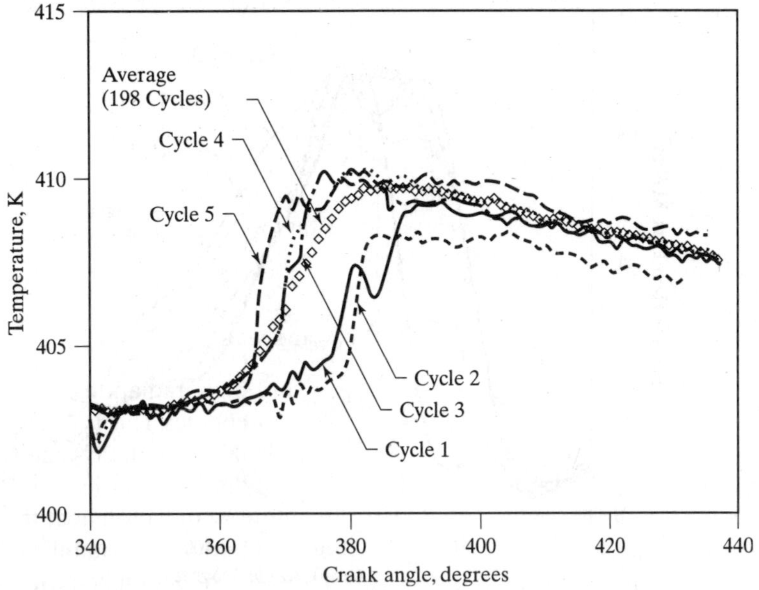

FIGURE 10-8

Temperature variation of five consecutive cycles recorded by sensor in wall of combustion chamber of an SI engine. Also shown is the average temperature for 198 cycles as a function of crank angle, 360° being top-dead-center. The engine has bore of 10.47 cm and stroke of 9.53 cm, and was operating at 1500 RPM with an equivalence ratio of 0.87. Reprinted with permission from *Journal of Heat Transfer* © ASME, [2].

flux (Fig. 10-9). Not only are there time variations at any given point, but there are spacial variations from one point to another on the combustion chamber wall.

Heat transfer occurs in all four strokes of a cycle, ranging from very high fluxes to low fluxes and even to zero or heat flow in the reverse direction (i.e., heat flow from the walls to the gas mixture in the cylinder). In a naturally aspirated engine, heat flow can be in either direction during the intake stroke, at a given cylinder location. During the compression stroke, the gases heat up and a heat flux to the walls results. Maximum temperature and maximum heat flux occur during combustion and then decrease during the power and exhaust strokes. For engines equipped with a supercharger or turbocharger, the intake gases are at a higher temperature, and the corresponding heat flux during intake is greater and into the walls.

Cooling difficulties caused by the protrusions of the spark plug, fuel injectors, and valves through the cylinder walls have been discussed. Another major cooling problem is the face of the piston. This surface is exposed to the hot combustion process but cannot be cooled by the coolant in the engine water jacket or an external finned surface. For this reason, the piston crown is one of the hotter points in an engine. One method used to cool the piston is by splashing or spraying lubricating oil on the back surface of the piston crown. In addition to being a lubricant, the oil then also serves as a coolant

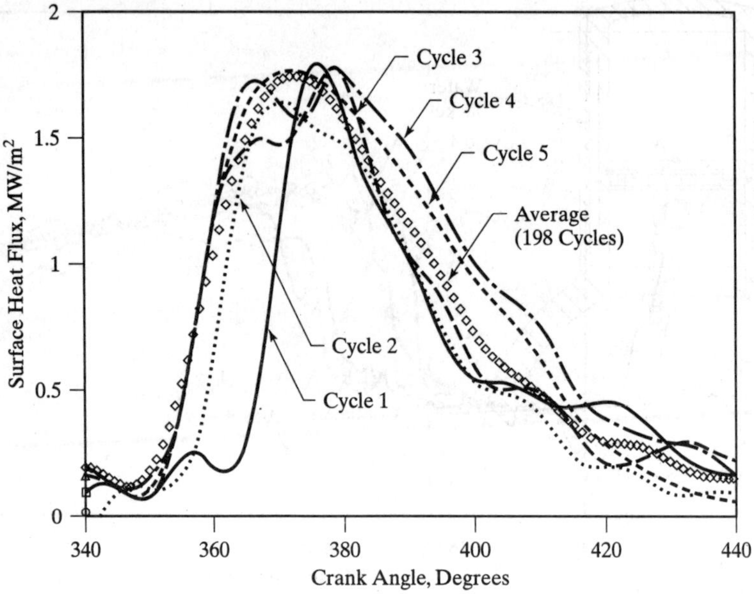

FIGURE 10-9

Variation in measured heat transfer rate at combustion chamber wall for five consecutive cycles of an SI engine operating at 1500 RPM with an equivalence ratio of 0.87. Also shown is the average value of 198 cycles as a function of crank angle. Reprinted with permission from *Journal of Heat Transfer* © ASME, [2].

fluid. After absorbing energy from the piston, the oil flows back into the oil reservoir in the crankcase, where it is again cooled. Heat is also conducted from the piston face, but thermal resistance for this is quite high. The two conduction paths available are (1) down the connecting rod to the oil reservoir, and (2) through the piston rings to the cylinder walls and into the coolant in the surrounding water jacket (Fig. 10-10). Thermal resistance through the piston body and connecting rod is low because they are made of metal. However, there is high resistance where these connect together at the wrist pin because of the lubricant film between the surfaces. This is also true where the connecting rod fastens to the crankshaft through lubricated surfaces. The oil film between surfaces needed for lubrication and wear reduction constitutes a large thermal resistance and a poor conduction path.

Aluminum pistons generally operate 30°–80°C cooler than cast-iron pistons due to their higher thermal conductivity. This reduces knock problems in these engines, but can cause greater thermal expansion problems between dissimilar materials. Many modern pistons have a ceramic face and operate at a higher steady-state temperature. Ceramic has poor heat conduction properties but can tolerate much higher temperatures. Some very large engines have water-cooled pistons.

To avoid thermal breakdown of the lubricating oil, it is necessary to keep the cylinder wall temperatures from exceeding 180°–200°C. As lubrication technology improves the quality of oils, this maximum allowable wall temperature is being raised. As

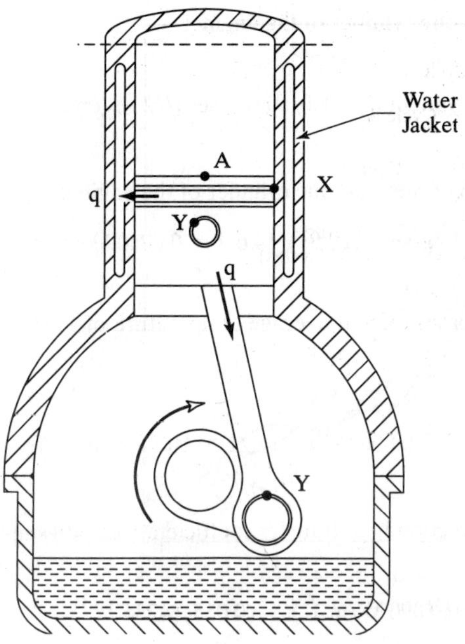

FIGURE 10-10

Cooling of piston. The face of a piston (A) is one of the hotter surfaces in a combustion chamber. Cooling is done mainly by convection to the lubricating oil on the back side of the piston face, by conduction through the piston face, by conduction through the piston rings in contact with the cylinder walls, and by conduction down the connecting rod to the oil reservoir. High conduction resistance occurs because of lubricated surfaces at cylinder walls (*X*) and rod bearings (*Y*).

an engine ages, deposits slowly build up on the walls of the cylinders, due to impurities in the air and fuel, imperfect combustion, and lubricating oil in the combustion chamber. These deposits create a thermal resistance and cause higher wall temperatures. Excessive wall deposits also slightly decrease the clearance volume of the cylinder and cause a rise in the compression ratio.

Some modern engines use heat pipes to help cool internal hot regions that are inaccessible to normal cooling by conduction or coolant flow. With one end of the heat pipe in the hot interior of the engine, the other end can be in contact with the circulating coolant or exposed to external air flow.

Example Problem 10-1

A 3.0 liter, 5-cylinder, 4-stroke cycle SI engine, with a volumetric efficiency of 82%, operates at 3000 RPM using gasoline with a lambda value of 0.91. Bore and stroke are related as $S = 1.08$ B. At a certain point in the engine cycle, the gas temperature in the combustion chamber is $T_g = 2100°\text{C}$ while the cylinder wall temperature is $T_w = 190°\text{C}$. Calculate the approximate convection heat transfer rate to the cylinder wall at this instant.

Use Eq. (2-8) to find bore:

$$V_d = (\pi/4)B^2S = [(3000 \text{ cm}^3)/(5 \text{ cylinders})] = (\pi/4)(B^2)(1.08\,B)$$
$$B = 8.91 \text{ cm} = 0.0891 \text{ m}$$

The area of one piston face is

$$A_p = (\pi/4)B^2 = (\pi/4)(0.0891 \text{ m})^2 = 0.006235 \text{ m}^2$$

Equation (2-71) gives the mass flow rate of air into one cylinder of the engine:

$$\begin{aligned}\dot{m}_a &= \eta_v \rho_a V_d N/n \\ &= (0.82)(1.181\ \text{kg/m}^3)[(0.003\ \text{m}^3/\text{cycle})/(5\ \text{cylinders})](3000/60\ \text{rev/sec})/(2\ \text{rev/cycle}) \\ &= 0.01453\ \text{kg/sec}\end{aligned}$$

Equations (2-55) and (2-58) give the mass flow rate of fuel into one cylinder of the engine:

$$\dot{m}_f = \dot{m}_a/(\text{AF})_{\text{act}} = \dot{m}_a/[\lambda(\text{AF})_{\text{stoich}}] = (0.01453\ \text{kg/sec})/[(0.91)(14.6)] = 0.00109\ \text{kg/sec}$$

Use Eq. (10-7) to find Reynolds Number:

viscosity μ_g and thermal conductivity k_g of gases (air) at average temperature of 1145°C are obtained from ref. [63]

$$\begin{aligned}\text{Re} &= [(\dot{m}_a + \dot{m}_f)B]/(A_p\mu_g) \\ &= \{[(0.01453) + (0.00109)\text{kg/sec}](0.0891\ \text{m})\}/[(0.006235\ \text{m}^2)(5.21 \times 10^{-5}\ \text{kg/m-sec})] \\ &= 4284\end{aligned}$$

Equation (10-8) gives the Nusselt number and convection heat transfer coefficient (ref. [40] suggests $C_1 = 0.035$ and $C_2 = 0.80$):

$$\text{Nu} = h_g B/k_g = C_1(\text{Re})^{C2} = h_g(0.0891\ \text{m})/(0.090\ \text{W/m-K}) = (0.035)(4284)^{0.80}$$

$$h_g = 28.44\ \text{W/m}^2\text{-K}$$

Equation (10-6) gives convection heat transfer rate at combustion chamber wall at this instant:

$$\dot{q} = h_g(T_g - T_w) = (28.44\ \text{W/m}^2\text{-K})(2373 - 463)\text{K} = \underline{54{,}320\ \text{W/m}^2 = 54.32\ \text{kW/m}^2}$$

10.5 HEAT TRANSFER IN EXHAUST SYSTEM

To calculate heat losses in an exhaust pipe, normal internal convection flow models can be used with one major modification: Due to the pulsing cyclic flow, the Nusselt number is about twice that which would be predicted for the same mass flow in the same pipe at steady flow conditions [82] (see Fig. 10-11). Heat losses from the exhaust system affect emissions and turbocharging.

Pseudo-steady-state exhaust temperatures of SI engines are generally in the range of 400°–600°C, with extremes of 300°–900°C. Exhaust temperatures of CI engines are lower due to their greater expansion ratio and are generally in the range of 200°–500°C.

Some automobile engines and large stationary engines have exhaust valves with hollow stems containing sodium. These act as heat pipes and are very effective in removing heat from the hot face area of the valve. Whereas solid stems remove heat by conduction only, heat pipes use a phase change cycle to remove a much greater amount of energy, up to 4000 W/cm^2 of surface area. Liquid sodium is vaporized in the hot end of the hollow valve stem and then is condensed back to liquid at the cooler end. Because of the large transfer of energy during a phase change, the *effective heat conduction* in the stem will be many times greater than pure conduction. Sodium is used as the

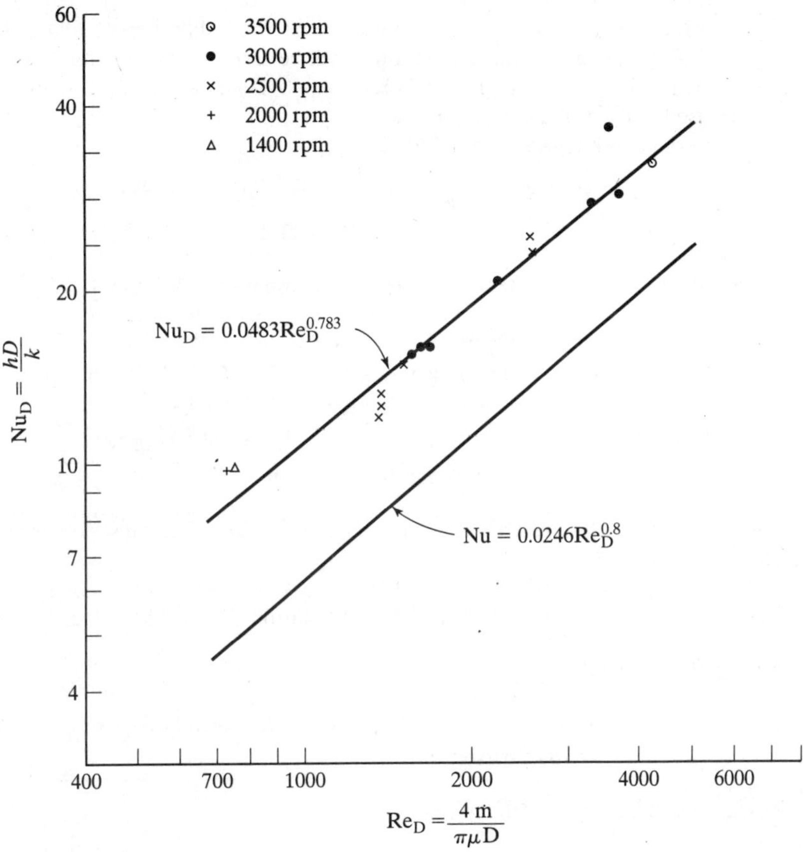

FIGURE 10-11

Average Nusselt number as a function of Reynolds number in the exhaust flow of a reciprocating IC engine (top curve). The cyclic pulsing that occurs in the exhaust increases the Nusselt number and convection heat transfer in the exhaust pipe by a factor of about two over steady-state flow conditions of equal mass flow rate in the same pipe (bottom curve). Reprinted with permission from SAE Paper No. 790309 © 1979 SAE International, [199].

working fluid because of its thermal properties, and its melting point of about 98°C (208°F) [223].

Example Problem 10-2

On the engine in Example Problem 8-1, the exhaust manifold and pipe leading from the engine to the catalytic converter can be approximated as a 1.8-m length of pipe with ID = 6.0 cm and OD = 6.5 cm. Volumetric efficiency of the engine at 3600 RPM is $\eta_v = 93\%$, the air–fuel ratio AF = 15:1, and the average wall temperature of the exhaust pipe is 200°C. Calculate the approximate temperature of the exhaust gas entering the catalytic converter.

Heat transfer equations from any standard textbook can be used; [63] will be used here. From Example Problem 8-1, the temperature of the exhaust gas leaving the engine is T_1 = 756 K = 483°C. As a first approximation, it will be assumed that the temperature loss in the exhaust pipe is ΔT = 100 K, or T_2 = 656 K = 383°C. As in air-standard analysis, air property values are used to approximate exhaust gas.

Average bulk temperature of gas is

$$T_{\text{BULK}} = (T_1 + T_2)/2 = (756 + 656)/2$$
$$= 706 \text{ K} = 433°\text{C}$$

Air property values, evaluated at average bulk temperature from [63]:

density	$\rho = 0.499\ kg/m^3$
kinematic viscosity	$\nu = 6.72 \times 10^{-5}$ m^2/sec
thermal conductivity	$k = 0.0526$ W/m-K
specific heat	$c_p = 1076$ J/kg-K
Prandtl number	$Pr = 0.684$

Equation (2-71) gives the mass flow rate of air. Mass flow of exhaust equals air plus fuel:

$$\dot{m}_{\text{ex}} = [\eta_v \rho_a V_d N/n](16/15)$$
$$= [(0.93)(1.181)(0.0064)(3600/60)/2](16/15) = 0.225 \text{ kg/sec}$$

Average flow velocity is

$$u = \dot{m}_{\text{ex}}/\rho A = (0.225 \text{ kg/sec})/(0.499 \text{ kg/m}^3)[(\pi/4)(0.06 \text{ m})^2]$$
$$= 159.5 \text{ m/sec}$$

Reynolds number for flow in a pipe is

$$\text{Re} = \text{ud}/\nu = (159.5 \text{ m/sec})(0.06 \text{ m})/(6.72 \times 10^{-5} \text{ m}^2/\text{sec}) = 142{,}411$$

Using the Dittus–Boelter equation for the Nusselt number of interior turbulent flow in a pipe,

$$\text{Nu} = 0.023 \text{ Re}^{0.8} \text{ Pr}^{0.3} = (0.023)(142{,}411)^{0.8}(0.684)^{0.3} = 272$$

This is multiplied by 2 because of the pulsed exhaust flow:

$$\text{Nu} = (2)(272) = 544$$

Convection heat transfer coefficient is

$$h = \text{Nu}(k/d) = (544)(0.0526 \text{ W/m-K})/(0.060 \text{ m}) = 477 \text{ W/m}^2\text{-K}$$

Convection heat transfer from exhaust gas to pipe walls is

$$\dot{Q} = hA(T_{\text{bulk}} - T_{\text{wall}}) = (477 \text{ W/m}^2\text{-K})[\pi(0.06 \text{ m})(1.8 \text{ m})](706 - 473)\text{K}$$
$$= 37{,}709 \text{ W.}$$

This gives a temperature drop in the exhaust flow between the engine and catalytic converter of

$$\Delta T = \dot{Q}/\dot{m}_{\text{ex}} c_p = (37{,}709 \text{ W})/(0.225 \text{ kg/sec})(1076 \text{ J/kg-K}) = 156°$$

Temperature of exhaust gas entering the catalytic converter is

$$T_2 = T_1 - \Delta T = 756 \text{ K} - 156 = 600 \text{ K} = 327°\text{C}$$

A second iteration is now done using these values for T_2 and ΔT in the calculations. This gives the temperature of the exhaust gas entering the catalytic converter:

$$\underline{\Delta T = 138°} \quad \underline{T_2 = 618 \text{ K} = 345°\text{C}}$$

10.6 EFFECT OF ENGINE OPERATING VARIABLES ON HEAT TRANSFER

Heat transfer within engines depends on so many different variables that it is difficult to correlate one engine with another. These variables include the air–fuel ratio, speed, load, brake mean effective pressure, spark timing, compression ratio, materials, and size. The subsections that follow provide general comparisons of some of these variables.

Engine Size

If two geometrically similar engines of different size (displacement) are run at the same speed, and all other variables (temperature, AF, fuel, etc.) are kept as close to the same as possible, the larger engine will have a greater absolute heat loss but will be more thermally efficient. If the temperatures and materials of both engines are the same, heat loss fluxes to the surroundings per unit area will be about the same, but the absolute heat loss of the larger engine will be greater due to its larger surface areas.

A larger engine will generate more output power and will do this at a higher thermal efficiency. As linear size goes up, volume increases on the order of linear dimension cubed. If one engine is 50% larger in linear size, its displacement will be on the order of $(1.5)^3 = 3.375$ larger. With similar mixture properties, the larger engine will therefore combust about 3.375 times the fuel of the smaller engine and will release 3.375 times the amount of thermal energy. Surface area, on the other hand, is proportional to length squared, and the larger engine will have only 2.25 times the surface area and consequent heat loss of the smaller engine. Energy generated goes up with length cubed, while heat losses go up with length squared. This makes the larger engine more efficient if everything else is the same.

This reasoning can be extended to more than just absolute size when designing an engine. What is desirable for good thermal efficiency is a combustion chamber with a high volume–to-surface area ratio. This is one reason why a modern overhead valve engine is more efficient than older valve-in-block L head engines that had large combustion chamber surface areas. This also says that a cylinder with a single, simple, open combustion chamber will have less percentage heat loss than one with a split dual chamber that has a large surface area.

Engine Speed

As engine speed is increased, gas flow velocity into and out of the engine goes up, with a resulting rise in turbulence and convection heat transfer coefficients. This increases heat transfer occurring during the intake and exhaust strokes and even during the early part of the compression stroke.

During combustion and power stroke, gas velocities within the cylinder are fairly independent of engine speed; being controlled instead by swirl, squish, and combustion motion. The convection heat transfer coefficient and, thus, convection are therefore fairly independent of engine speed at this time. Radiation, which is important only during this portion of the cycle, is also independent of speed. Rate of heat transfer (kW) during this part of the cycle is therefore constant, but because the time of the cycle is less at higher speed, less heat transfer per cycle (kJ/cycle) occurs. This gives the engine a higher thermal efficiency at higher speed. At higher speeds, more cycles per unit time occur, but each cycle lasts less time. The net result is a slight rise in heat transfer loss with time (kW) from the engine. This is partly due to the higher heat losses for part of the cycle, but is mostly due to the higher steady-state (pseudo-steady-state) losses that the engine establishes at higher speeds. Mass flow of gas through an engine increases with speed, with a net result of less heat loss per unit mass (kJ/kg) (i.e., higher thermal efficiency).

All steady-state temperatures within an engine go up as engine speed increases, as shown in Fig. 10-12.

Heat transfer to the engine coolant increases with higher speed

$$\dot{Q} = hA(T_w - T_c) \tag{10-10}$$

where

h = convection heat transfer coefficient, which remains about constant
A = surface area, which remains constant

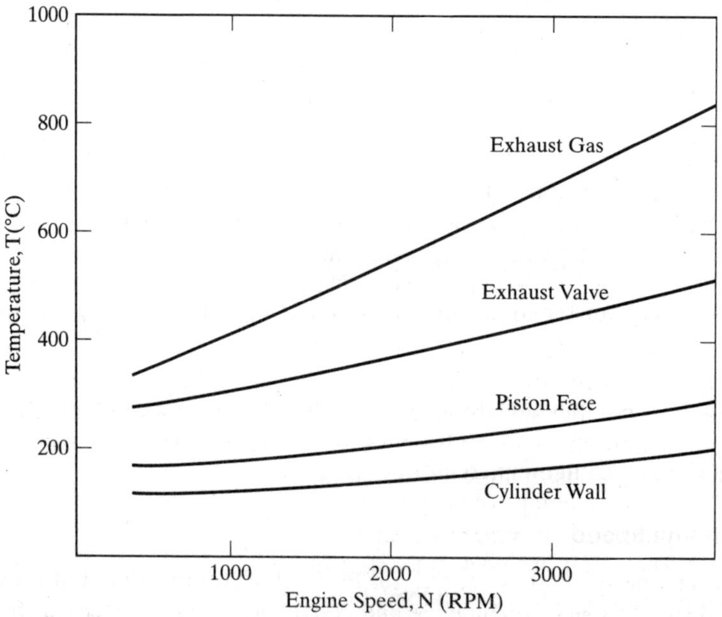

FIGURE 10-12

Engine temperatures as a function of engine speed for a typical SI engine.

T_c = coolant temperature, which remains about constant

T_w = wall temperature, which increases with speed

To stay at the same steady-state temperature as engine speed is increased, more heat must be transferred to the surroundings from the coolant in the automobile radiator heat exchanger.

At higher engine speeds, there is less time per cycle. Combustion occurs over about the same engine rotation (burn angle) at all speeds, so the time of combustion is less at higher speeds (Fig. 7-6). This means less time for self-ignition and knock. However, there is also less time for heat transfer per cycle, which means the engine runs hotter, and a hotter engine has a greater knock problem. The result of this is that some engines have a greater knock problem at higher speeds while other engines have less knock problem at higher speeds.

For part of the exhaust blowdown process, there will be sonic velocity through the exhaust valve, and the flow rate will be choked and independent of engine speed. At higher speeds, this causes the blowdown process to last over a larger engine rotation angle and results in hotter exhaust valves and ports. The hotter engine temperatures do increase sonic velocity slightly, which in turn increases the flow rate slightly. Gases in the exhaust system are hotter at higher engine speeds.

Load

As the load on an engine is increased (going uphill, pulling a trailer), the throttle must be opened further to keep the engine speed constant. This causes less pressure drop across the throttle and higher pressure and density in the intake system. Mass flow rate of air and fuel, therefore, goes up with load at a given engine speed. Heat transfer within the engine also goes up by

$$\dot{Q} = hA\Delta T \tag{10-11}$$

where

h = convection heat transfer coefficient

A = surface area at any point

ΔT = temperature difference at that point

The heat transfer coefficient is related to Reynolds number by

$$h \propto \mathrm{Re}^C \tag{10-12}$$

where C is a constant, usually on the order of 0.8. Reynolds number is proportional to the mass flow rate $\dot{m}$, so the time rate of heat transfer increases with $\dot{m}^{0.8}$. Density of fuel into the engine increases as $\dot{m}$, so energy into the engine increases as $\dot{m}$.

The percentage of heat loss goes down slightly as engine load increases (kJ/cycle). This quite often is offset by engine knock, which occurs most often in an engine under load. The result of knock is localized high temperature and high heat transfer. Engine temperatures increase with load. Figure 10-12 would be very similar if the X coordinate of speed at constant load were replaced with load at constant speed.

CI engines are run unthrottled, and total mass flow is almost independent of load. When speed or load is increased and more power is needed, the amount of fuel injected is increased. This increases the total mass flow in the latter part of each cycle only very slightly, on the order of 5%. This means that the convection heat transfer coefficient within the engine is fairly independent of engine load.

At light loads, less fuel is injected and burned, creating a cooler steady-state temperature. This decreases the corresponding heat transfer. At heavy load, more fuel is injected and burned, and the resulting steady-state temperature is higher. This causes a greater convective heat transfer. Combustion of the richer mixture at heavy load also creates a larger amount of solid carbon soot. This, in turn, further increases heat transfer by radiation, solid carbon being a good radiator. The amount of fuel and, consequently, the amount of energy released per cycle goes up with load. The percent of heat loss, therefore, changes very little with load in a CI engine.

Spark Timing

More power and higher temperatures are generated when the spark setting is set to give maximum pressure and temperature at about 5° to 10° aTDC. These higher peak temperatures will create a higher momentary heat loss, but this will occur over a shorter length of time. With spark timing set either too early or too late, combustion efficiency and average temperatures will be lower. These lower temperatures will give less peak heat loss, but the heat losses will last over a longer length of time and the overall energy loss will be greater. Higher power output is thus gained with correct ignition timing. Late ignition timing extends the combustion process longer into the expansion stroke, resulting in higher exhaust temperature and hotter exhaust valves and ports.

Fuel Equivalence Ratio

In an SI engine, maximum power is obtained with an equivalence ratio of about $\phi = 1.1$. This is also when the greatest heat losses will occur, with lower losses when the engine runs either leaner or richer. The greatest heat loss as a percentage of energy in $m_f Q_{HV}$, will occur at stoichiometric conditions, $\phi = 1.0$. An engine requires the highest fuel octane number when operating at stoichiometric conditions. Lower octane can be tolerated when the engine is running rich.

Evaporative Cooling—Water Injection

As fuel is vaporized during intake and start of compression, evaporative cooling lowers the intake temperature and raises intake density. This increases the volumetric efficiency of the engine. Fuels with high latent heats, such as alcohols, have greater evaporative cooling and generally make for cooler running engines. If an engine is operated fuel rich, evaporation of the excess fuel will lower cycle temperatures.

During World War II, water injectors were installed on the intake systems of high performance reciprocating engines of warplanes. As the water evaporates it increases evaporative cooling, which increases power because of higher volumetric efficiency. In recent years, the technology of adding water to the intake systems of engines is again being used. This is done in both automobile engines and in large ship and stationary engines. In addition to increasing volumetric efficiency and power, this is also done to decrease the generation of NOx by reducing cycle temperatures [200, 211]. Water can be added by one of three methods: (1) injection of water into the incoming air, either in the intake system or directly into the combustion chamber; (2) emulsifying water with the fuel; or (3) using high humidity inlet air. Of these methods, the direct injection of water seems to be the most practical, and is used in a number of existing systems. Water storage and concern of the water freezing are possible problems, especially with road vehicles. Mixing water with the fuel in an emulsion can cause possible mixing, storage, and/or injector problems. Large volumes of high humidity air can be difficult to provide, and can cause corrosion problems in the intake system.

Saab Automobile Company has been experimenting with water injection to improve fuel economy at high speed and fast acceleration. To avoid the need for an additional water source, fluid is taken from the windshield washer reservoir. Antifreeze solution and washer additives do not seem to harm the engine. High-speed fuel consumption has been reduced by 20–30% [70].

Example Problem 10-3

A large supercharged aircraft engine generates 900 kW at 3600 RPM when operating with air and gasoline at a fuel equivalence ratio $\phi = 1.05$. After supercharging and fuel addition, air enters the engine at 65°C. Gasoline can be approximated as isooctane.

Calculate how much the air is cooled by evaporative cooling when the fuel vaporizes.

The reaction for stoichiometric combustion is

$$C_8H_{18} + 12.5\,O_2 + 12.5(3.76)\,N_2 \rightarrow 8\,CO_2 + 9\,H_2O + 12.5(3.76)\,N_2$$

At an equivalence ratio of 1.05,

$$C_8H_{18} + (12.5/1.05)\,O_2 + (12.5/1.05)(3.76)\,N_2 \rightarrow$$
$$(8/1.05)\,CO_2 + (9/1.05)\,H_2O + 0.05\,C_8H_{18} + (12.5/1.05)(3.76)\,N_2$$

Evaporative cooling for one kgmole of fuel is

$$Q_{evap} = \Delta H_{air} = N_a M_a c_p \Delta T$$
$$Q_{evap} = N_f M_f h_{fg} = N_a M_a c_p \Delta T$$

where

N_a = number of moles of air
N_f = number of moles of fuel
M_a = molecular weight of air
M_f = molecular weight of fuel
h_{fg} = heat of vaporization of fuel
c_p = specific heat
T = temperature

$$(1\text{ kgmole})(114\text{ kg/kgmole})(290\text{ kJ/kg}) =$$
$$[(12.5/1.05)(4.76)\text{ kgmoles}](29\text{ kg/kgmole})(1.005\text{ kJ/kg-K})\Delta T$$
$$\Delta T = 20°\text{C}$$

Example Problem 10-4

Water injection is added to the engine in Example Problem 10-3, with 0.25 kg of water injected for each kg of fuel used. Let the heat of vaporization of water h_{fg} = 2350 KJ/kg. Calculate:

1. approximate inlet air temperature when water injection is used
2. approximate engine power with water injection

(1) For one mole of fuel, the mass of fuel will be

$$m_f = N_f M_f = (1 \text{ kgmole})(114 \text{ kg/kgmole}) = 114 \text{ kg}$$

The mass of water will be

$$m_w = (0.25)(114) = 28.5 \text{ kg}$$

Evaporative cooling from water is

$$\begin{aligned} m_w h_{fg} &= N_a M_a c_p \Delta T \\ (28.5 \text{ kg})(2350 \text{ kJ/kg}) &= [(12.5/1.05)(4.76) \text{ kgmoles}] \\ &\quad \times (29 \text{ kg/kgmole})(1.005 \text{ kJ/kg-K})\Delta T \\ \Delta T &= 41°\text{C} \end{aligned}$$

Temperature of the air entering the engine after water evaporation is

$$T_a = 65 - 41 = \underline{24°\text{C}}$$

(2) As an approximation, power produced in an engine can be assumed to be about proportional to inlet air density. Fuel is added in proportion to the mass of inlet air, and a percentage of this fuel energy is converted to power output. Inlet air density is inversely proportional to inlet air temperature. Power with water injection is, therefore,

$$\begin{aligned} (\dot{W})_{\text{with}} &= (\dot{W})_{\text{without}}(T_{\text{without}}/T_{\text{with}}) = (900 \text{ kW})[(338 \text{ K})/(297 \text{ K})] \\ &= \underline{1024 \text{ kW}} \end{aligned}$$

This value, however, is reduced somewhat when some of the inlet air is displaced by water vapor. For each mole of inlet fuel, there are $(12.5/1.05)(4.76) = 56.67$ moles of inlet air. With water injection, there are also $(0.25)(114)/(18) = 1.583$ moles of water vapor. Engine output power is then

$$\begin{aligned} (\dot{W})_{\text{output}} &= (\dot{W})_{\text{with}}[(N_a)/(N_a + N_{\text{vapor}})] \\ &= (1024 \text{ kW})[(56.67)/(56.67 + 1.583)] = \underline{996 \text{ kW}} \end{aligned}$$

Increase in power from water injection is

$$\%\Delta\dot{W} = [(996 - 900)/(900)](100) = \underline{10.7\%}$$

Inlet Air Temperature

Increasing inlet air temperature to an engine results in a temperature increase over the entire cycle, with a resulting increase in heat losses. A 100°C increase in inlet temperature will cause a 10–15% increase in heat losses. Increasing cycle temperatures also increases the chance of knock. Turbocharged or supercharged engines generally have higher inlet air temperatures due to compressive heating. Many systems have aftercooling to reduce air temperature before it enters the engine cylinders.

Coolant Temperature

Increasing the coolant temperature of an engine (hotter thermostat) results in higher temperatures of all cooled components. There is little change in the temperatures of the spark plugs and exhaust valves. Indicated thermal efficiency would be higher, but there is a potential for a greater knock problem in hotter engines.

Engine Materials

Different materials used in the manufacture of cylinder and piston components result in different operating temperatures. Aluminum pistons, with their higher thermal conductivity, generally operate about 30°–80°C cooler than equivalent cast-iron pistons. Ceramic-faced pistons have poor thermal conductivity, resulting in very high temperatures. This is by design, with the ceramic being able to tolerate the higher temperature. Ceramic exhaust valves are sometimes used because of their lower mass inertia and high temperature tolerance.

Compression Ratio

Changing the compression ratio of an engine changes the heat transfer to the coolant very little. Increasing the compression ratio decreases heat transfer slightly up to about $r_c = 10$. Increasing the compression ratio above this value increases heat transfer slightly [58]. There is about a 10% decrease in heat transfer as the compression ratio is raised from 7 to 10. These changes in heat transfer occur mainly because of the combustion characteristics that change as the compression ratio is raised (e.g., flame speed, gas motion, etc.). The higher the compression ratio, the more expansion cooling will occur during the power stroke, resulting in cooler exhaust. CI engines, with their high compression ratios, generally have lower exhaust temperatures than SI engines. Piston temperatures generally increase slightly with increasing compression ratio.

Knock

When knock occurs, the temperature and pressure are raised in very localized spots within the combustion chamber. This rise in local temperature can be very severe and, in extreme cases, can cause surface damage to pistons and valves.

Swirl and Squish

Higher swirl and squish velocities result in a higher convection heat transfer coefficient within the cylinder. This results in better heat transfer to the walls.

10.7 AIR-COOLED ENGINES

Many small engines and some medium-sized engines are air cooled. This includes most small-engine tools and toys like lawn mowers, chain saws, model airplanes, etc. This allows both the weight and price of these engines to be kept low. Some motorcycles, automobiles, and aircraft have air-cooled engines, also benefitting from lower weight.

Air-cooled engines rely on a flow of air across their external surfaces to remove the necessary heat to keep them from overheating. On vehicles like motorcycles and aircraft, the forward motion of the vehicle supplies the air flow across the surface. Deflectors and ductwork are often added to direct the flow to critical locations. The outer surfaces of the engine are made of good heat-conducting metals and are finned to promote maximum heat transfer. Automobile engines usually have fans to increase the air-flow rate and direct it in the desired direction. Lawn mowers and chain saws rely on free convection from their finned surfaces. Some small engines have exposed flywheels with air deflectors fastened to the surface. When the engine is in operation, these deflectors create air motion that increases heat transfer on the finned surfaces.

It is more difficult to get uniform cooling of cylinders on air-cooled engines than on liquid-cooled engines. The flow of liquid coolants can be better controlled and ducted to the hot spots where maximum cooling is needed. Liquid coolants also have better thermal properties than air (e.g., higher convection coefficients, specific heats, etc.). Figure 10-13 shows how cooling needs are not the same at all locations on an engine surface. Hotter areas, such as around the exhaust valve and manifold, need greater cooling and a larger finned surface area. Cooling the front of an air-cooled engine which faces the forward motion of the vehicle is often much easier and more efficient than cooling the back surface of the engine. This can result in temperature differences and thermal expansion problems.

When compared with liquid-cooled engines, air-cooled engines have the following advantages: (1) lighter weight, (2) less costly, (3) no coolant system failures (e.g., water pump, hoses), (4) no engine freeze-ups, and (5) faster engine warm-up. Disadvantages of air-cooled engines are that they (1) are less efficient, (2) are noisier, with greater air flow requirements and no water jacket to dampen noise, and (3) need a directed air flow and finned surfaces.

Standard heat transfer equations for finned surfaces can be used to calculate the heat transfer off of these engine surfaces.

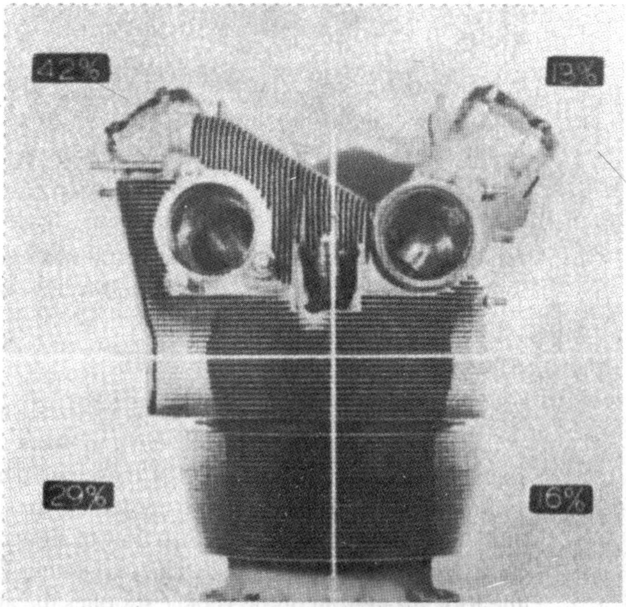

FIGURE 10-13

Variation of heat losses from the fins of an air-cooled aircraft engine. Seventy-one percent of the heat losses occur on the hotter side of the cylinder, containing the exhaust valve. The engine shown was used on a number of different aircraft, including the six-engine B-36 bomber. Reprinted with permission from SAE Paper 500197 © 1950 SAE International, [106].

10.8 LIQUID-COOLED ENGINES

The engine block of a water-cooled engine is surrounded with a water jacket through which coolant liquid flows (Fig. 10-14). This allows for a much better control of heat removal at a cost of added weight and the need for a water pump. The cost, weight, and complexity of a liquid coolant system makes this type of cooling very rare on small and/or low-cost engines.

Very few water-cooled engines use just water as the coolant fluid in the water jacket. The physical properties of water make it a very good heat transfer fluid, but it has some drawbacks. Used as a pure fluid it has a freezing point of 0°C, unacceptable in northern winter climates. Its boiling temperature, even in a pressurized cooling system, is lower than desired, and without additives it promotes rust and corrosion in many materials. Most engines use a mixture of water and ethylene glycol, which has the heat transfer advantages of water but improves on some of the physical properties. Ethylene glycol ($C_2H_6O_2$), often called **antifreeze**, acts as a rust inhibitor and a lubricant for the water pump, two properties not present when water is used alone. When added to water, it lowers the freezing temperature and raises the boiling temperature, both desirable consequences. This is true for mixtures with ethylene glycol concentrations from a very small amount up to about 70%. Due to a unique temperature–concentration–phase relationship, the freezing temperature again rises at high concentrations. The desirable heat transfer properties of water are also lost at high concentrations. Pure ethylene glycol should not be used as an engine coolant.

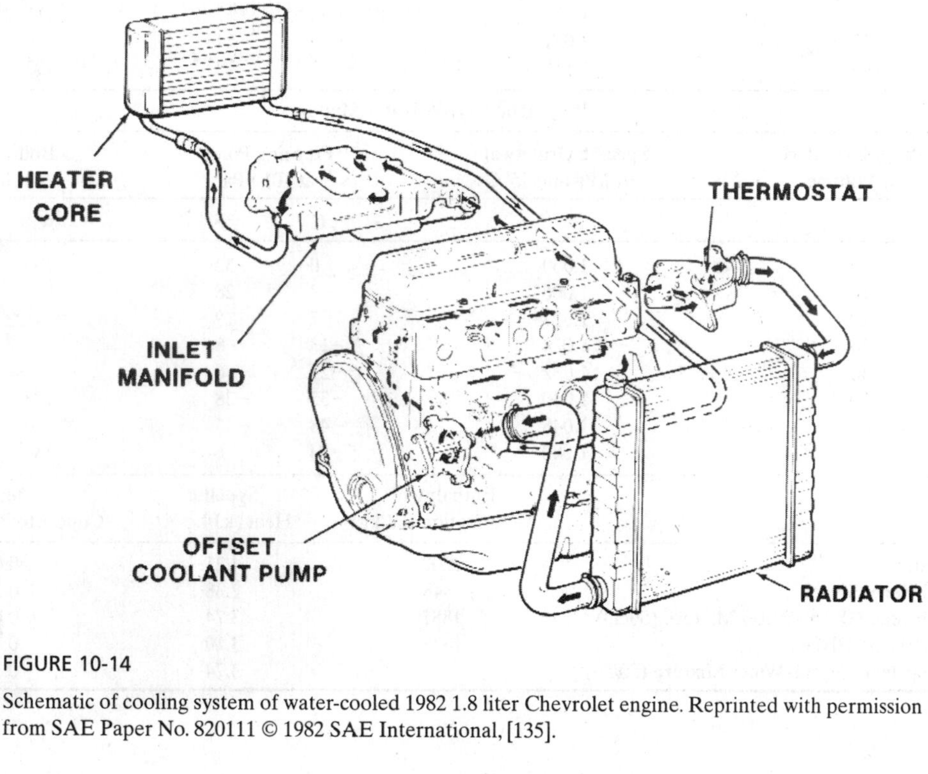

FIGURE 10-14

Schematic of cooling system of water-cooled 1982 1.8 liter Chevrolet engine. Reprinted with permission from SAE Paper No. 820111 © 1982 SAE International, [135].

Ethylene glycol is water soluble and has a boiling temperature of 197°C and a freezing temperature of −11°C in pure form at atmospheric pressure. Table 10-1 gives properties of ethylene glycol–water mixtures. When ethylene glycol is used as an engine coolant, the concentration with water is usually determined by the coldest weather temperature that is expected to be experienced.

Engine coolant cannot be allowed to freeze. If it does, it will not circulate through the radiator of the cooling system and the engine will overheat. A more serious consequence is caused when the water in the coolant expands on freezing and cracks the walls of the water jacket or water pump. This destroys the engine. Even in climates where there is no danger of freezing water, some ethylene glycol should be used because of its

TABLE 10-1 Properties of Antifreeze Solutions

Ethylene Glycol–Water Mixtures

% Ethylene Glycol by Volume	Specific Gravity at 101 kPa and 15°C	Freezing Point at 101 kPa		Boiling Point at 101 kPa	
		°C	°F	°C	°F
0	1.000	0	32	100	212
10	1.014	−4	24		
20	1.029	−9	15		
30	1.043	−16	3		
40	1.056	−25	−14		
50	1.070	−38	−37	111	231
60	1.081	−53	−64		
100	1.119	−11	12	197	386

Propylene Glycol–Water Mixtures

% Propylene Glycol by Volume	Specific Gravity at 101 kPa and 15°C	Freezing Point at 101 kPa		Boiling Point at 101 kPa	
		°C	°F	°C	°F
0	1.000	0	32	100	212
10	1.006	−2	28		
20	1.017	−7	19		
30	1.024	−13	8		
40	1.032	−21	−6		
50	1.040	−33	−28	108	225
60	1.048	−48	−55		
100	1.080	−14	6	188	370

	Enthalpy of Vaporization (kJ/kg)	Specific Heat (kJ/kg-K)	Thermal Conductivity (W/m-K)
Water	2202	4.25	0.69
Ethylene Glycol	848	2.38	0.30
Ethylene Glycol–Water Mixture (50/50)	1885	3.74	0.47
Propylene Glycol	1823	3.10	0.15
Propylene Glycol–Water Mixture (50/50)		3.74	0.37

better thermal and lubricating properties. In addition to good thermal properties, a coolant should have the following characteristics.

1. Chemically stable under conditions of use
2. Nonfoaming
3. Noncorrosive
4. Low toxicity
5. Nonflammable
6. Low cost

Most commercial antifreezes satisfy these requirements. Many of them are basically ethylene glycol with small amounts of additives.

A hydrometer is used to determine the concentration of ethylene glycol when it is mixed with water. The specific gravity of the mixture is determined by the height at which the calibrated hydrometer floats. Charts such as Table 10-1 can be used to determine the concentration needed. Most of these hydrometers are used by service station attendants who have no engineering training. For this reason they are usually not calibrated in concentration, but only in freezing temperature of the total water–ethylene glycol mixture. Most commercial antifreeze products (Prestone, Zerex, Dex-Cool, etc.) are basically ethylene glycol, and the same calibrated hydrometer can be used for all of these.

Some commercial engine coolants (Sierra, etc.) use propylene glycol (C_4H_8O) as the base ingredient. It is argued that, when coolant systems leak or when the coolant becomes aged and is discarded, these products are less harmful to the environment than ethylene glycol. A far lesser amount of these products is sold in the United States than those containing ethylene glycol.

HISTORIC—ANTIFREEZE

There have been some serious mistakes made when the few antifreeze products using something other than ethylene glycol as the base ingredient have appeared on the market. With a specific gravity different from that of ethylene glycol, a special hydrometer with a different calibration must be used when testing concentrations of these antifreezes. Most service stations use hydrometers calibrated for ethylene glycol. If these are used by mistake to test antifreezes made of other base materials, an incorrect freezing temperature is obtained. This can lead to unexpected freezing of the coolant mixture and a destroyed engine with a cracked block. This has happened a number of times.

The coolant system of a typical automobile engine is shown in Fig. 10-15. Fluid enters the water jacket of the engine, usually at the bottom of the engine. It flows through the engine block, where it absorbs energy from the hot cylinder walls. The flow passages in the water jacket are designed to direct the flow around the outer surfaces of the cylinder walls and past any other surface that needs cooling. The flow is also directed through any other component that may need heating or cooling (e.g., heating of the intake manifold or cooling of the oil reservoir). The flow leaves the engine block containing a high specific enthalpy because of the energy it absorbed in engine cooling. Exit is usually at the top of the engine block.

Enthalpy must now be removed from the coolant flow so that the circulation loop can be closed and the coolant can again be used to cool the engine. This is done by the use of a heat exchanger in the flow loop called a **radiator**. The radiator is a honeycomb heat exchanger with hot coolant flowing from top to bottom exchanging energy with cooler air flowing from front to back, as shown in Fig. 10-15. Air flow occurs because of the forward motion of the automobile, assisted by a fan located behind the radiator and either driven electrically or off the engine crankshaft. The cooled engine coolant exits the bottom of the radiator and reenters the water jacket of the engine, completing a closed loop. A water pump that drives the flow of the coolant loop is usually located between the radiator exit and engine block entrance. This pump is either electric or mechanically driven off the engine. Some early automobiles had no water pump and relied on a natural convection thermal flow loop.

Air leaving the automobile radiator is further used to cool the engine by being directed through the engine compartment and across the exterior surfaces of the engine. Because of the modern aerodynamic shape of automobiles and the great emphasis on cosmetics, it is much more difficult to duct cooling air through the radiator and engine compartment. Much greater efficiency is needed in rejecting energy with the modern

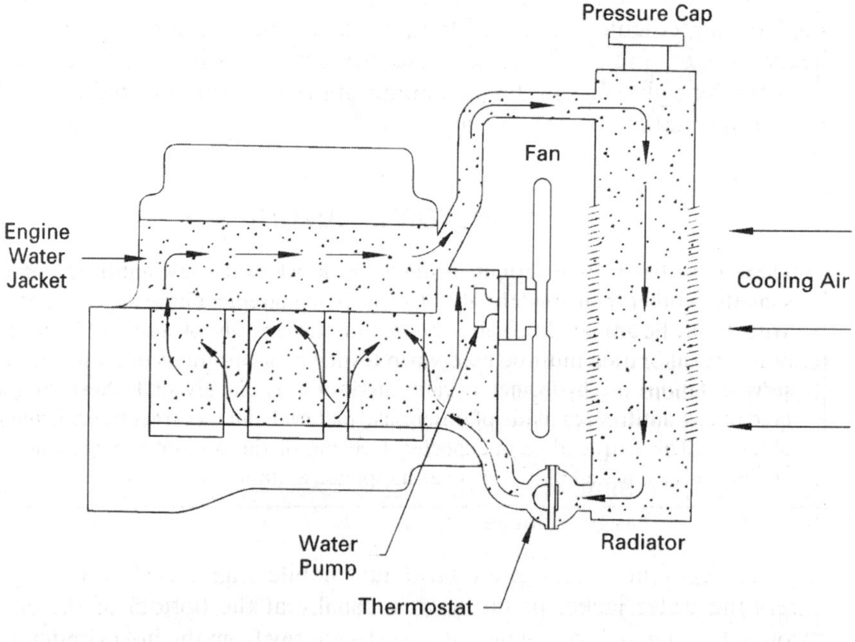

FIGURE 10-15
Radiator of a liquid-cooled engine used to remove heat from the coolant loop of the engine. A radiator is a liquid-to-air heat exchanger generally mounted in front of the engine on an automobile. Liquid flow is supplied by the engine water pump, while air flow is a result of the forward motion of the automobile, assisted by one or more fans. Adapted from [81].

radiator heat exchanger. Modern engines are designed to run hotter and thus can tolerate a lower cooling air-flow rate. Steady-state temperature of the air within the engine compartment of a modern automobile is on the order of 125°C.

To keep the coolant fluid temperature from dropping below some minimum value, and thus keeping the engine operating at a higher temperature and efficiency, a **thermostat** is installed in the coolant loop, usually at the engine flow entrance. A thermostat is a thermally activated go–no go valve. When the thermostat is cold, it is closed and allows no fluid flow through the main circulation channel. As the engine warms up, the thermostat also warms up, and thermal expansion opens the flow passage and allows coolant circulation. The higher the temperature, the greater the flow passage opening, with the greater resulting coolant flow. The coolant temperature is, therefore, controlled fairly accurately by the opening and closing of the thermostat. Thermostats are manufactured for different coolant temperatures, depending on engine use and climate conditions. They generally come in ratings from cold (140°F) to hot (240°F).

Coolant loops of older automobiles operated at atmospheric pressure using mostly water. This limited overall coolant temperatures to about 180°F (83°C), allowing for a safety margin to avoid boiling. In order to increase engine operating temperature for better efficiency, it was necessary to increase coolant temperature. This was done by pressurizing the coolant loop and adding ethylene glycol to the water. The ethylene glycol raised the boiling temperature of the fluid as shown in Table 10-1. Pressurizing the system further raises the boiling temperature of the fluid regardless of the concentration of ethylene glycol. Normal coolant system pressures are about 200 kPa absolute.

It is desirable for the coolant to remain mostly liquid throughout the flow loop. If boiling occurs, a small mass of liquid becomes a large volume of vapor, and steady-state mass flow becomes almost impossible to sustain. By using ethylene glycol in a pressurized system, high temperatures can be achieved without large-scale boiling. Localized boiling in small hot spots does occur within the engine water jacket. This is good. The very hottest spots within the engine (either momentary or almost steady state) require the greatest heat removal and cooling. The phase change that is experienced when boiling occurs at these local hot spots absorbs a large amount of energy and supplies the necessary large cooling at these spots. The circulating convection flow carries the resulting vapor bubbles away from the hot spots back into the main stream of the coolant. Here they condense back into liquid due to the cooler fluid temperature, and bulk flow is not interrupted.

As hot engine coolant leaves the engine block, it can be used to heat the passenger compartment of an automobile, when desired. This is done by routing a portion of the coolant flow through an auxiliary system that supplies the hot side of a small liquid-to-air heat exchanger. Outside or recirculated air is heated as it passes through the other half of the heat exchanger and is ducted into the passenger compartment and/or onto the cold windows for defrosting. Various manual and automatic controls determine the flow rates of the air and coolant to supply the desired warming results.

The small diesel engines of some vehicles are so efficient that they do not supply enough waste heat to adequately heat the passenger compartment under some operating conditions (e.g., idling at a stoplight). These vehicles sometimes have an electric resistance auxiliary heater to use at these times. These heaters will become more practical with 42-volt electrical systems. Another auxiliary method used on some of these

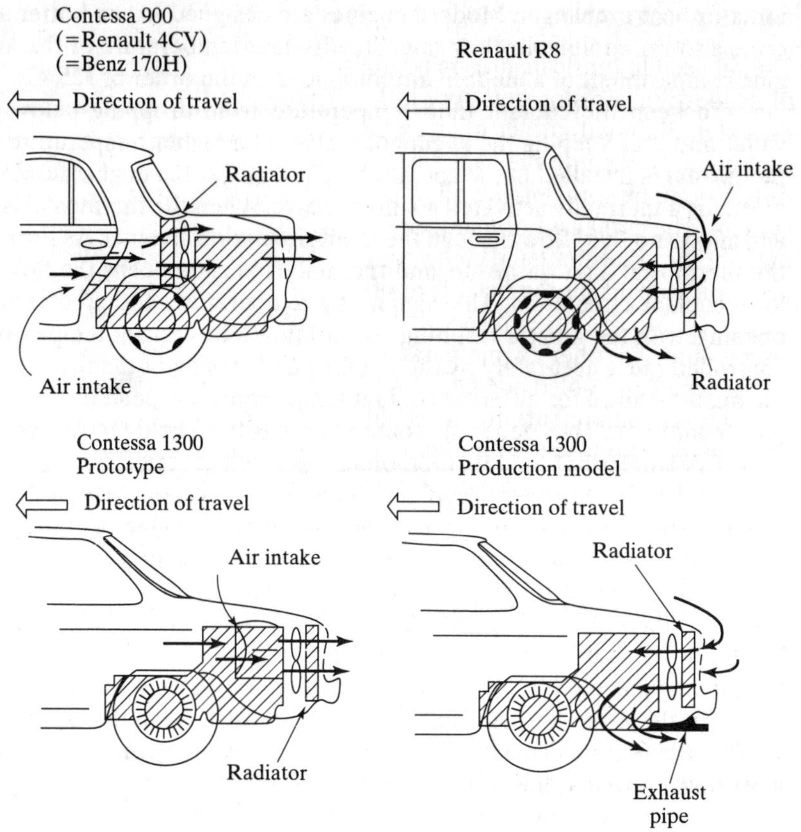

FIGURE 10-16

Various methods used to cool rear-mounted liquid-cooled engines. Reprinted with permission from *The Romance of Engines* by T. Suzuki, © 1997 SAE International, [227].

vehicles is a viscous heater. A viscous heater is a pump that churns a fluid to generate heat through friction, one such system using silicon gel [161]. The pump can be driven electrically or belted to the crankshaft and controlled on–off, as needed, by the EMS.

HISTORIC—ENGINE COOLANTS

Very early automobile engines were either air cooled or water cooled. At first, it was common practice to drain the water from water-cooled engines in cold weather and to store the car through the winter. Two of the first liquid antifreezes that were used were alcohols and kerosene. These allowed use of the automobile in cold weather, but great care was needed to avoid coolant system leaks. These liquids are combustible, and many automobiles burned up when leaking antifreeze came in contact with the hot engine and exhaust system.

10.9 OIL AS A COOLANT

The oil used to lubricate an engine in operation also helps to cool the engine. Because of its location, a piston gets very little cooling from the coolant in the water jacket or the external finned surface of an engine. To help cool the piston face, one of the hottest surfaces in the engine, the back surface of the piston crown, is subjected to a flow of oil. This is done by spraying the oil in pressurized systems or by splash in nonpressurized systems. The crankcase of many engines also serves as the oil reservoir, and the movement of the crankshaft and connecting rods splash oil over all exposed surfaces. The oil acts as a coolant on the back face of the piston crown as it absorbs energy and then runs back into the larger reservoir. Here, it mixes with the cooler oil and dissipates this energy into the other engine parts. This splash oil cooling of the piston is very important in small air-cooled engines as well as in automobile engines.

Other components are also cooled by oil circulation, either by splash or by pressurized flow from the oil pump. Oil passages through internal components like the camshaft and connecting rods offer the only major cooling these parts are subjected to. As the oil cools the various components, it absorbs energy and its temperature rises. This energy is then dissipated to the rest of the engine by circulation and eventually gets absorbed in the engine coolant flow.

Some high-performance engines have an oil cooler in their lubricant circulation system. The energy absorbed by the oil as it cools the engine components is dissipated in the oil cooler, which is a heat exchanger cooled by either engine coolant flow or external air flow.

10.10 ADIABATIC ENGINES

A small increase in brake power output can be gained by decreasing the heat losses from an engine cylinder. About 30% of available energy is converted to useful work (thermal efficiency), and this is done near TDC during combustion and the following expansion stroke, which encompasses about one-fourth of the total engine cycle. On the other hand, heat transfer occurs over the entire 720° of the cycle. Therefore only about one-fourth of the saved energy is available when output work is being generated, and only about 30% of this is utilized. If a 10% decrease of heat loss energy were accomplished over the cycle (a major accomplishment), only a fraction of this would appear as added crankshaft output:

$$\% \text{ Power Gained} = (10\%/4)(0.30) = 0.75\%$$

Most of the reduced heat loss energy ends up in the enthalpy of the exhaust. There would also be a higher steady-state temperature of internal engine components.

In recent years, so-called **adiabatic engines** have appeared on the market. They are not truly adiabatic (no heat losses) but do have greatly reduced heat loss from the combustion chambers. They usually have no coolant jacket or finned surfaces, and the only heat losses are from natural convection off the exterior surface. This results in much hotter engine components and some gain in brake power output.

Advances in material technology allow for engine components to operate at much higher temperatures without mechanical or thermal failure. This is because of better heat treating and alloying of metals and advancements in ceramics and composites. The development of flexible ceramic materials which could withstand the mechanical and thermal shocks that occured within an engine was a major breakthrough in the 1980s. These materials are now commonly found in modern engines, especially at the highest temperature spots, such as piston face and exhaust port. A common material found in adiabatic engines is silicon nitride (Si_3N_4). Because they have no cooling system (water pump, water jacket, finned surfaces, etc.), adiabatic engines can be made smaller and lighter than conventional engines. Vehicles can be made more aerodynamic with a lower drag coefficient because there is no radiator. This also gives greater flexibility in engine location and positioning.

All engine components of an adiabatic engine, including the cylinder walls, operate at higher temperatures. These heat the incoming air mixture quicker than in a conventional engine. This reduces the volumetric efficiency of the engine, which, in turn, deletes some of the brake power increase gained from less heat loss. The higher cylinder temperature during the compression stroke also raises the pressure and reduces the net work output of the cycle by increasing the compression work input.

Adiabatic engines are all compression ignition. They cannot be used as spark ignition engines, because the hot cylinder walls would heat the air–fuel mixture too quickly and knock would be a major problem. Another problem created by the hot cylinder walls, which are on the order of 800 K, is thermal breakdown of the lubricating oil. Better oils have been developed that can tolerate the conditions in present-day engines, but lubrication technology will need to continue to advance to keep up with increasing engine demands. One solution that is being considered and developed is the use of solid lubricants.

10.11 SOME MODERN TRENDS IN ENGINE COOLING

A number of different methods of cooling an engine are being tested and developed. These include engines with dual water jackets operating at two different coolant temperatures. A higher thermal efficiency can be obtained by the flexibility this allows in controlling engine temperatures. The coolant around the engine block is operated hotter, which reduces oil viscosity and lowers friction between pistons and cylinder walls. Coolant around the engine head is kept cooler to reduce knocking and to allow for a higher compression ratio. A variation of this is a cooling system that uses both a liquid coolant and a gaseous coolant in a dual-flow *water jacket*. Various other cooling systems take advantage of the large heat transfer that occurs during a phase change by operating as a two-phase flow at saturation conditions.

At least one company is working to reduce cylinder size and weight on small engines by building them with neither cooling fins nor a water jacket. The lubricating oil is used to cool the cylinders by piping it through circumferential passages built into the cylinder walls. Not only has this provided adequate cooling in test engines, but it also resulted in a more uniform temperature distribution. An oil cooler system would be required with this type of engine.

Some General Motors automobiles offer a safety feature on their engines in case of a leak in the coolant system. The automobile is capable of safely driving a long distance at moderate speed with no coolant in the cooling system. This is possible by firing only four of the eight cylinders at any given time. The four cylinders that do not get fuel, and do not fire, continue to pump air, which cools the engine enough to keep it from overheating. Each set of four cylinders cycles between firing for a period of time and pumping cooling air at other times.

10.12 THERMAL STORAGE

Some vehicles are equipped with a *thermal battery* that can be used to preheat an engine and automobile. A thermal battery takes waste heat from the engine coolant during operation and stores about 500 to 1000 W-hr (1800–3600 kJ) for as long as three days or more. Various methods to do this have been tried and used. The most common system stores energy by use of a liquid–solid phase change occurring in a water–salt crystal mixture. The stored energy can then be used in cold-weather starting to preheat the engine, preheat the catalytic converter, and/or heat the passenger compartment and defrost the car windows. Preheating can commence within a few seconds.

A number of different systems and materials have been tried with various success. One early system uses about 10 kg of $Ba(OH)_2 8H_2O$ as the base material. This has a latent heat (solid-to-liquid) of 89 W-hr/kg and a melting point of 78°C. The salt–water mixture of this system is contained within hollow fins fastened on the interior of a cylindrical flow chamber through which engine coolant flows. The exterior wall of the cylinder is insulated with super-high vacuum insulation that restricts heat losses to three watts or less when the surrounding temperature is at −20°C [79].

When the engine is running under normal conditions, hot coolant is ducted through the thermal battery and liquifies the salt–water mixture. This is done with energy that would otherwise be rejected in the automobile radiator, so there is no operating cost for the system. When the engine is turned off and coolant flow stops, the liquid–salt solution will very slowly change phase to solid as it cools. This phase change will take about three days because of the superinsulated wall of the container. As the salt solution changes phase, the temperature in the container remains at 78°C. To later recover the energy for use in preheating, an electric pump circulates the now cold coolant through the thermal battery, where it initiates a liquid-back-to-solid phase change in the salt–water mixture. When this occurs, the latent heat is absorbed by the coolant, which leaves the battery at a temperature near 78°C. The coolant can then be ducted to the engine or the catalytic converter for preheating or can be used to heat the passenger compartment. One or all of these are possible with proper piping and controls.

A preheated engine starts quicker with less wear and wasted fuel. The warmed cylinder walls and intake manifold promote better fuel evaporation, and combustion starts quicker. Also, the overrich intake mixture used to start a cold engine can be lessened, saving fuel and causing less emissions. The engine lubricating oil also gets preheated, which greatly reduces its viscosity. This allows for faster engine turnover with the starting motor and quicker engine starting. This also allows for quicker and better early oil distribution, which reduces engine wear. Preheating an alcohol-fueled engine is especially important. Because of its high latent heat, it is difficult to evaporate enough alcohol to get a cold engine started. This is one of the serious disadvantages of alcohol fuel, and engine preheating reduces the problem.

Chapter 9 explained how a major percentage of emissions occurs when a cold engine is started and before the catalytic converter reaches operating temperature. Preheating the catalytic converter is, therefore, one very effective way of reducing emissions. This can be done to a limited extent with a thermal battery.

Another way the stored energy of the thermal battery can be used is to pipe the heated coolant through the heat exchanger of the passenger compartment heater system. The heater can then immediately be turned on, with the air flow directed into the passenger compartment and/or onto the windows for deicing.

For the first 10 seconds of operation, a thermal battery can supply 50–100 kW. The system can be started when the ignition key is inserted or even when the car door is opened. Effective preheating occurs within 20–30 seconds. This compares to several minutes for effective heating of the engine, catalytic converter, and passenger compartment on nonpreheated automobiles.

Various systems supply different percentages of the stored energy, and in different sequences, to the various uses. Some systems have greater flexibility and changeability than others.

Even if all stored energy is delivered to the passenger compartment, the engine will warm up quicker. This is because no engine heat will be diverted to the cabin heater system. This may be important in large trucks as laws are being considered in some areas that require driver compartment heating before the driver is allowed to get into the vehicle. When the thermal battery is being recharged after the engine has heated up, there is no loss of cabin heater efficiency, since the engine supplies more than enough energy for both purposes.

Most thermal battery systems have a mass of about 10 kg and can supply 500–1000 W-hr as their temperature drops from 78°C to 50°C. Total energy depletion will normally take 20–30 minutes, depending on the engine and surrounding temperatures. The battery can be located either in the engine compartment or somewhere else in the automobile, such as the trunk. Mounting it in the engine compartment results in the least piping and greatest efficiency, but space limitations may not allow this.

The greatest benefit from thermal batteries will probably be in those automobiles that are used for city driving. Many city trips are short enough that the catalytic converter never reaches operating temperature, resulting in high exhaust emissions. Reducing startup emissions in densely populated areas is most important because of the large number of automobiles and other polluting systems, as well as the large number of people affected. Emissions from limited-range dual-powered automobiles that are being developed for city commuting will be greatly reduced with thermal storage. These automobiles, which are propelled by an electric motor most of the time and only use their small IC engine when extra power or range is needed, operate their engines in an on–off mode. The engine is started only when needed and therefore operates most often only for short periods of time after start-up, a highly polluting method. With a thermal battery, energy can be stored while the engine operates and then used to preheat the engine before the next start-up.

Although there is no direct running cost in charging a thermal battery, there will be a very slight cost in a higher fuel consumption due to the added mass of the vehicle contributed by the battery. A 10-kg battery is only 1% of the mass in a 1000-kg automobile. This, however, will cause a very slight steady-state increase in fuel consumption even on long-distance travel, which gets no benefit from the battery.

Thermal batteries are designed to last the lifetime of the automobile.

Example Problem 10-5

The fluid in a thermal battery system of an automobile consists of 6.2 lbm of salt solution which changes liquid–solid phase at 175°F. When the automobile engine is operating, steady-state engine coolant temperature is 215°F, and the coolant flow rate through the thermal battery is 0.025 lbm/sec. Because it is superinsulated, there is a steady heat loss from the thermal battery of only 11 BTU/hr. After the automobile has sat for some time, the engine coolant is at a temperature of 75°F, and the salt solution in the thermal battery is 60% liquid and 40% solid by mass. Immediately after engine start-up, coolant enters the thermal battery at a temperature of 75°F and exits at 165°F. Water property values can be used for engine coolant. Property values of the salt solution are:

$$\text{latent heat of solid–liquid phase change } h_{if} = 125 \text{ BTU/lbm}$$
$$\text{specific heat of salt in liquid phase } c_p = 0.220 \text{ BTU/lbm-°F}$$
$$\text{specific heat of salt in solid phase } c_p = 0.084 \text{ BTU/lbm-°F}$$

Calculate:

1. how long thermal battery can supply coolant at 165°F after engine start-up
2. time to cool thermal battery from 215°F to 75°F after engine has stopped

(1) Battery can supply coolant at 165°F until all liquid salt has turned to solid: energy to change phase $= Q_{cp} = [(6.2 \text{ lbm})(0.60)](125 \text{ BTU/lbm}) = 465.0 \text{ BTU}$
Rate of energy transfer to heat coolant flow is

$$\dot{Q} = \dot{m}c_p\Delta T = (0.025 \text{ lbm/sec})(1 \text{ BTU/lbm-°F})(165 - 75)\text{°F} = 2.25 \text{ BTU/sec}$$

Time which thermal battery can supply coolant at 165°F is

$$t = (465.0 \text{ BTU})/(2.25 \text{ BTU/sec}) = \underline{207 \text{ sec} = 3 \text{ min } 27 \text{ sec}}$$

(2) Heat loss to cool liquid from 215°F to 175°F is

$$Q_{\text{liq}} = mc_p\Delta T = (6.2 \text{ lbm})(0.220 \text{ BTU/lbm-°F})(215 - 175)\text{°F} = 54.56 \text{ BTU}$$

Heat loss during a phase change is

$$Q_{pc} = mh_{\text{if}} = (6.2 \text{ lbm})(125 \text{ BTU/lbm}) = 775 \text{ BTU}$$

Heat loss to cool solid from 175°F to 75°F is

$$Q_{\text{solid}} = mc_p\Delta T = (6.2 \text{ lbm})(0.084 \text{ BTU/lbm-°F})(175 - 75)\text{°F} = 52.08 \text{ BTU}$$

Total heat loss to cool from 215°F to 75°F is

$$Q_{\text{total}} = (54.56) + (775) + (52.08) = 881.64 \text{ BTU}$$

Time to cool from 215°F to 75°F is

$$t = (881.64 \text{ BTU})/(11 \text{ BTU/hr}) = \underline{80.15 \text{ hr} = 3 \text{ days } 8 \text{ hr}}$$

10.13 SUMMARY

Combustion temperatures in the cylinders of IC engines can reach values of 2700 K and higher. Without adequate cooling, temperatures of this magnitude would quickly destroy engine components and lubricants. If cylinder walls are allowed to exceed temperatures above 200°C, material failures will occur and most lubricating oils will break down. To keep the cylinders from overheating, they are surrounded with a water jacket on liquid-cooled engines or a finned surface on air-cooled engines. On the other hand, to obtain maximum efficiency from an engine, it is desirable to operate it as hot as possible. With improvements in materials and lubrication technology, modern engines can operate much hotter than engines of a few years ago.

Heat removed from engine cylinders is eventually rejected to the surroundings. Unfortunately, by keeping the engine from overheating with heat transfer to the surroundings, a large percent of the energy generated within the engine is wasted, and the brake thermal efficiency of most engines is on the order of 30–40%.

Because of the low profile of a modern automobile, cooling air flow is much more restricted and greater heat transfer efficiency is necessary.

Innovative cooling systems are being developed, but at present most automobile engines are liquid cooled using a water–ethylene glycol solution. Most small engines are air cooled because of the requirements of weight, cost, and simplicity.

PROBLEMS

10.1 An in-line, six-cylinder, 6.6-liter, four-stroke cycle SI engine with multipoint port fuel injection operates at 3000 RPM, with a volumetric efficiency of $\eta_v = 89\%$. The intake manifold runners can be approximated as round pipes with an inside diameter of 4.0 cm. Inlet temperature to the manifold is 27°C.

Calculate:

(a) Average velocity and mass flow rate of air to each cylinder, using inlet temperature to evaluate properties. [m/sec, kg/sec]

(b) Reynolds number in the runner to cylinder #1, which is 40 cm long (use standard interior pipe flow equations).

(c) Temperature of the air entering cylinder #1 if the runner wall temperature is constant at 67°C. [°C]

(d) Wall temperature needed for the runner to cylinder #3, such that the cylinder air inlet temperature is the same as that in cylinder #1. Runner length for cylinder #3 is 15 cm. [°C]

10.2 The engine in Problem 10-1 is converted to throttle body fuel injection with the fuel injected at the inlet end of the manifold. Forty percent of the fuel evaporates in the intake manifold runners, which cools the air by evaporative cooling. Wall temperatures remain the same.

Calculate:

(a) Temperature of the air entering cylinder #1 if the fuel is stoichiometric gasoline. [°C]

(b) Temperature of the air entering cylinder #1 if the fuel is stoichiometric ethanol. [°C]

10.3 The engine in Problem 10-2b has bore and stroke related by $S = 0.90\ B$. Using inlet conditions to evaluate properties, calculate the Reynolds number as defined by Eq. (10-7).

10.4 The engine in Problem 10-3 has an exhaust pipe between the engine and catalytic converter that can be approximated as a round pipe, 1.5 m long and 6.5 cm inside diameter. Exhaust temperature leaving the engine is 477°C, and the average wall temperature of the exhaust pipe is 227°C. Calculate the exhaust temperature entering the catalytic converter. [°C]

10.5 An automobile cruises at 55 mph using a brake power of 20 kW. The engine, which is represented by Fig. 10-1, runs at a speed of 2000 RPM.

Approximate:

(a) Power lost in the exhaust flow. [kW]

(b) Power lost to friction. [kW]

(c) Power dissipated in the coolant system. [kW]

10.6 An engine represented by Fig. 10-1 has a coolant system with a flow rate of 25 gal/min and a thermostat that controls coolant flow temperature into the engine at 220°F. The engine produces 30 bhp at 2500 RPM as the automobile travels at 30 MPH. Frontal area of the radiator is 4.5 ft^2, and a fan increases air flow velocity through the radiator by a factor of 1.1.

Calculate:

(a) Coolant temperature as it exits the engine. [°F]

(b) Air temperature leaving the radiator if the ambient temperature is 75°F. [°F]

10.7 A certain automobile model is offered with two engine options. The two engines are identical V8s with different displacements; one is 320 $in.^3$ and the other is 290 $in.^3$. The engines are run at identical speeds, temperatures, and operating conditions.

Calculate:

(a) Rough approximation of what percent greater (or less) will be the indicated thermal efficiency of the larger engine versus the smaller. [%]

(b) Approximation of what percent greater (or less) will be the total heat transfer to the coolant fluid of the larger engine versus the smaller. [%]

10.8 A service station attendant mixes a water–antifreeze solution so that the mixture will have a freezing point of −30°C. The antifreeze used is ethylene glycol, but by mistake the attendant uses a hydrometer calibrated for propylene glycol. Calculate the actual freezing temperature of the mixture. [°C]

10.9 A thermal storage battery consists of 10 kg of a salt solution that changes phase at 80°C and has the following properties:

$$\text{solid to liquid latent heat} = 80 \text{ W-hr/kg}$$
$$\text{specific heat as liquid} = 900 \text{ J/kg-K}$$
$$\text{specific heat as solid} = 350 \text{ J/kg-K}$$

The battery container is superinsulated with a heat loss rate of 3 W (assume constant). Engine coolant flows through the battery at a rate of 0.09 kg/sec, and has a temperature of 110°C when the engine is running. Coolant can be approximated as water.

Calculate:

(a) How long it takes for the battery solution to reach 80°C after the engine is turned off. [hr]

(b) How long the battery solution remains at 80°C as it changes phase. [hr]

(c) How long until the battery cools to an ambient temperature of 10°C. [hr]

(d) How long the battery can supply coolant at 80°C when the engine is started at a temperature of 20°C. Assume the battery solution starts as all liquid at 80°C, and the engine coolant enters at 20°C and leaves at 80°C. [min.]

10.10 Calculate the length of time the automobile in Example Problem 10-5 has been inoperative in order to reach the condition described (i.e., salt solution in thermal battery is 60% liquid and 40% solid). Surroundings are at 75°F. [hr]

10.11 Coolant (water) flow through an automobile engine has a flow rate of 20 gallons per minute and removes 1000 BTUs per minute from the engine. The water enters the engine at a temperature of 200°F. The radiator on the automobile has a frontal area of 4 ft^2 and an air flow velocity of 50 ft/sec through it. (for water 1 gal = 8.4 lbm)

Calculate:

(a) Temperature of water as it exits engine. [°F]

(b) Change of air temperature as it passes through radiator. [°F]

10.12 A large four-stroke cycle stationary V12 diesel engine is to be used as part of a cogeneration system by using the exhaust energy to generate steam. The engine, which has a 14.2-cm bore and 24.5-cm stroke, runs at 980 RPM with a volumetric efficiency $\eta_v = 96\%$. Air–fuel ratio is AF = 21:1. Steam is generated by running the exhaust through one side of a gas–steam heat exchanger.

Calculate:

(a) Energy made available to generate steam if the exhaust temperature decreases from 577°C to 227°C as it passes through the heat exchanger. [kW]

(b) Saturated steam vapor that can be generated if steam enters the heat exchanger as saturated liquid at 101 kPa. The heat exchanger efficiency is 98% and, for water at 101 kPa, $h_{fg} = 2257$ kJ/kg. [kg/hr]

10.13 During combustion, there is a momentary heat flux through the wall of the combustion chamber at a certain spot equal to 67,000 BTU/hr-ft^2. Gas temperature in the cylinder at this time is 3800°R, and the convection heat transfer coefficient within the cylinder is 22 BTU/hr-ft^2-°R. Coolant temperature is 185°F. Thermal conductivity of the 0.4-inch-thick cast-iron cylinder wall is 34 BTU/hr-ft-°R.

Calculate:

(a) Inside surface temperature of the cylinder wall. [°F]

(b) Surface temperature on the coolant side of the cylinder wall. [°F]

(c) Convection heat transfer coefficient on the coolant side of the cylinder wall. [BTU/hr-ft^2-°R]

10.14 Two engines have cylinders that are geometrically the same in size and shape. The cylinders of engine A are surrounded with a normal water jacket filled with a water–ethylene glycol solution. The cylinders of engine B are insulated, making this an adiabatic engine. Other than temperatures, the engines are operated with the same steady-state conditions (as much as possible). (a) Which engine has higher volumetric efficiency? Why? (b) Which engine has higher thermal efficiency? Why? (c) Which engine has hotter exhaust? Why? (d) Which engine would be more difficult to lubricate? Why? (e) Which engine would be a better SI engine? Why?

DESIGN PROBLEMS

10.1D Design a thermal storage system for an automobile. The system is to be used to preheat the oil and the catalytic converter, and to warm the passenger compartment. Determine the size and materials of a thermal battery. Determine flow rates versus time, and draw a flow diagram schematic. Explain the sequence of events when the automobile is started, using approximate energy flows and temperatures.

10.2D Design an engine cooling system that uses two separate water jackets. Give the fluids used, flow rates, temperatures, and pressures. Show the flow diagram and pumps on a schematic drawing of the engine.

CHAPTER 11

Friction and Lubrication

"The motors and water coolers are placed up front, usually over or nearly over the front axle, the gears and fuel in the rear and the weight of the passengers in the middle. The net results of this change are long wheel bases, low center of gravity, angle-iron frames, plain springs, running gears without reaches, the comfortable tonneau body, freedom from vibration, good traction, great hill-climbing qualities, almost total absence of slip, and easy access to all parts, the motor being covered only by a detachable metal hood or bonnet."

Resume of the New York Automobile Show

The Automobile Magazine (January 1902)

This chapter examines the friction that occurs in an engine and the lubrication needed to minimize this friction. **Friction** refers to the forces acting between mechanical components due to their relative motion and to the forces on and by fluids when they move through the engine. A percentage of the power generated within the engine cylinders is lost to friction, with a reduction in the resulting brake power obtained off the crankshaft. Accessories that are run off the engine also reduce crankshaft output and are often classified as part of the engine friction load.

11.1 MECHANICAL FRICTION AND LUBRICATION

When two solid surfaces are in contact in an engine, they will touch each other at the roughness high spots of the surfaces, as shown magnified in Fig. 11-1. The smoother the surfaces are machined (on a macroscopic level), the lower will be the surface high points (microscopic) and the less will be the average distance separating them. If one surface is moved relative to the other, the high points will come into contact and will resist motion (friction) (see Fig. 11-1a). Points of contact will become hot, sometimes to the point of

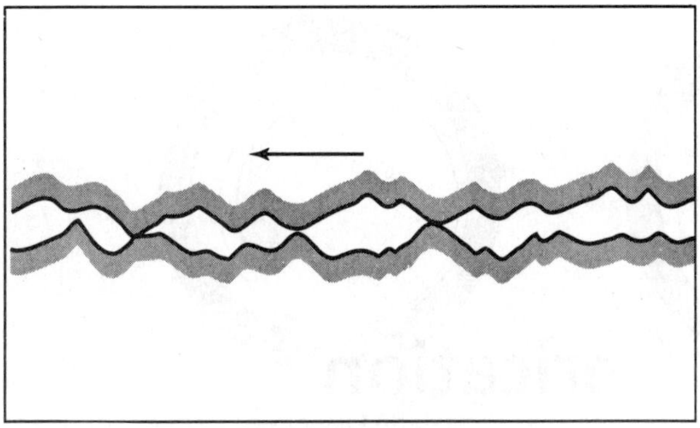

(a)

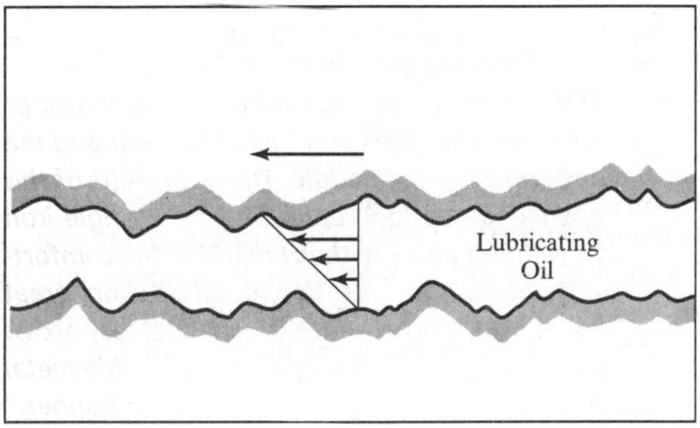

(b)

FIGURE 11-1

Motion between engine components, highly magnified to show surface roughness. (a) Dry or nonlubricated surface showing friction caused by high spots. (b) Lubricated surface showing reduction of friction by hydraulic floating.

trying to weld together. To greatly reduce resistance of surface-to-surface motion, lubricating oil is added to the space between the surfaces. Lubricating oil adheres to the solid surfaces, and when one surface moves relative to the other, oil is *dragged* along with the surface. The oil holds the surfaces apart and one surface hydraulically *floats* on the other surface. The only resistance to relative motion is the shearing of fluid layers between the surfaces, which is orders of magnitude less than that of dry surface motion. Three important characteristics are needed in a lubricating fluid:

1. It must adhere to the solid surfaces.
2. It must resist being squeezed out from between the surfaces, even under the extreme forces experienced in an engine between some components.
3. It should not require excessive force to shear adjacent liquid layers. The property that determines this is called viscosity and is addressed later in the chapter.

Bearings offer a unique lubrication problem because one surface (race) surrounds the other surface (shaft). When an engine is not in operation, gravity pulls the shaft in any bearing (crankshaft, connecting rod, etc.) down and squeezes out the oil film between the two surfaces (Fig. 11-2a). In operation, the combination of a rotating

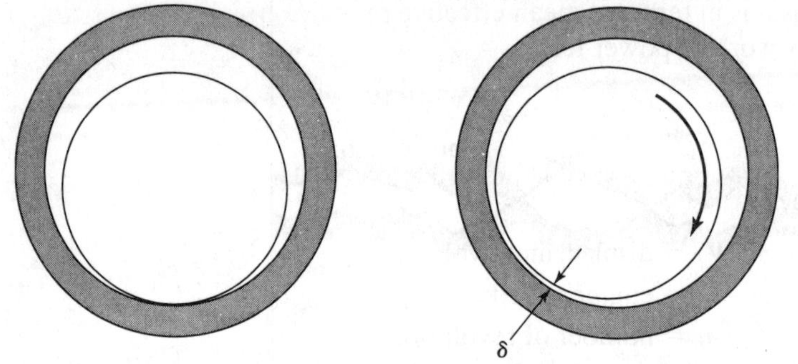

(a) Not Turning (b) Lubricated

FIGURE 11-2

Lubrication of bearings. (a) Nonrotating, lubricating oil is squeezed out and surface contacts surface. (b) Rotating, oil film is dragged by moving surface and surfaces are separated by thin layer of fluid.

shaft, viscous effects, and dynamic forces from various directions results in hydraulic floating of the shaft, offset slightly from center (Fig. 11-2b). The position and thickness of the minimum oil film in the bearing will depend on tolerances, load, speed, and oil viscosity. It will be on the order of $2\ \mu (1\ \mu = 10^{-6}\,\text{m})$ for the main bearings in an engine. For additional analysis, the reader is referred to the many books available on dynamic lubrication of bearings.

11.2 ENGINE FRICTION

Friction can be classified as a loss using power terms:

$$\dot{W}_f = (\dot{W}_i)_{\text{net}} - \dot{W}_b \tag{11-1}$$

where: $(\dot{W}_i)_{\text{net}} = (\dot{W}_i)_{\text{gross}} - (\dot{W}_i)_{\text{pump}}$
subscript

f = friction
i = indicated
b = brake

Friction can also be classified in terms of work:

$$w_f = (w_i)_{\text{net}} - w_b \tag{11-2}$$

with mechanical efficiency defined as

$$\eta_m = \dot{W}_b/\dot{W}_i = w_b/w_i \tag{11-3}$$

Because of the different sizes of engines operating at many different speeds, probably the most meaningful method of classifying and comparing friction and engine losses is in terms of mean effective pressure. Mean effective pressure can be related to any work or power term

$$\text{work } W = (\text{mep})V_d \tag{11-4}$$

$$\text{power } \dot{W} = (\text{mep})V_d(N/n) \tag{11-5}$$

where

V_d = displacement volume
N = engine speed
n = number of revolutions per cycle

Using frictional work or frictional power and rearranging, these become:

$$\text{fmep} = W_f/V_d \tag{11-6}$$

$$\text{fmep} = \dot{W}_f/[V_d(N/n)] \tag{11-7}$$

$$\text{fmep} = \text{imep} - \text{bmep} \tag{11-8}$$

In some analyses [e.g., 40], the mep concept is expanded to include all work and power inputs and outputs of an engine. Various mep terms and the work they correspond to include:

amep—work to drive auxiliaries such as the power steering pump
bmep—work done by the engine crankshaft
cmep—work to power the supercharger or turbocharger compressor
fmep—work lost to internal friction and to drive necessary engine equipment such as the oil pump
gmep—gross work—indicated work of compression and expansion strokes
imep—net work generated in the combustion chambers
mmep—work needed to motor an engine (see p. 415)
pmep—pumping work—indicated work of the exhaust and intake strokes
tmep—work recovered from the exhaust gas in a turbocharger turbine

These are related as follows:

$$\text{fmep} = \text{imep} - \text{bmep} - \text{amep} - \text{cmep} + \text{tmep} \tag{11-9}$$

$$\text{imep} = \text{gmep} - \text{pmep} \tag{11-10}$$

Assuming amep = 0 and cmep = tmep, it follows that

$$\text{fmep} = \text{imep} - \text{bmep} \tag{11-11}$$

Friction mep can quite accurately be related to engine speed by the empirical equation

$$\text{fmep} = A + BN + CN^2 \tag{11-12}$$

where

N = engine speed
A, B, C = empirical constants related to a specific engine

The first term on the right side of Eq. (11-12) (A = constant) is sometimes called boundary friction. It occurs between components of the engine where there is not enough lubrication to hydraulically separate totally the motion of one surface from another. Metal-to-metal contact occurs between the piston rings and cylinder walls at TDC and BDC, and in heavily loaded bearings of the crankshaft. Periodic metal-to-metal contact occurs when heavily loaded surfaces move at low speeds or undergo sudden acceleration and direction changes. When this happens, the lubrication is squeezed out and there is a momentary lack of hydraulic floating. Places where this happens include bearings of the crankshaft and connecting rods, the piston ring–cylinder wall interface at TDC and BDC, and in most components at start-up.

The second term on the right side of Eq. (11-12) is proportional to engine speed and relates to the hydraulic shear that occurs between many lubricated engine components. Shear force per unit surface area is given as

$$\tau_s = \mu(dU/dy) = \mu(\Delta U/\Delta y) \tag{11-13}$$

where

μ = dynamic viscosity of lubricating oil
(dU/dy) = velocity gradient between surfaces
ΔU = velocity difference between adjacent surfaces
Δy = distance between adjacent surfaces

For a given viscosity (temperature) and geometry, the velocity term ΔU is proportional to the speed N.

The third term of Eq. (11-12) is related to engine speed squared. This term accounts for the losses from turbulent dissipation in the intake and exhaust flows. Dissipation equates to the square of mass velocity, which, in turn, relates to engine speed. Constants A, B, and C must be determined for the operating conditions of a given engine.

An empirical equation similar to Eq. (11-12), but replacing engine speed with average piston speed $\overline{U}_p$, can be written as

$$\text{fmep} = A' + B'\overline{U}_p + C'\overline{U}_p^2 \tag{11-14}$$

The constants of the two equations are related as

$$\begin{aligned} A' &= A \\ B' &= B/2S \\ C' &= C/4S^2 \end{aligned} \tag{11-15}$$

where S is the piston stroke.

The magnitude of friction mean effective pressure (or friction power, or friction work) is on the order of 10% of the net indicated mean effective pressure (or net $\dot{W}_i$, or net w_i) at WOT. This increases to 100% at idle, when no brake power is taken off the crankshaft. A turbocharged engine will generally have a lower percent friction loss. This is due to the greater brake output realized, while absolute friction remains about the same. Most power lost to friction ends up heating the engine oil and coolant. Total engine friction can readily be determined from Eq. (11-1) by measuring indicated power and brake power. Indicated power is found by integrating the pressure–volume areas of a cycle from an indicator diagram generated from pressure sensors in the combustion chamber. Brake power is directly measured by connecting the crankshaft output to a dynamometer.

It is much more difficult and less accurate to divide total friction into parts to determine the percentage of the total contributed by various engine components. One of the best ways to do this is to **motor** the engine (i.e., drive an unfired engine with an external electric motor connected to the crankshaft). Many electric dynamometers are capable of doing this, making them an attractive type of dynamometer. Engine power output is measured with an electric dynamometer by running a generator off the engine crankshaft and measuring the electric load imposed on the generator. The generator is designed as a dual system, which also allows it to be used as an electric motor that can drive the connected IC engine. When an engine is motored, the ignition is turned off and no combustion takes place. Engine rotation is provided and controlled by the connected electric motor, with the resulting cycle much like that shown in Fig. 11-3. Unlike a fired engine, both the compression–expansion loop and the exhaust–intake loop of this cycle represent negative work on the cylinder gases. This work is provided through the crankshaft from the electric motor.

The power needed to be supplied by the electric motor to rotate the engine is

$$\dot{W}_m = \dot{W}_f + \dot{W}_g + \dot{W}_p \tag{11-16}$$

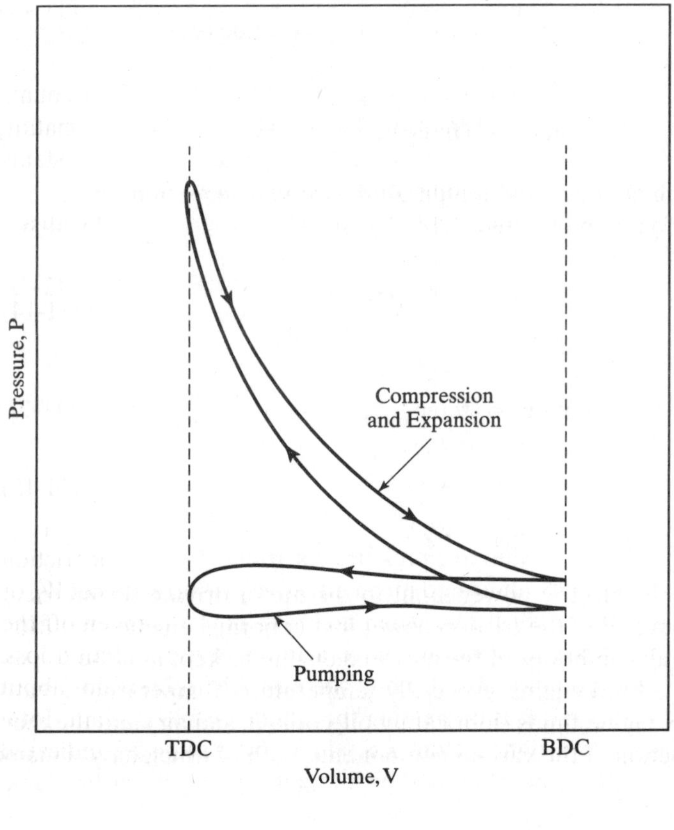

FIGURE 11-3

P-V diagram of motored engine. Intake and compression strokes are similar to those in the cycle of a fired engine. With no combustion to raise pressure, the expansion stroke reverses the intake stroke and there is very little blowdown.

where

$\dot{W}_m$ = power to motor the engine

$\dot{W}_f$ = friction power

$\dot{W}_g$ = gross indicated power (compression and expansion)

$\dot{W}_p$ = pumping indicated power (exhaust and intake)

In terms of mean effective pressure, this becomes

$$(\text{mmep}) = (\text{fmep})_m + (\text{gmep})_m + (\text{pmep})_m \tag{11-17}$$

where the subscript m indicates motoring condition.

Because there is no combustion occurring in a motored engine, normal combustion pressure rise does not occur and the expansion stroke just reverses the compression stroke. There is essentially no gross indicated work or power, and

$$(\text{gmep})_m = 0 \tag{11-18}$$

If the motored engine is operated at WOT, then the pump work loop is almost zero, and

$$(\text{pmep})_m = 0 \tag{11-19}$$

From Eq. (11-17), we get

$$\text{mmep} \approx (\text{fmep})_m \tag{11-20}$$

If all parameters such as speed and temperature, are kept consistent in the motored engine as would be found in the fired engine, then

$$(\text{fmep})_m \approx (\text{fmep})_{\text{fired}} \tag{11-21}$$

and

$$\text{mmep} \approx (\text{fmep})_{\text{fired}} \tag{11-22}$$

and

$$\dot{W}_m \approx \dot{W}_f \tag{11-23}$$

Thus, by measuring the electric power input to the motor driving the engine, a good approximation is obtained of the friction power lost in normal engine operation.

It is imperative that all conditions of the motored engine be kept as close as possible to the conditions of a fired engine, especially temperature. Temperature greatly affects the viscosity of the engine fluids (lubricating oil, coolant, and air) and the thermal expansion and contraction of the various components, both of which have a major influence on engine friction. The oil must be circulated at the same rate and temperature (viscosity) as in a fired engine. Air and engine coolant flow should be kept as consistent as possible (i.e., with the same throttle setting and pump rates).

The normal way of measuring friction in an engine by the motoring method is to first run the engine in a normal fired mode. When the engine has reached a steady-state condition with all temperatures, it is turned off and immediately tested using the electric motor. For a very brief period of time, the engine temperatures will be almost the same as with a fired engine. This will quickly change because no combustion is occurring, and the engine will start to cool off. There will be some differences immediately. Even with all other temperatures being correct, the exhaust flow will be quite different. Hot combustion products that make up the exhaust flow in a fired engine are approximated with much cooler air in a motored engine. At best, motored engine test results give a close approximation of engine friction.

As friction in an engine is being tested by the motoring method, engine components can be removed to determine how much they individually contribute to total friction. For instance, the engine can be motored with and without the valves connected. The difference in power required gives an approximation to the friction of the valves. A problem with this is the difficulty of keeping engine temperatures near normal operating temperatures when the engine is partially dismantled. Figure 11-4 gives typical results for the percent of friction contributed by various engine components.

The components that contribute a major part of total friction are the pistons and piston rings. Figure 11-5 shows the friction forces on a typical piston assembly as it goes through one cycle. The forces are greatest near TDC and BDC, where the piston momentarily stops. When there is no relative motion between the piston and cylinder walls, the oil film between these surfaces gets squeezed out by the high forces between them. When the piston then starts a new stroke, there is very little lubricant between surfaces, and some metal-to-metal contact occurs, with resulting high friction forces. As the piston gains speed over the lubricated cylinder wall surface, it drags a film of oil with it and hydraulic floating occurs. This is the most effective form of lubrication between moving surfaces, and friction forces are minimized.

It can be seen in Fig. 11-5 that there is even a small measurable friction force at TDC and BDC where the piston velocity is considered zero. This shows that there are deflections in the connecting components and stretching or compression of the piston occurring at these points due to mass inertia and high acceleration rates. This is the reason the maximum allowable average piston speed is about 5 to 20 m/sec for all engines regardless of size. With speeds higher than these, there would be a danger of structural failure, with too small of a safety margin for the materials in the piston assemblies of most engines (i.e., iron and aluminum).

The magnitude of the friction forces is about the same for the intake, compression, and exhaust strokes. It is much higher during the expansion stroke, reflecting the higher pressure and forces that occur at that time.

The piston assemblies of most engines contribute about half of the total friction and can contribute as much as 75% at light loads. The piston rings alone contribute about 20% of total friction. Most pistons have two compression rings and one or two oil rings. The second compression ring reduces the pressure differential that occurs

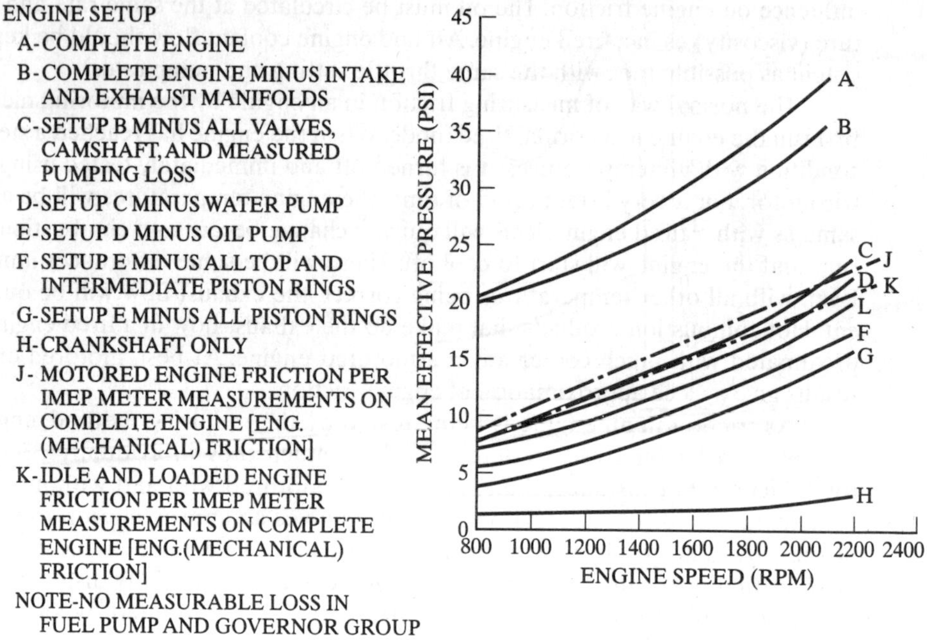

FIGURE 11-4

Friction losses for various engine components as measured by motoring of the engine. All losses, which are given in terms of fmep (psia), increase with increasing engine speed. Reprinted with permission from SAE Paper No. 730150 © 1973 SAE International, [20].

across the first compression ring during combustion and power stroke. To reduce friction, the trend has been to make compression rings thinner, some engines having rings as thin as 1 mm. Oil rings distribute and remove an oil film on the cylinder walls and sustain no pressure differential. All rings are spring loaded against the walls, which results in high friction forces.

Adding an additional compression ring can add about 10 kPa to fmep of an engine. Increasing the compression ratio by one will increase fmep by about 10 kPa. Increasing the compression ratio also requires heavier bearings on the crankshaft and connecting rods and may require an additional piston compression ring.

The valve train of an engine contributes about 25% of total friction, crankshaft bearings contribute about 10% of total, and engine-driven accessories contribute about 15% of total.

In Figs. 11-6 and 11-7, friction mean effective pressure (≈mmep) is plotted as a function of engine speed. Engine speed of the ordinate (X axis) can be replaced with average piston speed without changing the shape of the curves. When data are generated to make curves like these, a Reynolds number is defined in terms of an average piston speed such that

$$\text{Re} = \overline{U}_p B/\nu \tag{11-24}$$

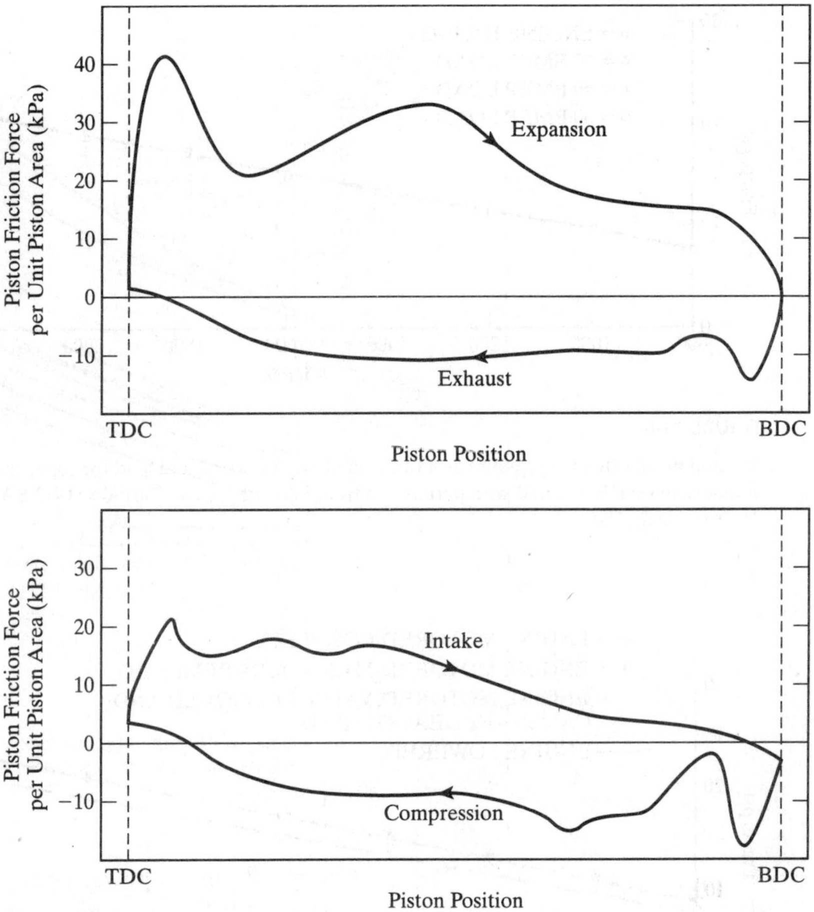

FIGURE 11-5

Friction force on piston and piston rings through one engine cycle. Greater friction during the expansion stroke reflects the higher pressure and forces during that part of the cycle. Adapted from [80].

where

B = bore

ν = kinematic viscosity of lubricating oil

Data from engines of different sizes can be compared at the same piston speed and temperature if the kinematic viscosity of the lubricating oil is adjusted to be proportional to the cylinder bore B (i.e., if B/ν is kept constant). When this is done, the ordinate variable of piston speed can be replaced with the Reynolds number, resulting in the same curves. This is limited by the amount the kinematic viscosity of an oil can be adjusted without affecting the lubrication of the engine.

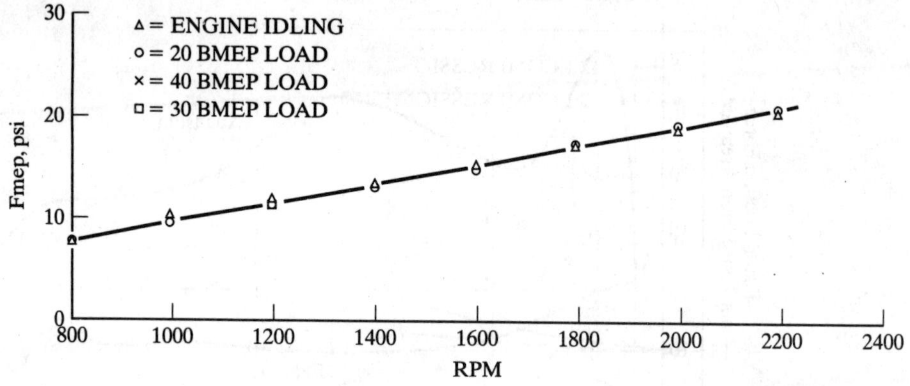

FIGURE 11-6

Friction mean effective pressure as a function of engine speed and load for six-cylinder CI engine. Pressure in psia. Reprinted with permission from SAE Paper No. 730150 © 1973 SAE International, [158].

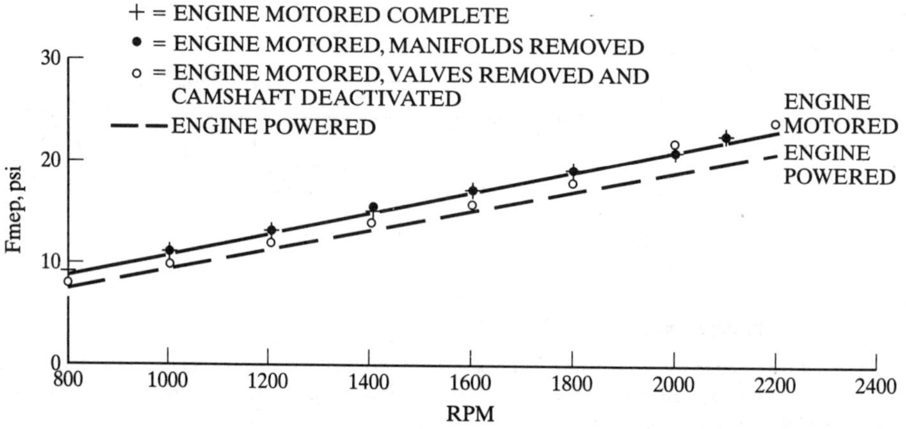

FIGURE 11-7

Comparison of friction mean effective pressure of a motored engine vs. fired engine. Engine is a six-cylinder CI engine. Pressure in psia. Reprinted with permission from SAE Paper No. 730150 © 1973 SAE International, [158].

Example Problem 11-1

A five-cylinder, in-line engine has an 8.15-cm bore, a 7.82-cm stroke, and a connecting rod length of 15.4 cm. Each piston has a skirt length of 6.5 cm and a mass of 0.32 kg. At a certain engine speed and crank angle, the instantaneous piston speed is 8.25 m/sec, and clearance between the piston and cylinder wall is 0.004 mm. SAE 10W-30 motor oil is used in the engine, and at the temperature of the piston–cylinder interface, the dynamic viscosity of the oil is 0.006 N-sec/m^2. Calculate the friction force on one piston at this condition.

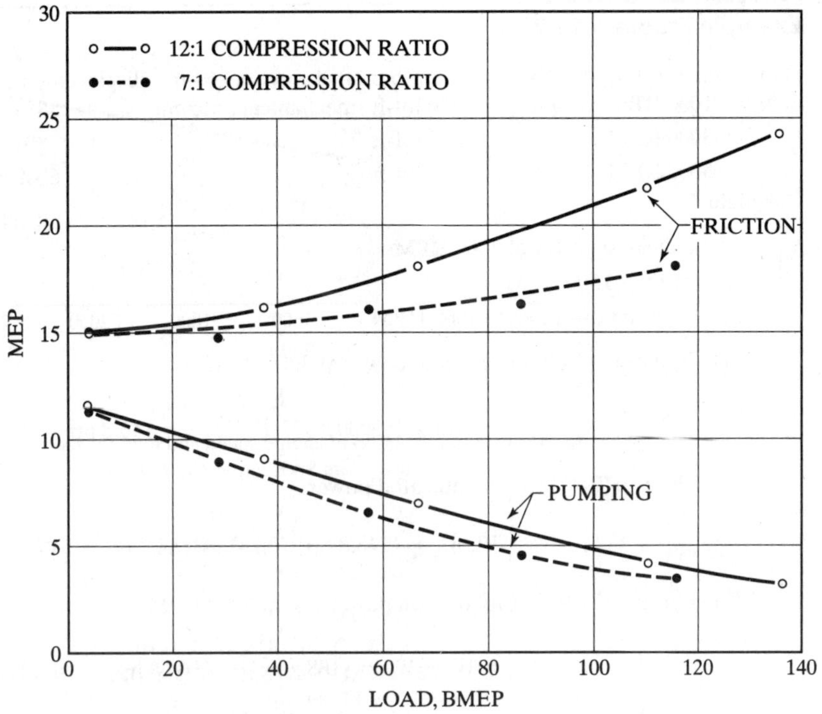

Friction and pumping work versus load at two compression ratios

FIGURE 11-8

Friction mean effective pressure (fmep) and pumping mean effective pressure (pmep) as a function of brake mean effective pressure (bmep) for two compression ratios. Engine is a four-cylinder, 3.25 liter, SI engine, with 9.53 cm bore and 11.40 cm stroke, operating at 1600 RPM. Pressures in psia. Reprinted with permission from SAE Transactions © 1958 SAE International, [177].

Using Eq. (11-13),

$$\tau_s = \mu(dU/dy) = \mu(\Delta U/\Delta y)$$
$$= (0.006 \text{ N-sec/m}^2)[(8.25 \text{ m/sec})/(0.000004 \text{ m})] = 12{,}375 \text{ N/m}^2$$

Contact area between the piston and cylinder wall:

$$A = \pi B(\text{height}) = \pi(0.0815 \text{ m})(0.065 \text{ m}) = 0.0166 \text{ m}^2$$

Friction force on the piston:

$$F_f = \tau_s A = (12{,}375 \text{ N/m}^2)(0.0166 \text{ m}^2) = \underline{205 \text{ N}}$$

Example Problem 11-2

Dynamometer data of a five-cylinder, four-stroke cycle, 260-in.^{3} displacement, CI engine gives:

at N = 1000 RPM	torque τ = 230 lbf-ft	mechanical efficiency η_m =	88%
3000 RPM	257 lbf-ft		78%
5000 RPM	224 lbf-ft		62%

Calculate:

1. indicated power at 3000 RPM
2. fmep at 1500 RPM
3. friction power lost at 4000 RPM

(1) Equation (2-82) gives brake power at 3000 RPM:

$$\dot{W}_b = [(3000)(257)]/(5252) = 146.8 \text{ hp}$$

Use Eq. (2-47) to find indicated power:

$$\dot{W}_i = \dot{W}_b/\eta_m = (146.8 \text{ hp})/(0.78) = \underline{188.2 \text{ hp} = 140.3 \text{ kW}}$$

(2) Use Eq. (2-49) to find friction power lost at 3000 RPM:

$$\dot{W}_f = \dot{W}_i - \dot{W}_b = (188.2 \text{ hp}) - (146.8 \text{ hp}) = 41.4 \text{ hp}$$

Equation (2-90) gives fmep at 3000 RPM:

$$\text{fmep} = [(396{,}000)(41.4)(2)]/[(260)(3000)] = 42.0 \text{ psia}$$

Use Eqs. (2-82), (2-47), (2-49), and (2-90) to find fmep at 1000 RPM:

$$\dot{W}_b = [(1000)(230)]/(5252) = 43.8 \text{ hp}$$
$$\dot{W}_i = (43.8)/(0.88) = 49.8 \text{ hp}$$
$$\dot{W}_f = (49.8) - (43.8) = 6.0 \text{ hp}$$
$$\text{fmep} = [(396{,}000)(6.0)(2)]/[(260)(1000)] = 18.3 \text{ psia}$$

Use Eqs. (2-82), (2-47), (2-49), and (2-90) to find fmep at 5000 RPM:

$$\dot{W}_b = [(5000)(224)]/(5252) = 213.3 \text{ hp}$$
$$\dot{W}_i = (213.3)/(0.62) = 344.0 \text{ hp}$$
$$\dot{W}_f = (344.0) - (213.3) = 130.7 \text{ hp}$$
$$\text{fmep} = [(396{,}000)(130.7)(2)]/[(260)(5000)] = 79.6 \text{ psia}$$

Use Eq. (11-12) to find fmep at 1000 RPM, at 3000 RPM, and at 5000 RPM:

$$\text{fmep} = A + BN + CN^2$$

At 1000 RPM: $18.3 = A + B(1000) + C(1000)^2$

At 3000 RPM $42.0 = A + B(3000) + C(3000)^2$

At 5000 RPM $79.6 = A + B(5000) + C(5000)^2$

Solve for A, B, and C:

$$A = 11.656 \quad B = 0.0049 \quad C = 1.738 \times 10^{-6}$$

Now use this equation to solve for fmep at 1500 RPM:

$$\text{fmep} = (11.656) + (0.0049)(1500) + (1.738 \times 10^{-6})(1500)^2 = \underline{22.9 \text{ psia} = 158 \text{ kPa}}$$

(3) Use Eq. (11-12) with the solved values of A, B, and C to find fmep at 4000 RPM:

$$\text{fmep} = (11.656) + (0.0049)(4000) + (1.738 \times 10^{-6})(4000)^2 = 59.1 \text{ psia}$$

Use Eq. (2-90) to find the friction power lost at 4000 RPM:

$$\dot{W}_f = [(59.1)(260)(4000)]/[(396{,}000)(2)] = \underline{77.6 \text{ hp} = 57.9 \text{ kW}}$$

Engine Accessories

There are many engine and automobile accessories powered off the crankshaft which reduce the brake power output of the engine. Some of these are continuous (fuel pump, oil pump, supercharger, engine fan), and some operate only part of the time (power brake pump, air conditioner compressor, emission control air pump, power steering pump). When an engine is motored to measure friction, it has been found that three essential accessories (water pump, fuel pump, and alternator) can account for as much as 20% of total friction power. The fuel pump and water pump on many older engines were driven mechanically off the crankshaft. Most modern engines have electric fuel pumps and some have electric water pumps. The power to drive these comes from the alternator, which in turn is driven off the engine crankshaft. Most engines have a cooling fan that draws external air through the radiator and blows it through the engine compartment. Some are powered by direct mechanical linkage to the crankshaft. As engine speed goes up, fan speed also goes up. Power needed to drive an air fan goes up as fan speed cubed, so power requirements can get high at higher engine speeds. Higher engine speeds often mean higher automobile velocity, which is when fan cooling is not necessary. At high automobile velocity, enough air is forced through the radiator and engine compartment to adequately cool the engine just by the forward motion of the car. A fan is not needed. To save power, some fans are driven only when their cooling effect is needed. This can be done with mechanical or hydraulic linkage that disconnects at higher speeds or at cooler temperatures (i.e., with a centrifugal or thermal clutch). Most fans are electrically driven and can be turned on with a thermal switch only when needed. Automobiles with air conditioners often require a larger fan due to the added cooling load of the AC condenser.

11.3 FORCES ON PISTON

Figure 11-9 shows the forces that act on a piston. The centerline of the cylinder is used as the X axis, with positive being down in the direction of piston motion during the

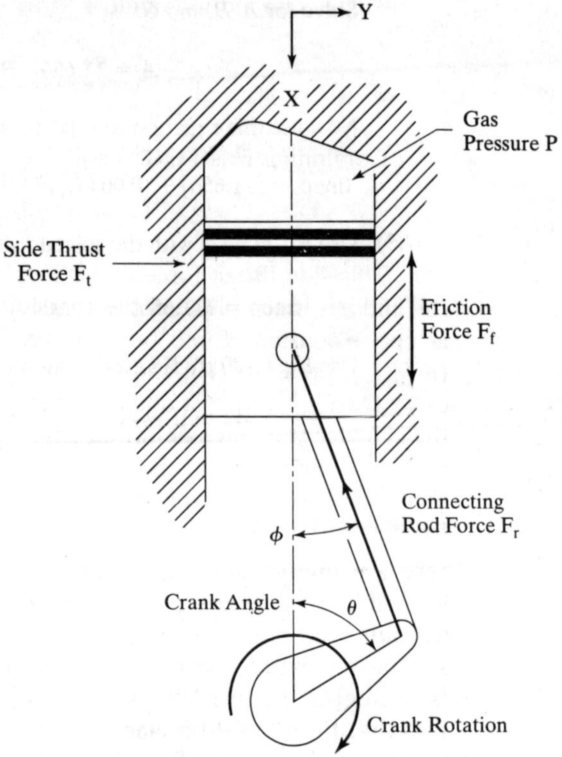

FIGURE 11-9

Force balance on a piston. Side thrust force is a reaction to the connecting rod force and is in the plane of the connecting rod. When the piston passes BDC, the side thrust force switches to the other side of the cylinder. The connecting rod force and the resulting side thrust force are greatest during the power stroke, and this is called the major thrust side. Lesser forces during the exhaust stroke occur on the minor thrust side. The friction force is in the direction opposite to the piston motion and changes direction after TDC and BDC.

power stroke. The Y axis is in the radial direction outward, with zero at the centerline. A force balance in the X direction gives

$$\sum F_x = m(dU_p/dt) = -F_r \cos\phi + P(\pi/4)B^2 \pm F_f' \qquad (11\text{-}25)$$

where

ϕ = angle between the connecting rod and centerline of the cylinder
m = mass of the piston
(dU_p/dt) = acceleration of the piston
F_r = force of the connecting rod
P = pressure in the combustion chamber
B = bore
F_f = friction force between the piston and cylinder walls

the sign on the friction force term depends on the crank angle θ:
− when $0° < \theta < 180°$
+ when $180° < \theta < 360°$

There is no motion in the Y direction, so a force balance gives

$$\sum F_y = 0 = F_r \sin\phi - F_t \tag{11-26}$$

Combining Eqs. (11-25) and (11-26) gives the side thrust force on the piston as

$$F_t = [-m(dU_p/dt) + P(\pi/4)B^2 \pm F_f]\tan\phi \tag{11-27}$$

This side thrust force is the Y-direction reaction to the force in the connecting rod and lies in the plane of the connecting rod. From Eq. (11-27) it can be seen that F_t is not a constant force but changes with piston position (angle ϕ), acceleration (dU_p/dt), pressure (P), and friction force (F_f), all of which vary during the engine cycle. During the power and intake strokes, the side thrust force will be on one side of the cylinder (the left side for an engine rotating as shown in Fig. 11-9) in the plane of the connecting rod. This is called the **major thrust side** of the cylinder because of the high pressure during the power stroke. This high pressure causes a strong reaction force in the connecting rod, which in turn causes a large side thrust reaction force. During the exhaust and compression strokes, the connecting rod is on the other side of the crankshaft and the resulting side thrust reaction force is on the other side of the cylinder (the right side in Fig. 11-9). This is called the **minor thrust side** due to the lower pressures and forces involved and is again in the plane of the connecting rod. The side thrust forces on the piston are less in planes turned circumferentially away from the plane of the connecting rod, reaching a minimum in the plane at a right angle to the connecting rod plane. There will still be a small force reacting to the spring-loaded piston rings.

The side thrust force also varies with crank angular position as the piston moves back and forth in the cylinder. Thus, there is continuous variation both in the circumferential direction and along the length of the cylinder from TDC to BDC. One result of this force variation is the variation in wear that occurs on the cylinder walls. The greatest wear occurs in the plane of the connecting rod on the major thrust side of the cylinder. Significant, but less, wear will occur on the minor thrust side. This wear will also vary along the length of the cylinder on both sides. Additional wear to various degrees will occur in the other rotation planes and at various distances along the length of the cylinder. As an engine ages, this wear can become significant in some spots. Even if the cross section of an engine cylinder is perfectly round when the engine is new, wear will erode this roundness with time [125].

To reduce friction, modern engines use pistons that have less mass and shorter skirts. Less mass lowers the piston inertia and reduces the acceleration term in Eq. (11-27). Shorter piston skirts reduce rubbing friction because of the smaller surface contact area. With closer tolerances between piston and cylinder wall, skirts can be made as short as about 60% of bore diameter. In addition, parts of the skirt can be removed to lower piston mass. Fewer and smaller piston rings are common compared with earlier engines, but these also require closer manufacturing tolerances. In some engines, the wrist pin is offset from center by 1 or 2 mm towards the minor thrust side of the piston. This reduces the side thrust force and resulting wear on the major thrust side.

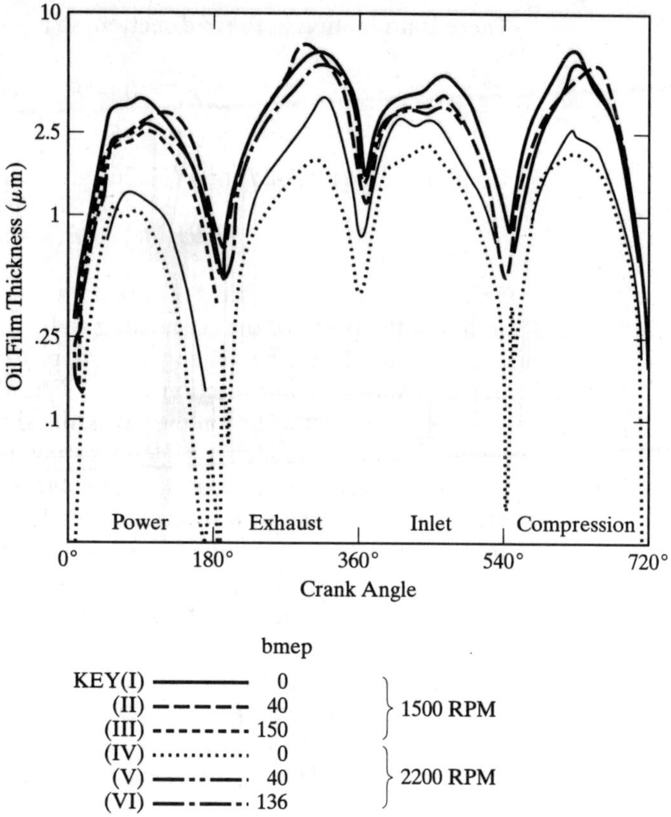

FIGURE 11-10

Oil film thickness between piston compression ring and cylinder wall as a function of speed and load (bmep). Film is minimum when the piston stops at TDC and BDC and oil gets squeezed out. When the piston moves in the opposite direction, the oil film is again dragged between the surfaces, reaches a maximum thickness at maximum piston speed, and again decreases with decreasing piston speed. Data are for the major thrust side of a cylinder in a Perkins AT6-354 diesel engine. Reprinted with permission, © Institute of Mechanical Engineers, London, UK, [3].

The philosophy of some manufacturers is to reduce friction by having a shorter stroke. However, for a given displacement this requires a larger bore, which results in greater heat losses due to the larger cylinder surface area. Greater flame travel distance also increases knock problems. This is why most medium-sized engines (automobile engines) are close to square, with $B \approx S$.

Figure 11-10 shows how the oil film thickness between the piston and cylinder wall varies with speed during an engine cycle for one circumferential position of the piston. Piston friction force will be proportional to oil viscosity, engine speed, and imep.

On some large CI engines, piston side thrust force is eliminated by adding a secondary sliding mechanism called a **crosshead** (Fig. 11-11). The crosshead is contained in an extension of the cylinder and is connected to the crankshaft by the connecting rod. The piston is connected to the crosshead by a secondary connecting rod which stays parallel with the cylinder walls. This eliminates side forces on the piston, transferring them to the crosshead, and reducing wear on the main cylinder walls. This system adds mass, height, and more mechanism to the engine, and is rarely found on smaller or vehicle engines [150].

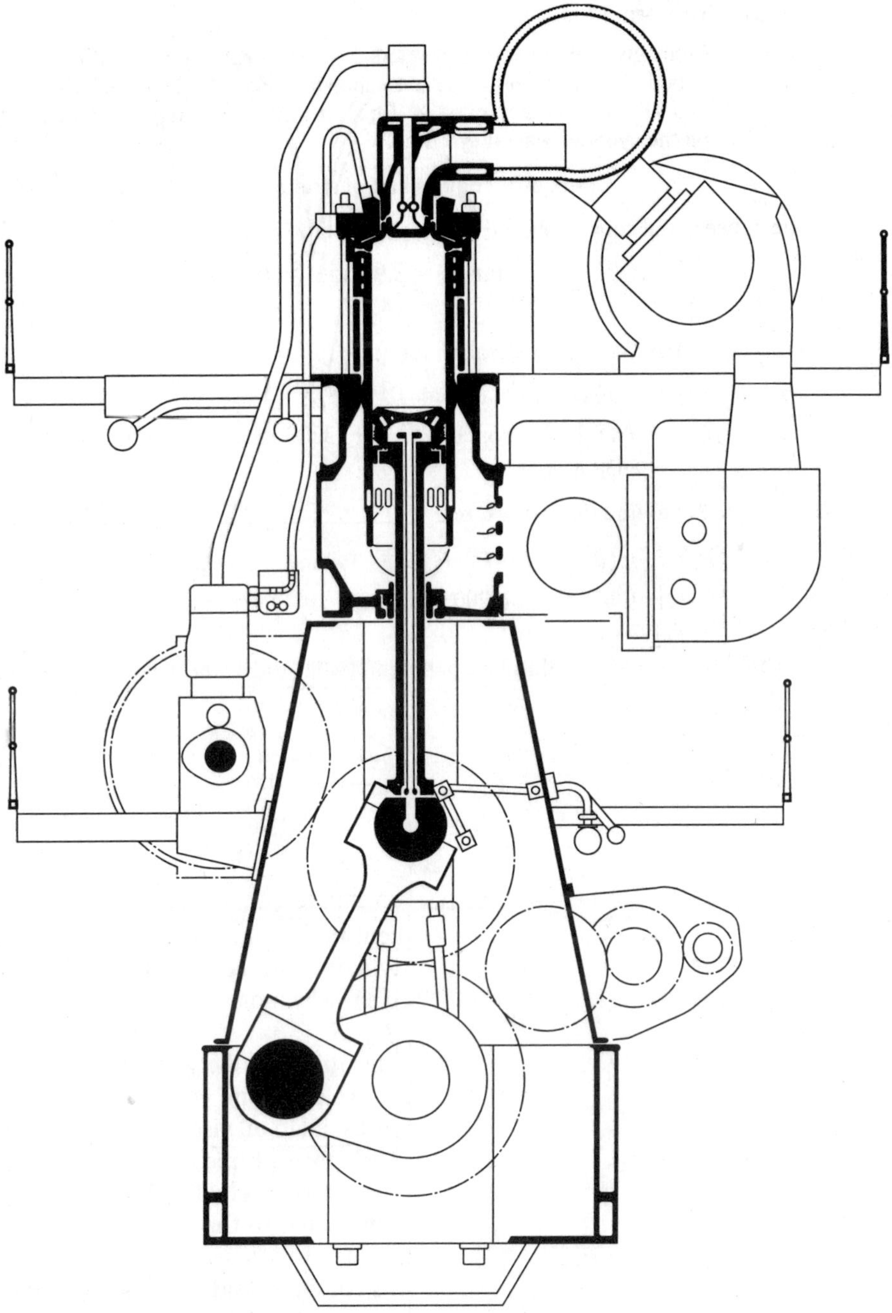

FIGURE 11-11

Cross-sectional view of large Sulzer RTA62 two-stroke cycle CI engine. The pistons of this engine are connected to the crankshaft through a secondary crosshead mechanism which eliminates side thrust forces on the pistons. The crosshead is connected to the crankshaft by connecting rods and thus experiences the side thrust forces. Reprinted with permission from SAE Paper 851219 © 1985 SAE International, [198].

Example Problem 11-3

At the conditions given in Example Problem 11-1, the engine performs a power stroke in the cylinder described and the crank angle is as shown in Fig. 11-12. At this point, pressure in the cylinder is 3200 kPa and the compressive force in the connecting rod is 8.1 kN. Calculate the thrust force on the cylinder wall at this time.

$$\text{Crank offset equals half of stroke length} = 3.91 \text{ cm.}$$

Angle between the connecting rod and the centerline of the cylinder:

$$\tan\phi = 3.91/15.4 = 0.2539$$
$$\phi = 14.25°$$

Use Eq. (11-25) to find the acceleration term:

$$\begin{aligned} m(dU_p/dt) &= -F_r \cos\phi + P(\pi/4)B^2 - F_f \\ &= -(8.1 \text{ kN})\cos(14.25°) + (3200 \text{ kPa})(\pi/4)(0.0815 \text{ m})^2 - (205 N) \\ &= 8638\ N \end{aligned}$$

Use Eq. (11-27) to find the thrust force:

$$\begin{aligned} F_t &= [-m(dU_p/dt) + P(\pi/4)B^2 - F_f]\tan\phi \\ &= [-(8638 \text{ N}) + (3200 \text{ kPa})(\pi/4)(0.0815 \text{ m})^2 - (205 \text{ N})]\tan(14.25°) \\ &= \underline{1994 N} \end{aligned}$$

This force would be in the plane of the connecting rod on the major thrust side of the cylinder.

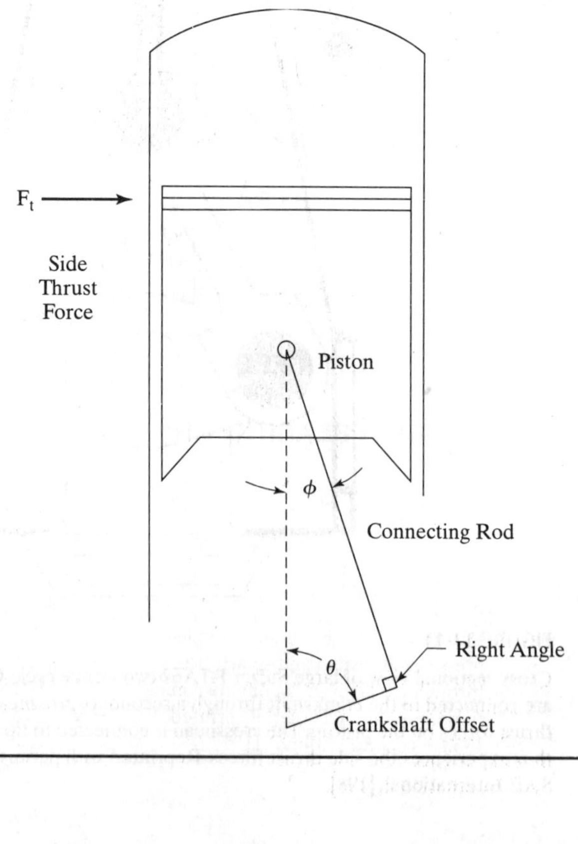

FIGURE 11-12

11.4 ENGINE LUBRICATION SYSTEMS

There are three basic types of oil distribution systems used in engines: splash, pressurized, or a combination of these.

The crankcase is used as the oil sump (reservoir) in a splash system, and the crankshaft rotating at high speed in the oil distributes it to the various moving parts by splash; no oil pump is used. All components, including the valve train and camshaft, must be open to the crankcase. Oil is splashed into the cylinders behind the pistons and onto the back of the piston crowns, acting both as a lubricant and a coolant. Many small four-stroke cycle engines (lawn mowers, golf carts, etc.) use splash distribution of oil.

An engine with a pressurized oil distribution system uses an oil pump to supply lubrication to the moving parts through passages built into the components (Fig. 11-13). A typical automobile engine has oil passages built into the connecting rods, valve stems, push rods, rocker arms, valve seats, engine block, and many other moving components. These make up a circulation network through which oil is distributed by the oil pump. In addition, oil is sprayed under pressure onto the cylinder walls and onto the back of

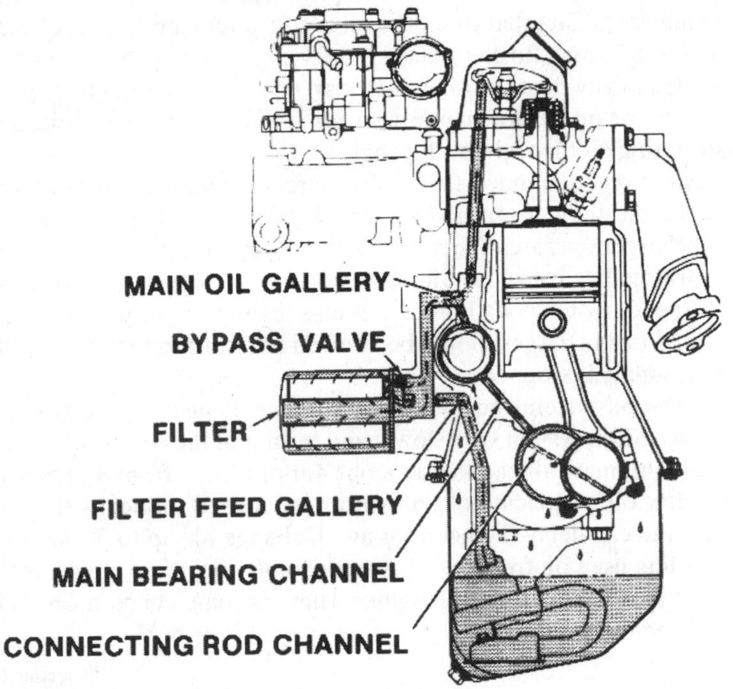

FIGURE 11-13

Lubrication of a 1980s automobile engine consisting of a combination of a pressurized system and splash. The oil pump distributes oil under pressure through passages in the engine components. Systems usually include filtration and sometimes an oil cooler. Reprinted with permission from SAE Paper 820111 © 1982 SAE International, [135].

the piston crowns. Most automobiles actually use dual distribution systems, relying on splash within the crankcase in addition to the pressurized flow from the oil pump. Most large stationary engines also use this kind of dual system. Most aircraft engines and a few automobile engines use a total pressurized system with the oil reservoir located separate from the crankcase. These are often called **dry sump** systems (i.e., the crankcase sump is *dry* of excess oil). Aircraft do not always fly level, and uncontrolled oil in the crankcase may not supply proper lubrication or oil pump input when the plane banks or turns. A diaphragm controls the oil level in the reservoir of a dry sump system, assuring a continuous flow into the oil pump and throughout the engine. Some automobile engines with overhead camshafts have a secondary oil sump in the engine head to supply the cam and valve mechanism.

Oil pumps can be electric or mechanically driven off the engine. Pressure at the pump exit is typically about 300 to 400 kPa. If an oil pump is driven directly off the engine, some means should be built into the system to keep the exit pressure and flow rate from becoming excessive at high engine speeds.

A time of excess wear is at engine start-up before the oil pump can distribute proper lubrication. It takes a few engine cycles before the flow of oil is fully established, and during this time, many parts are not properly lubricated. Adding to the problem is the fact that often the oil is cold at engine start-up. Cold oil has much higher viscosity, which further delays proper circulation. A few engines have oil preheaters that electrically heat the oil before start-up. Some engines have preoilers that heat and circulate the oil before engine start-up. An electric pump lubricates all components by distributing oil throughout the engine.

It is recommended that turbocharged engines be allowed to idle for a few seconds before they are turned off. This is because of the very high speeds at which the turbocharger operates. When the engine is turned off, oil circulation stops and lubricated surfaces begin to lose oil. Stopping the oil supply to a turbocharger operating at high speed invites poor lubrication and high wear. To minimize this problem, the engine and turbocharger should be allowed to return to low speed (idle) before the lubrication supply is stopped.

The oil systems on some large vehicle engines are designed to remove oil from the circulation system very slowly and burn it in the combustion chambers. Makeup oil is then automatically added on a one-for-one basis from a separate oil reservoir filled with new oil. Replacing old oil with new unused oil keeps the oil circulation system clean, and can delay the need for an oil change for up to 50,000 miles [166]. Instead of discarding used oil from an oil change, some ship companies mix this oil with fuel and burn it in the large CI ship engines. There is some concern for an increase of ash and wear when this is done.

11.5 TWO-STROKE CYCLE ENGINES

Many small engines and some experimental two-stroke cycle automobile engines use the crankcase as a compressor for the inlet air. Automobile engines that do this generally have the crankcase divided into several compartments, with each cylinder having its own separate compressor. These engines cannot use the crankcase as an oil sump, and an alternate method must be used to lubricate the crankshaft and other components in the crankcase. In these engines, oil is carried into the engine with the inlet air in much the same way as the fuel. When the fuel is added to the inlet air, usually with a carburetor, oil particles as well as fuel particles are distributed into the flow. The air flow then enters the crankcase, where it is compressed. Oil particles carried with the air lubricate the surfaces they come in contact with, first in the crankcase and then in the intake runner and cylinder.

In some systems (model airplane engines, marine outboard motors, etc.), the oil is premixed with the fuel in the fuel tank. In other engines (automobiles, some golf carts, etc.), there is a separate oil reservoir that feeds a metered flow of oil into the fuel supply line or directly into the inlet air flow. Fuel-to-oil ratio ranges from 30:1 to 400:1, depending on the engine. Some modern high-performance engines have controls that regulate the fuel–oil ratio, depending on engine speed and load. Under conditions of high oil input, oil sometimes condenses in the crankcase. Up to 30% of the oil is recirculated from the crankcase in some experimental automobile engines. It is desirable to get at least 3000 miles per liter of oil used. Most small, lower-cost engines have a single average oil input setting. If too much oil is supplied, deposits form on the combustion chamber walls and valves will stick (if there are valves). If too little oil is supplied, excess wear will occur and the piston can *freeze* in the cylinder. Engines that add oil to the inlet fuel obviously are designed to use up oil during operation. This oil also contributes to HC emissions in the exhaust due to valve overlap and poor combustion of the oil vapor in the cylinders. New oils that also burn better as fuel are being developed for two-stroke cycle engines.

Some experimental two-stroke cycle automobile engines and other medium- and large-size engines use an external supercharger to compress inlet air. These engines use pressurized/splash lubrication systems similar to those on four-stroke cycle engines with the crankcase also serving as the oil sump.

Example Problem 11-4

A four-cylinder, two-stroke cycle engine, with a 2.65-liter displacement and crankcase compression, is running at 2400 RPM and an air–fuel ratio of 16.2:1. At this condition, the trapping efficiency is 72%, relative charge is 87%, and the exhaust residual from the previous cycle in each cylinder is 7%. Oil is added to the intake air flow such that the input fuel-to-oil ratio is 50:1. Calculate:

1. rate of oil use
2. rate of unburned oil added to the exhaust flow

(1) A time rate form of Eq. (5-26) gives the total charge trapped in the engine with time:

$$\dot{m}_{tc} = V_d \rho_a \lambda_{rc} N/n$$
$$= (0.00265\ \text{m}^3/\text{cycle})(1.181\ \text{kg/m}^3)(0.87)(2400/60\ \text{rev/sec})/(1\ \text{rev/cycle})$$
$$= 0.1089\ \text{kg/sec}$$

With 7% exhaust residual, only 93% of the trapped charge is air–fuel:

$$\dot{m}_{mt} = (0.1089\ \text{kg/sec})(0.93) = 0.1013\ \text{kg/sec}$$

Use Eq. (5-24) to find the mass of air and fuel ingested:

$$\dot{m}_{mi} = \dot{m}_{mt}/\lambda_{te} = (0.1013 \text{ kg/sec})/(0.72) = 0.1407 \text{ kg/sec}$$

With AF = 16.2, the rate of fuel input is

$$\dot{m}_f = (0.1407 \text{ kg/sec})/(17.2) = 0.00818 \text{ kg/sec}$$

Oil input

$$\dot{m}_{oil} = (0.00818 \text{ kg/sec})/(50)$$
$$= \underline{0.000164 \text{ kg/sec} = 0.59 \text{ kg/hr} \approx 0.67 \text{ L/hr}}$$

(2) Of the oil input, 72% is trapped in the cylinders and burned, and 28% enters the exhaust during valve overlap.

Mass of unburned oil in the exhaust

$$\dot{m}_{oil} = (0.59 \text{ kg/hr})(0.28) = \underline{0.17 \text{ kg/hr}}$$

11.6 LUBRICATING OIL

The oil used in an engine must serve as a lubricant, a coolant, and a vehicle for removing impurities. It must be able to withstand high temperatures without breaking down and must have a long working life. The development trend in engines is toward higher operating temperatures, higher speeds, closer tolerances, and smaller oil sump capacity. All of these require improved oils compared with those used just a few years ago. Certainly, the technology of the oil industry has to continue to improve along with the technology growth of engines and fuel.

Early engines and other mechanical systems were often designed to consume the lubricating oil as it was used, requiring a continuous input of fresh oil. The used oil was either burned up in the combustion chamber or allowed to fall to the ground. Just a couple of decades back, the tolerances between pistons and cylinder walls was such that engines burned some oil that seeped past the pistons from the crankcase. This required a periodic need to add oil and frequent oil changes due to blowby contamination of the remaining oil. HC levels in the exhaust were high because of the oil in the combustion chamber. A rule in the 1950s and 1960s was to have an oil change in an automobile every 1000 miles.

Modern engines run hotter, have closer tolerances which keep oil consumption down, and have smaller oil sumps due to space limitations. They generate more power with smaller engines by running faster and with higher compression ratios. This means higher forces and a greater need for good lubrication. At the same time, many manufacturers now suggest changing the oil every 6000 miles. Not only must the oil last longer under much more severe conditions, but new oil is not added between oil changes. Engines of the past that consumed some oil required periodic makeup oil to be added. This makeup oil mixed with the remaining used oil and improved the overall lubrication properties within the engine. Some modern high-performance automobiles (Mercedes, Corvette) have sensors in the oil sump that monitor oil level, age, temperature, etc. [191]. These systems tell the operator when the oil has degraded to a point that an oil change is required. These vehicles can sometimes go 40,000 km (25,000 miles) or two years between oil changes.

The oils in modern engines must operate over an extreme temperature range. They must lubricate properly from the starting temperature of a cold engine to beyond the extreme steady-state temperatures that occur within the engine cylinders. They must not oxidize on the combustion chamber walls or at other hot spots such as the center crown of the piston or at the top piston ring. Oil should adhere to surfaces so that they always lubricate and provide a protective covering against corrosion. This is often called *oiliness*. Oil should have high film strength to assure no metal-to-metal contact even under extreme loads. Oils should be nontoxic and nonexplosive. Lubricating oil must satisfy the following needs:

1. **Lubrication**
   It must reduce friction and wear within the engine. It improves engine efficiency by reducing friction forces between moving parts.
2. **Coolant**
3. **Removal of contaminants**
4. **Enhancement of ring seal and reduction of blowby**
5. **Slow corrosion**
6. **Stability over a large temperature range**
7. **Long life span**
8. **Low cost**

The base ingredients in most lubricating oils are hydrocarbon components made from crude oil. These are species with larger molecular weights obtained from the distillation process (Table 11-1). Various other components are added to create a lubricant that will allow for maximum performance and life span of the engine. These additives include the following:

1. **Antifoam agents**
   These reduce the foaming that would result when the crankshaft and other components rotate at high speed in the crankcase oil sump.
2. **Oxidation inhibitors**
   Oxygen is trapped in the oil when foaming occurs, and this leads to possible oxidation of engine components. One such additive is zinc dithiophosphate.
3. **Pour-point depressant**

Table 11-1 Hydrocarbon Components in Lubricating Oil

	Number of Carbon Atoms in Component	
	Range	Average Number
SAE 10	25–35	28
SAE 20	30–80	38
SAE 30	40–100	41

Adapted from [56].

4. Antirust agents
5. Detergents
   These are made from organic salts and metallic salts. They help keep deposits and impurities in suspension and stop reactions that form varnish and other surface deposits. They help neutralize acid formed from sulfur in the fuel.
6. Antiwear agents
7. Friction reducers
8. Viscosity index improvers

Viscosity

Lubricating oils are generally rated using a viscosity scale established by the Society of Automotive Engineering (SAE). Dynamic viscosity is defined from the equation

$$\tau_s = \mu(dU/dy) \tag{11-13}$$

where

τ_s = shear force per unit area
μ = dynamic viscosity
(dU/dy) = velocity gradient

The higher the viscosity value, the greater is the force needed to move adjacent surfaces or to pump oil through a passage. Viscosity is highly dependent on temperature, increasing with decreasing temperature (Fig. 11-14). In the temperature range of engine operation, the dynamic viscosity of the oil can change by several orders of magnitude. Oil viscosity also changes with shear, du/dy, decreasing with increasing shear. Shear rates within an engine range from very low values to extremely high values in the bearings and between piston and cylinder walls. The change of viscosity over these extremes can be several orders of magnitude. The following viscosity *grades* are commonly used in engines:

SAE 5
SAE 10
SAE 20
SAE 30
SAE 40
SAE 45
SAE 50

The oils with lower numbers are less viscous and are used in cold-weather operation. Those with higher numbers are more viscous and are used in modern high-temperature, high-speed, close-tolerance engines. Oils become more viscous with age, because the components with lower molecular weights evaporate quicker.

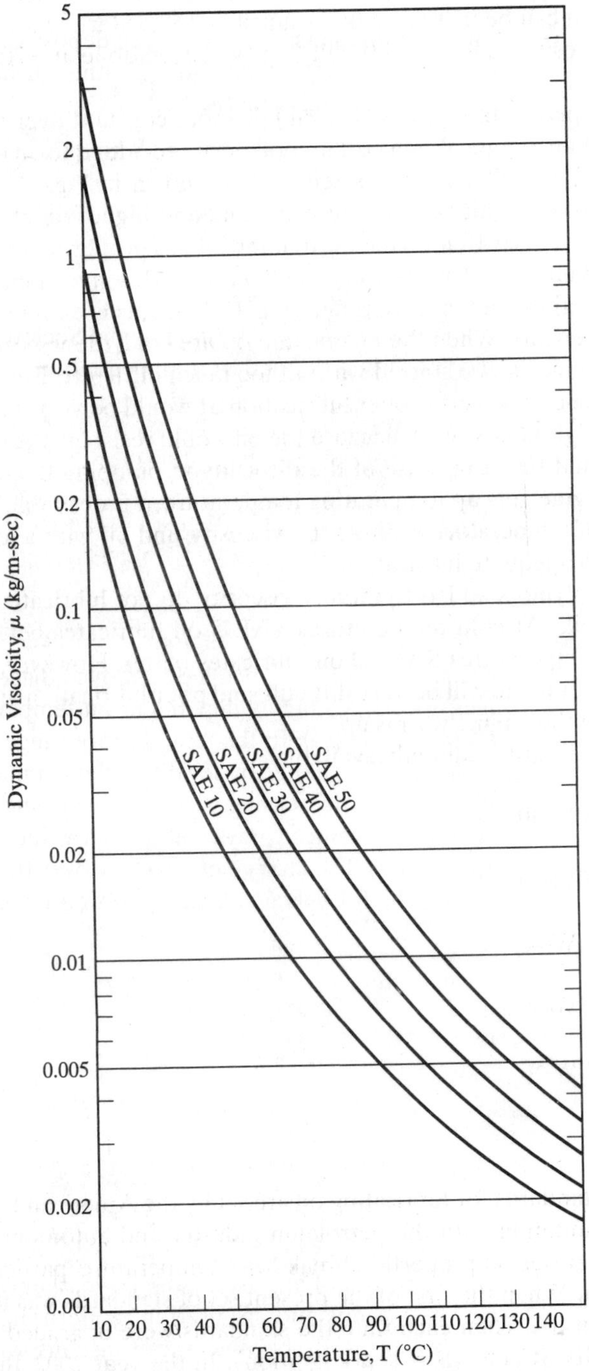

FIGURE 11-14

Dynamic viscosity as a function of temperature for common engine oils. Adapted from [113].

If oil viscosity is too high, more work is required to pump it and to shear it between moving parts. This results in greater friction work and reduced brake work and power output. Fuel consumption can be increased by as much as 15%. Starting a cold engine lubricated with high-viscosity oil is very difficult (e.g., an automobile at −20°C or a lawn mower at 10°C).

Multigrade oil was developed so that viscosity would be more constant over the operating temperature range of an engine. When certain polymers are added to an oil, the temperature dependency of the oil viscosity is reduced, as shown in Fig. 11-15. These oils have low-number viscosity values when they are cold and higher numbers when they are hot. A value such as SAE 10W-30 means that the oil has properties of 10 viscosity when it is cold (W = winter) and 30 viscosity when it is hot. This gives a more constant viscosity over the operating temperature range (Fig. 11-15). This is extremely important when starting a cold engine. When the engine and oil are cold, the viscosity must be low enough that the engine can be started without too much difficulty. The oil flows with less resistance and the engine gets proper lubrication. It would be very difficult to start a cold engine with high-viscosity oil, because the oil would resist engine rotation and poor lubrication would result because of the difficulty in pumping the oil. On the other hand, when the engine gets up to operating temperature, it is desirable to have a higher viscosity oil. High temperature reduces the viscosity, and oil with a low viscosity number would not give adequate lubrication.

Some studies show that polymers added to modify viscosity do not lubricate as well as the base hydrocarbon oils. At cold temperatures SAE 5 oil lubricates better than SAE 5W-30, and at high temperatures SAE 30 oil lubricates better. However, if SAE 30 oil is used, starting a cold engine will be very difficult, and poor lubrication and very high wear will result before the engine warms up.

The following multigrade oils are commonly available:

SAE 5W-20	SAE 10W-40
SAE 5W-30	SAE 10W-50
SAE 5W-40	SAE 15W-40
SAE 5W-50	SAE 15W-50
SAE 10W-30	SAE 20W-50

Lubricating Oil Standards

In the United States, voluntary standards for lubricating oil are set by the American Petroleum Institute (API), in consultation with the petroleum industry and automobile makers. These standards cover various oil properties: breakdown temperature, particle suspension, lubricating ability, etc. When the first of the present set of standards was issued it was given the designation SA. Then each time the standards were upgraded a new letter grade was assigned: SB, SC, etc. (S = spark ignition). In the year 2002, the

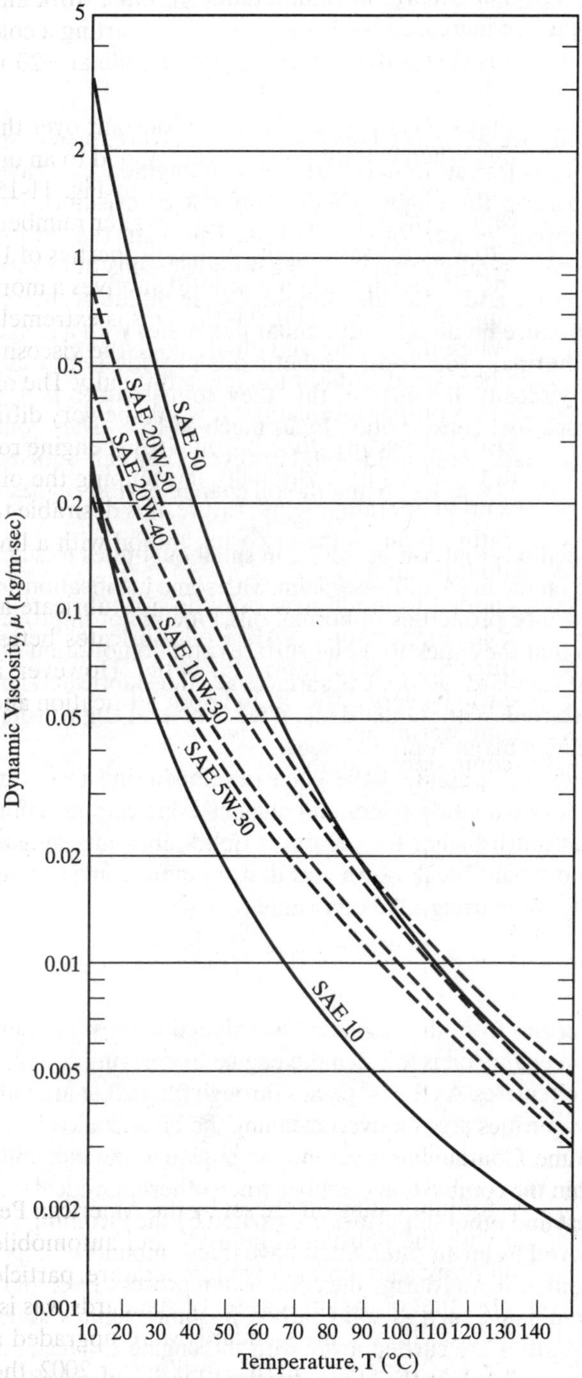

FIGURE 11-15

Dynamic viscosity as a function of temperature for common multigrade engine oils. Adapted from [113].

standards were up to SL. Standards for oils intended for use in diesel engines were given a similar set of letter designations: CA, CB, etc. (C = compression ignition). In 2002, the diesel standards were up to CH-4.

Synthetic Oils

A number of synthetically made oils are available that give better performance than those made from crude oil. They are better at reducing friction and engine wear, have good detergency properties which keep the engine cleaner, offer less resistance for moving parts, and require less pumping power for distribution. The main reason for this better performance is that the base synthetic material is a homogeneous fluid with similar molecules of the same structure and molecular weight. Crude oil lubricant, on the other hand, has a base fluid structure made up of dissimilar molecules with a range of molecular weights. With good thermal properties, synthetic oils provide better engine cooling and less variation in viscosity. Because of this, they contribute to better cold-weather starting and can reduce fuel consumption by as much as 15%. These oils cost several times as much as those made from crude oil. However, they can be used longer in an engine, with 24,000 km (15,000 miles) being the oil change period suggested by most manufacturers.

Various oil additives and special oils that can be added in small quantities to standard oils in the engine are available on the market. These claim, with some justification, to improve the viscous and wear resistance properties of normal oils. One major improvement that some of them provide is that they stick to metal surfaces and do not drain off when the engine is stopped, as most standard oils do. The surfaces are thus lubricated immediately when the engine is next started. With standard oils, it takes several engine rotations before proper lubrication occurs, a major source of wear.

Solid lubricants, such as powdered graphite, have been developed and tested in some engines. These are attractive for adiabatic engines and engines using ceramic components, which generally operate at much higher temperatures. Solid lubricants remain functional at high temperatures that would break down and destroy more conventional oils. Distribution is a major difficulty when using solid lubricants.

11.7 OIL FILTERS

A filtration system to remove impurities from the engine oil is included in most pressurized oil systems. One of the duties of engine oil is to clean the engine by carrying contaminant impurities in suspension as it circulates. As the oil passes through filters that are part of the flow passage system, these impurities are removed, cleaning the oil and allowing it to be used for a greater length of time. Contaminants get into an engine in the incoming air or fuel, or can be generated within the combustion chamber when other-than-ideal stoichiometric combustion occurs. Dust and other impurities are carried by the incoming air. Some, but not all, of these are removed by an air filter. Fuels have trace amounts of impurities such as sulfur, which create contaminants during the combustion process. Even pure fuel components form some contaminants, such as solid carbon in some engines under some conditions. Many engine impurities are carried away with the engine exhaust, but some get into the interior of the engine, mainly in the blowby process. During blowby, fuel, air, and combustion products are forced past the pistons into the crankcase, where they mix with the engine oil. Some of the water vapor in the exhaust products condenses in the crankcase, and the resulting liquid water adds to the contaminants. The gases of blowby pass through the crankcase and are routed back into the air intake. Ideally, most of the contaminants are trapped in the oil, which then contains dust, carbon, fuel particles, sulfur, water droplets, and many other impurities. If these were not filtered out of the oil, they would be spread throughout the engine by the oil distribution system. Also, the oil would quickly become *dirty* and lose its lubricating properties, resulting in greater engine wear. Instead of an oil filter, Cummins Diesel uses a centrifuge on some of their large engines [166]. The oil impurities are forced to the outer edge of the centrifuge where they are removed.

Flow passages in a filter are not all the same size but usually exist in a normal bell-shaped size distribution (Fig. 11-16). This means that most larger particles will be filtered out as the oil passes through the filter, but a few as large as the largest passages will get through.

The choice of filter pore size is a compromise. Better filtration will be obtained with smaller filter pores, but this requires a much greater flow pressure to push the oil through the filter. This also results in the filter becoming clogged quicker and requiring earlier filter cartridge change. Some filter materials or material of too small a pore size can even remove some additives from the oil. Filters are made from cotton, paper, cellulose, and a number of different synthetic materials. Filters are usually located just downstream from the oil pump exit.

As a filter is used, it slowly becomes saturated with trapped impurities. As these impurities fill the filter pores, a greater pressure differential is needed to keep the same flow rate. When this needed pressure differential gets too high, the oil pump limit is reached and oil flow through the engine is slowed. The filter cartridge should be replaced before this happens. Sometimes, when the pressure differential across a filter gets high enough, the cartridge structure will collapse and a hole will develop through

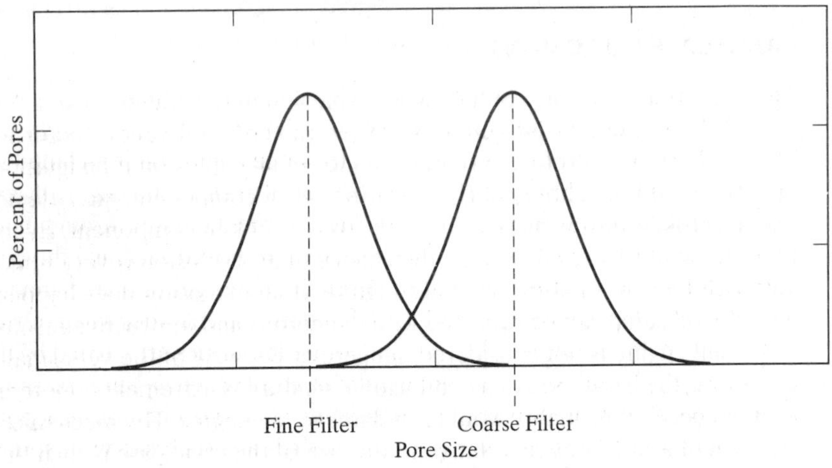

FIGURE 11-16

Pore size distribution for common filters.

the cartridge wall. Most of the oil pumped through the filter will then follow the path of least resistance and flow through the hole. This *short circuit* will reduce the pressure drop across the filter, but the oil does not get filtered.

There are several ways in which the oil circulation system can be filtered:

1. *Full-flow oil filtration.* All oil flows through the filter. The filter pore size must be fairly large to avoid extreme pressures in the resulting large flow rate. This results in some larger impurities in the oil.
2. *Bypass oil filtration.* Only part of the oil leaving the pump flows through the filter, the rest bypassing it without being filtered. This system allows the use of a much finer filter, but only a percentage of the oil gets filtered during each circulation loop.
3. *Combination.* Some systems use a combination of full-flow and bypass. All the oil first flows through a filter with large pores and then some of it flows through a second filter with small pores.
4. *Shunt filtration.* This is a system using a full-flow filter and a bypass valve. All oil at first flows through the filter. As the filter cartridge dirties with age, the pressure differential across it needed to keep the oil flowing increases. When this pressure differential gets above a predetermined value, the bypass valve opens and the oil flows around the filter. The filter cartridge must then be replaced before filtering will again occur.

HISTORIC—FILTER CARTRIDGES

During the 1950s, one manufacturer sold an accessory oil filtration system that could be added onto an automobile engine. The system consisted of a filter cartridge canister that was bolted onto the oil distribution system. The filter cartridge that was to be used with this canister was a standard roll of toilet paper.

11.8 CRANKCASE EXPLOSIONS

There is a remote chance of having an explosion in the crankcase of a reciprocating engine [152]. The crankcase contains oxygen (air) and oil vapor and/or fuel vapor from blowby. There is, therefore, a remote chance of an explosion if an ignition source develops. This could be a hot spot (bad bearing, worn component, etc.), flame getting past a piston (broken piston ring), or a spark from a broken component. Even when oxygen, fuel, and ignition are all present, the chance of an explosion is very remote because the very rich fuel–air mixture will fail to ignite. If an explosion does happen in a large engine, the oil sump can be ruptured with dangerous and destructive results. An explosion in a small engine is not considered dangerous. Because of the very small volume of the crankcase, the likelihood of a combustible mixture is extremely remote, and if an explosion did occur, it probably would not damage the engine. The mechanical structure and strength of a small engine relative to the size of the crankcase is such that an explosion probably would not even be noticed, and the engine would continue to run. International safety codes consider safe any engine with less than 6.1 m^3 of crankcase volume or bores less than 0.2 m. Engines larger than this require explosion relief valves.

Explosion relief valves are designed to release the pressure buildup caused by a crankcase explosion without damaging the engine or injuring anyone near the engine [164]. A well-designed valve will open, direct the flow of hot gases away from operators, arrest the flame, and immediately close without allowing air to flow back into the crankcase. Most valves will open when a pressure pulse in the range of 5 to 20 kPa (0.7 to 3 psia) is detected. The engine will usually continue to run, but should be stopped to determine the cause of explosion. Engines should be allowed to cool before they are opened; the onrush of air into the hot crankcase could be very dangerous.

11.9 SUMMARY AND CONCLUSIONS

Brake power output from an engine is less than the power generated in the combustion chambers, due to engine friction. Two types of friction occur that result in dissipation and loss of useful power. Mechanical friction between moving parts is a major source of engine power loss, with piston motion in the cylinders causing a large percentage of this loss. Fluid friction occurs in the intake and exhaust systems, in flow through valves, and because of motion within the cylinders. Operation of engine accessories, although not friction in the normal sense, is often included as part of the engine friction load. That is because the accessories are powered, directly or indirectly, off the engine crankshaft and reduce final crankshaft output power.

To minimize friction and to reduce engine wear, lubrication systems are a major required facet of any engine. Oil distribution can be accomplished by using a pressurized system supplied by a pump, as with automobile engines, or by splash distribution, as on many small engines. In addition to lubricating, engine oil helps to cool the engine and is a vehicle for removing engine contaminants.

PROBLEMS

11.1 The connecting rod in Fig. 11-12 experiences a force of 1000 N in the position shown during the power stroke of a four-cylinder, four-stroke cycle SI engine operating at 2000 RPM. Crankshaft offset equals 3.0 cm and connecting rod length equals 9.10 cm. Calculate:

(a) Side thrust force felt in the cylinder wall at this moment. [N]
(b) Distance the piston has traveled from TDC. [cm]
(c) Engine displacement if $S = 0.94\ B$. [L]
(d) Side thrust force felt in the cylinder wall when the piston is at TDC. [N]

11.2 Why do cylinders in IC engines get *out-of-round* as the engine is operated for a long period of time? Why is wear on the cylinder walls not the same along the length of the cylinder? Theoretically, why is piston frictional force equal to zero at TDC and BDC? In actuality, why is piston frictional force not equal to zero at TDC and BDC?

11.3 A six-cylinder IC engine has a 6.00-cm bore, a 5.78-cm stroke, and a connecting rod length of 11.56 cm. In the power stroke of the cycle for one cylinder at a crank position of 90° aTDC, the pressure in the cylinder is 4500 kPa and the sliding friction force on the piston is 0.85 kN. Piston acceleration at this point can be considered zero.

Calculate:

(a) Force in the connecting rod at this point. [kN] Is it compressive or tensile?
(b) Side thrust force on the piston at this point. [kN] Is it on the major thrust side or the minor thrust side?
(c) Side thrust force on the piston at this point if the wrist pin is offset 2 mm to reduce the side thrust force. (Assume rod force and friction force are the same as above.) [kN]

11.4 A V6, two-stroke cycle SI automobile engine has a 3.1203-inch bore and 3.45-inch stroke. The pistons have a height of 2.95 inches and diameter of 3.12 inches. At a certain point during the compression stroke piston speed in one cylinder is 30.78 ft/sec. The lubricating oil on the cylinder walls has a dynamic viscosity of 0.000042 lbf-sec/ft^2. Calculate the friction force on the piston under this condition. [lbf]

11.5 A four-cylinder, four-stroke cycle, 2.8-liter, opposed-cylinder SI engine has brake mean effective pressure and mechanical efficiency as follows:

at 1000 RPM	bmep = 828 kPa	η_m = 90%
2000 RPM	bmep = 828 kPa	η_m = 88%
3000 RPM	bmep = 646 kPa	η_m = 82%

Calculate:

(a) Brake power at 2000 RPM. [kW]
(b) Friction mean effective pressure at 2500 RPM. [kPa]
(c) Friction power lost at 2500 RPM. [kW]

11.6 A 110-in.^{3}-displacement, six-cylinder SI automobile engine operates on a two-stroke cycle with crankcase compression and throttle body fuel injection. With AF = 17.8 and the engine running at 1850 RPM, the automobile cruises at 65 mph and gets 21 miles per gallon of gasoline. Oil is added to the inlet air at a rate such that input fuel-to-oil ratio is 40:1. Relative charge is 64% and the exhaust residual from the previous cycle is 6%. Combustion efficiency η_c = 100% and the density of gasoline ρ_g = 46.8 lbm/ft^3.
Calculate:

(a) Rate of oil use. [gal/hr]
(b) Trapping efficiency of the engine. [%]
(c) Rate of unburned oil added to the exhaust flow. [gal/hr]

11.7 When a supercharger is installed on the four-stroke cycle SI engine with a compression ratio r_c = 9.2:1, the indicated thermal efficiency at WOT is decreased by 6%. Mass of air in the cylinders is increased by 22% when operating at the same speed of 2400 RPM. Engine mechanical efficiency stays the same, except that 4% of the brake crankshaft output is needed to run the supercharger.
Calculate:

(a) Indicated thermal efficiency without a supercharger. [%]
(b) Indicated thermal efficiency with a supercharger. [%]
(c) Percent increase of indicated power when a supercharger is installed. [%]
(d) Percent increase of brake power when a supercharger is installed. [%]

DESIGN PROBLEMS

11.1D Design a two-stroke cycle SI engine with crankcase compression that uses a conventional oil distribution system (i.e., a pressurized system with an oil pump and an oil reservoir in the crankcase).

APPENDIX A

Appendix

TABLE A-1 Thermodynamic Properties of Air

Temperature K	Temperature °C	c_p (kJ/kg-K)	c_v (kJ/kg-K)	$k = c_p/c_v$	Gas Constant $R = c_p - c_v$ (kJ/kg-K)
273	0	1.004	0.717	1.40	0.287
298	25	1.005	0.718	1.40	0.287
300	27	1.005	0.718	1.40	0.287
500	227	1.029	0.742	1.39	0.287
850	577	1.108	0.821	1.35	0.287
1000	727	1.140	0.853	1.34	0.287
1500	1227	1.210	0.923	1.31	0.287
2000	1727	1.249	0.962	1.30	0.287
2500	2227	1.274	0.987	1.29	0.287
3000	2727	1.291	1.004	1.29	0.287

Temperature °R	Temperature °F	c_p (BTU/lbm-°R)	c_v (BTU/lbm-°R)	$k = c_p/c_v$	Gas Constant $R = c_p - c_v$ (BTU/lbm-°R)	(ft-lbf/lbm-°R)
492	32	0.240	0.171	1.40	0.069	53.33
537	77	0.240	0.171	1.40	0.069	53.33
1000	540	0.249	0.180	1.38	0.069	53.33
1500	1040	0.264	0.195	1.35	0.069	53.33
2000	1540	0.277	0.208	1.33	0.069	53.33
2500	2040	0.287	0.218	1.32	0.069	53.33
3000	2540	0.293	0.224	1.31	0.069	53.33
3500	3040	0.298	0.229	1.30	0.069	53.33
4000	3540	0.302	0.233	1.30	0.069	53.33
4500	4040	0.305	0.236	1.29	0.069	53.33
5000	4540	0.307	0.238	1.29	0.069	53.33
5500	5040	0.309	0.240	1.29	0.069	53.33

TABLE A-2 Properties of Fuels

Fuel		Molecular Weight	Heating Value HHV (kJ/kg)	Heating Value LHV (kJ/kg)	Stoichiometric $(AF)_s$	Stoichiometric $(FA)_s$	Octane Number MON	Octane Number RON	Heat of Vaporization (kJ/kg)	Cetane Number
gasoline	C_8H_{15}	111	47300	43000	14.6	0.068	80–91	92–99	307	
light diesel	$C_{12.3}H_{22.2}$	170	44800	42500	14.5	0.069			270	40–55
heavy diesel	$C_{14.6}H_{24.8}$	200	43800	41400	14.5	0.069			230	35–50
isooctane	C_8H_{18}	114	47810	44300	15.1	0.066	100	100	290	
methanol	CH_3OH	32	22540	20050	6.5	0.155	92	106	1147	
ethanol	C_2H_5OH	46	29710	26950	9.0	0.111	89	107	873	
methane	CH_4	16	55260	49770	17.2	0.058	120	120	509	
propane	C_3H_8	44	50180	46190	15.7	0.064	97	112	426	
nitromethane	CH_3NO_2	61	12000	10920	1.7	0.588			623	
heptane	C_7H_{16}	100	48070	44560	15.2	0.066	0	0	316	
cetane	$C_{16}H_{34}$	226	47280	43980	15.0	0.066			292	100
heptamethylnonane	$C_{12}H_{34}$	178			15.9	0.063				15
α-methylnaphthalene	$C_{11}H_{10}$	142			13.1	0.076				0
carbon monoxide	CO	28	10100	10100	2.5	0.405				
coal (carbon)	C	12	33800	33800	11.5	0.087				
butene-1	C_4H_8	56	48210	45040	14.8	0.068	80	99	390	
triptane	C_7H_{16}	100	47950	44440	15.2	0.066	101	112	288	
isodecane	$C_{10}H_{22}$	142	47590	44220	15.1	0.066	92	113		
toluene	C_7H_8	92	42500	40600	13.5	0.074	109	120	412	
hydrogen	H_2	2	141800	120000	34.5	0.029		90		

TABLE A-3 Chemical Equilibrium Constants

Log_{10} K_e for equilibrium constants as defined by Eq. (4-4) for the following reactions:

(A) $H_2 \rightarrow 2H$
(B) $N_2 \rightarrow 2N$
(C) $O_2 \rightarrow 2O$
(D) $CO_2 \rightarrow CO + \frac{1}{2}O_2$
(E) $H_2O \rightarrow OH + \frac{1}{2}H_2$
(F) $\frac{1}{2}N_2 + \frac{1}{2}O_2 \rightarrow NO$
(G) $CO_2 + H_2 \rightarrow CO + H_2O$

T(K)	(A)	(B)	(C)	(D)	(E)	(F)	(G)	T(K)
298.15	−71.30	−159.7	−81.28	−45.05	−46.10	−15.15	−4.950	298.15
300	−70.83	−158.7	−80.73	−44.74	−45.80	−15.07	−4.905	300
400	−51.73	−117.4	−58.91	−32.43	−33.48	−11.14	−3.215	400
500	−40.26	−92.63	−45.82	−25.03	−26.09	−8.784	−2.193	500
600	−32.62	−76.12	−37.10	−20.10	−21.16	−7.210	−1.506	600
700	−27.16	−64.33	−30.86	−16.57	−17.64	−6.086	−1.014	700
800	−23.06	−55.48	−26.19	−13.92	−15.00	−5.243	−0.642	800
900	−19.88	−48.60	−22.55	−11.86	−12.95	−4.587	−0.352	900
1000	−17.33	−43.10	−19.64	−10.21	−11.31	−4.063	−0.120	1000
1100	−15.18	−38.54	−17.21	−8.843	−9.922	−3.633	0.040	1100
1200	−13.40	−34.75	−15.20	−7.739	−8.784	−3.275	0.152	1200
1300	−11.89	−31.54	−13.49	−6.802	−7.821	−2.972	0.251	1300
1400	−10.60	−28.79	−12.03	−6.004	−6.996	−2.712	0.330	1400
1500	−9.474	−26.41	−10.76	−5.315	−6.280	−2.487	0.397	1500
1600	−8.492	−24.32	−9.657	−4.711	−5.654	−2.290	0.456	1600
1700	−7.626	−22.48	−8.680	−4.175	−5.102	−2.116	0.510	1700
1800	−6.856	−20.85	−7.811	−3.697	−4.611	−1.962	0.560	1800
1900	−6.168	−19.39	−7.033	−3.268	−4.172	−1.824	0.608	1900
2000	−5.548	−18.07	−6.334	−2.879	−3.777	−1.699	0.652	2000
2100	−4.987	−16.88	−5.701	−2.527	−3.419	−1.587	0.692	2100
2200	−4.477	−15.79	−5.125	−2.207	−3.094	−1.484	0.729	2200
2300	−4.012	−14.81	−4.600	−1.917	−2.797	−1.391	0.761	2300
2400	−3.585	−13.90	−4.118	−1.652	−2.525	−1.306	0.788	2400
2500	−3.192	−13.06	−3.675	−1.412	−2.274	−1.227	0.810	2500
2600	−2.830	−12.29	−3.266	−1.194	−2.043	−1.154	0.828	2600
2700	−2.495	−11.58	−2.887	−0.995	−1.829	−1.087	0.840	2700
2800	−2.183	−10.92	−2.536	−0.813	−1.631	−1.025	0.849	2800
2900	−1.893	−10.30	−2.208	−0.646	−1.446	−0.967	0.855	2900
3000	−1.622	−9.729	−1.903	−0.491	−1.273	−0.913	0.859	3000
3100	−1.369	−9.191	−1.617	−0.347	−1.111	−0.863	0.863	3100
3200	−1.131	−8.686	−1.349	−0.208	−0.960	−0.815	0.869	3200
3300	−0.908	−8.213	−1.097	−0.073	−0.818	−0.771	0.881	3300
3400	−0.698	−7.767	−0.860	0.062	−0.684	−0.729	0.900	3400
3500	−0.501	−7.346	−0.637	0.202	−0.558	−0.690	0.929	3500

Adapted from [69].

TABLE A-4 Conversion Factors for Engine Parameters

Length	1 cm = 0.01 m = 10 mm = 0.394 in. = 0.0328 ft 1 m = 100 cm = 1000 mm = 3.281 ft = 39.37 in. 1 km = 0.6214 mile = 3281 ft 1 in. = 0.0833 ft = 0.0254 m = 2.54 cm 1 ft = 0.3048 m 1 mile = 5280 ft = 1609 m = 1.609 km
Area	1 cm^2 = 0.0001 m^2 = 0.155 $in.^2$ 1 m^2 = 10000 cm^2 = 10.76 ft^2 = 1550 $in.^2$ 1 $in.^2$ = 0.00694 ft^2 = 6.45 cm^2 1 ft^2 = 144 $in.^2$ = 0.00929 m^2 = 92.9 cm^2
Volume	1 cm^3 = 0.001 L = 0.061 $in.^3$ 1 L = 0.001 m^3 = 1000 cm^3 = 61.2 $in.^3$ = 0.264 gal 1 m^3 = 1000 L = 35.32 ft^3 = 10^6 cm^3 1 $in.^3$ = 0.000574 ft^3 = 16.39 cm^3 = 0.01639 L 1 ft^3 = 1728 $in.^3$ = 7.481 gal = 28.32 L = 0.02832 m^3 1 gal = 0.1337 ft^3 = 231 $in.^3$ = 3.785 L
Specific Volume	1 m^3/kg = 16.018 ft^3/lbm 1 ft^3/lbm = 0.0624 m^3/kg
Density	1 kg/m^3 = 0.0624 lbm/ft^3 1 lbm/ft^3 = 16.02 kg/m^3
Mass	1 kg = 2.205 lbm 1 lbm = 0.4536 kg
Force	1 N = 0.225 lbf 1 lbf = 4.448 N
Energy or Work	1 kJ = 0.948 BTU = 737.6 ft-lbf 1 BTU = 778.2 ft-lbf = 1.055 kJ 1 ft-lbf = 0.00129 BTU = 0.00136 kJ
Power	1 kW = 1.341 hp = 3412 BTU/hr 1 hp = 2545 BTU/hr = 550 ft-lbf/sec = 0.7457 kW
Torque	1 N-m = 0.738 lbf-ft 1 lbf-ft = 1.355 N-m
Pressure	1 kPa = 0.145 psia 1 psia = 6.895 kPa 1 atm = 101.35 kPa = 14.7 psia
Velocity	1 m/sec = 3.60 km/hr = 3.281 ft/sec = 2.237 MPH 1 km/hr = 0.2778 m/sec = 0.6214 MPH 1 ft/sec = 0.682 MPH = 0.3048 m/sec = 1.097 km/hr 1 MPH = 1.467 ft/sec = 1.609 km/hr

TABLE A-4 (*cont.*) Conversion Factors for Engine Parameters

Acceleration	1 m/sec^2 = 3.281 ft/sec^2 1 ft/sec^2 = 0.305 m/sec^2
Rotational Speed	1 RPM = 0.0167 rev/sec = 0.1047 radians/sec 1 rev/sec = 60 RPM = 2π radians/sec
Mass Flow Rate	1 kg/sec = 2.205 lbm/sec 1 lbm/sec = 0.4536 kg/sec
Specific Energy or Heating Value	1 kJ/kg = 0.4299 BTU/lbm 1 BTU/lbm = 2.326 kJ/kg
Specific Heat or Gas Constant	1 kJ/kg-K = 0.2388 BTU/lbm-°R = 185.8 ft-lbf/lbm-°R 1 BTU/lbm-°R = 778.2 ft-lbf/lbm-°R = 4.1868 kJ/kg-K 1 ft-lbf/lbm-°R = 0.001285 BTU/lbm-°R = 0.005382 kJ/kg-K
Heat Transfer Coefficient	1 W/m^2-K = 0.1761 BTU/hr-ft^2-°R 1 BTU/hr-ft^2-°R = 5.678 W/m^2-K
Thermal Conductivity	1 W/m-K = 0.5778 BTU/hr-ft°-R 1 BTU/hr-ft-°R = 1.731 W/m-K
Dynamic Viscosity	1 kg/m-sec = 0.672 lbm/ft-sec 1 lbm/ft-sec = 1.488 kg/m-sec
Shear Force	1 N/m^2 = 0.0209 lbf/ft^2 1 lbf/ft^2 = 47.84 N/m^2
Mass Moment of Inertia	1 kg-m^2 = 23.74 lbm-ft^2 1 lbm-ft^2 = 0.0421 kg-m^2
Angular Momentum	1 kg-m^2/sec = 23.74 lbm-ft^2/sec 1 lbm-ft^2/sec = 0.0421 kg-m^2/sec

References

[1] Abthoff, J., H. Schuster, H. Langer, and G. Loose, "The Regenerable Trap Oxidizer—An Emission Control Technique for Diesel Engines," SAE paper 850015, 1985.

[2] Alkidas, A. C. and J. P. Myers, "Transient Heat–Flux Measurements in the Combustion Chamber of a Spark Ignition Engine," *Journal of Heat Transfer*, ASME Trans., vol. 104, pp. 62–67, 1982.

[3] Allen, D. G., B. R. Dudley, J. Middletown, and D. A. Panka, "Prediction of Piston Ring–Cylinder Bore Oil Film Thickness in Two Particular Engines and Correlation with Experimental Evidences," *Piston Ring Scuffing*, p. 107, London: Mechanical Engineering Pub. Ltd., 1976.

[4] Amann, C. A., "Control of the Homogeneous-Charge Passenger Car Engine—Defining the Problem," SAE paper 801440, 1980.

[5] Amann, C. A., "Power to Burn," *Mechanical Engineering*, ASME, vol. 112, no. 4, pp. 46–54, 1990.

[6] Amsden, A. A., T. D. Butler, P. J. O'Rourke, and J. D. Ramshaw, "KIVA—A Comprehensive Model for 2-D and 3-D Engine Simulations," SAE paper 850554, 1985.

[7] Amsden, A. A., J. D. Ramshaw, P. J. O'Rourke, and J. K. Dukowicz, "KIVA—A Computer Program for Two- and Three-Dimensional Fluid Flows with Chemical Reactions and Fuel Sprays," report LA-10245-MS, Los Alamos National Laboratory, 1985.

[8] "A Stirling Briefing," NASA, Cleveland: Lewis Research Center, March 1987.

[9] "A Survey of Variable Valve Actuation," Automotive Engineering, vol. 98, no. 1, pp. 29–33, 1990, SAE International.

[10] Austen, A. E. W. and W. T. Lyn, "Relation Between Fuel Injection and Heat Release in a Direct-Injection Engine and the Nature of the Combustion Processes," *Proc. Institute of Mechanical Engineers*, pp. 47–62, 1960.

[11] *Automotive Engineering*, a monthly publication by SAE International.

[12] Birch, S., "NGVs in the U.K.," *Automotive Engineering*, vol. 102, no. 3, pp. 26–27, 1994, SAE International.

[13] Birch, S., "Thermo Accumulator," *Automotive Engineering*, vol. 100, no. 2, pp. 85–86, 1992, SAE International.

[14] Birch, S., "Two-Stroke Power," *Automotive Engineering*, vol. 100, no. 8, pp. 45–47, 1992, SAE International.

[15] Birch, S., J. Yamaguchi, A. Demmler, and K. Jost, "Honda's Oval-Piston Mega-Bike," *Automotive Engineering*, vol. 100, no. 6, pp. 46–47, 1992, SAE International.

[16] Borgnake, C., V. S. Arpaci, and R. J. Tabaczynski, "A Model for the Instantaneous Heat Transfer and Turbulence in a Spark Ignition Engine," SAE paper 800287, 1980.

[17] Bracco, F. V., "Modeling of Engine Sprays," SAE paper 850394, 1985.

[18] Brandstatter, W., R. J. R. Johns, and G. Wigley, "The Effect of Inlet Port Geometry on In-Cylinder Flow Structure," SAE paper 850499, 1985.

[19] Brooks, D., "Development of Reference Fuel Scales for Knock Rating," *SAE Journal*, vol. 54, no. 8, August 1946.

[20] Brown, W. L., "The Caterpillar imep Meter and Engine Friction," SAE paper 730150, 1973.

[21] Butler, T. D., L. D. Cloutman, J. K. Dukowicz, and J. D. Ramshaw, "Multidimensional Numerical Simulation of Reactive Flow in Internal Combustion Engines," *Proc. Energy Combustion Science*, vol. 7, pp. 293–315, 1981.

[22] Cameron, K., "NR750," *Cycle World*, pp. 30–35, Jan. 1992.

[23] Catania, A. E., C. Dongiovanni, and A. Mittica, "Further Investigation into the Statistical Properties of Reciprocating Engine Turbulence," *JSME International Journal*, vol. 35, pp. 255–265, 1992.

[24] Catania, A. E. and A. Mittica, "Autocorrelation and Autospectra Estimation of Reciprocating Engine Turbulence," *Trans. ASME, J. Eng. Gas Turbines Power*, vol. 112, 1990.

[25] Catania, A. E. and A. Mittica, "Extraction Techniques and Analysis of Turbulence Quantities from In-Cylinder Velocity Data," *Trans. ASME, J. Eng. Gas Turbines Power*, vol. 111, 1989.

[26] Catania, A. E. and A. Mittica, "Induction System Effects on Small-Scale Turbulence in a High-Speed Diesel Engine," *Trans. ASME, J. Eng. Gas Turbines Power*, vol. 109, 1987.

[27] Caton, J., "A Brief Review of Coal-Fueled Engines," *Internal Combustion Engine Division Newsletter*, ASME, Summer 1995.

[28] Chapman, M., J. M. Novak, and R. A. Stein, "Numerical Modeling of Inlet and Exhaust Flows in Multi-Cylinder Internal Combustion Engines," *Flows in Internal Combustion Engines*, ASME, 1982.

[29] Cummins, C. L. Jr, *Internal Fire*, SAE International Inc., 1989.

[30] Demmler, A., "Smog-Treating Catalyst," *Automotive Engineering*, vol. 103, no. 8, p. 32, 1995, SAE International.

[31] *Diesel and Gas Turbine Worldwide*, a monthly publication by Diesel and Gas Turbine Publications.

[32] Dinsdale, S., A. Roughton, and N. Collings, "Length Scale and Turbulence Intensity Measurements in a Motored Internal Combustion Engine," SAE paper 880380, 1988.

[33] Douard, A. and P. Eyzat, "DIGITAP—An On-Line Acquisition and Processing System for Instantaneous Engine Data—Applications," SAE paper 770218, 1977.

[34] Duck, G. E., H. Beyer, and A. Mierbach, *Piston Ring Manual*, GOETZE-AG, Germany, 1977.

[35] "Eddy Current Dynamometer Series W," paper L3220/3e, Schenck Company, 1995.

[36] "Electronic Valve Timing," *Automotive Engineering*, vol. 99, no. 4, pp. 19–24, 1991, SAE International.

[37] "Engine Mounts and NVH," *Automotive Engineering*, vol. 102, no. 7, pp. 19–23, 1994, SAE International.

[38] "ER Fluid Engine Mounts," *Automotive Engineering*, vol. 101, no. 2, pp. 52–55, 1993, SAE International.

[39] "Evolution of the Automobile Engine Development," The Civic Report, Honda Motor Company, Inc., 1978.

[40] Ferguson, C. R., *Internal Combustion Engines*. New York: Wiley, 1986.

[41] Fiedler, R. A., "General Motors Internal Combustion Engine Simulation Program," Geode, vol. 67, pp. 7–8, 1991, University of Wisconsin—Platteville.

[42] Gatowski, J. A., E. N. Balles, K. M. Chun, F. E. Nelson, J. A. Ehchian, and J. B. Heywood, "Heat Release Analysis of Engine Pressure Data," SAE paper 841359, *SAE Trans.*, vol. 93, 1984.

[43] Gatowski, J. A., J. B. Heywood, and C. Deleplace, "Flame Photographs in a Spark-Ignition Engine," *Combustion and Flame*, vol. 56, pp. 71–81, 1984.

[44] "Generator Gas," SERI, U. S. Department of Energy, EG-77-C-01-4042, 1979.

[45] Givens, L., "A Technical History of the Automobile," *Automotive Engineering*, vol. 98, nos. 6–8, SAE International Inc.

[46] "Global Warming, Fuels, and Passenger Cars," *Automotive Engineering*, vol. 99, no. 2, pp. 15–18, 1991, SAE International.

[47] Glover, A. R., G. E. Hundleby, and O. Hadded, "An Investigation into Turbulence in Engines Using Scanning LDA," SAE paper 880378, 1988.

[48] Goodsell, D. L., *Dictionary of Automotive Engineering*. SAE International Inc., 1995, 2nd ed.

[49] Gordon, S. and B. J. McBride, "Computer Program for the Calculation of Complex Chemical Equilibrium Composition, Rocket Performance, Incident and Reflected Shocks, and Chapman–Jouquet Detonations," NASA publication SP-273, 1971.

[50] Gorr, E. and H. S. Hilbert, "The Future of Two-Stroke Engines in Street Bikes," *Motorcyclist*, pp. 32–34, Nov. 1992.

[51] Gosman, A. D., "Computer Modeling of Flow and Heat Transfer in Engines, Progress and Prospects," COMODIA '85, Tokyo, Japan, 1985, pp. 15–26.

[52] Gosman, A. D., "Multidimensional Modeling of Cold Flows and Turbulence in Reciprocating Engines," SAE paper 850344, 1985.

[53] Gosman, A. D. and R. J. R. Johns, "Computer Analysis of Fuel–Air Mixing in Direct-Injection Engines," SAE paper 800091, 1980.

[54] Gosman, A. D., Y. Y. Tsui, and A. P. Watkins, "Calculation of Three Dimensional Air Motion in Model Engines," SAE paper 840229, 1984.

[55] Gosman, A. D., Y. Y. Tsui, and A. P. Watkins, "Calculation of Unsteady Three-Dimensional Flow in a Model Motored Reciprocating Engine and Comparison with Experiment," Fifth International Turbulent Shear Flow Meeting, Cornell Univ., 1985.

[56] Gruse, W. A., *Motor Oils: Performance and Evaluation*. New York: Van Nostrand Reinhold, 1967.

[57] "Heated Catalytic Converter," *Automotive Engineering*, vol. 102, no. 9, 1994, SAE International.

[58] Heywood, J. B., *Internal Combustion Engine Fundamentals*. New York: McGraw-Hill, 1988.

[59] Hinze, J. O., *Turbulence*. New York: McGraw-Hill, 1975.

[60] Hires, S. D., A. Ekchian, J. B. Heywood, R. J. Tabaczynski, and J. C. Wall, "Performance and NOx Emissions Modeling of a Jet Ignition Pre-Chamber Stratified Charge Engine," SAE paper 760161, 1976.

[61] Hires, S. D., R. J. Tabaczynski, and J. M. Novak, "The Prediction of Ignition Delay and Combustion Intervals for a Homogeneous Charge, Spark Ignition Engine," SAE paper 780232, *SAE Trans.*, vol. 87, 1978.

[62] Hoffman, H., "Development Work on the Mercedes-Benz Commercial Diesel Engine, Model Series 400," SAE paper 710558, 1971.

[63] Holman, J. P., *Heat Transfer*. New York: McGraw-Hill, 2002.

[64] "Hydrogen as an Alternative Automotive Fuel," *Automotive Engineering*, vol. 102, no. 10, pp. 25–30, 1994, SAE International.

[65] Ikegami, M., M. Shioji, and K. Nishimoto, "Turbulence Intensity and Spatial Integral Scale During Compression and Expansion Strokes in a Four-Cycle Reciprocating Engine," SAE paper 870372, 1987.

[66] Isshiki, Y., Y. Shimamoto, and T. Wakisaka, "Numerical Prediction of Effect of Intake Port Configurations on the Induction Swirl Intensity by Three-Dimensional Gas Flow Analysis," COMODIA 85, Tokyo, Japan 1985.

[67] JANAF Thermochemical Tables, 2nd ed., NSRDS-NBS37, U. S. National Bureau of Standards, 1971.

[68] "Japanese 'Miller-Cycle' Engine Development Accelerates," *Automotive Engineering*, vol. 101, no. 7, 1993, SAE International.

[69] Jones, J. B., and R. E. Dugan, *Engineering Thermodynamics*. Upper Saddle River, NJ: Prentice Hall, 1996.

[70] Jost, K., "Future Saab Engine Technology," *Automotive Engineering*, vol. 103, no. 12, 1995, SAE International.

[71] Jost, K., "NGV User's Guide," Parker Hannifin Corporation, 1994.

[72] Kajiyama, K., K. Nishida, A. Murakami, M. Arai, and H. Hiroyasu, "An Analysis of Swirling Flow in Cylinder for Predicting D. I. Diesel Engine Performance," SAE paper 840518, 1984.

[73] Keenan, J. H., J. Chao, and J. Kaye, *Gas Tables—International Version*, 2nd ed., Malabar, FL: Krieger, 1992.

[74] Kramer, A. S., "The Electric Motor that Killed the Electric Car," *Old Cars Weekly News and Marketplace*, Oct. 1994.

[75] Krieger, R. B. and G. L. Borman, "The Computation of Apparent Heat Release for Internal Combustion Engines," ASME paper 66-WA/DGP-4, 1966.

[76] Kummer, J. T., "Catalysts for Automobile Emission Control," *Prog. Energy Combustion Science*, vol. 6, pp. 177–199, 1981.

[77] Kuroda, H., Y. Nakajima, K. Sugihara, Y. Takagi, and S. Muranaka, "The Fast Burn with Heavy EGR, New Approach for Low NOx and Improved Fuel Economy," SAE paper 780006, 1978.

[78] Langworth, R. M., *The Complete Book of the Corvette*. Beckman House, 1987.

[79] "Latent Heat Storage," *Automotive Engineering*, vol. 100, no. 2, pp. 58–61, 1992, SAE International.

[80] Leary, W. A. and J. U. Jovellanos, "A Study of Piston and Piston–Ring Friction," NACA ARR-4J06, 1944.

[81] Liljedahl, J. B., W. M. Carleton, P. K. Turnquist, and D. W. Smith, *Tractors and Their Power Units*. New York: Wiley, 1979.

[82] Malchow, G. L., S. C. Sorenson, and R. O. Buckius, "Heat Transfer in the Straight Section of an Exhaust Port of a Spark Ignition Engine," SAE paper 790309, 1979.

[83] Maly, R., and M. Vogel, "Initiation and Propagation of Flame Fronts in Lean CH_4—Air Mixtures by the Three Modes of the Ignition Spark," in *Proc. Seventeenth International Symposium on Combustion*, The Combustion Institute, 1976, pp. 821–831.

[84] Matsui, K., T. Tanaka, and S. Ohigashi, "Measurement of Local Mixture Strength of Spark Gap of S. I. Engines," SAE paper 790483, *SAE Trans.*, vol. 88, 1979.

[85] Mattavi, J. N. and C. A. Amann, *Combustion Modeling in Reciprocating Engines*, Plenum Press, 1980, pp. 41–68.

[86] "Mazda Hydrogen-Fueled Rotary Development," *Automotive Engineering*, vol. 101, no. 6, pp. 61–65, 1993, SAE International.

[87] Meintjes, K., "A User's Guide for the General Motors Engine Simulation Program," GMR-5758, General Motors Research Laboratories, Warren, MI, 1987.

[88] "Methanol/Gasoline Blends and Emissions," *Automotive Engineering*, vol. 100, no. 5, pp. 17–19, 1992, SAE International.

[89] "Mitsubishi Variable Displacement and Valve Timing/Lift," *Automotive Engineering*, vol. 101, no. 1, pp. 99–100, 1993, SAE International.

[90] Moran, M. J., and H. N. Shapiro, *Fundamentals of Engineering Thermodynamics*. New York: Wiley, 2000.

[91] Morel, T. and N. N. Mansour, "Modeling of Turbulence in Internal Combustion Engines," SAE paper 820040, 1982.

[92] Newhall, H. K. and S. M. Shahed, "Kinetics of Nitric Oxide Formation in High-Pressure Flames," in *Proc. Thirteenth International Symposium on Combustion*, The Combustion Institute, 1971, pp. 381–390.

[93] Obert, E. F., *Internal Combustion Engines and Air Pollution*. New York: Harper and Row, 1973.

[94] O'Connor, L., "Clearing the Air with Natural Gas Engines," *Mechanical Engineering*, vol. 115, no. 10, pp. 53–56, 1993, ASME.

[95] O'Donnell, J., "Gasoline Allies," *Autoweek*, pp. 16–18, Feb. 1994.

[96] Olikara, C. and G. L. Borman, "A Computer Program for Calculating Properties of Equilibrium Combustion Products with Some Applications to I. C. Engines," SAE paper 750468, 1975.

[97] Oppel, F., *Motoring in America*. Castle Books, 1989.

[98] Pulkrabek, W. W. and R. A. Shaver, "Catalytic Converter Preheating by Using a Chemical Reaction," SAE paper 931086, 1993.

[99] Quader, A. A., "Why Intake Charge Dilution Decreases Nitric Oxide Emission From Spark Ignition Engines," SAE paper 710009, *SAE Trans.*, vol. 80, 1971.

[100] Ramos, J. I., *Internal Combustion Engine Modeling*. Hemisphere, 1989.

[101] Reed, D., "Compressed-Natural-Gas Vehicles," *Automotive Engineering*, vol. 103, no. 2, p. 269, 1995, SAE International.

[102] Rinschler, G. L. and T. Asmus, "Powerplant Perspectives," *Automotive Engineering*, vol. 103, nos. 4–6, 1995, SAE International.

[103] Rogowski, S. M., *Elements of Internal-Combustion Engines*. New York: McGraw-Hill, 1953.

[104] "*Rotary Engine Design: Analysis and Development*," SP-768, SAE International, 1989.

[105] Ruddy, B., "Calculated Inter-Ring Gas Pressures and Their Effect Upon Ring Pack Lubrication," *DAROS Information*, vol. 6, pp. 2–6, Sweden, 1979.

[106] Ryder, E. A., "Recent Developments in the R-4360 Engine," *SAE Quart. Trans.*, vol. 4, p. 559, 1950.

[107] *SAE Fuels and Lubricants Standards Manual*, SAE HS-23, 1993.

[108] Sakai, Y., H. Miyazaki, and K. Mukai, "The Effect of Combustion Chamber Shape on Nitrogen Oxides," SAE paper 730154, 1973.

[109] Schapertons, H. and F. Thiele, "Three-Dimensional Computations for Flowfields in DI Piston Bowls," SAE paper 860463, 1986.

[110] "Sensors and the Intelligent Engine," *Automotive Engineering*, vol. 99, no. 4, pp. 33–36, 1991, SAE International.

[111] Shampine, L. F. and Gordon, M. K., *Computer Solution of Ordinary Differential Equations*. Freeman, 1975.

[112] Shapiro, A. H., *The Dynamics and Thermodynamics of Compressible Fluid Flow*. New York: Ronald Press, 1953.

[113] Shigley, J. E. and L. D. Mitchell, *Mechanical Engineering Design*. New York: McGraw-Hill, 1983.

[114] Smith, J. R., R. M. Green, C. K. Westbrook, and W. J. Pitz, "An Experimental and Modeling Study of Engine Knock," Twentieth Symposium on Combustion, The Combustion Institute, Pittsburgh, PA, 1984.

[115] "Southern California Alternative-Fuel Projects," *Automotive Engineering*, vol. 103, no. 3, pp. 63–66, 1995, SAE International.

[116] Stone, R., *Introduction to Internal Combustion Engines*. SAE International Inc., 1992.

[117] Svehla, R. A. and B. J. McBride, "Fortran IV Computer Program for Calculation of Thermodynamic and Transport Properties of Complex Chemical Systems," NASA technical note, TND-7056, 1973.

[118] Tabaczynski, R. J., "Turbulence and Turbulent Combustion in Spark Ignition Engines," *Proc. Energy Combustion Science*, vol. 2, pp. 143–165, 1976.

[119] Tabaczynski, R. J., F. H. Trinker, and B. A. S. Shannon, "Further Refinement and Validation of a Turbulent Flame Propagation Model for Spark Ignition Engines," *Combustion and Flame*, vol. 39, pp. 111–122, 1980.

[120] Taylor, C. F., *The Internal Combustion Engine in Theory and Practice*. Cambridge, MA: M.I.T. Press, 1977.

[121] "The Changing Nature of Gasoline," *Automotive Engineering*, vol. 102, no. 1, pp. 99–102, 1994, SAE International.

[122] "The Road to Clean Air: Powered with Alternate Fuels," pamphlet by Wisconsin Alternate Fuels Task Force, 1994.

[123] Thomas, F. J., J. S. Ahluwalia, E. Shamah, and G. W. Van der Horst, "Medium-Speed Diesel Engines Part I: Design Trends and the Use of Residual/Blended Fuels," ASME paper 84-DGP-15, 1984.

[124] Thompson, K. D., "Applying Parametrics and AutoCad in Engine Design," *Cadence*, pp. 63–68, March 1988.

[125] Ting, L. L., and J. E. Mayer Jr., "Piston Ring Lubrication and Cylinder Bore Wear Analyses, Part II—Theory Verification," *Trans. ASME, J. Lub. Tech.*, pp. 258–266, 1974.

[126] Tizard and Pye, *Philosophical Magazine*, July 1922.

[127] Uzkan, T., *Flows in Internal Combustion Engines—II*, FED-vol. 20, pp. 39–46, ASME, New York, 1984.

[128] Uzkan, T., C. Borgnakke, and T. Morel, "Characterization of Flow Produced by a High-Swirl Inlet Port," SAE paper 830266, 1983.

[129] Uzkan, T., W. G. Tiederman, and J. M. Novak, *International Symposium on Flows in Internal Combustion Engines—III*, FED-vol. 28, pp. 125–134, ASME, New York, 1985.

[130] Valenti, M., "Alternate Fuels: Paving the Way to Energy Independence," *Mechanical Engineering*, vol. 113, no. 12, pp. 42–46, 1991, ASME.

[131] Valenti, M., "Insulating Catalytic Converters," *Mechanical Engineering*, vol. 117, no. 5, pp. 14–16, 1995, ASME.

[132] Valenti, M., "Pollution-Reducing Cars," *Mechanical Engineering*, vol. 117, no. 7, p. 12, 1995, ASME.

[133] "Variable Valve Actuation," *Automotive Engineering*, vol. 99, no. 10, pp. 12–16, 1991, SAE International.

[134] Wakisaka, T., Y. Shimamoto, and Y. Isshiki, "Three Dimensional Numerical Analysis of In-Cylinder Flows in Reciprocating Engines," SAE paper 860464. 1986.

[135] Walker, J. W., "The GM 1.8 Liter Gasoline Engine Designed by Chevrolet," SAE paper 820111, 1982.

[136] Wallace, T. F., "Buick's Turbocharged V-6 Powertrain for 1978," SAE paper 780413, 1978.

[137] Watkins, A. P., A. D. Gosman, and B. S. Tabrizi, "Calculation of Three Dimensional Spray Motion in Engines," SAE paper 860468, 1986.

[138] Wentworth, J. T., "Effects of Top Compression Ring Profile on Oil Consumption and Blowby with the Sealed Ring-Orifice Design," SAE paper 820089, 1982.

[139] Wise, D. B., *The Illustrated Encyclopedia of the World's Automobiles*. New York: A and W Publishers, 1979.

[140] Witze, P. O., "Measurements of the Spatial Distribution and Engine Speed Dependence of Turbulent Air Motion in an I. C. Engine," SAE paper 770220, 1977.

[141] Woehrle, W. J., "A History of the Passenger Car Tire: Part I," *Automotive Engineering*, vol. 103, no. 9, pp. 71–75, 1995.

[142] Yamada, T., T. Inoue, A. Yoshimatsu, T. Hiramatsu, and M. Konishi, "In-Cylinder Gas Motion of Multivalve Engine—Three Dimensional Numerical Simulation," SAE paper 860465, 1986.

[143] Yamaguchi, J., "Honda's Oval-Piston Mega-Bike," *Automotive Engineering*, vol. 100, no. 6, pp. 46–47, 1992.

[144] Arai, M., M. Tabata, and H. Hiroyasu, "Disintegrating Process and Spray Characteristics of Fuel Jet Injected by a Diesel Nozzle," SAE paper 840275, 1984.

[145] Ashley, A., "Fuel Cells Start to Look Real," *Automotive Engineering International*, vol. 109, no. 3, pp. 64–80, 2001, SAE International.

[146] "A Survey of Variable Valve Actuation," *Automotive Engineering International*, vol. 98, no. 1, pp. 29–33, 1990, SAE International.

[147] Auburn–Cord–Duesenberg Museum, Auburn, IN, private correspondence in 2002.

[148] Birch, S., "Combustion and Expansion at Saab," *Automotive Engineering International*, vol. 109, no. 3, pp. 46–48, 2001, SAE International.

[149] Birch, S., "New Saab and Citroen Technology at Geneva," *Automotive Engineering International*, vol. 108, no. 5, pp. 96–101, 2000, SAE International.

[150] Birch, S., "Two-Seaters from MG and Fiat," *Automotive Engineering International*, vol. 103, no. 6, pp. 51–53, 1995, SAE International.

[151] Birch, S., "Variations on a Theme by Saab," *Automotive Engineering International*, vol. 109, no. 4, pp. 54–57, 2001, SAE International.

[152] Blizzard, D. T., and W. N. Shade, "Crankcase Explosion Safety Revisited—Good Maintenance Practices Necessary for Prevention," *Diesel & Gas Turbine Worldwide*, pp. 10–14, March 1996, Diesel and Gas Turbine Publications.

[153] "BMW Returns to Formula One with V10," *Automotive Engineering International*, vol. 108, no. 3, pp. 42, 2000, SAE International,

[154] "BMW's Hydrogen Message," *Automotive Engineering International*, vol. 110, no. 5, 2002, SAE International.

[155] Bower, G. R., and D. E. Foster, "A Comparison of the Bosch and Zuech Rate of Injection Meters," SAE paper 910724, 1991, SAE International.

[156] Broge, J. L., "Revving up for Diesel," *Automotive Engineering International*, vol. 110, no. 2, pp. 40–49, 2002, SAE International.

[157] Broge, J. L., "Smog-Eating Technology," *Automotive Engineering International*, vol. 108, no. 2, pp. 141–142, 2000, SAE International.

[158] Brown, W. L., "The Caterpillar imep Meter and Engine Friction," SAE paper 730150, 1973.

[159] Buchholz, K., "Chevy Revs for 2002 IRL Season," *Automotive Engineering International*, vol. 110, no. 3, pp. 33–34, 2002, SAE International.

[160] "Butane/Propane Mixtures as Fleet Fuel," *Automotive Engineering International*, vol. 107, no. 12, pp. 41–44, 1999, SAE International.

[161] Carney, D., "Denso Heats Up Diesels," *Automotive Engineering International*, vol. 109, no. 7, pp. 47–48, 2001, SAE International.

[162] Carney, D., "Developments in Fuel Cells," *Automotive Engineering International*, vol. 110, no. 3, pp. 47–52, 2002, SAE International.

[163] Catania, A. E., C. Dongiovanni, and A. Mittica, "Further Investigation into the Statistical Properties of Reciprocating Engine Turbulence," *JSME International Journal*, series II, vol. 35, no. 2, pp. 255–265, 1992, JSME.

[164] Chellini, R., "Improved Design Explosion Relief Valves," *Diesel & Gas Turbine Worldwide*, p.46, March 2002, Diesel and Gas Turbine Publications.

[165] Clymer, F., and L. R. Henry, *Ford Model A Album*, Polyprints, 1960.

[166] Cummins Diesel, private communication, 2002.

[167] Czadzeck, G. H., "Ford's 1980 Central Fuel Injection System," SAE paper 790742, 1979.

[168] Date, T., S. Yagi, A. Ishizuya, and I. Fuji, "Research and Development of the Honda CVCC Engine," SAE paper 740605, 1974.

[169] DeAngelis, G., E. P. Francis, and L. R. Henry, *The Ford Model A*, Motor Cities Publishing Company, 1983.

[170] Dent, J. C., and P. S. Mehta, "Phenomenological Combustion Model for a Quiescent Chamber Diesel Engine," SAE paper 811235, 1981.

[171] "Emission Standards," DieselNet Website, 2002.

[172] "Emissions with Butane/Propane Blends," *Automotive Engineering International*, vol. 104, no. 11, pp. 49–54, 1996, SAE International.

[173] "Engine Tech 2001," *Diesel & Gas Turbine Worldwide*, pp. 45–48, Dec 2001, Diesel and Gas Turbine Publications.

[174] "EPA Approves Clean Diesel Measure," *Automotive Engineering International*, vol. 109, no. 2, p. 240, 2001, SAE International.

[175] "Federal and California Exhaust and Evaporative Emission Standards for Light-Duty Vehicles and Light-Duty Trucks," EPA Website, 2002.

[176] "Ford Invention Will Promote Natural Gas Vehicles," *Automotive Engineering International*, vol. 106, no. 6, p. 120, 1998, SAE International.

[177] Gish, R. E., J. D. McCullough, J. B. Retzloff, and H. T. Mueller, "Determination of True Engine Friction," SAE Trans., vol. 66, pp. 649–661, 1958, SAE International.

[178] Glockler, O., H. Knapp, and H. Manger, "Present Status and Future Development of Gasoline Fuel Injection Systems for Passenger Cars," SAE paper 800467, 1980.

[179] Greiner, M., P. Romann, and U. Steinbrenner, "BOSCH Fuel Injectors—New Developments," SAE paper 870124, 1987.

[180] Hamburg, D. R., and J. E. Hyland, "A Vaporized Gasoline Metering System for Internal Combustion Engines," SAE paper 760288, 1976.

[181] Hames, R. J., R. D. Straub, and R. W. Amann, "DDEC Detroit Diesel Electronic Control," SAE paper 850542, 1985.

[182] Hara, S., A. Hidaka, N. Tomisawa, M. Nakamura, and T. Todo, "Application of a Variable Valve Event and Timing System to Automotive Engines," SAE paper 2000-10-1224, 2000.

[183] Hardenberg, H. O., and F. W. Hase, "An Empirical Formula for Computing the Pressure Rise Delay of a Fuel from its Cetane Number and from the Relevant Parameters of Direct-Injection Diesel Engines," SAE paper 790493, 1979.

[184] "Harm Free Use of Diesel Additives," *Automotive Engineering International*, vol. 107, no. 7, pp. 84–88, 1999, SAE International.

[185] "Holley's Nitrous Oxide Systems (NOS) Brand," Holley Website, 2002.

[186] "Hydrogen as an Alternative Automobile Fuel," *Automotive Engineering International*, vol. 102, no. 10, 1994, SAE International.

[187] Jost, K., K. Buchholz, and R. Gehm, "Advances in Fuel-Cell Development," *Automotive Engineering International*, vol. 110, no. 6, pp. 67–71, 2002, SAE International.

[188] Jost, K., "Fuel Cell Autonomy," *Automotive Engineering International*, vol. 110, no. 2, pp. 35–37, 2002, SAE International.

[189] Jost, K., "Fuel-Stratified Injection from VW," *Automotive Engineering International*, vol. 109, no. 1, pp.63–65, 2001, SAE International.

[190] Jost, K., "Mercedes-Benz Launches Cylinder Cutout," *Automotive Engineering International*, vol. 107, no. 1, pp. 38–39, 1999, SAE International.

[191] Jost, K., "New Diesel V8 for S-Class," *Automotive Engineering International*, vol. 109, no. 1, pp. 78–80, 2001, SAE International.

[192] Kates, E. J., *Diesel and High Compression Gas Engines*, American Technical Society, 1954.

[193] Khan, I. M., G. Greeves, and C. H. T. Wang, "Factors Affecting Smoke and Gaseous Emissions from Direct Injection Engines and a Method of Calculation," SAE paper 730169, 1973.

[194] Li, C. H., "Piston Thermal Deformation and Friction Considerations," SAE paper 820086, 1982.

[195] Liou, T. M., M. Hall, D. A. Santavicca, and F. N. Bracco, "Laser Doppler Velocimetry Measurements in Valved and Ported Engines," SAE paper 840375, 1984.

[196] "Lotus Active-Valve Control," *Automotive Engineering International*, vol. 101, no. 1, pp. 97–98, 1993, SAE International.

[197] Lumley, John L., *Engines, An Introduction*, Cambridge University Press, 1999.

[198] Lustgarten, G. A., and S. Winterthur, "The Latest Sulzer Marine Diesel Engine Technology," SAE Technical Paper 851219, 1985.

[199] Malchow, G. L., S. C. Sorenson, and R. O. Buckius, "Heat Transfer in the Straight Section of an Exhaust Port of a Spark Ignition Engine," SAE paper 790309, 1979.

[200] "MAN B & W Announces IS Diesels," *Diesel & Gas Turbine Worldwide*, p.54, April 1999, Diesel and Gas Turbine Publications.

[201] Matekunas, F. A., "Modes and Measures of Cyclic Combustion Variability," SAE paper 830337, 1983.

[202] Matsui, Y., T. Kamimoto, and S. Matsuoka, "Formation and Oxidation Processes of Soot Particles in a D.I. Diesel Engine—An Experimental Study Via the Two-Color Method," SAE paper 820464, 1982.

[203] Matsumura, R., K. Higashiyama, and K. Kojima, "The Turbocharged 2.8L Engine for the Datsun 280ZX," SAE paper 820442, 1982.

[204] Matsuo, I., S. Nakazawa, H. Maeda, and E. Inada, "Development of a High-Performance Hybrid Propulsion System Incorporating a CVT," SAE paper 2000-01-0992, 2000.

[205] "Mayflower's Variable Engine Technology," *Automotive Engineering International*, vol. 110, no. 1, pp. 43–44, 2002, SAE International.

[206] "Mitsubishi Variable Displacement and Valve Timing/Lift," *Automotive Engineering International*, vol. 101, no. 1, pp. 99–100, 1993, SAE International.

[207] Moklegaard, L., A. G. Stefanopolulou, and J. Schmidt, "Transition from Combustion to Variable Compression Braking," SAE paper 011228, 2001, SAE International.

[208] "More than One Way to Stop a Truck," *Mechanical Engineering*, vol. 121, no. 11, pp. 34–36, 1999, ASME International.

[209] Mullins, P., "Future Emissions Control at Wartsila," *Diesel & Gas Turbine Worldwide*, pp. 32–35, May 2001, Diesel and Gas Turbine Publications.

[210] Mullins, P., "Lubricating Landfill Gas Engines," *Diesel & Gas Turbine Worldwide*, pp. 40–42, March 2000, Diesel and Gas Turbine Publications.

[211] Mullins, P., "Ro-Ro Vessels Will have Water Injection," *Diesel & Gas Turbine Worldwide*, pp. 28–32, Sept 1999, Diesel and Gas Turbine Publications.

[212] Mullins, P., "Shipbuilding in Asia," *Diesel & Gas Turbine Worldwide*, pp. 22–26, March 2001, Diesel and Gas Turbine Publications.

[213] "Natural Gas Fueling of Diesel Engines," *Automotive Engineering International*, vol. 104, no. 11, pp. 87–90, 1996, SAE International.

[214] "Next-Generation Power Sources," *Automotive Engineering International*, vol. 107, no. 9, p. 57, 1999, SAE International.

[215] "NOx Reduction at a Vermont Ski Resort," *Diesel & Gas Turbine Worldwide*, pp.14–16, July 2002, Diesel and Gas Turbine Publications.

[216] Pierik, R. J., and J. F. Burkhard, "Design and Development of a Mechanical Variable Valve Actuation System," SAE paper 2000-01-1221, 2000.

[217] Pischinger, M., W. Salber, F. van der Staay, H. Baumgarten, and H. Kemper, "Benefits of the Electromechanical Valve Train in Vehicle Operation," SAE paper 2000-01-1223, 2000.

[218] Ponticel, P., "High Time for Hybrids," *Automotive Engineering International*, vol. 110, no. 2, pp. 77–80, 2002, SAE International.

[219] "Saab Variable," www.saabnet.com, 2000.

[220] Seinosuke, H., A. Hidaka, N. Tomisawa, M. Nakamura, T. Todo, S. Takemura, and T. Nohara, "Application of a Variable Valve Event and Timing System to Automotive Engines," SAE paper 011224, 2001, SAE International.

[221] Sharke, P., "Power of 42," *Mechanical Engineering*, vol. 124, no. 4, pp. 40–42, 2002, ASME International.

[222] Shimizu, R., T. Tadokoro, T. Nakanishi, and J. Funamoto, "Mazda 4-Rotor Rotary Engine for the LeMans 24-Hour Endurance Race," SAE paper 920309, 1992.

[223] Silverstein, C. C., "Staying Cool," *Mechanical Engineering*, vol. 121, no. 7, pp. 64–65, 1999, ASME International.

[224] Smith, J., "Nitrous vs. Blowers," *Hot Rod*, pp. 60–62, Nov 1996.

[225] Stone, Richard, *Introduction to Internal Combustion Engines*, 2nd ed., 1992, SAE International.

[226] "The Stretch for Better Passenger Car Fuel Economy: A Critical Look, Part 1," *Automotive Engineering International*, vol. 106, no. 2, pp. 305–313, 1998, SAE International.

[227] Suzuki, T., *The Romance of Engines*, 1997, SAE International.

[228] "Toyota Prius: Best Engineered Car of 2001," *Automotive Engineering International*, vol. 109, no. 3, pp. 27–28, 2001, SAE International.

[229] Tsuchiya, K, and S. Hirano, "Characteristics of 2-Stroke Motorcycle Exhaust HC Emission and Effects of Air–Fuel Ratio and Ignition Timing," SAE paper 750908, 1975.

[230] "Two-Stroke Engine Technology," *Automotive Engineering International*, vol. 99, no. 7, pp. 11–14, 1991, SAE International.

[231] Uchiyama, H., T. Chiku, and S. Sayo, "Emission Control of Two-Stroke Automobile Engine," SAE paper 770766, 1977.

[232] Urciuoli, V., B. R. Mason, and M. Marcacci, "Simulation Techniques Applied to the Development of a 125cc 4-Stroke Scooter Engine," SAE paper 951819, 1995.

[233] "Urea Selective Catalytic Reduction," *Automotive Engineering International*, vol. 108, no. 11, pp.125–128, 2000, SAE International.

[234] "Valeo and Ricardo Team for 42-V Diesel Engine," *Automotive Engineering International*, vol. 109, no. 11, pp. 56–59, 2001, SAE International.

[235] "Variable Valve Actuation," *Automotive Engineering International*, vol. 89, no. 10, pp. 12–16, 1991, SAE International.

[236] Wilson, K., "Reducing Emissions: The Effects on Shipowners," *Diesel & Gas Turbine Worldwide*, pp.28–30, Sept 1996, Diesel and Gas Turbine Publications.

[237] Wong, C. L., and D. E. Steere, "The Effects of Diesel Fuel Properties and Engine Operating Conditions on Ignition Delay," SAE paper 821231, 1982.

[238] Wong, V. W., M. Stewart, G. Lundholm, and A. Hoglund, "Increased Power Density via Variable Compression/Displacement and Turbocharging Using the Alvar-Cycle Engine," SAE technical paper 981027, 1998, SAE International.

[239] Yamaguchi, J., "Opa - Toyota's New-Age Vehicle," *Automotive Engineering International*, vol. 108, no. 9, pp. 23–34, 2000, SAE International.

[240] Yu, R. C., and S. M. Shahed, "Effects of Injection Timing and Exhaust Gas Recirculation on Emissions from a D.I. Diesel Engine," SAE paper 811234, 1981.

[241] Zhao, F. Q., M. C. Lai, and D. L. Harrington, "A Review of Mixture Preparation and Combustion Control Strategies for Spark-Ignition Direct-Injection Gasoline Engines," SAE paper 970627, 1997.

Answers to Selected Review Problems

CHAPTER 1

5. **(a)** 6062, **(b)** 11,462
6. **(a)** 7991, **(b)** 4.22, **(c)** 38.41
8. **(a)** 80, **(b)** 4.5, **(c)** 67.5

CHAPTER 2

1. **(a)** 4.36×10^8, **(b)** 1.74×10^9, **(c)** 2.18×10^8
2. **(a)** 4703, 4.703, **(b)** 561, **(c)** 420, **(d)** 69.2
6. **(a)** 1429, **(b)** 886, **(c)** 543, **(d)** 64.6, 86.6, **(e)** 247
7. **(a)** 0.178, **(b)** 0.0185, **(c)** 54.1, **(d)** 39.6, 53.1
10. **(a)** 0.0407, **(b)** 145.5, **(c)** 153.5, **(d)** 32
12. **(a)** 27.8, **(b)** 8.47, **(c)** 1.32
13. **(a)** 11.56, **(b)** 429.8, **(c)** 964
15. **(a)** 5.33, **(b)** 0.00084, **(c)** 5.63×10^{-5}, **(d)** 4.22×10^{-7}
19. **(a)** 95.0, **(b)** 0.337, **(c)** 5.35, **(d)** 21.6

CHAPTER 3

1. **(c)** 1689, **(d)** 2502, **(e)** 1362, **(f)** 54.5
2. **(a)** 78.6, **(b)** 313, **(c)** 1311, **(d)** 15.0, **(e)** 192.4, **(f)** 91.1, **(g)** 26.2
3. **(a)** 856, **(b)** 3.2, **(c)** 34
4. **(a)** 58, **(b)** 455
7. **(a)** 60.3, **(b)** 2777, **(c)** 53.3, **(d)** 2580
10. **(a)** 599, **(b)** 3.6, **(c)** 708, **(d)** 825
13. **(c)** 1317, **(d)** −301, **(e)** 7.0, **(f)** 56.8
15. **(a)** 74.5, **(b)** 0, **(c)** 400.0
18. **(a)** 1339, **(b)** 16.33, **(c)** 12.3, **(d)** 0.59

CHAPTER 4

1. **(a)** 12.325, **(b)** 1.20, **(c)** 43.8, **(d)** 42.9
5. **(a)** 18.16, **(b)** 20, **(c)** 100, 0
7. **(a)** 8.63, **(b)** 58, **(c)** 59, **(d)** 99
10. **(a)** 1, **(b)** 0.596, **(c)** −39.2
13. **(a)** 102.5, **(b)** 96, **(c)** 9.22
16. **(a)** 0.1222, **(b)** 0.058, **(c)** 0.471, **(d)** 0.058, **(e)** 0.058
18. **(a)** 11.76, **(b)** 93.0
22. **(a)** 34.6, **(b)** −4.9
25. **(a)** 16.08, **(b)** 0.525
27. **(a)** 3507, 3507, **(b)** 0.856, **(c)** no dew point

CHAPTER 5

1. **(a)** 778, 2342, **(b)** 8.28, **(c)** 704, 1596
2. **(a)** 485, **(b)** 8.88, **(c)** 17
3. **(a)** 40, **(b)** +5.0, **(c)** 60
7. **(a)** 0.00242, **(b)** 8.49, **(c)** 6.01
9. **(a)** 2.224, **(b)** 1.016
14. **(a)** 1.535×10^6, **(b)** 0.360
16. **(a)** 52, **(b)** 5.6
17. **(a)** 1.642, **(b)** 0.2324, **(c)** 0.0130
18. **(a)** 0.936, **(b)** 0.711, **(c)** 0.955, **(d)** 0.745

CHAPTER 6

2. **(a)** 281, **(b)** 291, **(c)** 4.85
3. **(a)** 5.36, **(b)** 9.15, **(c)** 732.8
5. **(a)** 4.74, **(b)** 0.1075, **(c)** 1.28
6. **(a)** 4.04, **(b)** 18.3
8. **(a)** 0.0024, **(b)** 0.012, **(c)** 5

CHAPTER 7

1. **(a)** 0.0036, **(b)** 12.4° aTDC
2. **(a)** 24.23, **(b)** 26.6° bTDC, **(c)** 15.8° bTDC
7. **(a)** 17.2, **(b)** 0.849
8. **(a)** 70.15, **(b)** 1203, **(c)** 0.133, **(d)** 189

CHAPTER 8

2. **(a)** 1685, **(b)** 3.5, **(c)** 180, 8.8
3. **(a)** 565, **(b)** 774
6. **(a)** 0.0021, **(b)** 62.0, **(c)** 703

8. **(a)** 566, **(b)** 4.4, **(c)** 133, **(d)** 0.82
10. **(a)** 1003, **(b)** 0.000242, **(c)** 39.7
12. **(a)** 0.0361, **(b)** 0.0386, **(c)** 53, **(d)** 467

CHAPTER 9

4. **(a)** 7.99, **(b)** 57.7
5. **(a)** 1.72, **(b)** 1.61, **(c)** 11.7
7. **(a)** 0.0445, **(b)** 0.42, **(c)** 3.79×10^{15}
8. **(a)** 199, **(b)** 1.04, **(c)** 5.2
12. **(a)** 4.54, **(b)** 2.57
13. **(a)** 2652, **(b)** 17.9
15. **(a)** 0.014, **(b)** 0.0040, **(c)** 0.018
18. **(a)** 2.40, **(b)** 0.348, **(c)** 69.0
20. **(a)** 0.126, **(b)** 1.96×10^{7}
21. **(a)** 0.315

CHAPTER 10

1. **(a)** 19.47, 0.0289, **(b)** 49,825, **(c)** 32.2, **(d)** 134
2. **(a)** 24.4, **(b)** −2.5
6. **(a)** 227, **(b)** 81
8. **(a)** −13.5
9. **(a)** 25, **(b)** 267, **(c)** 314.7, **(d)** 2.12
10. 47.23
12. **(a)** 175.3, **(b)** 274

CHAPTER 11

3. **(a)** 12.27 comp., **(b)** 3.06 major, **(c)** 2.84
4. **(a)** 20.8
5. **(a)** 38.64, **(b)** 126.5, **(c)** 7.38
6. **(a)** 0.0774, **(b)** 86.3, **(c)** 0.0106

Index

B

C

D

E

P

TREVOR ANDERSON
KYLE ARMSTRONG
NOEL BÉGIN
ELISABETH BELLIVEAU
RICHARD BROWN
ERIC CAMERON
BRUNO CANADIEN
SHERRI CHABA
CHRIS CRAN
ALYSHA CREIGHTON
DAVEANDJENN
AMANDA FORBIS AND WENDY TILBY
MACKENZIE FRÈRE
SARAH FULLER
JASON DE HAAN AND MIRUNA DRAGAN
FAYE HEAVYSHIELD
TERRANCE HOULE
GARY JAMES JOYNES
KRISTOPHER KARKLIN
EMILY LUCE
ROBYN MOODY
ERIC MOSCHOPEDIS AND MIA RUSHTON
PAMELA NORRISH
GABRIELLE PARÉ
TARAS POLATIAKO
JEWEL SHAW
LAURA ST. PIERRE
LARISSA TIGGELERS
LAURA VICKERSON
JENNIFER WANNER
DONNA WHITE
MARIA WHITEMAN

2013 Alberta Biennial of Contemporary Art

THE NEWS FROM *HERE*

CURATED BY NANCY TOUSLEY

January 26 – May 5, 2013
Art Gallery of Alberta

Library and Archives Canada Cataloguing in Publication

Tousley, Nancy
2013 Alberta Biennial of Contemporary Art / Nancy Tousley.

Catalogue of an exhibition held at the Art Gallery of Alberta, Edmonton, Alta. from Jan. 26 - May 5, 2013.
ISBN 978-0-88950-170-6

1. Art, Canadian--Alberta--21st century--Exhibitions. I. Art Gallery of Alberta II. Title. III. Title: Two thousand thirteen Alberta Biennial of Contemporary Art.

N6546.A4T68 2013 709.7123074'712334 C2012-907168-4

art gallery of alberta

INTRODUCTION

Since its inception in 1996, the *Alberta Biennial of Contemporary Art* has maintained its focus on the presentation of work by Alberta artists. In this history, Alberta Biennial exhibitions have featured the work of over 125 artists, to which we are adding 36 in this year.

THE NEWS FROM HERE: The 2013 Alberta Biennial of Contemporary Art represents the diversity of Alberta's contemporary art practices and investigations, and provides a framework through which we can see how contemporary art is shaped by and shapes different perceptions of time and place. The exhibition has been curated by long-time Alberta resident, Nancy Tousley, who has watched and documented the Alberta art community since the late 1970s, making an immense contribution to our knowledge and understanding of art in the province as well as beyond. Her curatorial premise for the 2013 *Alberta Biennial*, looks at how artists today are both grounded in the place that they live and work, but also recognizing that they live and work globally and are connected to an art world that is international in its scope. I would like to extend a sincere thanks to Nancy Tousley for her thoughtful work in curating this year's exhibition. It has been a great pleasure to work with her over these last two years.

The Alberta Biennial is a unique opportunity for the AGA to celebrate and share the remarkable work of our province's artists. I would like to congratulate and thank all of them for the work that is represented in this exhibition and for their commitment to Alberta's cultural communities. Artists contribute to our culture in so many ways, as teachers, thinkers, inventors, commentators and critics, and they are integral to how we shape and create society.

I would like to express our gratitude to ATB Financial whose generous support enables us to continue our commitment to the presentation of the *Alberta Biennial of Contemporary Art* and to sharing the many different stories of art in the province. We also acknowledge the ongoing support of the City of Edmonton and the Edmonton Arts Council; the Province of Alberta and the Alberta Foundation for the Arts; and the Canada Council for the Arts. As well we are grateful for the support of all of our Members and Donors. With their assistance we are able to present this exhibition and accompanying publication, and are proud to able to continue our commitment to writing the history of art of this place we call Alberta.

Catherine Crowston
Executive Director and Chief Curator
Art Gallery of Alberta

THE NEWS FROM HERE
BY NANCY TOUSLEY

Biennials large and small are most often associated with particular places, usually cities, which are scattered around the world and play host to changing cohorts of international artists. At present, more than 200 biennials are held worldwide. Unless the art on display is site-specific, it needn't connect directly to the locations in which these high-stakes exhibitions are shown. Recent exceptions are the first two *Prospect New Orleans* biennials, unusual for the number of works that related specifically to the history and culture of a unique city. The many biennials named for cities are by and large international in scope. Less numerous, the institutional biennials, like that of the Whitney Museum of American Art in New York, often have a national mandate — in the Whitney's case, the United States — and they are framed by an institutional voice. The regional biennials, which are fewer still, might focus on a continent, like Europe, or that of a province, in which all of the art in an exhibition, like the *Alberta Biennial of Contemporary Art*, will have been made.

Smaller and quieter than, say, the international *Venice Biennale*, the oldest biennial of all, the regional biennial has an opportunity to do something that the others do not or cannot do. The regional context presents the possibility of making an exhibition with a strong connection to its location and the artists, geography and ethos of the place. The art on display can tell us not simply that this art has been made in this place by these artists, it can convey a sense of place. It can point to changes in the position of its artists vis-à-vis where they live and work, as well as to their expanded understanding of the global art world. Regional biennials can be, simultaneously, exhibitions about "here" and "there."

The News From Here: The 2013 Alberta Biennial of Contemporary Art attempts to situate our understanding of the practice of Alberta-based artists by looking outward and inward at the same time. It brings "The News" because that's what biennials do: they are temporal, ephemeral and potentially experimental; snapshots capturing something of a time- and date-stamped point of view. A biennial takes stock of things. The work on view is new, both in terms of its recent manufacture or, additionally, an artist's change in direction. The "From Here" puts the focus on the resonances of place, as well as the politics of being located (placed), a central issue of human geography. And it takes the position that whether or not it is immediately apparent in a given artwork, place is a significant factor in the making of art and art is a significant factor in the shaping of place.

The News From Here follows as closely as possible the idea that, by design, biennials should present art made in the previous two years. It showcases works made for the exhibition, freshly minted work not shown before, two film debuts and young artists who are making their first outings in a major public art gallery, as well as new and recent work by mid-career artists and art world veterans. Since the first *Alberta Biennial* held in 1996, this inter-generational make-up has been one of the exhibition's characteristic traits.

The 2013 edition takes as its starting point what, for want of a better term, I will call post-regionalism, in order to point to a new frontier of geographical context, whose reach moves beyond traditional notions of regionalism. The idea grew, somewhat unexpectedly, out of the studio visits I made to artists in the province. My sample, taken from Fort McMurray in the north to Standoff in the south, admittedly was small; I reviewed 164 submissions and, out of these, made 62 studio visits. Even so I was struck by something I had not seen before, to such an extent, during the 35 years I have lived in Calgary and written about artists in Alberta and their work. The physical, social, political, historical and cultural geography of Alberta is appearing once again in her art — in the foreground and in the background — in the imagery, subject matter, content, metaphorical constructs, symbolism and critical issues that it presents.

Clifford Foard Robinson
Alberta Scene, n.d.
woodcut 4/25
Art Gallery of Alberta Collection, purchased with funds from the Soper Endowment

This reappearance reaches deeper and goes beyond the representations of landscape and genre painting of earlier years, whose dominance as art forms in Alberta began to give way to modernist abstraction, pop, conceptual and experimental modes of art making during the 1970s. Although it might seem counter-intuitive to say so, by re-positioning themselves in relationship to place artists have sloughed off the old narratives of regionalism — which were imposed from the outside during a time not so long ago when "regional" was a pejorative term equal to "parochial" — as though they were shedding an old skin. The discourse now has shifted from regionalism to place.

"This is the most straightforward and common definition of place — a meaningful location," writes the British geographer Tim Cresswell, who examines the complexities in what seems to be a common-sense term.[1] Place is concrete, a thing, and, "As well as being located and having a material visual form, places must have some relationship to humans and the human capacity to produce and consume meaning." Importantly, Cresswell continues, if place as a thing can be "obscure and hard to grasp," place is "also a way of seeing, knowing and understanding the world."[2]

Today the most innovative and ambitious artists in the province display an identification with and commitment to and considerations of place and the particulars of daily lived experience. One might say they are "in place." At the same time, they are partaking of the inter-connectedness and speedy communications of a global economy. They express place in the visual and technological vocabularies of international contemporary art, which they feel free to renovate and recast to suit the needs of a changed context. And because place is a social and cultural construction, they both reflect it and participate in the shaping and reshaping of its meanings.

Canada governs and regulates the country, and funds artists, based on political, economic and cultural regions. This might be one reason why the discourse of regionalism has been hard to shake. Human geographers began to look beyond regional geography for the specific yet always changing territory of place in the early

1970s, coincident with the beginning of post-modernism. Into the 1980s, many artists in Alberta felt isolated, overlooked and envious. The most ambitious moved away or yearned, openly or inwardly, to be somewhere else, like Toronto or New York. Serious art discourse travelled from the centres of the art world to the periphery, sometimes in the guise of visiting critics and artists, but not the other way around. This, of course, was not dialogue at all, but in some cases the very nearly colonizing dominance of ideas that arrived from outside the local or even the national parameters.

Writing in the mid 1970s, in an essay in his book *Radical Regionalism*, published in 1981, University of Calgary cultural historian George Melnyk proposed, "The metropolis-hinterland theory of Canadian-American relations can just as easily be applied to East-West relations in Canada."[3] In what amounted to a political and cultural manifesto, he argued for self-determination, recognition of difference and a redefinition of regionalism "to give it a major role in freeing the West of neo-colonialism."[4] Now, even as the political and economic status of Alberta increases in importance nationally and internationally, her artists are challenging the parochial myopia of eastern centres, as artists have been doing for decades in the newer western centres of Vancouver and Los Angeles. Their outlook is global, yet they see the world from the vantage point of a particular place and actively participate in the construction of its identity.

In his introduction to Melnyk's manifesto, George Woodcock observes, "Regionalism is one of those concepts that, because it is so close to the changing impact of history on the enduring structure of geography, has constantly to be rethought and relived."[5] Adopting Melnyk's term, curator Ihor Holubizky re-examined issues of regionalism and its relationship to Canadian contemporary art in 2006 with his exhibition *Radical Regionalism*. To Melnyk's insistence on self-determination, Holubizky, who looks at several contributions to the subject of regionalism, adds "we must examine the particular in all of its complexities and diversity in order 'to comprehend a world of locales without center or periphery.' "[6] Richard Rhodes, in his essay for the 2010 *Alberta Biennial, Timeland*, wrote about the relationships of the local and the global. "The encapsulating scale of . . . globalism subsumes the idea of local identity within a new information-fed internationalism . . . . Yet, at the same time, the local has also taken on a new significance . . . which functions as a life-blood of diversity within the global framework."[7]

Place, in which the renewed interest is international, is not just here, but there and everywhere. Places are the nodes on the global network. Rethinking the centre-periphery model of regionalism in 2013, in the decentered globalized world that Holubizky and Rhodes talked about, it seems that while regions might continue to have relevance for governance and funding agencies, they are no longer pertinent to the discourses surrounding the making of art. The many reasons for this include the vast changes in travel, technology and ways of communicating that since the 1990s have given birth to personal computers, smartphones, tablets and the various forms of social media. In 2013, artists in Alberta are cosmopolitans who actively travel the world; study abroad; establish international networks; show locally, nationally and abroad, where mainstream, DIY and artist-run culture offer a variety of opportunities;

and participate in international artist residencies, which have mushroomed in more than 70 countries during the past 20 years.

Yet they resist the standardizations and homogenization of globalism. They make their work by attending to the textures and specificities of life where they live. The present moment is characterized by the active engagement by artists with the phenomenal and social realities of place, space and time — a specific geography, history, memory, and psychic and cultural inscription. Shifting the discourse from the outmoded, even punitive, notions of regionalism to an acknowledgment of the centrality and relevance of place establishes a new subject-position for artists that calls for recognition. I want to acknowledge this in *The News From Here*.

Several of the works in *The News From Here* were conceptualized in far-flung parts of the world. Jason de Haan and Miruna Dragan talked out their plans for *The Wood and Wave Each Other Know* (2011) during a year in Beirut, while Dragan attended a residency. Terrance Houle wrote the proposal for a film featuring his Blackfoot ancestor, *isstahpikssi (eestah-peek-see) — Ghost* (2013), based on the French filmmaker Chris Marker's famous *La Jetée* (1962), during a residency in Bogota, Colombia. The drawings of Elisabeth Belliveau's *Entre Chien et Loup* (2010-2012) comprise a kind of travel diary: the graphite horizons of cities at dusk, which include Sackville, NB and Grande Prairie, where she lives and works; Manhattan and Sleepy Hollow, NY; Bologna, Venice and Rome.

Taras Polataiko, a conceptualist who immigrated to Canada from Ukraine in 1990, is a truly transnational artist. His practice moves back and forth between Lethbridge and Ukraine, and his most recent performance work, *The Sleeping Beauty Project* (2012), mounted in Kiev, caught the attention of the world press. In a series of "Self-Portraits," which turns on darkly humorous Gogolesque reversals, Polataiko takes the role of the outsider and makes of himself an un-sophisticate as he parodies artists to whom he also offers homage, constructing photographs that are multilayered commentaries on artists, the art world and the history of his medium.

Top:
Illingworth Kerr, *Discovery On A Mountain Slope*, 1948, oil on panel
Art Gallery of Alberta Collection, gift of the Artist, 1982
Bottom:
Maria Whiteman, *Touching: Grizzly Bear #2*, 2011

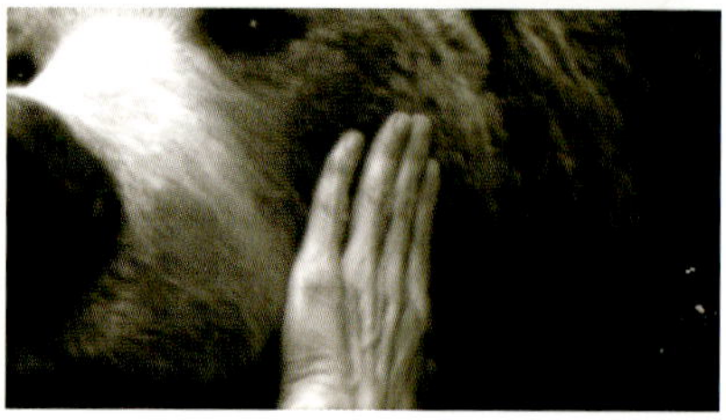

Pissing into the Wind (2010), the "Self-Portrait" in this biennial, celebrates and parodies, with the perfect touch of ironic self-deprecation, the constructed photography of certain Vancouver artists, like Jeff Wall and Rodney Graham, while the title, which describes exactly what the artist is doing on the side of a mountain, is a slang phrase for "wasting your time." The works in *The News From Here* present an assemblage of images, art objects, actions and ideas that are separate, individual, multilayered and diverse, but, in concert, might be taken as the beginnings of a thought about how to construct a provisional visual poetics of place by making an exhibition.

A model might include, as the space of this exhibition does, the natural environment or physical landscape, notions of the sublime, the influence of space, flora and fauna, people, stories, investigations of identity and the self, personal desire and transformations, domestic space, remembered and invented histories, the framing of Canadian art, ecological issues, intimations of mortality, death, and mythologies that begin with the stuff of the everyday only to transcend it through the imagination. Within the exhibition, place materializes through the subjectivity of artists and the representations in artworks. The construction of place can be only partial, it's true, and not every viewer will see it the same way, just as the artists in the exhibition speak with a multiplicity of voices and present diverse points of view.

Top:
Carl Rungius, *Mount Rundle*, n.d., oil on canvas
Art Gallery of Alberta Collection, purchased in 1925
Bottom:
Sara Fuller, *Experiment in Landscape No. 1*, 2011

The physical landscape, perhaps the first aspect of place that comes to many people's minds, appears in *The News From Here* always in some relationship to culture. Jason de Haan and Miruna Dragan's video installation, *The Wind and Wave Each Other Know* (2011), places the viewer in the centre of a moving 360° panorama of northern Alberta's boreal forest seen through the windows of a fire lookout tower. The view is both intimate and vast, enclosed and open to space. Each time the camera, mounted on a turntable, makes a complete revolution, it passes the lookout, Daniel Bosch, a self-taught musician who is playing improvised music, mixing phrases of Bach and pop songs, on a cello. The handcarved instrument is really just the fingerboard and neck; the small cabin acts as its resonating chamber. As the cellist plays the clouds flow past and the sky grows darker with the advancing dusk.

In the carefully worked-out melding of these elements, de Haan and Dragan create a visual and aural pastoral that speaks of folkways and the rustic life at the same time that it invokes the sublime in nature through an actual physical landscape. Sarah Fuller points to the sublime in *Experiment in Landscape, No.1* (2012), a set of three short videos, acted out in moonlight in front an iconic topographical feature of the Rockies at Banff, the oft painted and photographed Mt. Rundle. Fuller walks into the shot framing a snow-covered meadow below the mountain and attempts a headstand, which she does in every place she travels to, almost as a ritual, as a way of being present in the moment. Her goal is to integrate herself into the landscape. When she wipes out, the result is comical. When she achieves the headstand and holds it, she is not only in the moment, she is fully "in place".

Being "in place" connotes belonging and identifying with place. Noel Bégin's installation, *A Decombinant Diapositive Verisimilitude Leaning Precarious Against the Verdance* (2011), imagines a nighttime fête champêtre, with references to Watteau, Fragonard and Manet, acted out in a Calgary back lane. He mythologizes the ordinary, anchored in place but lifted above the everyday. DaveandJenn mythologize history, the events of their own lives and their being-in-place even more fantastically and extravagantly in a wondrous grotesque entitled *-TheBindingLine-* (2012), a three-headed skeletal creature whose recombinant DNA shapes it into a hybrid of painting and sculpture. On its neck, it carries the circular image of a dark, teeming forest. One thinks of swampy jungles and the Mesozoic Era.

Chris Cran, on the other hand, draws attention to the landscape as a background teaser as he cheerfully and obscenely mocks the extravagant bigness of prairie space and prairie boasting in *king!* (2012), a three-part painting-cum-relief sculpture. Playing on painting-as-window, he hides a big-sky landscape by revealing it in fragments as seen through the windows of three S,M,L Victorian-baroque style frames. The partially visible slogan "***king big!" — perhaps the exclamation of a gob-smacked city boy — and the bespectacled farmer in the landscape play with the idea that the man is king of all he surveys.

Robyn Moody and Mackenzie Frere extrapolate from natural forces in the prairie environment, in which each grew up. Moody builds a kinetic cascade of lighted neon tubes, accompanied by a melancholic organ-drone, in *Wave Interference* (2011), to embody a light wave that addresses both the operations of nature and the distrust of science. Frere weaves *Noise* (2012), whose intricate, non-repeating, zigzag patterns he designed by using random numbers generated by lightning, which he found on a web site. Both works touch on the characteristically Canadian theme of the mediation of nature by technology. Abstract painters Richard Brown and Larissa Tiggelers respond to empty open space. Each locates a single image/object on a colour field that implies vastness. Tiggelers's "characters" float against a single colour that could be thought of as sky, while Brown's paint objects sit on a ruler-straight horizon line that differentiates sky and ground. Both artists are adept colourists.

Works that reach far back into time in one locale are deeply seated in place. Faye HeavyShield's *Currency* (2012) came about after she attended the slaughter and butchering of a buffalo, a huge animal, whose meat was given to family and friends. The impact of witnessing this event in the present made a deep impression. Her work speaks of ancient spiritual traditions, modern mediums of exchange and consumption. In her offering of "meat" —represented as "coins" on which heads and tails are hide and flesh — piled onto a ceramic platter, the "feast" is reduced to thin, flat photographic images. With this depleted currency, she contrasts plenty and poverty, then and now.

Representing something or someone as being "out of place" from a position of immersion "in place" can be a powerful artistic strategy. Workers living in the vast pre-fab camps of booming Ft. McMurray, and working in the Alberta oil sands,

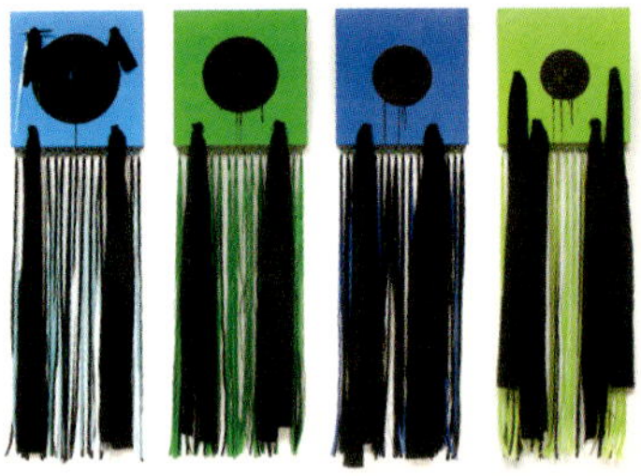

Top:
Bruno Canadien, *Pipe Dreams*, 2011
Bottom:
Orest Semshichen, *Oil company maintenance road near Brooks, Alberta, June 1989*,
1989, silver gelatin print
Art Gallery of Alberta Collection, gift of Orest Semchishen

inhabit Kristopher Karklin's *Camp Life* (2011). He makes this controversial territory seem dreamlike, rumored and far away, as if its inhabitants were indeed out of place. His method of dropping photographs of his subjects into constructed model environments enhances the feeling of the unreality, ennui and alienation of camp life. Bruno Canadien points both to the sacredness of oil, part of the earth's bounty, and the dangers of pipelines running across native lands in a four-part painting, *Pipe Dreams* (2011), from his *Freedom Fighter Series*. Combining modernist tropes with multi-coloured satin pow-wow ribbons and beads, he symbolizes bitumen, water and grass, natural substances that can be endangered if one of them, bitumen, is taken "out of place."

Trevor Anderson's gay great-uncle, the protagonist of Anderson's inventive film musical, *The Man That Got Away* (2011), filmed on the spiral ramp of an Edmonton parking garage, is out of place as a gay man in 1950s Canada. Terrance Houle displaces his ancestor, Isstahpikssi, from mid-19th century Alberta to set him adrift, via the alchemical vehicles of photography and film, as a time traveller who recounts the history of First Nations people's relations with Europeans. Looking back to the late-19th century, in their animated film *Wild Life* (2011), Amanda Forbis and Wendy Tilby tell the sad tale of an English remittance man who comes to live in Alberta and fails to understand or adapt to his new environment. The repeat image of a woman with bare, arching arms, weeps floods of lead-sinker tears and patterns Donna White's Victorian-inspired *Despair Wallpaper* (2012). A powerful image, one that points simultaneously to history and the present, she is powerless and out of place in a patriarchal society that traps her and isolates her in domestic life.

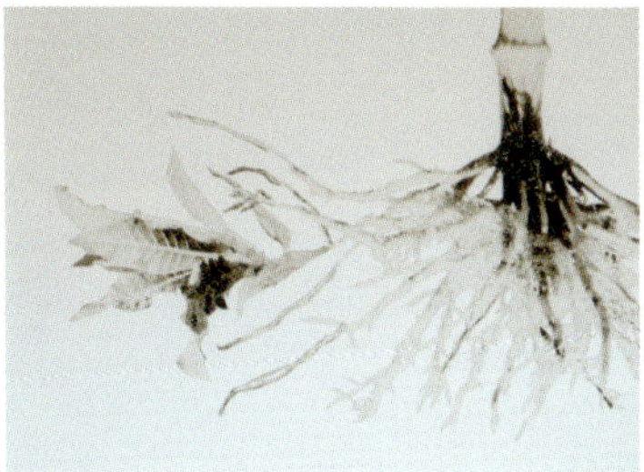

Top:
James McLaren Nicoll, *Seed Buds*, 1979, lithograph on paper
Art Gallery of Alberta Collection, purchased with funds from the Soper Endowment
Bottom:
Jennifer Wanner, *Herbacentrice*, 2010-2012

Kyle Armstrong's lushly visual experimental film, *Magnetic Reconnection* (2012) juxtaposes stunning images of the aurora borealis with the ruins of a military facility and industrial junk that litters the North. (Although he shot the footage in Manitoba, Alberta's North would yield similar phenomena.) Genetically altered seeds, out of place in nature, grow into fierce, cannibalistic mutations of familiar plants, like tomatoes, canola and corn, in Jennifer Wanner's spectral, poetic stop-motion animation, *Herbacentrice* (2012). Jewel Shaw steps away from a troubled northern Alberta childhood to create an alternative space. In a series of etchings entitled, *Investigations of the Dog* (2011), she visualizes this place as a snow-covered landscape strewn with magical animals, a Bible, objects and bits of text, through which she refers allegorically to historical personal narratives.

A persona (homeless person, survivalist?) has fashioned the makeshift dwellings in huge pipes with lighted interiors that dot the darkening panoramic landscape of Laura

St. Pierre's *06.21* (2012), set on the margins of urban space, and the rough-hewn shack in Sherri Chaba's *The Paradox of Plenty* (2012), which inverts the relationship of inside and outside by sheltering miniature landscape dioramas in its interior. Both of these works envision homelessness, the ill effects of the oil industry and the breakdown of society in some not too distant future. Also a figure of displacement is the animal Other, embodied in Maria Whiteman's visceral video stills from the *Touching: Grizzly Bear Series* (2011), in which the artist caresses taxidermy as a gesture of mourning for the loss of contact with the animal body and decreasing animal populations in the wild.

Death, whose presence is intimated in HeavyShield's *Currency*, is no stranger to Canadian art. Pamela Norrish weaves her own *Burial Suit* (2012), made entirely of glass beads. The accurately detailed, ceremonial, white T-shirt and jeans are an intimation of mortality, tailored to fit her at her present size. She is a young woman. Eric Cameron's *Thanatos II* (2012), composed of 225 of his *Dipped Paintings* suspended from the ceiling in a grid, presents a colorful inverted garden of "flowers," each an object made of paint with a Remembrance Day poppy at its core. Cameron, who is in his 70s, was inspired to make the work which he titled with the Greek word for "peaceful death," after a serious illness led him to believe he would die. He found the idea "agreeable."

Alysha Creighton literally rises upwards on her toes and disappears, removing herself bodily from place, in her video, *Ascension* (2011), but the accent is on transcendence rather than death. Transcendence and transformation are again figured on the female body in Elisabeth Belliveau's *Go so we may see (lady of gold arms doom)* (2012), an animation that isolates scenes from five films that depict moments of transformation in the lives of the female protagonists. Similarly, Gabrielle Pare's "performed" photo etchings, *The Pulling Body*, *On the Precipice* and *The Slow Creep of Repercussion* (2011) are animated by the uncanny movements of a woman, the artist, who acts out the throes of an inner struggle between reason and the relinquishment of control in the anonymous space of an Edmonton hotel room.

Actual performance has a part to play in *The News From Here* in the work of collaborators Eric Moschopedis and Mia Rushton and Gary James Joynes. For their place-specific work, *Hunter, Gatherer, Purveyor* (2012-13), Moschopedis and Rushton gathered indigenous and cultivated plants from Edmonton gardens in selected neighbourhoods, which they dried and made into tea to serve at the Biennial opening at the Art Gallery of Alberta. The duo uses the social engagement of serving and drinking tea to focus on economics, ecology and place "by tasting class difference and geography." Joynes, who is a sound and installation artist, electronic musician and singer, presents a dramatic live cinema performance, which breaks the fourth wall of the theatre as part of the Biennial film program, which includes the works of Forbis and Tilby, Anderson and Armstrong.

The Art Gallery of Alberta itself can be thought of as a place, which is marked by Laura Vickerson's *Luxe* (2013), a site-specific work in which she upholsters a curved steel architectural projection on the building's façade with a red, Victorian-style damask fabric. The contrast with the façade draws immediate attention, feminizes, softens and adds bright colour to the steel, glass and zinc of the Gallery's award-winning architecture, and moves into the space of the atrium to claim a womb-like nook. Emily Luce's *The Cardiff-Miller House* (2013), is a work built to the specifications of the tiny-house movement, a miniature of a house in Lethbridge once lived in and still owned by artists Janet Cardiff and George Bures Miller, artists who now are known internationally. It is fully functional and filled with miniature art.

In *The Cardiff-Miller House*, the AGA meets its opposite as a place, an architectural structure that is downscale, domestic and mobile, whose presence is temporary. While the tiny house speaks of process — achieving a small ecological footprint, living simply — the public art gallery speaks of permanence, collections, preservation, presentation and place. Its very name locates it in place. Among its own processes is the *Alberta Biennial*, a biannual exhibition that holds a mirror to the contemporary art of a region, held in an institution named for a western province, a region yet open to the world. What better venue could there be for a biennial that in its present manifestation seeks to sound the end of regionalism?

Geographers tell us that place is a process; it is built on what came before yet it is always changing. The Art Gallery of Alberta is a place where the past and present can be brought together to show us where we have been and to point to where we might be going. Its exhibitions are both local and global, showing us where "here" is in relation to "there." If this Biennial can be thought of as an embodiment of place, one that surely will change in its next manifestation, it situates a place within a place, like nesting boxes. It shows us something of where we are now. That is *The News From Here*.

Endnotes

1. **Tim Cresswell**, *Place: A Short Introduction* (Malden, MA: Blackwell Publishing, 2004), 7.

2. Ibid., 11.

3. **George Melnyk**, "On Cultural Regionalism," in *Radical Regionalism* (Edmonton, AB: NeWest Press, 1981), 17.

4. **Melnyk**, *Radical Regionalism*, op. cit., 83.

5. **George Woodcock**, "Introduction," in Melnyk, *Radical Regionalism*, op. cit., 6.

6. **Ihor Holubizky**, "Radical Regionalism, Local Knowledge and Making Places," in *Radical Regionalism: Local Knowledge and Making Places* (London, ONT: Museum London, 2006), 38.

7. **Richard Rhodes**, *Timeland* (Edmonton, AB: Art Gallery of Alberta, 2010), 14.

THE ARTISTS

TREVOR ANDERSON
KYLE ARMSTRONG
NOEL BÉGIN
ELISABETH BELLIVEAU
RICHARD BROWN
ERIC CAMERON
BRUNO CANADIEN
SHERRI CHABA
CHRIS CRAN
ALYSHA CREIGHTON
DAVEANDJENN
AMANDA FORBIS AND WENDY TILBY
MACKENZIE FRÈRE
SARAH FULLER
JASON DE HAAN AND MIRUNA DRAGAN
FAYE HEAVYSHIELD
TERRANCE HOULE
GARY JAMES JOYNES
KRISTOPHER KARKLIN
EMILY LUCE
ROBYN MOODY
ERIC MOSCHOPEDIS AND MIA RUSHTON
PAMELA NORRISH
GABRIELLE PARÉ
TARAS POLATIAKO
JEWEL SHAW
LAURA ST. PIERRE
LARISSA TIGGELERS
LAURA VICKERSON
JENNIFER WANNER
DONNA WHITE
MARIA WHITEMAN

TREVOR ANDERSON

My great-uncle Jimmy was the black sheep from a side of the family we already didn't talk about, so it took some digging to discover this hidden story. Finally another uncle told me everything he knew, and then he passed away. As far as I was aware, I was suddenly the only person left who knew anything about it. I had to decide how best to share what I knew about Jimmy's life. The answer seemed clear: *The Man That Got Away* had to be a musical.

Trevor Anderson
The Man That Got Away, 2012
HD Video
Running time: 25'16"
Courtesy of Dirt City Films, Edmonton

Illustrated: Production stills

KYLE ARMSTRONG

In March of 2012 Canadian filmmaker Kyle Armstrong travelled to Churchill, Manitoba to capture images of the northern lights. Using a one of a kind purpose built digital IMAX resolution camera Armstrong captured astonishingly detailed images of the aurora borealis on nine of ten possible nights. The camera utilizes two precisely calibrated fast wide angle prime lenses allowing for a field of view of nearly 90° at a 4:3 aspect ratio captured at over 6K resolution, and a very large and highly sensitive digital sensor capable of ultrahigh ISO settings allowing for frame rates previously impossible.

During the day, Armstrong and co-cinematographer and auroral scientist Dr. Trond Trondsen explored the landscape surrounding the town of Churchill. They went to great lengths in arctic conditions to shoot super8, 16mm and DSLR time-lapse footage of *Miss Piggy*, a C46 aircraft which crashed just short of the airport in 1979, the *Ithaca IV*, a Greek steamship trapped on the shores of Hudson Bay since the 1950s, and Fort Churchill, the American-operated rocket range active during the 1960s.

Armstrong manipulated footage, altering and decaying the super-8 and 16mm film with bleach, dyes, paints, scratches and other methods, and photographing each frame individually. Armstrong then edited the resulting footage as a meditation on the contrast between mankind's attempts to conquer the natural world and the regenerative power of nature.

Magnetic Reconnection features an original score by Grammy Award winning musician Jim O'Rourke (Sonic Youth, Gastr Del Sol, Wilco, Werner Herzog's Grizzly Man) and is narrated by singer/songwriter/actor Will Oldham (Bonnie "Prince" Billy, Old Joy, Matewan).

Kyle Armstrong
Magnetic Reconnection, 2012
Super-8mm film, 16mm film, DSLR, digital IMAX
Running time: 12'30"
Courtesy of the Artist

Illustrated: Production stills

NOEL BÉGIN

Created during the events of the Arab Spring in 2011, *Verisimilitude* was designed to draw parallels between the lavish and wasteful lifestyle led by the average First-World person, the extravagant lives of North African royalty and the frivolous excesses of patrons of the Rococo period in art, which contributed to the rise of the French Revolution.

The installation is also a critical response to a current goal of digital artists to create interactive works. In contrast to them, it accomplishes its effects with the use of a simple and defunct analogue technology. The 22 photographs that comprise the mural-size cinematic image collage were recorded as slides, each of which is projected by one of 22 slide projectors, while the ambient audio track plays as a continual loop.

The slide projectors are arranged in a row, at roughly waist height, along the opposite wall. Each projector contributes one element to the large image. As viewers move through the installation, they simultaneously deconstruct the piece by blocking the projections or parts of them, and implicate themselves in the narrative by casting their shadows into the image while they are in turn projected upon. In so doing, they reveal other readings or visual layers of the projected composite image.

One slide image also exists as a 76 cm x 51 cm (30" x 20") archival photographic print, which allows the viewer to experience one image in isolation and offers a key to the installation and how the images are combined.

The squeaky, cricket-like chirping sound, recorded when a projector was in need of lubrication, plays through a set of speakers, striking a convincing nocturnal tone at the same time that it suggests the projection equipment might fail.

Noel Bégin
A Decombinant Diapositive Verisimilitude Leaning Precarious Against the Verdance, 2011
Chrome diapositives, 22 slide projectors, paper, ambient audio recording.
Commissioned by Andrea Mann
Courtesy of the Artist

Illustrated: Installation view

ELISABETH BELLIVEAU

My interdisciplinary practice braids stories, history and everyday life, employing intuition, curiosity and material exploration.

Entre Chien et Loup is a French expression that translates roughly as "in-between dog and wolf". The expression describes the moment at dusk when it is difficult to differentiate the form of a dog from a wolf — it is the witching hour. The horizons drawn for this series describe the transition that marks the end of day and beginning of night. Each drawing chronicles various places I have found myself in the world over the past three years. These drawings attempt to arrest a moment of light and time.

The animation *Go so we may see (lady of gold arms doom)* is a collection of vignettes, based on scenes extracted from films that feature walking cycles of women in film that are significant to the narrative. Through stop-motion animation, moments of emotional, psychological or physical transformation in walking cycles are isolated and removed from the context and *mise en scene* of the film. The walking sequence is then hand drawn and animated. The animation is not rotoscoped (no tracing), each cell is hand drawn with black plasticine on a light-box.

The films quoted in *Go so we may see (lady of gold arms doom)* include: *Belle de Jour*, *Picnic at Hanging Rock*, *Bridget Jones' Diary*, *Poltergeist*, *Persona* and *The Children's Hour*.

The title of the work, *Go so we may see (lady of gold arms doom)* is from a poem by Sappho as translated by Anne Carson in the book *If Not, Winter — Fragments of Sappho*. Carson's translations are arranged with brackets to indicate the missing fragments of papyrus and to interpret the spacing and rhythm of Sappho's words. Below is how this poem appears in Carson's book.

"6

so
]

]
]
]
]

Go   [
so we may see [
]
lady

of gold arms [
]
]
]doom
]"

Elisabeth Belliveau
Go so we may see (lady of gold arms doom), 2012
Animation: plasticine, paper, digital video
Running time: 3'
Courtesy of the Artist

Entre Chien et Loup, 2010-2012
20 drawings
Graphite on found paper, MDF and Plexiglass
Courtesy of the Artist

Illustrated: Production still from *Go so we may see (lady of gold arms doom)*

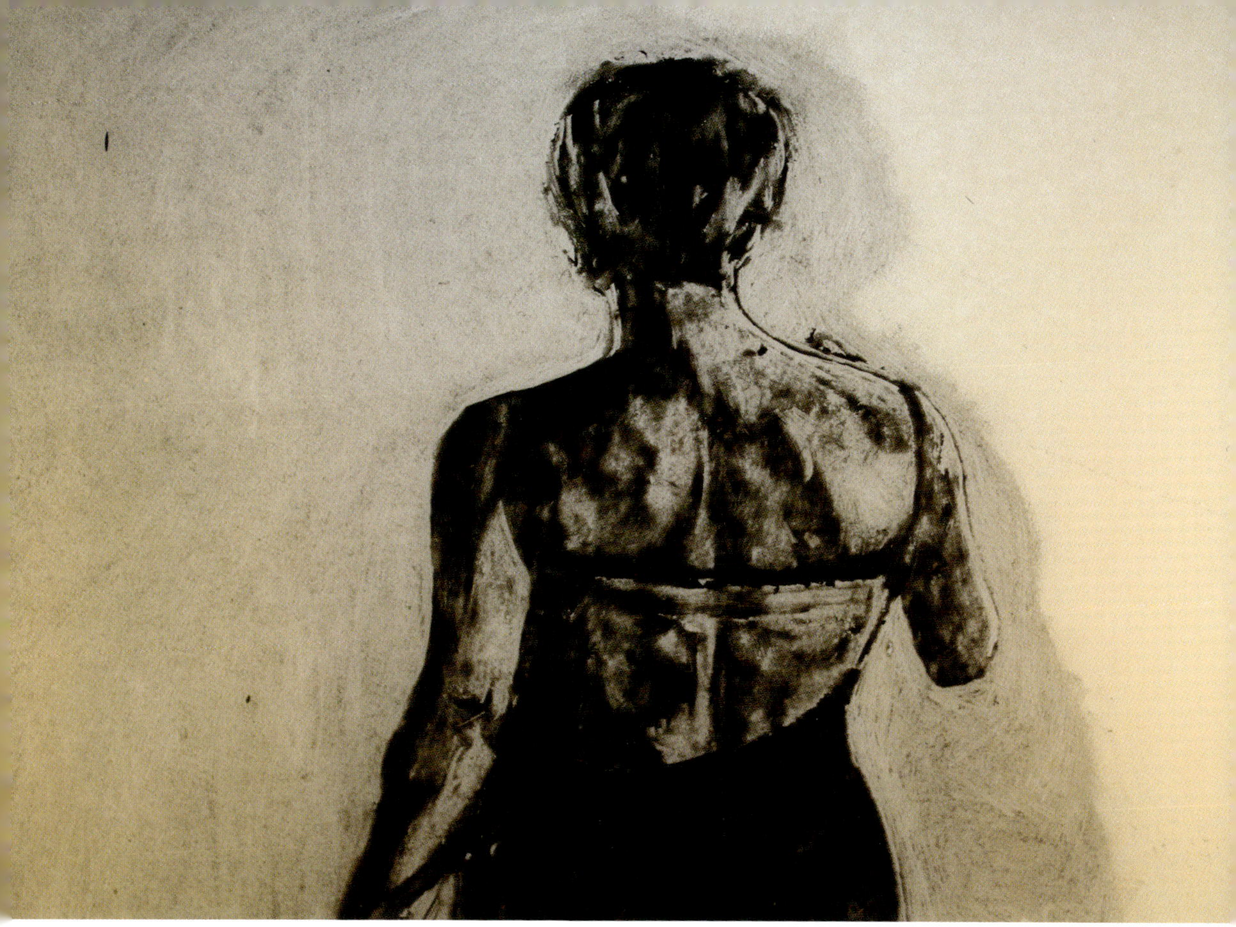

RICHARD BROWN

I want to think about the most elusive of my feelings and experiences, those that strike me as strange, funny, warm, vaguely worrisome, contradictory. I'm not really interested in separating these from my various and possibly more superficial obsessions with music, art, books, records and bicycles. When I start to put these thoughts together they come as shapes, lines and colours, mostly because I've recently become interested in idiosyncratic abstraction, using a loose and hermetic vocabulary of forms borrowed from architecture, public sculpture, electric guitars and elements from comic books, among other things.

The things in these paintings aren't feelings themselves, but a recollection of something felt, at a moment in time, in a particular place, among friends or sometimes alone, now remote and inaccessible.

Richard Brown
New Jersey, 2010-2012
Oil on canvas
50 paintings
Courtesy of the Artist

Illustrated: Detail

ERIC CAMERON

Thanatos

My *Thanatos*, a group of 225 *Dipped Paintings*, was made by repeatedly dipping Remembrance Day poppies into cans of latex paint in a variety of colours. The poppies are first sewn onto spools of fish wire, which are then suspended from the open beams in my basement ceiling, with empty cans or cardboard boxes placed underneath to catch the drips of paint. After several dippings the slightly convex downward-facing fronts of the poppies begin to yield to deepening concavities. At the stage where the concavities become just deep enough to start to create air pockets that will soon prevent the paint reaching the upper surface of the concavity, the colours break up into completely unpredictable and often quite spectacularly complex configurations, that are equally remarkable for their strangeness and for their clarity of definition of form. This is the stage that holds most interest and excitement for me and that brings the dipped paintings into most meaningful relation with my major projects since graduation more than half a century ago, all of which depend for their final resolution on my loss of control over the materials with which I was working. As controlling mind yields to the overwhelming power of natural forces, activated matter manifests its own unthinking propensities.

My choice of the title *Thanatos*, the Greek word for peaceful death, relates to a period of illness some fifteen years ago. At its onset, I thought I might be dying and found the experience remarkably agreeable. In their suspended state, the works themselves may invoke more sinister connotations of suicidal death by hanging.

Eric Cameron
Thanatos II, 2012
Remembrance Day poppies, latex paint, filament
Courtesy of TrépanierBaer Gallery, Calgary

Illustrated: Installation detail

BRUNO CANADIEN

My work is primarily focused on addressing issues surrounding the intersection of First Nation/Tribal sovereignty, resource exploitation and environmental concerns.

Through a language that strives to reflect my personal visual history within our larger complex and blended North American history, from Plains and Dene visual traditions to Modernism to mass media to powwow, I use collage, adornment, painting and drawing to present evidence of contemporary Indigenous resistance in my series of paintings entitled the *Freedom Fighter Series*. Just as First Nations' struggles are ongoing, the *Freedom Fighter Series* is an ongoing and evolving work, an attempt at honouring Nations and People.

Currently, I am exploring the resistance mounted against Alberta's oil and gas industry by aboriginal communities in western Canada and the United States in regards to the effects of the Athabasca Oilsands development, which is located within the MacKenzie/Peace watershed, upstream from my home community of Fort Providence.

Pipe Dreams presents a view of bitumen as a dangerous and a sacred thing, while touching on concerns First Nations have expressed over the transport of oil across their territories. The black circles can be viewed as oil deposits encased in the blues and greens of the earth, or as cross sections of pipes or pipelines. The cloth and ribbons represent grass and water, a celebration of life, and the hopes and nightmares of First Nations who live in proximity to oil and gas operations.

Bruno Canadien
Pipe Dreams, 2011
From the *Freedom Fighter Series*
Acrylic, graphite, cloth, beads, ribbon, hardware, on canvas.
Courtesy of the Artist

Photo: John Dean

SHERRI CHABA

What would the future look like in a post petroleum centered society? Having grown up in Alberta, in a region impacted by the presence and extraction of oil, I have been challenged for many years by this question. It now underlies much of my recent work.

As a third generation landowner, a member of a rural community and someone whose livelihood has depended on oil, I have developed an appreciation of the diverse and competing values — the political, historical, economic, social and environmental — that position each of us in the debate on natural resources. In my work, I express these values through installations that often meld industrial materials with historical artifacts. I am inspired by the aesthetics of materials and in particular worn items that embody their histories.

While I draw on my own experience with oil, I also recognize the complexity of experiences through which others view my work. My installations invite people to share their experiences and reflect on them in a wider context, as a way of moving beyond the personal. My hope is that as viewers bring their knowledge and vision into the space, they will develop a deeper understanding of where they situate themselves within 21st century dialogues of ecological, economic and social sustainability.

The work I have created for the *2013 Alberta Biennial* challenges viewers to reflect on social and environmental issues that impact us locally, but also are relevant nationally and internationally.

Sherri Chaba
The Paradox of Plenty, 2012
Steel, wire, animal stretchers, artifacts, mirrors, viewing devices, taxidermy
Courtesy of the Artist

Illustrated: Installation views

CHRIS CRAN

In the mid-nineties I began a new body of work. I was used to making paintings by this procedure: have an idea, figure out the technical procedure, make the work. One day a question posed itself. Does my hand have its own intelligence? It was followed by a thought: I dunno! Ignoring my higher intelligence, I began by approaching a canvas without specific intention. My hand began at the upper left corner and drew an interior, asymmetrical framing device that countered the stability of the canvas' rectangle. I then responded to that first gesture and then again and again until the painting seemed done. As I continued with this approach to making work I noticed a repeat of this framing device even into hundreds of works. Over time this evolved into many other versions of the interior frame. I began to think of it as an "archetype". At some point I made a work that echoed a frame I had grown up with in my family home and I realized that this frame might be the source of my "archetype". The frame in question is an ornate, gold leafed, slightly asymmetrical beauty that holds a watercolour of a woman done by an English painter in 1830. Several years ago I had the idea to take that framed work and have a 3D scan made of the entire object. Having done this I then had three different sizes (large, medium and small) made in wood by a CNC process in preparation to make silicone molds and ultimately plastic objects upon which I could paint. The image comes from a painting I made in 2005 for an exhibition called *The McIntyre Ranch Project*. The year before I was on an artist's residency at the Ranch which is the biggest in Alberta. In fact it was ***king big! which is where the painting got its title. Along with the lettering and blue sky there was added a floating close-cropped head taken from an ad in a late 1940s Canadian magazine devoted to agriculture. That painting was the source of the three-part *king!* that is in this exhibition. This is the first version of what will be yet another new body of work.

Chris Cran
king! 2012
Oil and acrylic on polyurethane plastic structures
Courtesy of TrépanierBaer Gallery, Calgary and Clint Roenisch Gallery, Toronto

ALYSHA CREIGHTON

This work explores a poetics of the body, a language of gesture, a space beyond words. I am interested in the way that a single moment, the moment of suspension, becomes charged with significance; a fleeting metaphor for that which escapes us, that which can barely be grasped. The subtle language of the body destabilizes. The moment of tension presents itself, withdraws, leaving behind the experience of dissonance and disbelief that surfaces in a space beyond words.

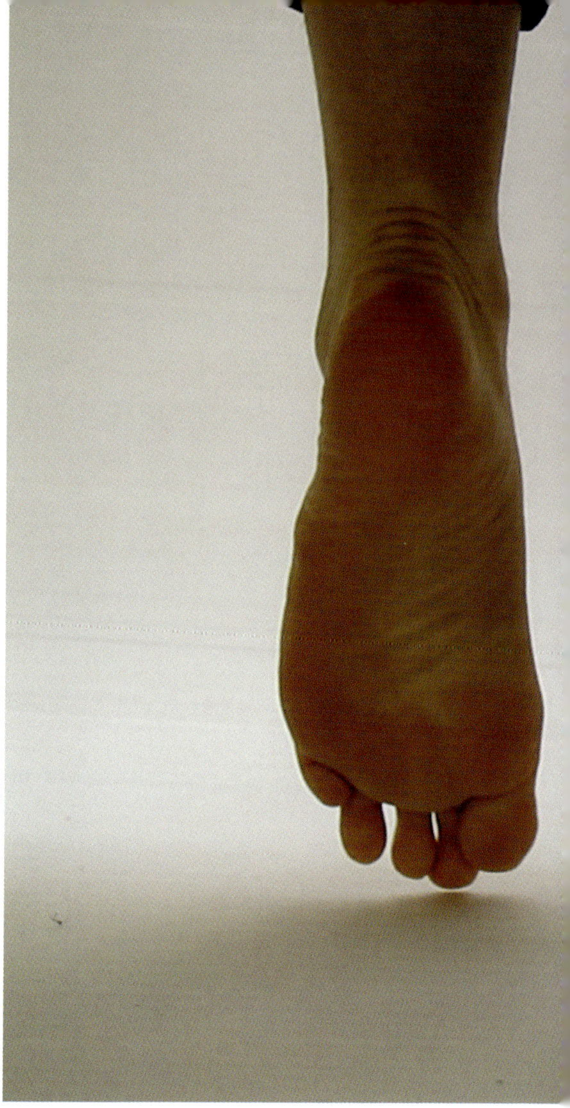

Alysha Creighton
Ascension, 2011
Digital video
Running time: 2'24"
Courtesy of the Artist

Illustrated: Production still

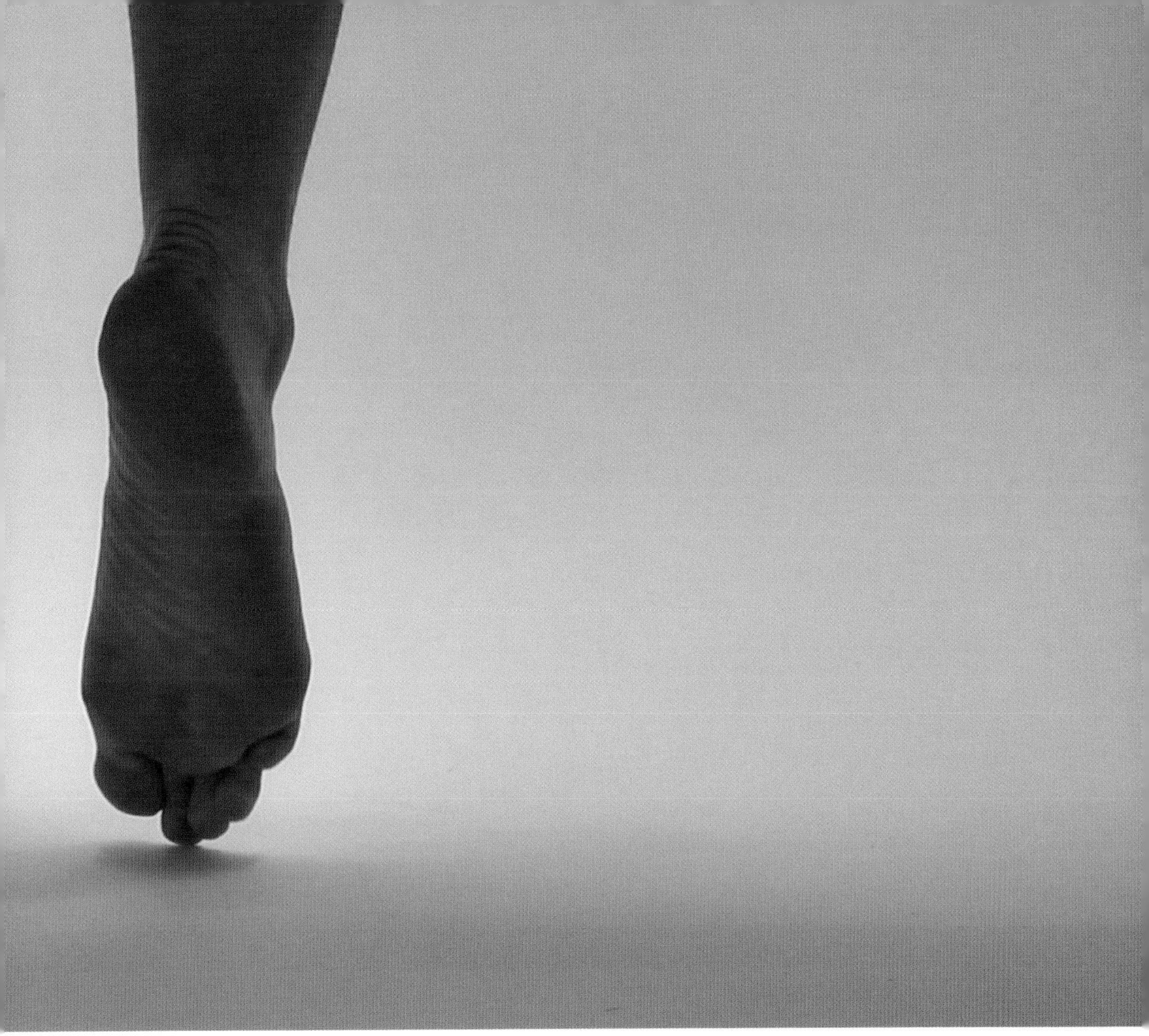

DAVEANDJENN

Art, Make-Believe, History and Hearsay

DaveandJenn have a weakness for private spectacles and inner landscapes. Often autobiographical, their work is the culmination of an almost endless amount of information and meaning gleaned from their day-to-day lives, experiences and research.

Accumulation and deconstruction factors heavily into their modus operandi, the original narratives stripped down to their essential parts and then built up again as fantastical mythology.

Their process involves painting on multiple layers of resin, slowly working up the final image. Many times the world they depict is past a point of no return but the characters within it, be they monsters or not, often have the appearance of revelers in a new world order. This is a landscape with a mind of its own, where humans are an afterthought and others have risen to claim their place.

-TheBindingLine- is a sculptural painting that depicts the alchemic transformation of a place and the necessary migration of those that once called it home. The world reforms itself even as it disintegrates.

DaveandJenn
-TheBindingLine-, 2012
Resin, acrylic, oil, bronze and mixed media
Courtesy of TrépanierBaer Gallery, Calgary

Illustrated: Studies and work in progress

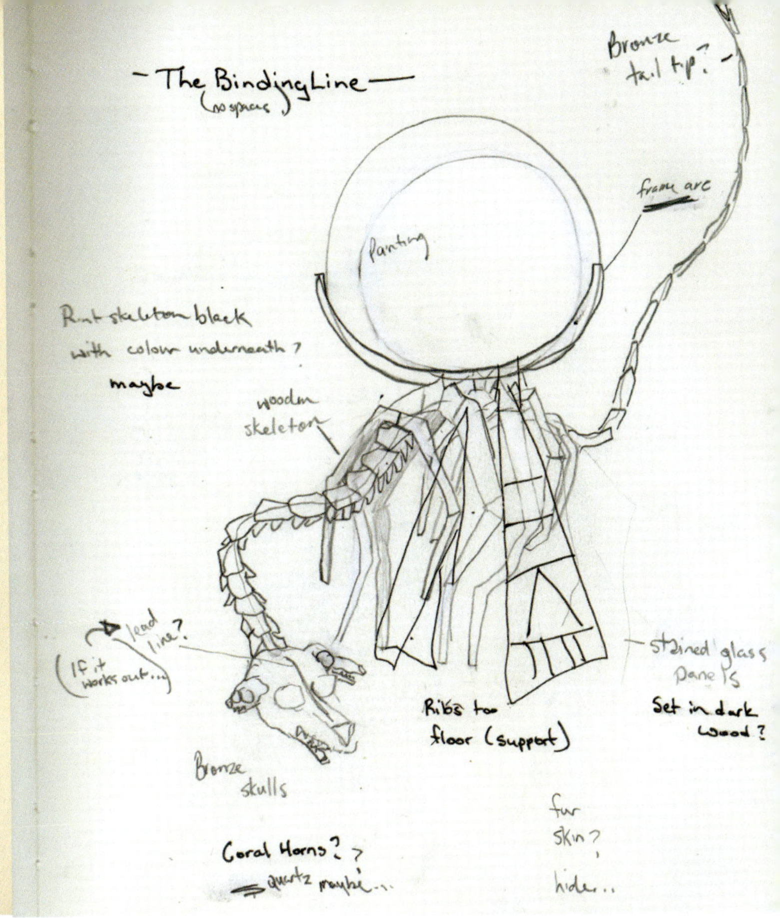
The BindingLine
(no spaces)
Bronze tail tip?
frame arc
Painting
Paint skeleton black with colour underneath ?
maybe
wooden skeleton
lead line?
(If it works out...)
stained glass panels
Ribs to floor (support)
Set in dark wood ?
Bronze skulls
Coral Horns? ?
quartz maybe...
fur
skin?
hide..

AMANDA FORBIS AND WENDY TILBY

The starting points for our films are best described as small, emotionally resonant moments which gradually coalesce into a concept or narrative. They are usually imagined observations and often have a strong sound component (such as sirens, water drops or a specific piece of music).

We avoid making proper scripts or storyboards. Conceptualizing is more a combination of writing, drawing, painting and gathering. The story and technique evolve simultaneously and haphazardly by much trial and error. As in documentary filmmaking, it is a process of distillation where order is found from the chaos created.

A recurring theme in our work is the tension between solitude and connectedness. *When the Day Breaks* is about the irony of isolation in an urban context - that it is our physical proximity to others that compels us to isolate ourselves. In *Wild Life*, the protagonist's isolation is very real. He too is wistful – but he longs for the comfort and safety of the city and the company of others.

Though we have worked in a variety of techniques, our images are most often painterly. We love close-ups and extreme long-shots. We don't consider traditional animation drawing skills to be the most crucial element of animation filmmaking. Idea, a sense of timing and aesthetic coherence are all more important to us. We strive for aesthetic unity by limiting the palette and seeking strong visual (and aural) motifs.

Our influences are found as much in painting, short stories, music, science and experimental/documentary films as in animated film. We do, nonetheless, continue to be inspired to make animation because of its completeness as an art form - and the control we have of every frame.

Amanda Forbis and Wendy Tilby
Wild Life, 2011
Animation, gouache on paper
Running time: 13'30"
Courtesy of The National Film Board of Canada

Illustrated: Production still

MACKENZIE FRÈRE

My studio practice is rooted in a contemplative approach to cloth construction using natural materials and plant-sourced colour. I favour compositional strategies based on random numerical sequences, measurements taken from my own body, and antique winding tools made irregular with time and wear. In combination, these methods produce complexity and variations that echo patterns found in nature and the irregular symmetries of the body. As I engage with layers of complexity driven by factors over which I have little control, intervals emerge in which apparently random sequences coalesce into visible repetitions or patterns.

In my recent *Noise* series random sequences and computer-generated bitmaps are interleaved creating both structural (specific to the interlacement of threads) and visual noise that interferes with the perceptual process. The online random number generator used in these works is driven by the atmospheric noise of lightning strikes. This link to natural phenomena is compelling for me as it seems to conflate the digital and the natural. Ultimately the source of the integers I utilize to compose this work is immaterial to its intention as I am more interested in drawing one's attention to the process of perception itself. In *Noise* I am intrigued that hints of repetition, or the occasional alignment of apparently scattered blocks of colour may draw us to construct pattern and make meaning in that which we perceive.

There is a gestural logic to the rhythmic and cumulative processes of cloth construction in which the ancient human impulse toward pattern, order and cohesion is embedded as much into our idea of cloth as it may be in the cloth itself. The intimacy of cloth made by hand, and our centuries-old familiarity with it make cloth not only related to the body, but itself (an)other body. In the studio, I am compelled by the ways in which cloth may intimate human presence and absence (often simultaneously) and have begun to explore the notion of embodiment as it relates to the praxis of making cloth by hand. As a material aggregation of dedicated gestures, breath and thought through an interval of time cloth tacitly communicates the immaterial and enmeshes the intangible.

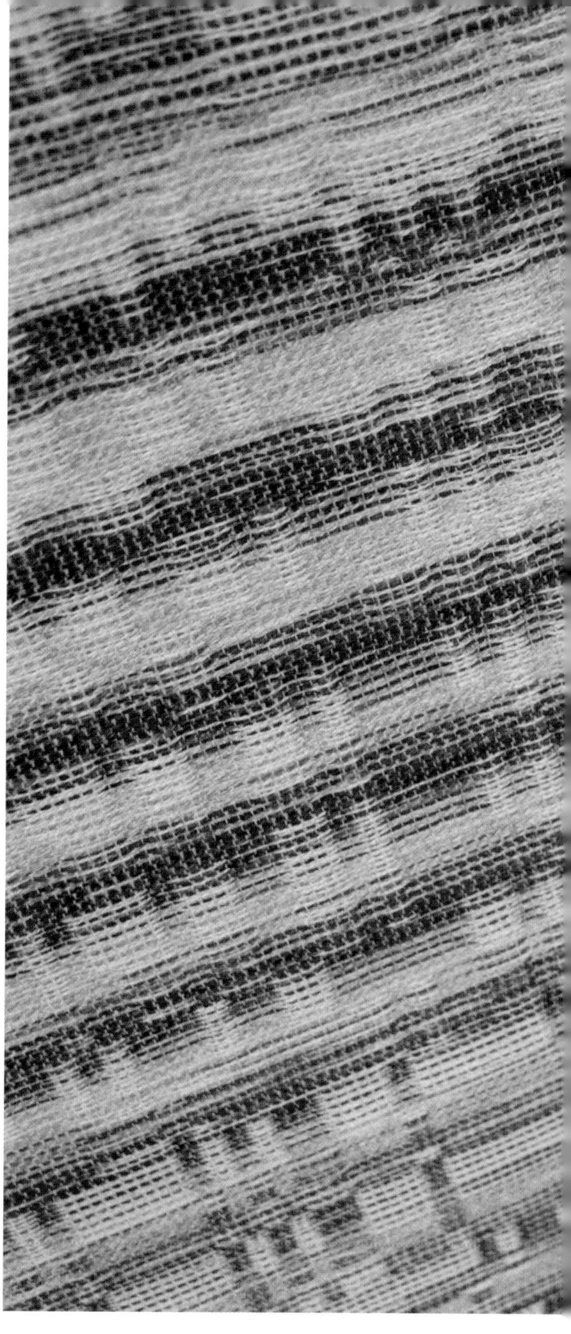

Mackenzie Frère
Noise 120429, 2012
Silk, linen, hemp, indigo
Courtesy of the Artist

SARAH FULLER

Experiment in Landscape, No.1 is an investigation into constructed landscape, narrative and performance. Using the backdrop of Mt. Rundle under a full moon, I attempt to describe a moment of suspension in nature; suspension of belief and of physical reality.

I am interested in shooting the mountains at night because of the paper cutout effect rendered upon them by moonlight. My act of entering the scene to do a headstand is intended as an interruption to the surreal nature of the landscape, but also an attempt for me to gain a different perspective on the landscape through performing a physical gesture that requires concentration and focus.

The work is as much about the language of photography as it is about video, as I try to reconcile the construct of the tradition of monumental landscape photography with the act of being a body in space in nature.

Sarah Fuller
Experiment in Landscape, No.1, 2012
HD video
Running time: 8'9"
Courtesy of the Artist

Illustrated: Production stills

JASON DE HAAN AND MIRUNA DRAGAN

Daniel Bosch is a wildfire lookout in Northern Alberta, Canada. Every day, from April to October, he looks out upon the treetops from the eight by eight foot cabin of the tallest tower in the province. Dan is also a self-taught musician. While living in the woods, he crafted a cello from a solid block of spruce and then taught himself to play it. Because the cello was too big to fit in the cage of the tower's 120 vertical steps, Dan made a body-less version, allowing him to fit it into the lookout and practice during his many hours inside the tower's cabin. By wedging the practice cello between the edge of a small worktable and the cabin's octagonal fiberglass cupola, Dan discovered that he could more than compensate for the instrument's lack of a body. The cabin itself becomes the resonant chamber and the tower becomes the instrument within which the cellist plays. Now, the lookout tower broadcasts Dan's music into the immensity of the landscape and the trees become his audience.

The Wood and Wave Each Other Know is a video work documenting this activity; the camera was set on a modified turntable in the centre of the cupola to pan constantly at one rotation every 100 seconds. A bookwork containing peripheral images accompanies the video.

Jason de Haan and Miruna Dragan
The Wood and Wave Each Other Know, 2011
HD video installation
Running time: 38'9"
Courtesy of the Artists and Clint Roenisch Gallery, Toronto

Illustrated: Production still

FAYE HEAVYSHIELD

I look to the art that exists in language, community and land.

Regarding "currency"

Blackfoot is my first language and though I've known and spoken English since the age of 6, there are times when I still feel or hear the oddity of some words. For instance, currency. While I recognize the definition (exchange or units thereof) I am aware also of its relative, current (describing a flow or the present).

First sight may offer a shortcut to or a commentary on a segment in the history of niitsitapii/real people ... read: life with buffalo. A platter of coins, their heads/tails sides bearing glimpses of beastly viscera and fur. The images, material and processes gave up other things as I spent time with them.

Exchange is translation; the paper circles are bits of photographs, flat, light weight and distant from real. I think of tender.

Faye HeavyShield
Currency, 2012
Digital print on paper, acrylic compound, ink, ceramic platter, plinth
Courtesy of the Artist

Illustrated: Print details and installation sketch

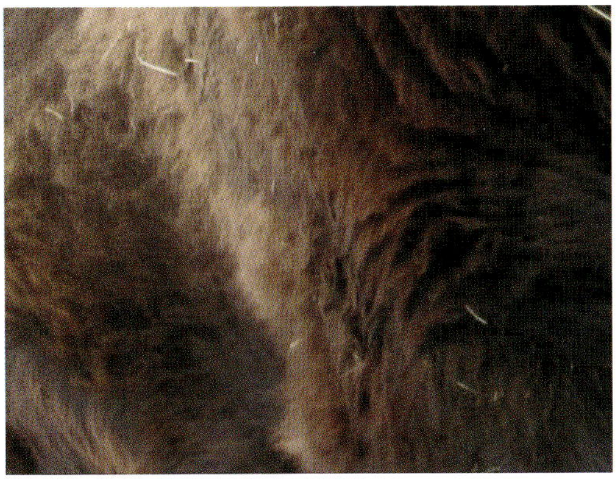

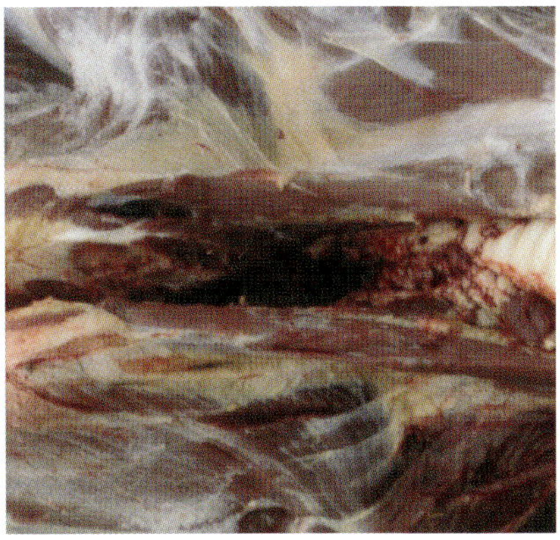

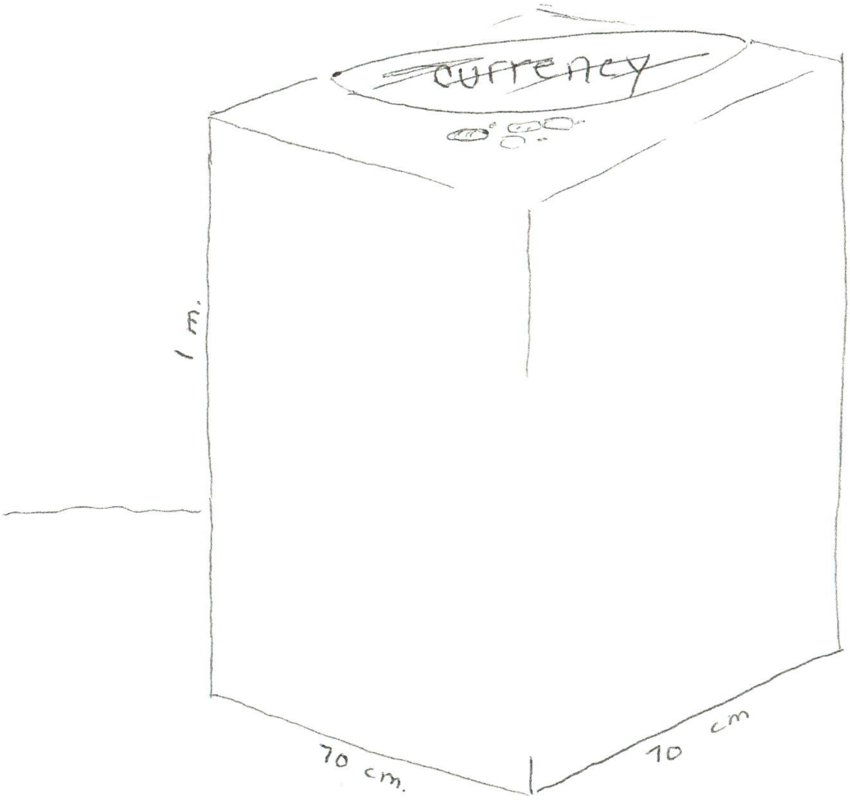

plinth:
1m H
70 cm. sq.

Shallow dish approx 50 cm diam.

fhs-currency 2012

TERRANCE HOULE

Historically, First Nations people have had close ties to the technology of photography and moving images such as film, although they viewed it with apprehension in the belief that it would capture or take one's soul. In my own family, there are photographs of well-dressed and stoic ancestors that go back more than 100 years. The notion that one's image or soul can be captured by mediums that are commonplace in our own day and age led me to the idea of time travel.

The story of *isstahpikssi (eestah-peek-see) — Ghost* is based on the fictional experience of a Native man who is asked to have his photograph taken by an unknown artist on the plains of Canada. With a willingness to explore something new, the Native man agrees to the photo and finds himself captured by the alchemy of the technology.

He sets forth as a time traveler, using photography and film as dead-reckoning devices as the technology develops. Never being able to look at his image for fear his soul will be forever lost to him, the Native man is tied to the technology. As he travels through time, he sees his people transformed as the onslaught of Western culture reverses and despises the initial precious image it captured so long ago.

I use photography, pinhole photography, archival documentary and film to create a silent, black-and-white 16mm film that uses the idea of film, Celluloid, as the alchemy in the history of the capturing of the First Nations peoples' image as icon.

Terrance Houle
Film Still #1 for *isstahpikssi (eestah-peek-see) — Ghost*, 2012
Inkjet print

Film Still #2 for *isstahpikssi (eestah-peek-see) — Ghost*, 2012 (illustrated)
Inkjet print

Trailer for *isstahpikssi (eestah-peek-see) — Ghost*, 2012
16mm film transferred to HD video
Running time: 2" (approx.)

isstahpikssi (eestah-peek-see) — Ghost, 2012
16mm film transferred to HD video
Approx. running time: 10'

Courtesy of the Artist and the Calgary Society of Independent Filmmakers, $100 Film Festival Commission Project

GARY JAMES JOYNES

Peregrination is a live cinema work reflecting on the nature of sound and its visual and material nature. A meditation on my visual experimentation using historical scientific instruments and anachronistic musical technology. *Peregrination* will combine compositional sound elements and improvisational use of recorded footage of live play of the wavedriver machinery I use to create in the studio. For the *2013 Biennial* I will bring the early phase of my career in Live Cinema together with my most recent experiments in cymatics by composing and triggering my waveform animations live in the AGA theatre and accompanying these visual sound wave animations with layered vocals and overtone singing. The performance will run through movements of varying intensity rising to a dynamic chordal crescendo.

All elements of my *Frequency Painting* work connect sound and visual definitions of sound. This new work expands these senses with aspects of space and time in my most ambitious Live Cinema work to date. My explorations and experimentations in art are focused on the active process of listening to, looking at, and feeling sound; slowing down time and linking sensoria to allow for immersion in a whole body experience. *Peregrination* (Live Cinema) will work to offer new opportunities to the audience for the understanding of the importance of deep listening and viewing life, and the wonderful journey that this brings us.

Gary James Joynes
Peregrination, 2013
Live cinema: visual/sound performance with video projection, audio and layered vocals
Running Time: 30'–35' (variable)
Courtesy of the Artist and dc3 Art Projects

Illustrated: Video still from *Peregrination*

KRISTOPHER KARKLIN

An integral component of my work is my investigation into the relationship between the living space and its occupant, where the exchange that occurs is intimate and secure; it is also temporary and homogenous. This idea is manifested in my work where the spaces that I re-create are minimal to the point where there is almost nothing, except the occupant. Due to the sparseness of the environments, the viewer is drawn into the space, and interprets the piece in relation to their own past experiences.

My most recent work is in response to living and working for the past 19 months in the Alberta tar sands of Fort McMurray. Here I continue to explore personal recollections on time, environment and experience, focusing on the static quality of life in a northern Canadian work camp.

The process of my work is done through the meticulous sculpting of miniature models. These are labour-intensive handmade reflections of the narrative. The process of building the miniatures models deal with the same problems one engages if they are painting or drawing; the color, composition, relation of objects, perspective and size. When the model is completed it is photographed. The last element of my process is spent photographing a human model. The human figure is then composited into the photo of the miniature setting, emphasizing the contrast between the figure and the space. Once finished the situation is presented as a large-scale photograph.

Kristopher Karklin
Jack & Jill Room (morning), 2011
From the series *Camp Life* (Illustrated top)
Inkjet print
Courtesy of Skew Gallery, Calgary

Jack & Jill Room (evening), 2011
From the series *Camp Life*
Inkjet print
Collection of the Alberta Foundation for the Arts

Ping Pong Room, 2011
From the series *Camp Life* (Illustrated bottom)
Inkjet print
Courtesy of Skew Gallery, Calgary

EMILY LUCE

The project is a multi-part project comprised of miniaturized but inhabitable replicas of famous (or infamous) art houses. *The Cardiff-Miller House*, a small-scale re-creation of the house in Lethbridge owned by artists Janet Cardiff and George Bures Miller, is mounted on a trailer enabling it to be moved around, easily parked, and open to the public. In the future, other art sites in Canada will be re-created in the same miniaturized but inhabitable format, including the Agnes Martin Family Homestead in Macklin, Saskatchewan, the Emily Carr House in Victoria, BC, and the Maud Lewis House in Digby + Halifax, Nova Scotia.

Concept : Domesticity in the Art World, Sustainable Architecture (and Living,) Community/Collaborative Aspects, and the Pleasure of Tiny.

I live in a suite in a house in Lethbridge owned by the artists Janet Cardiff and George Bures Miller. At one time, they lived here, as have a slew of other Canadian artists who invited people in, made their work, and were a part of the arts community over the years. People brought their stuff, and sometimes left it in the basement. A painting belonging to my partner, Rodney Sayers, and abandoned in his studio in Halifax, was left in a closet here by artists we knew who lived here two tenants before us. To our surprise and horror we found it the day we moved into the middle suite six years later. Throughout the house there is a collection of 'artist repairs,' slightly left of centre solutions to problems. Because of this, the house has started to establish somewhat of a mythic quality for itself; locals are curious about who is living in the suites; everyone has an opinion about how it should be repaired. It's a place that's right on the brink of a certain history—someday it could become a significant art site. Then again, it could be sold and renovated by someone who wants to take advantage of its large footprint and strategic location.

I am interested in this tipping point in the building's history. In my project, I explore the qualities of this building in the present moment, and its domestic presence in Canadian art history. The work is an 8.5' x 16' miniature portrait of this house, on a trailer, so that the specific history of this space up to now can be recorded, engaged, and presented to the public.

Emily Luce, with collaborators Rodney Sayers and Alan Maten
The Cardiff/Miller House, 2012
Construction materials

Elevation: Miniature Replica and *Cardiff-Miller House*, 2012
Vinyl diagram

Elevation: Floorplans, 2012
Vinyl diagram

House Numbers Array and Details, 2012
Vinyl diagram
Courtesy of the Artist

Illustrated: Artist's Rendering

ROBYN MOODY

What I thought about while building *Wave Interference*.

The elecromagnetic spectum consists of waves of electromagnetic radiation, from low frequency radio waves to high frequency gamma rays. A sliver within this spectrum is visible light, and it is only its frequency that is different from the rest. We are typically oblivious to these waves until we have some detector (such as our eyes) to alert us to their presence. That we have the ability, through the use of these organic detectors on our faces, to effortlesly filter a portion of this mess of frequencies into a coherent image is a fact worthy of awe. Outside of this range (but keeping in mind they are exactly the same sorts of waves), other cleverly designed detectors are needed. For a long time the only detector that appeared in everyday life was the Radio, and by tuning the frequency to a particular wavelength, we became aware of that wave's existence.

Though it seems obvious when I say it, it is intriguing to think that the normally invisible waves - those that require clever devices other than our senses to detect - do not come into existence at the moment we tune into them. The waves from every radio station, mobile phone call, wifi signal (and countless others) are around us whether we are detecting them or not (just as waves of visible light illuminate the world beyond the tiny detectors we have on our faces). So it is worth noting that: (1) We cannot easily escape these waves, and (2) Point 1 doesn't particularly matter.

As an increasing number of devices allow us to detect and make use of these frequencies, a new fear of waves has arisen. A modern hysteria, fueled by conspiritorial websites and hearsay, has arisen around the waves emitted by wifi, mobile phones and power lines. Though no credible evidence exists to support these fears, and much exists to discredit them, believers quickly adopt conspiratorial ideas as support, involving scores of corrupt scientist henchmen and impossibly elaborate shady government coverups.

Believing these fantasies may seem like harmless fun. Who wouldn't want to live the excitement of James Bond or the X-files? But these are not documentaries, and their perceived crossover into reality becomes especially perplexing and disheartening in the form of popular denial of impartial scientific evidence, and in the unquestioning and fervent adoption of baseless rumour. This is a behaviour that applies not only to innocuous waves, but also to medical science, public health initiatives, important environmental issues, and even fear of our food. It appears that, like the witch trials, an accusation is enough to convict, and no amount of rational argument will convince otherwise. Chasing ghosts is far too exciting.

******

Any image can be made to elicit any emotional response based on the soundtrack that accompanies it. This is a fact that filmmakers have exploited for years; a soundtrack can make an image cheerful, or nostalgic, or (in a minor key) induce feelings of melancholy or dread. *Wave Interference* combines a beautiful and elegant vision of a wave of light with a gradually changing melancholic soundtrack; just as deniers of science manage to eke out a thriving existence by imposing feelings of fear and dread upon the beauty, and the awe inspiring nature, of nature.

Robyn Moody
Wave Interference, 2012
Fluorescent lamps, PVC, electronics, wood, organ
Courtesy of the Artist

Illustrated: Installation views
Photos: J. Guzzo Desforges

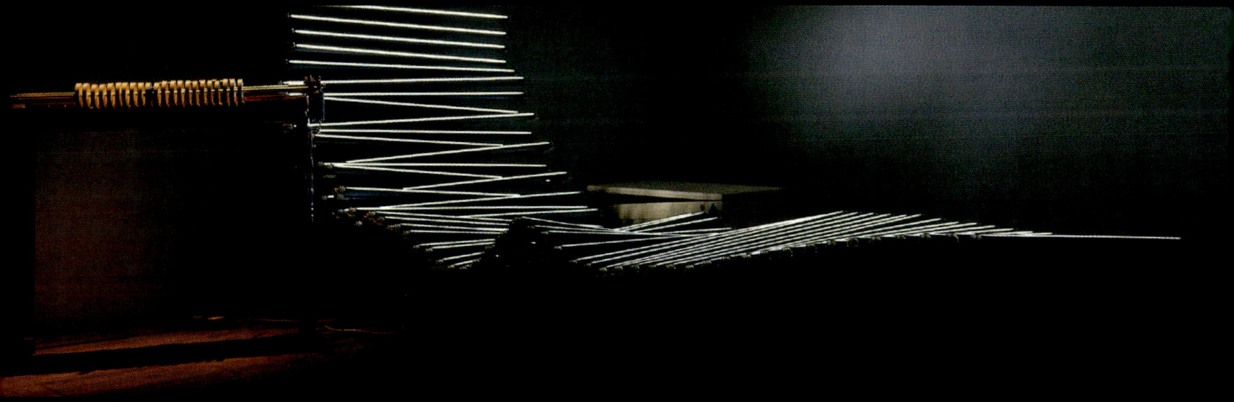

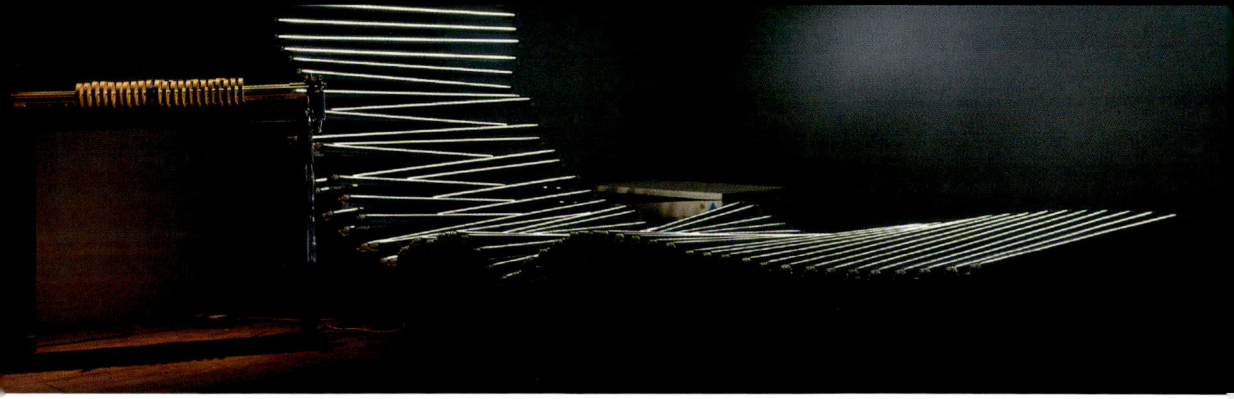

ERIC MOSCHOPEDIS & MIA RUSHTON

Hunter, Gatherer, Purveyor—a project in art and social engagement—brings together elements of craft, performance and civic participation to create edible art objects made from local vegetation and civic water sources—hot tea! Composed of three primary public actions, *Hunter, Gatherer, Purveyor* includes: a) the finding and collecting of fruits, grasses, flowers, bark, roots and other edible plant life that grow in specific neighbourhoods in a city; b) the processing of these ingredients into community-specific teas and c) the distribution of free naturally flavoured teas to participants/ gallery-goers for consumption during the *Biennial* opening. Throughout the highly public and performative process Moschopedis and Rushton will be dressed as hybrid park warden/ vendors and will act as naturalist and huckster, hunter and gatherer.

Through a series of performative gestures *Hunter, Gatherer, Purveyor* responds to the social and geographical context of a city and the distribution of the city's indigenous and planted flora. By creating a performance for a highly specific and intimate site—the participant's mouth—this project seeks to reintroduce a discussion about economics, ecology, and place by tasting class difference and geography. What better way to reimagine the City of Edmonton than to warm up on a cold winter night by drinking and tasting a hot tea made from flowers found in an affluent community, vegetables located in a derelict garden, or berries sourced along the North Saskatchewan river?

A playful, but critically aimed project, *Hunter, Gatherer, Purveyor* considers how discussions of class, ecology and spatial relations can become of "perennial" concern after thirty years of neoliberal policies that have sought to "uproot" and detoxify the body politic of critical discourse. Our answer to how best create a healthy civic body: eat the city and it's surroundings! With *Hunter, Gatherer, Purveyor* we question capitalist systems of exchange and capitalist social relationships—how are these determined by local vegetation; and how are class differences revealed by the types of flora found in different neighbourhoods within a city? With *Hunter, Gatherer, Purveyor* we are seeking to introduce a process of "digesting" capitalism and the city and thereby retoxifying the body politic as having a proclivity towards social and spatial justice.

It is our intention to provide a temporary (it's going to be eaten after all!) reminder of the city and neighborhoods as a source of sustenance, but also a class dividing entity. We will ask how collecting flora in neighbourhoods and communities can reveal how different groups of people have been distributed. Do well-groomed gardens and cosmetic plants represent a leisure class, as opposed to a working class district that might only have weeds and apple trees? By sharing the natural resources we have collected, we are not only rehearsing a utopic future where food is sourced locally and spatial relations socially just, but we are also looking to mimic the non-hierarchical and generous governance structures inherent in traditional hunter-gatherer societies. In this regard, the found flora—when transformed into tasty hot teas—becomes not only a present day geographical and social representation of Edmonton, but signifies an alternative economy.

Eric Moschopedis and Mia Rushton
Hunter, Gatherer, Purveyor, 2013
Found vegetation, mixed-media cart, bicycle, drawing, performance and social engagement
Courtesy of the Artists

Illustrated: Work in progress
Photo (top): Lori Rushton

PAMELA NORRISH

After death, I wish to be buried wearing a beaded t-shirt and pair of jeans for transcendence to my afterlife.

Throughout history, beads have been connected to religion. Whether pagan or orthodox, "[b]eads have frequently been enlisted as symbolic repositories of sacred knowledge, been deemed to have curative powers, served as fee for passage to the afterlife, and been used as prompters to insure the proper conduct of ritual and prayer."[1] Even though I am not particularly religious, I bead representations of quotidian objects to suggest an alternate universe, an extraordinary beaded universe, where the everyday can transcend mediocrity. The ancient Egyptians "believed that to insure comfort in the afterlife, the dead must be surrounded with the paraphernalia of daily life. Quantities of beads were buried with their owner ... Occasionally, delicately beaded funeral pieces were also included."[2]

I can only hope that there is an eternal party waiting for me at the end of my life to which I will wear this beaded outfit, but even if I am left alone in the dark I will probably need the rest.

Notes:

1. **Dubin**, Lois Sherr. *The History of Beads: from 30,000 B.C. to the Present*. New York: Harry N. Abrams, Inc., 1987. p. 17.
2. **Dubin**, Lois Sherr. *The History of Beads: from 30,000 B.C. to the Present*. New York: Harry N. Abrams, Inc., 1987. p. 42.

This project was funded by The Canada Council for the Arts through the Alberta Creative Development Initiative.

Special thanks to Mentoring Artists for Women's Art, Winnipeg and Artscape Gibraltar Point, Toronto.

Pamela Norrish
Outfit for the Afterlife, 2012
Glass beads and nylon thread
Courtesy of the Artist

Illustrated: Detail of work in progress
Photo: Ward Bastian

GABRIELLE PARÉ

I have a violence in me that is hot as deathblood. I can kill myself or – I know it now - even kill another. I could kill a woman, or wound a man. I think I could. I gritted to control my hands, but had a flash of bloody stars in my head as I stared that sassy girl down, and a blood longing to fly at her & tear her to bloody beating bits.
–Sylvia Plath, *The Unabridged Journals*

I am captivated by the things we do with no good reason. As rational beings, we have the ability to make choices that go beyond the needs and desires of the body. What I am compelled by, and what has driven this body of work, is the fact that such an inner struggle between reason and desire can storm over from within us and that reason does not always prevail. We identify this as a loss of control, or as a "letting go". Through various media I am exploring the way lived experiences transcend the body. It is in this transcendence that a person is the most open and vulnerable, and the least in control.

In photographing the body in motion within a staged environment, and subsequently layering and drawing onto the image, I am creating an experience that is anchored upon the body, though not contained by it. By denying the body the integrity of its edges I am refusing the figure its self-rule. In doing so, I present the viewer with a figure who has willingly surrendered control. I would like the viewer to feel compelled to question the ownership of their own actions, and the reason or lack of reason behind them.

Gabrielle Paré
The Pulling Body, 2011
The Slow Creep of Repercussion, 2011
On the Precipice, 2011 (illustrated)
Photo-etching and chine collé

Courtesy of the Artist

TARAS POLATAIKO

Taras Polataiko has developed a significant body of painting and photography, concurrent with installations and performances. This output is split between formalism and social concerns, between art and life, and between realities and fictions. A number of these works are connected with the artist's life and his origins. In 1994 he visited Chernobyl, from which he returned contaminated. While present within him, this damage is nonetheless invisible. Since then, his works have been haunted by the indescribable, the undetectable, as well as by the history of art and the ghosts of certain artists of the recent and less recent past.

In his paintings and photographs, he deals with history of the medium. Pointing up the reflections, simulacrums and shams, Polataiko develops a critique of perception that makes the viewers suspicious of the images and of what they are seeing.

In his installations, where the context and environment surrounding cultural institutions are used as a means to create situations in which life and art converge, Polataiko develops projects that favor interactions. These works form a critique of the society of control and give concrete shape to the imperceptible, just as they speak to us about contamination and transformation, about what spreads and seeps in.

Taras Polataiko
Pissing Into the Wind, 2010
From the "Self-Portraits" series
Archival chromagenic print
Courtesy Barbara Edwards Contemporary, Toronto

JEWEL SHAW

The drawn line has become a symbolic representation of existence in my work. Similar to writing in a journal, drawing and re-drawing a spare set of found objects has become a daily ritual in my studio practice. Through tracing and transfer processes, I am documenting memories and experiences that may be real or imagined, literal or metaphoric. The objects drawn, such as animal muzzles, bridles, snouts, tools and old machinery are meant to expose personal feelings of containment, memory and loss while pictorial/compositional decisions reveal internal landscapes where ideas of identity and place are examined.

It is my intent to create works that allow for illusion and an open-ended quality as well as provoke and invite viewers to speculate on implied narratives as well as possible symbolic associations with the objects rendered within the compositions. Animal emissaries such as the horse, dog and rook are silent witnesses to our humanity and our progress as a society. Through drawing, I listen for their stories stitching fiction with my grandmother's stories to create my personal narrative.

These vestiges of memories, objects and time are intertwined and become united within the picture plane and become a personal archive where political and social issues can be played out.

Animals real or unreal derived from an old naturalist text

Fragile, vulnerable, trauma, lost untold stories

Lore: transmission of cultural material, cultural memory

Iconography, pictographic

Experiences passed down to me

Jewel Shaw
On Ground, 2011
Archival digital print

Investigations of the Dog i, 2011
Investigations of the Dog ii, 2011
Investigations of the Dog iii, 2011 (illustrated)
Copper etching, digital print, chin collé

Gossip's Bridle, 2011
Digital and screenprint

Scold's Bridle, 2011
Digital and screenprint

Courtesy of the Artist

tar Sands

LAURA ST. PIERRE

Urban Vernacular is a sculptural/photographic project that imagines a series of cobbled-together architectures located on the fringes of urban space. Plastic bags, deli containers, Styrofoam packaging — the detritus of consumer culture — become construction materials, tenuously assembled and decorated by an unseen hand. Located in urban "junkscapes", such as vacant lots or alleys behind big-box stores, the structures encroach upon existing architectural elements and absorb nearby debris. They are short-lived, the only remaining proof of their existence is a large-format, panoramic photograph, titled with the date of capture. The constructions refer back to older forms of shelter, and evoke vernacular architectures that might have been built from earth, sod or snow. However, unlike igloos or mud huts, which are built with natural materials, these DIY structures are assembled from our most accessible and plentiful material: trash.

While the camera records the constructions and their urban surroundings, it fails to capture the builder at work. Although the scavenger's origins and motivations are enigmatic, "(e)xamining their space while they are out (...) allows the viewer to gather clues. This is a clever character, someone who scrabbles together a shelter from materials at hand, but not without a certain panache, a sense of home décor, an eye for colour, a knack for decoration, and an inclination to preen. Someone playful? poor? criminal? mad?".[1]
Whatever the scavenger's story might be, the structures s/he builds simultaneously create and inhabit a landscape much like our own: a no-man's land located somewhere between the resplendence and the absurdity of consumer culture.

This body of work explores the relationship between sculpture and photography in several ways. First, the scale of the panoramic photographs creates an immersive experience for the viewer that refers back to an embodied experience of installation and architecture. Secondly, in the translation of three-dimensional construction to two-dimensional image, mundane objects and spaces are transformed through lighting and composition, "making visible, things that cease to be of any consequence once the picture has been snapped."[2] Furthermore, the juxtaposition of the structures with the urban landscape creates a semiotic shift whereby the objects used to build the structures cease to be trash or art materials and instead become construction materials, emphasizing the underlying narrative of the scavenger. Finally, the solidity, bulk and permanence of sculpture are abandoned in favor of the ephemeral image, underlining the fragility and mutability of the constructions and the urban landscape. After the scene is photographed, architectural elements could be altered or destroyed, or the structures could continue to grow from a never-ending supply of trash.

Laura St. Pierre
06.21, 2012
Inkjet print on self-adhesive polypropylene
Courtesy of the Artist

Notes:

1. **Katherine Ylitalo**, essay for *Urban Vernacular,* exhibition catalogue published by the Art Gallery of Alberta, 2010.
2. **Sarah Hamill**, "The World is Flat: Photography and the Matter of Sculpture", *Camerawork*, v36 no1, Spring/Summer 2009.

The artist would like to thank the Canada Council for the Arts for their generous support.

LARISSA TIGGELERS

Larissa Tiggelers' paintings explore her fascination with colour relationships, the composition of form, and the 'push and pull' of spatial perspective. The elements of shape, form, colour, and texture are used as tools to manipulate and fracture space. Though her paintings appear measured and systematic, the works are highly experimental. This is evident in the play on perspective and the varying relationships between elements within the work.

Larissa Tiggelers
Untitled a through k, 2011-2012
Acrylic paint on wood panel
Courtesy of the Artist

LAURA VICKERSON

Unusual sites have always been an evocative trigger for my artistic practice. I am challenged both by the physical peculiarities of a space as well as it's history of use. When I work this way I attempt to address both.

Luxe is designed to interact with the flowing neo-baroque architecture that moves from the inside to the outside of the building. During my site visit, I noticed a structure to the right as one walks into the main lobby of the gallery: under the stairs. It was reminiscent of a streamlined airplane wing with its hard, shiny metal surface. On closer inspection, I realized that this form continued through the window providing an exterior presence. It is this part of the architecture that I have upholstered with a Victorian-inspired damask fabric.

The red covering softens and feminizes the structure contrasting the historical with the contemporary. The choice to make a piece, outside of the formal exhibition space attempts to address the gallery itself: offering alternative possibilities for siting work and perhaps even suggesting the building as a collaborator.

I would like to thank Aran McCormick and Kasia Koralewska for their assistance and expertise with this project.

Laura Vickerson
Luxe, 2013
Jacquard upholstery fabric
Courtesy of the Artist

Illustrated: Concept sketch and fabric sample

JENNIFER WANNER

The essentialist position of nature as "truth" and art's "duty" to closely observe and translate it as such, is no longer regarded as viable for contemporary artists. Instead of simply re-presenting a Romantic failure, my botanical watercolour paintings and video works serve as critiques of our continued human drive for dominance over nature, as well as our complacency towards that dominance. These works act as propositions for both nature (or "wildlife"), cinema and botanical art in a time of virtual, high-speed systems, influenced by Jean Baudrillard's "simulacrum" (where the ability to distinguish between "the model and the real" has been lost among the mediations of cultural constructs), and Paul Virilio's theory of "substitution" (in which "virtuality destroys reality").

Floral still-life painting has a history of working in a manner reminiscent of collage. Seventeenth century floral still-life painters would create idealised bouquets depicting flowers that would often bloom at different times of the year. These artists would generate their composite subjects from a variety of sources (sketches from life, books of botanical engravings and how-to-books). Both my watercolours and stop-motion animations are constructed from the Internet's image database of the most common genetically modified plant crops (corn, canola, soy, cotton, tomato, tobacco), which are downloaded, printed and carefully cut out. The videos as well as the preliminary collages for the watercolours are essentially "sliced" images that are "transferred" onto other images to create a new "modified" whole.

My art practice draws not only on art-historical conventions, but also on various scientific objective means of observing and representing the natural world – botany, genetics and the camera. By painting these new modified-plant forms in meticulous detail, I am mimicking the nineteenth-century Romanticist method of painting based on careful, minute observations to allow for "penetration into the very essence, the life force, of natural processes."[1] Yet the subject with which I am attempting to engage has no "essence," "life force" or "natural process" to be revealed. There is no "empirical reality" to attend to in order to achieve "artistic truth".

Both my watercolours and video works attempt to operate between two Romantic realms: fascination with mastery over natural processes and unease with what our technology might unleash. Ryan Bishop states, "As science became more enamored with not only finding the secrets of nature but of mastering them and turning the natural world to the will of human society, the concept of animation, converting dead tissue into a living being again, captured the imagination of artists and scientists in Europe in the late 18th and early 19th centuries."[2] This infatuation with the "simulation of life" still persists in contemporary technology-driven society, coupled with the "unconscious anxiety...[that] the very technologies that deliver humans this divine-like power might not always deliver what they promise."[3]

Notes:

1. **Cosgrove, Denis E.** *Social Formation and Symbolic Landscape*. Madison, Wisconsin: The University of Wisconsin Press, 1984.
2. **Bishop, Ryan.** "Animation/Re-animation." *Theory, Culture & Society* ns 2-3. 23 (March – May 2006): 346. Special Issue on Problematizing Knowledge. Goldsmiths College Library. April 18, 2008 <http//tcs.sagepub.com>.
3. Ibid

Jennifer Wanner
Herbacentrice, 2010-2012
Stop-motion animation
Running time: 6'2"
Courtesy of the Artist

Illustrated: Production still

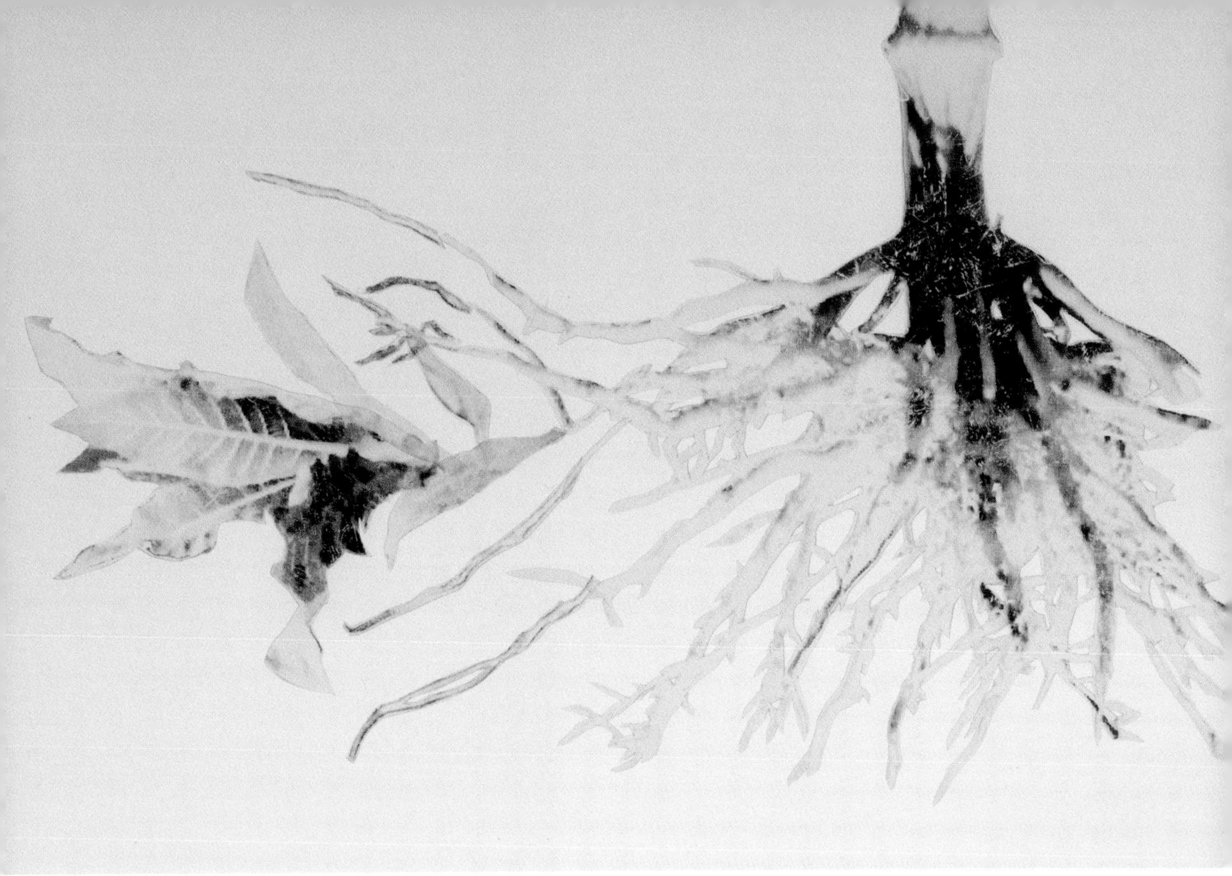

DONNA WHITE

My research interests have turned from looking at the past and the effects of patriarchy on women to the current moral attitudes in which North American women have entrenched in their own lives. I monitor current news media for articles covering the social, economic, legal, cultural and political status of all women. I am specifically interested in the actions of First World women on the rest of the world through behaviors such as consumption, political apathy and humanitarian disinterest.

Cultural critic and feminist philosopher Donna Haraway suggests that the potential for our future transformation as women is into a cyborg identity, a genderless identity existing in a social utopian future. Contrary to her theory, I define the cyborg as a contemporary complex woman who exists in a decidedly non-utopian reality. What Haraway and I agree on is the potential for the cyborg to shape her identity, and thus, her existence in the world.

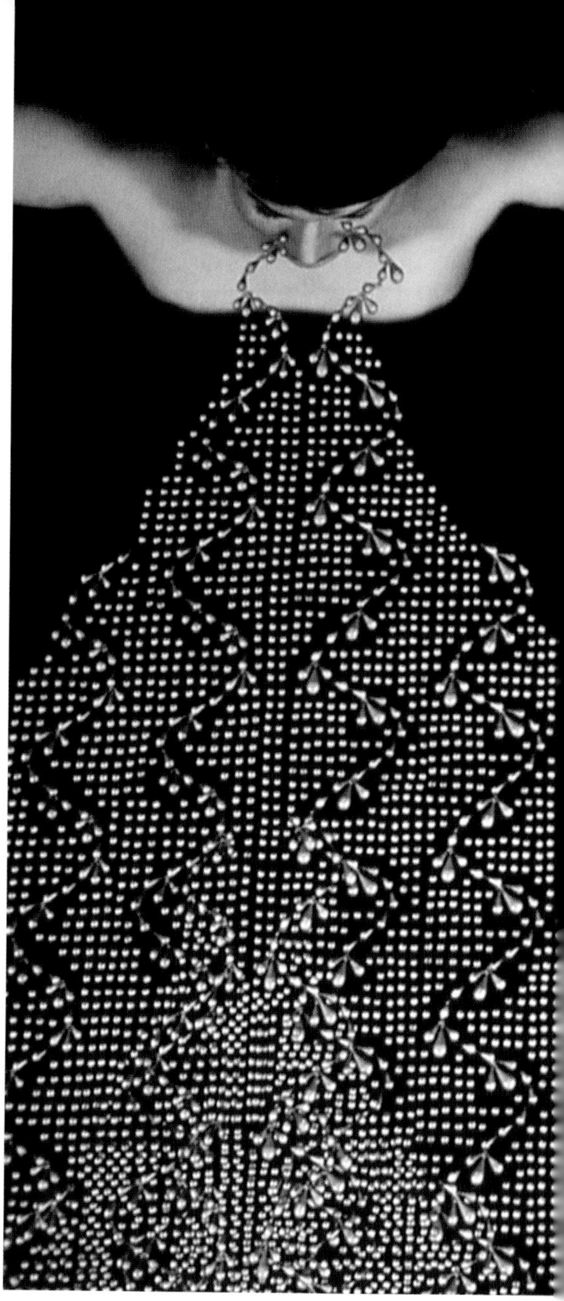

Donna White
Despair Wallpaper, 2012
Ink-jet print on adhesive-backed PVC film
Courtesy of the Artist

Illustrated: Detail of wallpaper

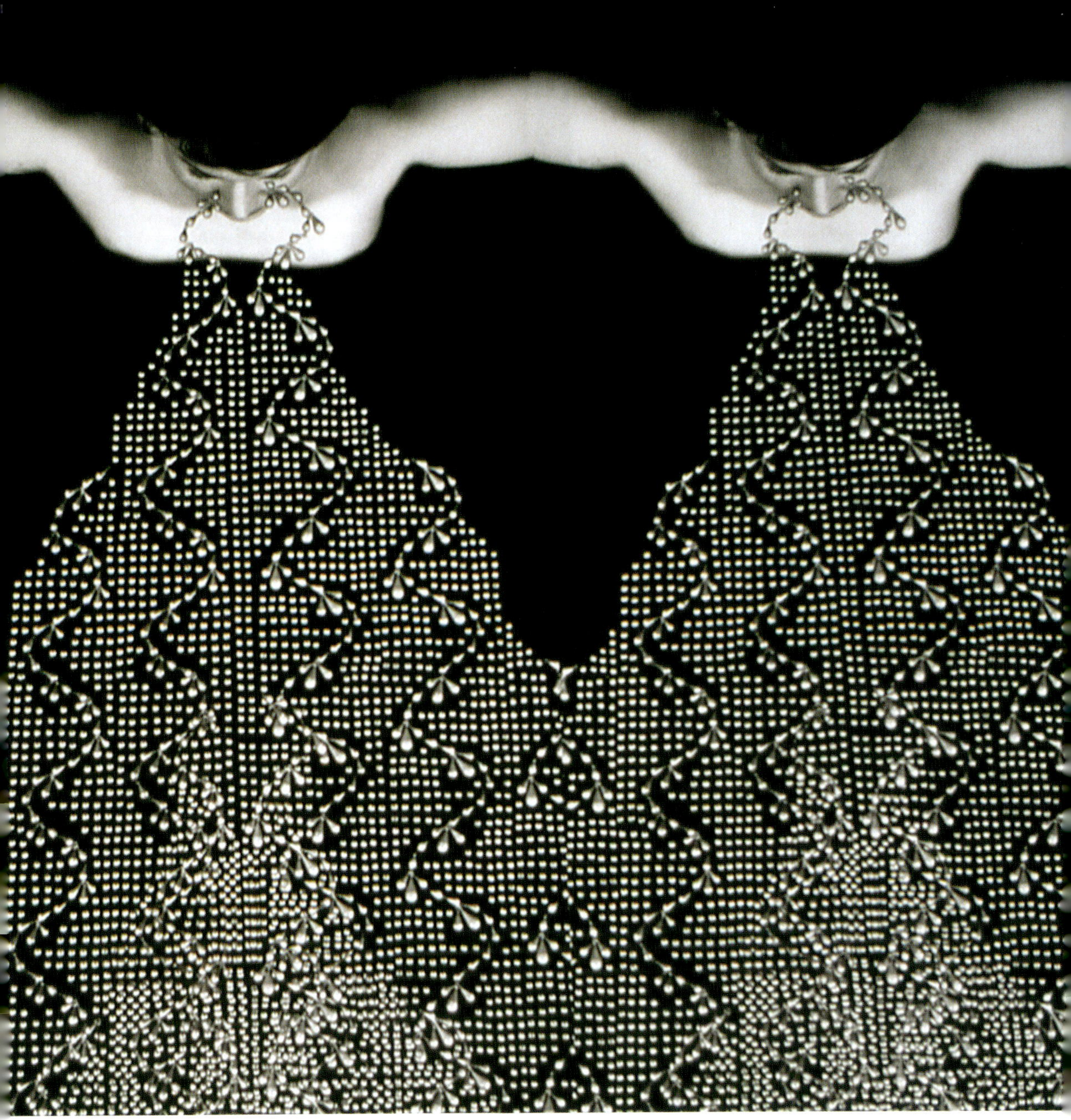

MARIA WHITEMAN

Visitors to our national parks arrive burdened with a desire that is unnerving, despite (or perhaps, because of) how common it has become. This is the desire to encounter the bodies of wild animals. For most, it is enough to catch a glimpse of an animal moving through the forest or escaping into the underbrush. For others, the pull of animal bodies is so powerful that they want contact: closer, closer still, moving forward in hope, fear, anticipation, expectation. All that is missing in the sometimes deadening conveniences of modern life is condensed into the possibility of a touch between two living, unlike species.

And what if such a touch became possible? These images offer evidence of contact between an animal body and human one. They are possible only because the animal bodies have become domesticated in one of the many ways we have developed to fix them in place and open them up to our knowledge systems. The images are of my hand slowly caressing mammals in the storage room of a natural history museum. In such places, even when the bodies of animals are no longer alive, the fact that they are placed in dioramas behind glass and shielded from our hands, makes us want to touch them.

Touching explores our desire for contact, investigating the intimacy of animal bodies. It does so in a way that mourns their death; it mourns, too, that gap in our present sense of ourselves that creates a need for a touch of the wild that is fast disappearing.

Maria Whiteman
Touching: Grizzly Bear #1, 2011 (illustrated top)
Touching: Grizzly Bear #2, 2011
Touching: Grizzly Bear #3, 2011 (illustrated bottom)
Touching: Grizzly Bear #4, 2011
From the *Touching: Grizzly Bear Series*
Video Stills on UPremium OHP Pictorico
Transparency Film

Courtesy of the Artist

Trevor Anderson

Trevor Anderson is a self-taught, independent filmmaker based in Edmonton. His work has been screened at many major international film festivals, including Sundance, Berlin and Toronto. His newest film, *The Man That Got Away* (2012), premiered at the Berlin International Film Festival (Berlinale) where it won the D.A.A.D. Short Film Prize. His previous films include *The High Level Bridge* (2010), which was included in the Sundance Institute Art House Project, where it was favourably reviewed by Roger Ebert, and which won a "Special Mention — Live Action Short Film" at the American Film Institute's AFI Fest in Los Angeles; *The Island* (2009), which won Best Short Film at the Pink Apple Festival in Zürich, Switzerland; and *Rock Pockets* (2007) which won the inaugural Lindalee Tracey Award at Hot Docs in Toronto, presented to "an emerging Canadian filmmaker working with passion, humour, a strong sense of social justice and a personal point of view".

In January 2012, Anderson was selected to study under master filmmaker Werner Herzog at his Rogue Film School. Anderson is also a stage director and playwright: he and Bryce Kulak wrote the award-winning children's musical *Nami Namersson: The Viking Who Liked to Name Things*. Anderson directed Edmonton's original, live improvised soap opera *Die-Nasty* for five years, co-directing the first two of those seasons with Stewart Lemoine. Aside from film and theatre, Anderson is co-founder of the Canadian rock 'n' roll band *The Wet Secrets*, whose song *Secret March* was listed in 2010 by Grant Lawrence of CBC Radio 3 as one of the "Best Songs of the Decade".

Kyle Armstrong

As a filmmaker, Kyle Armstrong focuses predominantly on short, non-narrative film. Working with super8 and 16mm, as well as lo-fi video, Armstrong frequently uses techniques that alter film by hand, employing bleaches, dyes, scratches, and paint to both his own footage and found films. His works have been screened at various galleries, during live-cinema performances and in traditional cinemas across Canada. His work is influenced by filmmakers as diverse as Stan Brakhage, Hollis Frampton, Bela Tarr, Alexander Sokurov, Guy Maddin and Andrei Tarkovsky.

Kyle Armstrong has been working with Guy Maddin on the acclaimed director's *Hauntings/ Seances* project as a "deputized" director, 2nd Unit Director and visual manipulator. Armstrong is currently an instructor at MacEwan University in the Design Studies program and the president of the Metro Cinema Society in Edmonton. In 2012, he was a recipient of the Lieutenant Governor of Alberta Emerging Artist Award.

Noel Bégin

Noel Bégin is an interdisciplinary artist and time-based media curator, who works spatially with objects, all manner of projected media, and performance. Bégin has exhibited and/or curated installation, performance and media art with the Art Gallery of Calgary, the Glenbow Museum, The Banff Centre, the Mountain Standard Time Performance Festival, the Art City Festival and EMMEDIA. He received an ACDI grant in 2009, and is an Artis-in-Residence for Calgary 2012. He has served on the boards and committees of several artist-run centres in Calgary.

Elisabeth Belliveau

Elisabeth Belliveau is an interdisciplinary artist and the published author of three graphic novels. Her work has been screened and exhibited internationally including venues in Cuba, Berlin, London, The Czech Republic, Sweden and across North America. She has attended residencies at The Banff Centre, Women's Studio Workshop, New York and the National Film Board of Canada, among others. Her work has been supported by the Andy Warhol Foundation for the Visual Arts, The New York State Council on the Arts, Canada Council for the Arts and Conseil des Arts et des Lettres du Québec.

Belliveau completed a BFA at Alberta College of Art & Design and an MFA at Concordia University in Montréal where she was awarded the William Blair Brucebo Scholarship in 2009.

Richard Brown

Richard Brown's practice consists of painting, drawing and stereo photography. He currently lives in Calgary, where he teaches in the Drawing Department of the Alberta College of Art & Design. He also plays in several bands, one with the artist, Rita McKeough.

Eric Cameron

Eric Cameron was born in 1935 in Leicester, England. He received his BA in painting from King's College, Durham University in Newcastle and his Diploma in Art History from the Courtauld Institute at the University of London. He is currently a professor of art at the University of Calgary. Cameron is known for his "Thick Paintings" as well as his "Process Paintings." "Thick Paintings" consist of objects that Cameron has painted with thousands of layers of gesso, changing everyday things into art objects that he says reveal the "mode of being in the world". "Process Paintings" are minimalist works that he produces by layering vertical and horizontal grids of tape. More recently, he has added the series of "Dipped Paintings," which he first showed in 2011. Cameron is the recipient of numerous prestigious awards, including the 1994 Gershon Iskowitz Prize and the 2004 Governor General's Award in Visual and Media Arts.

Bruno Canadien

Bruno Canadien has been working as an artist since graduating from the Alberta College of Art & Design's Painting Department in 1993. His work, such as the *Freedom Fighter Series* of paintings, brings about an awareness of the modern presence of First Nations, and native resistance, in the Americas. He exhibits steadily in Alberta, and his work is included in the collections of the Alberta Foundation for the Arts, the Glenbow Museum, the Red Deer Museum and the Indian Art Centre in Ottawa.

Bruno Canadien is Dene (of mixed heritage), born and raised in Denendeh. He is a member of the Deh Gah Got'ie Koe First Nation, a Deh Cho Region member of the Dene Nation. Bruno lives and works in the Calgary area, and enjoys spending much of his time outdoors in the traditional territories of the Blackfoot and Nakoda nations.

Sherri Chaba

Sherri Chaba received a Masters degree specializing in Drawing and Intermedia from the University of Alberta. She enjoys her work as a Lead Artist at the Nina Haggarty Centre for the Arts in Edmonton, an art center for people with developmental disabilities. Chaba has received numerous awards and grants for her work, including creation grants from the Canada Council, The Alberta Foundation for the Arts and the Edmonton Artists Trust Fund. Her works have been shown provincially and nationally and are in several permanent collections. She has lived most of her life in rural Alberta and has a strong personal connection to the land and its history, a connection that has been altered by industrial presence on her property. Her experience is reflected in works that merge the industrial with nature and recycled historical objects. She creates multimedia installations that draw on issues of ecological and social sustainability. Her work challenges viewers to imagine new ways of considering the environment and our relationship to it.

Chris Cran

Chris Cran graduated in 1979 from the Alberta College of Art & Design in Calgary. The recipient of numerous Canada Council grants, Cran has exhibited internationally at the 49th Parallel Gallery, New York City (1992) and in the *Fourth International Painting Biennial*, Cuenca, Ecuador (1994). In addition to his professional activities as a painter and teacher (ACAD), Cran has been actively involved with Calgary's internationally recognized One Yellow Rabbit Performance Theatre. His work may be found in numerous private and public collections including the National Gallery of Canada, the Glenbow Museum, the Nickle Arts Museum and The Mackenzie Art Gallery. Chris Cran is represented by TrépanierBaer Gallery in Calgary and Clint Roenisch Gallery in Toronto.

Alysha Creighton

Alysha Creighton was born and raised in Vancouver, BC. After completing her BA in Visual Art (great distinction) at Trinity Western University she moved east, to Edmonton, to pursue further study at the University of Alberta.

She is currently in her final year of her MFA studies in Drawing and Intermedia. Her current work explores the body, language and encounter through drawing, video and performance-based work. She has exhibited her work in galleries and screenings across North America, in Vancouver, Calgary, Ottawa, New York and the Los Angeles area.

DaveandJenn

Jennifer Saleik and David John Foy formed DaveandJenn in 2004. Together in 2006, they went on to earn their BFAs with distinction from the Alberta College of Art & Design.

They were shortlisted for the Royal Bank of Canada's Canadian Painting Competition in 2006 and 2009. In 2010, DaveandJenn were awarded the Lieutenant Governor of Alberta Biennial Emerging Artist Award and in 2011 were included on the long-list for the Sobey Art Award. They have exhibited in public spaces such as the Southern Alberta Art Gallery, the University of Lethbridge Art Gallery, the Glenbow Museum, the Massachusetts Museum of Contemporary Art, as well as Special Projects Spaces at the Toronto Art Fair and *NEXT: The Invitational Exhibition of Emerging Art* in Chicago.

Their work has been mentioned in publications including *Canadian Art*, *Border Crossings* Magazine and Magenta Publishing's *Carte Blanche 2: Painting*. DaveandJenn's work has been collected by the Alberta Foundation for the Arts, The Calgary Civic Art Collection as well as The Royal Bank of Canada. They have also received project grants and support from the Alberta Foundation for the Arts, the Lieutenant Governor of Alberta Arts Awards Foundation and the Canada Council for the Arts. DaveandJenn live and work in Calgary.

Amanda Forbis and Wendy Tilby

Amanda Forbis and Wendy Tilby met at the Emily Carr College of Art & Design in Vancouver, where they studied film and animation. They each went on to create their own works with The National Film Board of Canada before co-directing *When the Day Breaks* (1999), which received over 30 international awards including the Palme D'Or at Cannes and an Academy Award nomination. Wendy and Amanda have continued to collaborate on commercials (their spot for United Airlines, Interview, was nominated for an Emmy) and commissioned projects, including video projections for the Alberta Ballet. *Wild Life*, their latest short, has won several prizes and was nominated for an Academy Award in 2012.

Mackenzie Frère

Mackenzie Frère lives in Calgary, Alberta. His work has been exhibited in Canada, China, Japan, Korea and the United States. He holds a BFA from the Alberta College of Art & Design in Calgary and an MFA from NSCAD University in Halifax, Nova Scotia. He has written for several Canadian publications including *Artichoke* Magazine, *Craft, and Perception and Practice Vol III*. Frère currently teaches at the Alberta College of Art & Design where he has introduced Jacquard weaving. Recent studio-based projects include an artist residency at the Museum of Contemporary Craft in Portland, Oregon and ongoing research of Jacquard weaving at the Montreal Centre for Contemporary Textiles in Montreal, Quebec.

Sarah Fuller

Sarah Fuller was born in Winnipeg and works and lives in Banff. She earned a BFA from Emily Carr University in Vancouver in 2003, after completing her first two years of study at the University of Manitoba, School of Fine Arts. In the summer of 2011, she attended the XVII Advanced Course in Visual Arts *Dream Seminar II* with Susan Hiller at the Fondazione Antonio Ratti in Como, Italy.

Most recently, she has exhibited work at Les Territoires, Montreal, the Art Gallery of Alberta, Edmonton; Latitude 53 Gallery, Edmonton; the Illingworth Kerr Gallery, Calgary; and Three Walls Gallery, Chicago. Her work is held in public and private collections across Canada including the Canada Council Art Bank and the Alberta Foundation for the Arts and she has received support for her projects from the Canada Council for the Arts and the Alberta Foundation for the Arts. Fuller will be presenting work at Truck Gallery, Calgary, and the Odd Gallery, Dawson, YT in 2013.

Jason de Haan and Miruna Dragan

Jason de Haan works through numerous forms to recognize the potential in various systems, materials and sets of conditions. Recent solo exhibitions include the Southern Alberta Art Gallery, Lethbridge (2012); Clint Roenisch Gallery, Toronto (2012); with Miruna Dragan at The Khyber, Halifax (2011) and Museo de la Ciudad, Queretaro (2012). His work has been presented in group exhibitions including Oakville Galleries (2012); with Miruna Dragan at 98 Weeks Research Project at the *Thessaloniki Biennale* (2011); Nuit Blanche, Toronto (2011); MOLAF, Bergen (2011); Art Gallery of Alberta (2010); *OK Quoi?*, Sackville (2010) and Galleri Kling og Bang, Reykjavik (2009).

De Haan recently participated in the Arctic Circle 2012 residency and the residency *The Soiree Retreat: A bit of a Chekhovian Situation* at The Banff Centre, and spent a year living and working in Beirut. Upcoming solo exhibitions include the Kitchener Waterloo Art Gallery and the Art Gallery of Nova Scotia, Halifax. De Haan was recently shortlisted for the Sobey Art Award representing the Prairies and the North. He is currently pursuing his MFA at Bard College, New York and is represented in Canada by Clint Roenisch Gallery.

Miruna Dragan was born in Bucharest, has lived in New York, Los Angeles, Amsterdam, Beirut, the island of Paros, Greece, and is currently based in Calgary. Her work reflects themes of dispersion and transcendence through site-responsive acts. Working within and between the realms of intervention, video, fresco, drawing, and photography, Dragan has exhibited her projects internationally at Museo de la Ciudad in Queretaro, Mexico (2009 & 2012), Khyber ICA in Halifax (2011), 98 Weeks Research Project at the *Thessaloniki Biennale* (2011), Truck in Calgary (2010), and *Hearts of the New West: The 2012 Calgary Biennial*. She has also made art for situations outside of the art-world, attended numerous residencies worldwide, holds an MFA from the Rhode Island School of Design, and teaches in the Drawing program at the Alberta College of Art & Design.

Faye HeavyShield

Faye HeavyShield, a member of the Kainai First Nation, lives and works on the Blood Reserve in southern Alberta. Her work has been exhibited throughout Canada and the U.S.A. and is held in public and private collections throughout North America.

Terrance Houle

Terrance Houle is an internationally recognized, interdisciplinary media artist and a member of the Blood Tribe. Involved with Aboriginal communities all his life, he has traveled to reservations throughout North America participating in Powwow dancing along with native ceremonies. Houle utilizes performance, photography, video/film, music and painting in a practice that includes tools of mass dissemination such as billboards and vinyl bus signage.

A graduate of the Alberta College of Art & Design, Houle received his BFA in 2003. His groundbreaking art quickly garnered significant accolades and opportunities, including a 2003 invitation to participate in a Thematic Residency at The Banff Centre, which focused on 34 international indigenous people exploring issues of colonization and communion.

Houle received the 2006 Enbridge Emerging Artist Award presented at the Mayor's Luncheon for the Arts, City of Calgary. In 2008 Houle was a semi-finalist for the prestigious Sobey Art Award. His work has been exhibited across Canada, parts of the United States, Australia, Europe and England. His first major solo exhibition *GIVN'R*, presented his works in film, video, performance, installation, mixed media, and photography, made between 2003-2009. Terrance Houle lives and maintains his art practice in Calgary and is a founding member of the Indigeneity Artist Collective Society.

Gary James Joynes

Gary James Joynes, also known by his performing moniker "Clinker"(www.clinkersound.com), has been active in the electronic and experimental music performance community for many years. He has more recently embarked on a broader artistic exploration using his analog synthesizers and custom-made machines to create visual sound as photograph/drawing/sculpture hybrids along with immersive and unending installation environments for reflection and deep listening experiences.

His recent installation work *Ouroboros* had its world premiere at the Koffler Centre of the Arts international exhibition *Spin Off: Contemporary Art Circling the Mandala* in September 2011. His sound and visual installation *Frequency Painting: 12 Tones* premiered at the Latitude 53 Art Gallery in January 2011. *Soundbursting No. 1*, the most recent chapter in Joynes' visual catalog was shown at Edmonton's Harcourt House main gallery space from March 1 to April 14, 2012. Clinker's most recent Live Cinema work, *On the Other Side...*, was commissioned by the 2008 International Leonard Cohen Festival and has toured internationally.

Kristopher Karklin

Kristopher Karklin graduated from the Alberta College of Art & Design in 2007 with a BFA with distinction. His work has been exhibited at the University of Arts in Kyoto, Japan in 2009, as well as exhibitions in Alberta and Ontario. He is the recipient of several awards and scholarships including the ACAD Board of Governors Award in 2007, the Jason Lang Scholarship for two consecutive years, and the National In-Course Millennium Scholarship in 2005.

His work can be found in the collection of the Alberta Foundation for the Arts as well as private collections. In 2011 Karklin's work caught the attention of Sir Elton John, a recognized collector of important contemporary photography. The legendary musician has since become a significant collector of Karklin's art. Kristopher Karklin is represented in Canada by Skew Gallery.

Emily Luce

Emily Luce is an artist and graphic designer based in Lethbridge, Alberta. Her artwork and research address concerns of mass transportation, consumption, crowds and social action. She has presented her quirky, multidisciplinary, research-based visual work in the form of lectures, exhibitions, installations and gatherings in North America and Europe.

In addition to her studio and teaching practices, her work as President of the Board of Directors of DesignInquiry furthers her explorations into the complex nature of visual communication and the opportunities presented for building more meaningful relationships with our environment and with each other.

Robin Moody

Robyn Moody currently lives and works in Calgary, Alberta, where he is a sessional instructor in the department of Media Arts and Digital Technology at the Alberta College of Art & Design. Moody received a BFA from the University of Lethbridge in 2000, and an MFA from NSCAD University in 2006. He takes a whimsical and multifaceted approach to artmaking, incorporating electronics, mechanics, film, performance, installation, sculpture, or whatever a project requires.

Often humorous, often strikingly beautiful, and often hiding a dark secret, Moody's work explores (in whole or in part) the complex relationships between technological advances, human belief and interpretations of the world, and humanity's relationship with science, politics and nature.

In the past decade, Moody has shown his work regularly across Canada and in Europe. Notable examples include Kling and Bang (Reykjavik), La Fabbrica del Vapore (Milan), Kitchener Waterloo Art Gallery (Kitchener), PM galerie (Berlin), Lydgalleriet (Bergen), the AND festival (Liverpool) and Articule (Montréal). Upcoming exhibitions include the Whitewater Gallery (North Bay, ON), the Mendel Art Gallery (Saskatoon), and Eyelevel Gallery (Halifax). He has been twice nominated for the Sobey Art Award, in 2010 and 2012.

Eric Moschopedis and Mia Rushton

Eric Moschopedis is an award winning interdisciplinary performer, facilitator, educator, and curator. In 2008 Moschopedis completed his Master of Fine Arts in Interdisciplinary Studies at University of British Columbia, Okanagan. By combining a child-like curiosity with the scrutiny of an ethnographer, Moschopedis creates community-specific, relational and participatory works that invite audiences to become active collaborators in the creation of community. In addition to participatory works, Moschopedis maintains a performance practice that oscillates between staged performance, performance for video, installation, performative work, intervention, and walking, finding, and collecting.

Mia Rushton is a mixed media artist, interested in the intersection of craft and performance. Her practice includes elements of screen-printing, sewing, knitting, social practice and intervention art. A graduate of Alberta College of Art & Design, Rushton has shown her installations and participatory works in artist-run centres and festivals in Canada, the United States and Switzerland.

Pamela Norrish

Pamela Norrish studied at the Alberta College of Art & Design, Calgary, including one year abroad at the New York Studio Program at Parsons, The New School, New York, and the Ontario College of Art and Design, Toronto. Norrish has attended artist residencies at The Banff Centre, the Ledge Gallery, Calgary; Mentoring Artist's for Women's Art (MAWA), Winnipeg, and Artscape Gibraltar Point, Toronto.

Gabrielle Paré

Gabrielle Paré completed her Bachelor of Fine Arts at the University of Alberta in 2012. She has participated extensively in exhibitions and festivals all over Edmonton. Her work has also appeared internationally at the Wimbledon College of Art in London, England.

Paré creates artwork that focuses on the actions we take, and the ownership we claim to have over them. She explores these themes through painting, printmaking and performance.

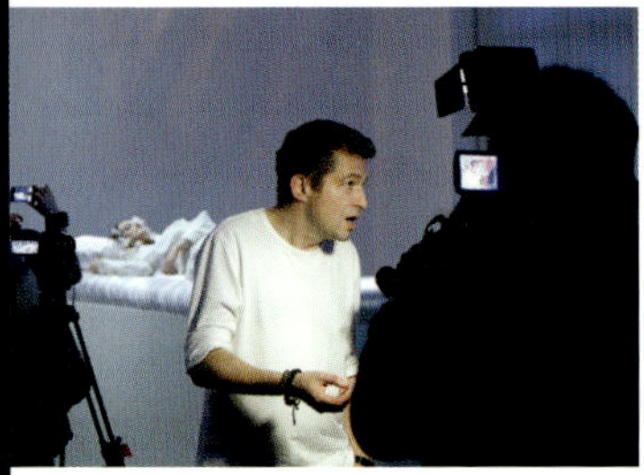

Taras Polataiko

Taras Polataiko was born in Chernivtsi, Ukraine. He received his BFA at the Moscow State Stroganov University of Fine and Industrial Arts, and his MFA at the University of Saskatchewan.

His work has been exhibited at the 25th *Sao Paulo Biennale of Contemporary Art*, Brazil; *Incheon Biennial*, South Korea; *5th Biennale of Video Art*, Tel Aviv, Israel; *Volta 5*, Basel, Switzerland; Antoni Tapies Foundation, Barcelona, Spain; Muse d'Art Contemporain de Montreal; The Power Plant, Toronto, and across Canada. Most recently, his *Sleeping Beauty Project* at the National Art Museum of Ukraine, in Kiev, Ukraine, attracted attention worldwide.

Jewel Shaw

Jewel Shaw earned a BFA, and a MFA in Printmaking from the University of Alberta, Edmonton. She was a participant artist in the thematic residency *Archive Restored*, and the Aboriginal Studio Work Study program in Visual Arts, The Banff Centre. Her work combines traditional printmaking, drawing and digital media, that considers themes of identity, memory and loss. Her work has been exhibited in Canada, the United States, and Japan. She lives and works in Red Deer.

Laura St. Pierre

Laura St. Pierre is a multi-disciplinary artist and educator. Her installations, sculptures, public works and photography have been exhibited throughout Canada, garnering critical attention in national and international publications. St.Pierre earned undergraduate degrees at the University of British Columbia and the University of Alberta, and completed a MFA degree at Concordia University in Montreal. She has been the recipient of numerous fellowships and scholarships as well as major arts grants from the Conseil des Arts et Lettres du Quebec, the Alberta Foundation for the Arts, and the Canada Council. St.Pierre's work is included in the permanent collections of the Art Gallery of Alberta and the Alberta Foundation for the Arts. She currently lives, works and teaches in Grande Prairie, Alberta.

Larissa Tiggelers

Calgary-based multidisciplinary artist Larissa Tiggelers received a BFA from the Alberta College of Art & Design in 2010. She maintains an active studio practice and contributes to the artist-run community in Calgary.

Laura Vickerson

Laura Vickerson is a mixed-media installation artist and educator who lives and works in Calgary. She currently teaches at the Alberta College of Art & Design. She has exhibited extensively in Canada as well as in the United States, Britain, Turkey, Poland and China.

Vickerson has produced site-specific installations for various international exhibitions and venues including the *Istanbul Biennial* (1997), *le Manifestation Internationale d'Art de Québec* (2000), *Biennial Internationale Du Lin de Portneuf*, Québec (2007) as well as *Fairytales and Factories*, a site-specific project produced by Locus+ for a decommissioned textiles mill in the Yorkshire Dales, Britain (1999). She has participated in artist residencies at The Banff Centre and the Sympo-Fibres International in Saint-Hyacinth, Québec (1999). She has been the recipient of numerous grants and awards through the Alberta Foundation for the Arts, the Canada Council for the Arts, Canada's Department of Foreign Affairs and International Trade and the Arts Council of England.

Jennifer Wanner

Jennifer Wanner is a Calgary-based artist who holds a BFA in Drama from the University of Calgary (1995), a BFA with distinction in Visual Arts from the Alberta College of Art & Design (1999), and an MFA from the University of Western Ontario (2009). In 2008 she was awarded a Social Sciences and Humanities Research Council of Canada Scholarship (SSHRC). Wanner has exhibited her work widely in Alberta and throughout Canada. Recent solo exhibitions include *Immuto* at the Dunlop Art Gallery in Regina and *Florilegium: Jennifer Wanner* at the Glenbow Museum in Calgary.

Wanner's work examines botanical art and nature cinema in a world dominated by "second nature" – a virtual simulation of pristine "first nature" wrought by the revolution in information, biotechnologies, and consumer culture.

Donna White

Donna White's most recent large-scale artwork was the 2008 public billboard project, *Prevailing Past*, for Stride Gallery, Calgary. Displayed during the Calgary Stampede, on the parade route, the three rotating images questioned the inherent associations of the white cowboy hat through the lenses of gender and history. Previous to this project, White completed her first billboard project in Saskatoon as part of her 2006 MFA graduation exhibition. This double billboard project marked the beginning of the inclusion of photographic processes into her sculpture.

White received her BFA (with distinction) in Interdisciplinary Studies at the Alberta College of Art & Design in 1995. Between periods of art study, she has worked as a director/curator in both artist-run centres and public art galleries in Alberta.

Maria Whiteman

Maria Whiteman is Assistant Professor of Drawing and Intermedia in Fine Arts at the University of Alberta. She teaches courses in mixed media, photography, video, drawing and installation art. Her current art practice explores themes such as art and science, relationships between industry, community and nature, and the place of animals in our cultural and social imaginary. In addition to her studio work, she conducts research in contemporary art theory and visual culture. She is currently working on *The Retreat* (co-editor; to be published by Autonomedia) and *Refiguring the Animal: Plasticity and Contemporary Art* (co-edited with Amanda Boetzkes; to be published by University of Minnesota Press). In 2011, Whiteman was the recipient of an Interdisciplinary Course Seminar Grant from the Kule Institute for Advanced Study. She is presently a scholar in the Canadian Institute for Research Computing in the Arts. In 2011, she had a solo exhibition at Latitude 53 in Edmonton. Whiteman is a co-director of the 2012 Banff Research in *Culture/ documenta 13* research residency.

LIST OF WORKS

Trevor Anderson
The Man That Got Away, 2012
HD video
Running time: 25'16"
Courtesy of Dirt City Films, Edmonton

Kyle Armstrong
Magnetic Reconnection, 2012
Super-8mm film, 16mm film, DSLR, digital IMAX
Running time: 12'30"
Courtesy of Dropper Pictures, Edmonton

Noel Bégin
A Decombinant Diapositive Verisimilitude Leaning Precarious Against the Verdance, 2011
Chrome diapositives, 22 slide projectors, paper, ambient audio recording.
336 cm x 853 cm x 488 cm
Commissioned by Andrea Mann
Courtesy of the Artist

Elisabeth Belliveau
Go so we may see (lady of gold arms doom), 2012
Animation: plasticine, paper, digital video
Running time: 3'

Entre Chien et Loup, 2010-2012
20 drawings, each 30 cm x 24 cm
Graphite on found paper, MDF and Plexiglass
Courtesy the Artist

Richard Brown
New Jersey, 2010-2012
Oil on canvas
50 paintings, each 30 cm x 30 cm; overall 170 cm x 415 cm
Courtesy the Artist

Eric Cameron
Thanatos II, 2012
A cluster of 225 Dipped Paintings, each approx. 15 cm x 5 cm
Remembrance Day poppies, latex paint, filament
Courtesy of TrépanierBaer Gallery, Calgary

Bruno Canadien
Pipe Dreams, 2011
From the *Freedom Fighter Series*
Acrylic, graphite, cloth, beads, ribbon, hardware, on canvas.
Four panels: each 137 cm x 41 cm x 5 cm
Courtesy the Artist

Sherri Chaba
The Paradox of Plenty, 2012
Steel, wire, animal stretchers, artifacts, mirrors, viewing devices, taxidermy
127 cm x 81.3 cm x 304.8 cm
Courtesy the Artist

Chris Cran
king! 2012
Oil and acrylic on polyurethane plastic structures
127 cm x 304.8 cm x 6.35 cm overall; three components, large 127 cm x 96.5 cm x 6.35 cm, medium 101.6 cm x 77.5 cm x 6.35 cm, and small 76.2 cm x 58.4 cm x 6.35 cm
Courtesy of TrépanierBaer Gallery, Calgary

Alysha Creighton
Ascension, 2011
Digital Video
Running time: 2' 24"
Dimensions variable
Courtesy the Artist

DaveandJenn
-TheBindingLine-, 2012
Resin, acrylic, oil, bronze and mixed media
165 cm x 145 cm x 60 cm
Courtesy of TrépanierBaer Gallery, Calgary

Amanda Forbis and Wendy Tilby
Wild Life, 2011
Animation, gouache on paper
Running time: 13'30"
Dimensions variable
Courtesy of The National Film Board of Canada

Mackenzie Frère
Noise 120429, 2012
Silk, linen, hemp, indigo
201 cm x 78 cm
Courtesy the Artist

Sarah Fuller
Experiment in Landscape, No.1, 2012
HD video
Running time: 8'9"
Dimensions variable
Courtesy the Artist

Jason de Haan and Miruna Dragan
The Wood and Wave Each Other Know, 2011
HD video installation
Running time: 38' 9"
Dimensions variable
Courtesy the Artists and Clint Roenisch Gallery

Faye HeavyShield
Currency, 2012
Digital print on paper, acrylic compound, ink, ceramic platter.
Each circular unit: 5 cm (2") diameter, .25 cm thick.
Number of units variable, minimum of 300.
Mounted on plinth 100 cm x 70 cm square.
Courtesy the Artist

Terrance Houle
Film Still #1 for *isstahpikssi (eestah-peek-see) — Ghost*, 2012
Inkjet print
75 cm x 100 cm (30" x 40")

Film Still #2 for *isstahpikssi (eestah-peek-see) — Ghost*, 2012
Inkjet print
75 cm x 100 cm (30" x 40")

Trailer for *isstahpikssi (eestah-peek-see) — Ghost*, 2012
16mm film transferred to HD video
Running time: 2" (approx.)

isstahpikssi (eestah-peek-see) — Ghost, 2013
16mm film transferred to HD video
Approx running time: 10'

Courtesy of the Artist and the Calgary Society of Independent Filmmakers.
$100 Film Festival Commission Project

Gary James Joynes
Peregrination, 2013
Live cinema: visual/sound performance with video projection, audio and layered vocals
Running Time: 30 – 35' (variable)
Courtesy of dc3 Art Projects

Kristopher Karklin
Jack & Jill Room (morning), 2011
From the series *Camp Life*
95 cm x 140 cm (38"x 56" sheet, 42"x 59.25" framed)
Inkjet print
Courtesy of Skew Gallery, Calgary

Jack & Jill Room (evening), 2011
From the series *Camp Life*
95 cm x 140 cm (38"x 56" sheet, 42"x 59.25" framed)
Inkjet print
Collection of the Alberta Foundation for the Arts

Ping Pong Room, 2011
From the series *Camp Life*
95 cm x 140 cm (38"x 56" sheet, 42"x 59.25" framed)
Inkjet print
Courtesy of Skew Gallery, Calgary

Emily Luce, with collaborators Rodney Sayers and Alan Maten
The Cardiff/Miller House, 2012
Construction materials
13' 2.5" x 20' x 8'6"

Elevation: Miniature Replica and *Cardiff-Miller House*, 2012
Vinyl diagram
Dimensions variable

Elevation: Floorplans, 2012
Vinyl diagram
Dimensions variable

House Numbers Array and Details, 2012
Vinyl diagram
Dimensions variable
Courtesy the Artist

Robyn Moody
Wave Interference, 2012
Fluorescent lamps, PVC, electronics, wood, organ
272 cm x 300 cm x 376 cm
Courtesy the artist

Eric Moschopedis and Mia Rushton
Hunter, Gatherer, Purveyor, 2013
Found vegetation, mixed-media cart, bicycle, drawing, performance and social engagement
Courtesy the Artist

Pamela Norrish
Outfit for the Afterlife, 2012
Glass beads and nylon thread
180 cm x 72 cm x 5 cm (71" x 28.5" x 2")
Courtesy the Artist

Gabrielle Paré
The Pulling Body, 2011
Photo-etching and chine collé
81.3 cm x 59.7 cm

The Slow Creep of Repercussion, 2011
Photo-etching and chine collé
81.3 cm x 59.7 cm

On the Precipice, 2011
Photo-etching and chine collé
81.3 cm x 59.7 cm
Courtesy the Artist

Taras Polataiko
Pissing Into the Wind, 2010
From the *Self-Portraits* series
Archival chromagenic print
180 cm x 115 cm
Courtesy Barbara Edwards Contemporary, Toronto

Jewel Shaw
On Ground, (2011)
Archival digital print
28.75 cm x 27.5 cm

Investigations of the Dog i, 2011
Copper etching, digital printing, chin collé
60 cm x 45 cm (sheet)

Investigations of the Dog ii, 2011
Copper etching, digital printing, chin collé
60 cm x 45 cm (sheet)

Investigations of the Dog iii, 2011
Copper etching, digital print, chin collé
60 cm x 45 cm (sheet)

Gossip's Bridle, 2011
Digital and screenprint
42.5 cm x 27.5 cm (sheet)

Scold's Bridle, 2011
Digital and screenprint
42.5 cm x 27.5 cm (sheet)
Courtesy the Artist

Laura St. Pierre
06.21, 2012
Inkjet print on self-adhesive vinyl
240 cm x 700 cm
Courtesy the Artist

Larissa Tiggelers
Untitled a through h, 2011-2012
Acrylic paint on wood panel
Each panel is 19 cm x 19 cm
Courtesy the Artist

Laura Vickerson
Luxe, 2013
Jacquard upholstery fabric
Exterior Structure:
2.9 m x 11 m x 1.85 m
Interior Structure:
2.9 m x 6.76 m x 1.85 m
Courtesy the Artist

Jennifer Wanner
Herbacentrice, 2010-2012
Stop-motion animation
Running time: 6'2"
Courtesy the Artist

Donna White
Despair Wallpaper, 2012
Ink-jet print on adhesive-backed PVC film
304.8 cm x 533.4 cm
Courtesy the Artist

Maria Whiteman
Touching: Grizzly Bear #1, 2011
From the *Touching: Grizzly Bear Series*
Video Stills on UPremium OHP Pictorico
Transparency Film
89 cm x 117 cm

Touching: Grizzly Bear #2, 2011
From the *Touching: Grizzly Bear Series*
Video Stills on UPremium OHP Pictorico
Transparency Film
89 cm x 117 cm

Touching: Grizzly Bear #3, 2011
From the *Touching: Grizzly Bear Series*
Video Stills on UPremium OHP Pictorico
Transparency Film
89 cm x 117 cm

Touching: Grizzly Bear #4, 2011
From the *Touching: Grizzly Bear Series*
Video Stills on UPremium OHP Pictorico
Transparency Film
89 cm x 117 cm
Courtesy the Artist

Nancy Tousley, winner of the 2011 Governor General's Award for Visual and Media Arts, is a nationally known senior art critic, arts journalist and independent curator, who received the award for her outstanding contribution.

Tousley was art critic of the *Calgary Herald* for more than 30 years and the first Critic-in-Residence at the Alberta College of Art & Design. She has been a contributing editor to *Canadian Art* magazine since 1986. Her writing has appeared in magazines such as *Artscanada*, *Vanguard*, *Parachute*, *Border Crossings* and *Canadian Art* since the early 1970s, and in more than 30 public art gallery and museum catalogues.

Tousley is also an accomplished curator, having organized exhibitions for the Brooklyn Museum, the Art Gallery of Ontario, the Glenbow Museum and the Art Gallery of York University, among others.

Her work has been recognized with many prestigious awards including the Board of Governors of the Alberta College of Art & Design's Award of Excellence (1997), the Ontario Association of Art Galleries award for best curatorial writing on contemporary art (1999 and 2001), the Canadian Museums Association's award for outstanding achievement in arts journalism (2002) and the medal of the Royal Canadian Academy of Art (2009).

Published in conjunction with the exhibition
THE NEWS FROM HERE: 2013 Alberta Biennial of Contemporary Art
Curated by Nancy Tousley

Art Gallery of Alberta
26 January – 5 May, 2013

2 Sir Winston Churchill Square
Edmonton, Alberta, T5J 2C1, Canada

ISBN: 978-0-88950-170-6

Editor: Nancy Tousley
Managing Editor: Catherine Crowston
Essay: Nancy Tousley
Photographs: Courtesy of the Artists unless otherwise noted
Catalogue Design: Pamela Clark
Typeface: Executive (www.optimo.ch)
Printed in Canada by Friesens Corporation

Front Cover: Eric Moschopedis and Mia Rushton, *Hunter, Gatherer, Purveyor*, 2013 (photo: Lori Rushton)
Back Cover and page 2: Jason de Haan and Miruna Dragan, *The Wood and Wave Each Other Know*, 2011 (screenshot detail)
Inside cover: Kristopher Karklin, *Jack & Jill Room (morning)*, 2011

art gallery of alberta